Oceanography
An Illustrated Guide

C.P. *Summerhayes* and **S.A. *Thorpe***
Southampton Oceanography Centre,
Empress Dock,
Southampton, UK

Foreword by *Robert D. Ballard*
Center for Marine Exploration
Woods Hole Oceanographic Institution,
Woods Hole, MA, USA

John Wiley & Sons
NEW YORK   TORONTO

The Contributors

All contributors are based at the Southampton
Oceanography Centre, apart from Dr A. Gebruk (P.P.
Shirshov Institute of Oceanography, Moscow, Russia),
and Dr C.M. Young (Harbor Branch Oceanographic
Institution, Ft Pierce, Florida, USA).

This edition is published and distributed in North and South America by
John Wiley & Sons, Inc., 605 Third Avenue, New York, NY 10156-0012, USA.

Library of Congress Cataloging-in-Publication Data
Oceanography: an illustrated text / [edited by] C.P. Summerhayes, S.A. Thorpe.
     352 pages. 19.4 × 26.1 cm.
  Includes bibliographical references and index.
  ISBN 0–470–23574–8 (hc)
   1. Oceanography.   I. Summerhayes, C.P.   II. Thorpe, S.A.
GC11.2.022 1996
551.48–dc20   96-38712
                                                  CIP

Copyright © 1996 Manson Publishing Ltd
ISBN 1–874545–38–3 (UK)

Text organisation and supervision: John Ormiston
Design and layout: Patrick Daly
Line artwork: Kate Davis
Proofreading and index: Michael Forder
Color separations by: Tenon & Polert, Hong Kong
Printed by: Grafos SA, Barcelona, Spain

Contents

Foreword

The latter half of the twentieth century has seen the science and technology of Oceanography progress from a primitive state of understanding to a highly sophisticated science, although many believe oceanography is still in its infancy.

It was not until 1960, for example, that oceanographers first recognized the size and significance of the Mid-Ocean Ridge, a great mountain range that stretches through the ocean basins of the world for a distance of 70,000 km and covers close to 23% of the Earth's total surface area. Even more ironic is the fact that astronauts walked on the surface of the Moon before Earth scientists explored the Ridge's rift valley for the first time, in 1973, using manned submersibles.

Following this lowly start, oceanographers have not only learned the significance of this great undersea mountain range to the genesis of oceanic crust, but have also discovered the existence of hydrothermal events along the axis of the Ridge, surrounded by important mineral deposits and exotic life forms that live independently of the Sun's life-supporting energy.

The discovery of volcanism and hydrothermal circulation within the Mid-Ocean Ridge not only had an impact on the biological and geological sciences, it also helped us to better understand the chemistry of the oceans. We now know, for example, that the entire volume of the world's oceans passes inside the Ridge and out every six to eight million years.

During this same period, we have seen the primitive sampling techniques of surface-towed dredges and plankton nets give way to manned submersibles and, more recently, to advanced remotely operated vehicle (ROV) systems. These ROV systems not only provide scientists with around-the-clock access to the deepest reaches of the ocean floor, but can also be operated from shore via satellites. Clearly, the information highway of the twenty-first century will enable scientists world-wide to have easy access to the oceans of our planet.

Despite this better understanding, few outside the field of Oceanography are aware of these new findings. It is with such thoughts in mind that the staff of the Southampton Oceanography Centre in the UK have joined forces to create this introductory book on Oceanography.

We believe that this book will help everyone better appreciate this exciting new field of science, especially the next generation who will most likely explore more of the ocean floor and learn more about its hidden secrets than all the previous generations combined.

Dr Robert D. Ballard
Director, Center for Marine Exploration
Woods Hole Oceanographic Institution,
Woods Hole, MA, USA

Prologue

This book is published to mark the occasion of a major event in Oceanography in Britain, the formation of the Southampton Oceanography Centre (SOC). This Centre brings together the Natural Environment Research Council's Institute of Oceanographic Sciences and Research Vessel Services, with its three ships, *Discovery*, *Charles Darwin*, and *Challenger*. In a fine new building alongside the Empress Dock in Southampton, the SOC is one of the largest institutions devoted to the study of the Earth and its oceans in Europe and in the world.

The book is a collection of contributions largely by the staff of the SOC and their colleagues. It reflects the range of their interests, and I hope that it conveys something of their excitement and enthusiasm for their subject.

All of us were saddened by the death of our eminent colleague John Swallow during the preparation of the book. We are sure that he would have approved of this venture, which aims to bring the science of the oceans he loved so much to a wider audience. The book thus marks the passing of an era dominated by great oceanographers, including John's American friend and colleague Henry Stommel, who also died quite recently. We hope that its publication marks the commencement of a new era, in which we can continue their work with equal vigour and honesty, and maintain the spirit of friendship and co-operation in which they worked.

It is therefore also a great pleasure and entirely appropriate to include a Foreword by Bob Ballard from the Woods Hole Oceanographic Institution, where Henry Stommel worked for so many years, and which John Swallow regarded as his second home.

There is, after all, only one ocean, and we must work together to understand its secrets.

Professor John G. Shepherd
Director, Southampton Oceanography Centre,
Southampton, UK

Preface

The original motivation for producing this book was a comment and question from a young student visiting the Department of Oceanography at the University of Southampton; 'I couldn't find anything in the bookshops to tell me what oceanography is, so what is it?' Our purpose here is to provide some answers, to explain the science, and to express some of the delight and the privilege we feel in being involved in it. Chapters are written in a variety of styles, but all, we hope, are at a level at which a science undergraduate should have no difficulty in understanding; the book contains a wealth of information and guides to further reading, which all should find of interest and informative. This book is not directed at any particular teaching course, but provides much which will supplement courses in environmental science.

So, what is Oceanography? Oceanography begins at the shoreline. It is the science of the oceans, their interaction with the atmosphere above and with the underlying sea-floor sediments and oceanic crust, their chemical and biological components, their physical properties and motion, their geology, their creation, past history, and development, their present state, and their future. Oceanography is founded on the basic scientific disciplines of biology, chemistry, geology, physics, and mathematics. Many of the problems addressed by oceanographers are interdisciplinary, so their solution demands a breadth of knowledge that crosses conventional scientific boundaries and requires multidisciplinary team collaboration. The science uses the range of facilities of all these basic sciences, for example advanced computers and the analytical laboratory instruments of biologists and chemists. Oceanography also relies heavily upon a range of technologies, for example computing, electronics, optics, and acoustics which, together with the engineering involved in the design and construction of winches and wires, enable the scientist to make observations and sample the remote ocean depths – it is this remoteness which makes Oceanography akin to Space Science. It requires elaborate and robust instruments to survive the hostile environment, one which by its very nature spans the Globe. Because the collection of material and measurements from the sea is essential, it is an expensive and high-risk science, using ships, aircraft, satellites, and submersibles. International collaboration is often necessary to provide the resources and ships for major experiments and programmes of research.

Oceanography is a science with application to areas of considerable economic importance; for example, to fisheries, to hydrocarbon or mineral extraction from the sea bed, to coastal protection, to tidal or wave power, and to waste disposal. It is of great relevance to defence and to the proper management and regulation of the use of the seas, especially when this may involve pollution or degradation of the marine environment. The global ocean, because it responds slowly to changes in temperature or atmospheric composition, acts as a flywheel to climate change, and the prediction of future climate is presently a major driving force affecting the direction of a significant part of the science.

In this book we describe, as individuals or with colleagues, those facets of the science in which we have found particular delight and satisfaction and for which we have special expertise and enthusiasm. It touches on many of the present areas of greatest activity in the science today and, novel in a book of this nature, it describes the equipment used in research, some of the methodology, and some of the applications of the science – the discovery and assessment of resources and ocean wealth, and the use of the oceans for the disposal of waste.

There is excitement, stimulus, and satisfaction in finding, often quite unexpectedly, that discoveries in one area of the science have important consequences for another. Oceanography is not the geography of the oceans, it does not have all the glamour of Jacques Cousteau's films or of diving in the warm waters of a coral reef – it is good, often applicable and useful, science, hard and mind-challenging work which provides great personal satisfaction – and a lot of fun.

The reader should keep in mind the toil, fatigue, and effort needed to gather information about the ocean – the heaving and pitching of a vessel in severe weather – and recognise that part of the oceanographer's reward and satisfaction derives from the occasionally successful (would that it were always so!) combat with a harsh, uncompromising and hostile environment. This book may be seen as a tribute to the dedication of those who have wrested that information from the sea.

We are most grateful to Dr Bob Ballard of the Woods Hole Oceanographic Institute, US, for the Foreword. Dr Ballard is well-known for his investigations of the deep ocean, for discovering the wreck of Titanic, and for his initiative in bringing the science of oceanography live by satellite links to school children.

Dr C.P. Summerhayes,
Professor S.A. Thorpe, FRS,
Southampton Oceanography Centre,
Southampton, UK

The Global Oceans

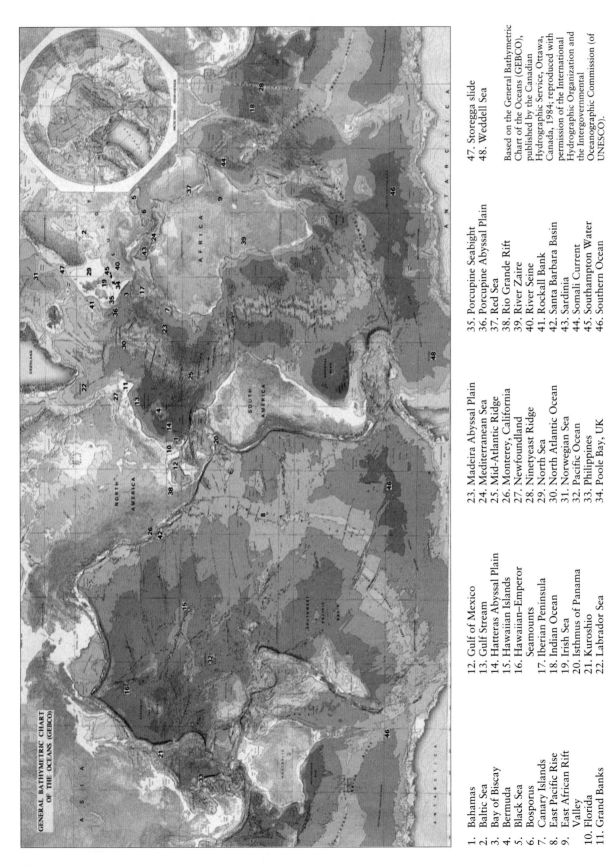

GENERAL BATHYMETRIC CHART
OF THE OCEANS (GEBCO)

1. Bahamas
2. Baltic Sea
3. Bay of Biscay
4. Bermuda
5. Black Sea
6. Bosporus
7. Canary Islands
8. East Pacific Rise
9. East African Rift
   Valley
10. Florida
11. Grand Banks

12. Gulf of Mexico
13. Gulf Stream
14. Hatteras Abyssal Plain
15. Hawaiian Islands
16. Hawaiian–Emperor
    Seamounts
17. Iberian Peninsula
18. Indian Ocean
19. Irish Sea
20. Isthmus of Panama
21. Kuroshio
22. Labrador Sea

23. Madeira Abyssal Plain
24. Mediterranean Sea
25. Mid-Atlantic Ridge
26. Monterey, California
27. Newfoundland
28. Ninetyeast Ridge
29. North Sea
30. North Atlantic Ocean
31. Norwegian Sea
32. Pacific Ocean
33. Philippines
34. Poole Bay, UK

35. Porcupine Seabight
36. Porcupine Abyssal Plain
37. Red Sea
38. Rio Grande Rift
39. River Zaire
40. River Seine
41. Rockall Bank
42. Santa Barbara Basin
43. Sardinia
44. Somali Current
45. Southampton Water
46. Southern Ocean

47. Storegga slide
48. Weddell Sea

Based on the General Bathymetric
Chart of the Oceans (GEBCO),
published by the Canadian
Hydrographic Service, Ottawa,
Canada, 1984; reproduced with
permission of the International
Hydrographic Organization and
the Intergovernmental
Oceanographic Commission (of
UNESCO).

The Geological Time Scale[a,b]

Era	Sub-era, Period, Sub-period		Epoch	Age (Myr)
Cenozoic	Quaternary Sub-era		Holocene	0.01
			Pleistocene	1.64
	Tertiary Sub-era		Pliocene	5.20
	Neogene Period		Miocene	23.3
	Palaeogene		Oligocene	35.4
			Eocene	56.5
			Palaeocene	65.0
Mesozoic	Cretaceous		Senonian	88.5
			Gallic	131.8
			Neocomian	145.6
	Jurassic Period		Malm	157.1
			Dogger	178.0
			Lias	208.0
	Triassic Period		Triassic 3	235.0
			Triassic 2	241.1
			Scythian	245.0
Palaeozoic	Permian Period		Zechstein	256.1
			Rotliegendes	290.0
	Carboniferous Period	Pennsylvanian Sub-period	Gzelian	295.1
			Kasimovian	303.0
			Moscovian	311.3
			Bashkirian	322.8
		Mississippian Sub-period	Serpukhovian	332.9
			Visean	349.5
			Tournaisian	362.5
	Devonian Period		Devonian 3	377.4
			Devonian 2	386.0
			Devonian 1	408.5
	Silurian Period		Pridoli	410.7
			Ludlow	424.0
			Wenlock	430.4
			Llandovery	439.0
	Ordovician Period		Ashgill	443.1
			Caradoc	463.9
			Llandeilo	468.6
			Llanvirn	476.1
			Arenig	493.0
			Tremadoc	510.0
	Cambrian Period		Merioneth	517.2
			St. David's	536.0
			Caerfai	570.0

a   After Harland, W.B., Armstrong, R.L., Cox, A.V., Craig, L.E., Smith, A.G., and Smith, D.G. (1990), *A Geologic Time Scale*. Cambridge University Press, Cambridge.

b   These Eras comprise the Phanaerozoic Eon. Preceding it is the PreCambrian, dating back to the origin of the Earth, at around 4600.0 M.

Standard International (SI) Units

Wherever possible the units used are those of the International System of Units known as SI. Oceanographers have traditionally used other units, such as the litre, which often cannot be avoided because of their common usage. Despite the recommendations periodically published by international committees as to what constitutes a standardised scientific terminology, agreement is still rather poor. Conversion between units often requires great care.

SI Unit Prefixes

Name	Symbol	Multiplying factor 10^N, N is given below
peta	P	15
tera	T	12
giga	G	9
mega	M	6
kilo	k	3
hecto	h	2
deca	da	1
deci	d	−1
centi	c	−2
milli	m	−3
micro	μ	−6
nano	n	−9
pico	p	−12
femto	f	−15
atto	a	−18

Commonly Used SI Units

SI units

Name	Symbol	Name	Equivalent cgs
Force	N	Newton	$kg\ m/s^2$
Pressure	Pa	Pascal	$kg/m\ s^2$
Energy/work	J	Joule	$kg\ m^2/s^2$
Power/energy flux	W	Watt	$kg\ m^2/s^3$
Irradiance	E/m^2s	Einstein/m^2s	mol photons/m^2s

The expression of gas concentrations is a particularly problematic area. The SI unit for pressure is the Pascal (1 Pa = 1 N/m^2). Although the bar (1 bar = 10^5 Pa) is also retained for the time being, it does not belong to the SI system. Various texts and scientific papers still refer to gas pressure in units of the torr (symbol: Torr), the bar, the conventional millimetre of mercury (symbol: mmHg), atmospheres (symbol: atm), and pounds per square inch (symbol: psi) – although these units will gradually disappear. Irradiance is also measured in W/m^2. Note; 1 mol photon = 6.02×10^{23} photons.

    The SI unit used for the amount of substance is the mole (symbol: mol), and for volume the SI unit is the cubic metre (symbol: m^3). It is technically correct, therefore, to refer to concentration in units of mol/m^3. However, because of the volumetric change that sea water experiences with depth, marine chemists prefer to express sea water concentrations in molal units, mol/kg.

CHAPTER 1:

How the Science of Oceanography Developed

M.B. Deacon

Early Ideas about the Sea

Oceanography is a young science with a long history. Scientists in the seventeenth and eighteenth centuries tried to study the sea, but were often frustrated by its sheer size and complexity and the practical difficulties involved. During the nineteenth century technological advances made systematic exploration of the deep sea possible for the first time, and oceanography became an independent scientific discipline. However, the greatest strides toward understanding the sea and its importance, both as a feature which governs the Earth as we know it and one that influences human activities in many ways, have been made during the twentieth century. Oceanography today is very different from what it was 50 years ago, let alone 200 years or more, so is there any good reason why anyone, apart from historians, should be interested in its past? Science is a continuum; as an activity it grows out of its past, even if that sometimes means rejecting outmoded ideas or unreliable data. A look at how oceanography has developed can be a valuable way of helping to understand the modern science, both in terms of a set of ideas and as an institution.

Primitive societies developed complex mythologies to explain the workings of the universe, and invoked deities to account for natural phenomena. By Greek and Roman times, however, philosophers were beginning to look for natural causes for things about them, from the movement of the heavens[8] to the waves of the sea[5]. Of these, Aristotle (in the fourth century BC) most influenced later European science. He wrote widely on natural science, as well as politics and philosophy, and his works contain much to interest oceanographers, including the first known observations on marine biology. We find him considering such diverse topics as how winds cause waves, water movements in straits, and the water balance of the ocean. Aristotle believed that the presence of water vapour in the atmosphere, the source of rain, was due to evaporation, principally from the sea. Rainfall supplied rivers and these flowed into the sea, so the level of the ocean was maintained. This seems obvious today, but other

suggestions were plausible given the state of knowledge at the time. Up to the end of the seventeenth century it was widely held that ground water was absorbed by the land directly from the sea, and that this formed the source of wells and springs. Then measurements made by the French scientist Edmé Mariotte, showing that the rainfall in the Paris area was, in fact, sufficient to account for the flow of water in the Seine, removed the objection that rainfall was insufficient to account for large rivers.

The Age of the Discoveries

To account for the movement of the heavens, Aristotle suggested that the Sun, Moon, and stars revolved around the Earth attached to concentric, crystalline (and therefore invisible) spheres. These, he said, derived their movement from an outermost sphere, which became known to medieval science as the *primum mobile*, or prime mover. When Columbus made his voyages of discovery across the Atlantic in the 1490s, he experienced a westward-flowing current in the tropics, and similar currents were also identified in the Pacific and Indian Oceans. Only in the Indian Ocean, north of the equator, where the proximity of the Asian land mass creates strong seasonal variations in winds and weather (the monsoons), is this flow subject to periodic reversals in direction, a fact already known to Arab geographers of the ninth century AD. The westward flow near the equator (there is actually an eastward-flowing equatorial counter-current dividing it into two streams, but this was not identified until the early nineteenth century) was thought by many sixteenth century writers to be due to motion transmitted to the Earth's fluid envelope by the *primum mobile*[3]. However, after Copernicus suggested that the Sun was at the centre of the Universe and that the Earth rotated around it, this explanation had to be adapted. The new version supposed that the westward movement, thought to exist throughout the oceans, although most marked near the equator, was due to inertia – as the Earth rotated daily on its axis, the sea lagged behind.

During the seventeenth century we find alternative explanations beginning to appear[4]. These were

1.1

TABULA GEOGRAPHICO-HYDROGRAPHICA MOTUS OCEANI, CURRENTES, ABYSSOS, MONTES IGNIVOMOS IN UNIVERSO ORBE INDICANS.

Figure 1.1 Chart showing ocean currents from *Mundus Subterraneus* by Athanasius Kircher[14], a Jesuit mathematician who taught at Rome. This chart was probably the first to attempt to show ocean currents in the major oceans, whose geographical limits were by then quite well-understood, except in the polar regions (where far more land is shown than actually exists). Kircher speculated that the ocean was connected with water masses in the interior of the Earth (his subterranean world) through openings in the sea floor. The siting of the abysses he shows on his map was not entirely fanciful, some being located on sites of reputed whirlpools. For example, Kircher supposed that water was sucked in at the Maelstrom – in the Atlantic off Norway – and flowed via a subterranean tunnel into the Baltic. This had no actual basis in fact, but modern oceanographers are finding that sea water under the high pressures that exist at the sea bed penetrates the rocks and sediments of the sea floor and migrates through them. At mid-ocean rift systems, this sea water, charged with chemicals leached from the rocks through which it has passed at high temperatures, is forced back into the ocean in a scenario that would, if suggested, have seemed as exotic as Kircher's abysses until relatively recently (see Chapter 10). (Courtesy of The Royal Society, London, England.)

mainly the work of Roman Catholic philosophers who, after Galileo's condemnation, were forbidden to express Copernican ideas. One of these, Athanasius Kircher[14] (*Figure 1.1*) suggested that as the Sun travels over the sea its heat evaporates the water below. This creates a depression in the sea surface so that currents flow in from either side to restore its level. The evaporated water falls as rain in higher latitudes, so that the circulation is maintained. Isaac Vos, a Dutch scholar who later settled in England, objected that this would actually lead to a current flowing eastward, from the part of the ocean which the Sun had not yet reached[34]. He agreed that it was the Sun's heat which was responsible for currents, but said it operated by expanding the water, so that the level of the sea rose slightly as the Sun moved across it. This was sufficient to cause a flow toward a lower level. Vos' theory was one of the most original and well-supported to appear up to that time (1663). He was less convincing when he tried to introduce effects from the Earth's annual rotation in its orbit around the Sun in order to explain tides as a by-product. The idea that surface currents are actually caused by winds occurs in seventeenth-century literature, but few scientists took it seriously at that time.

The cause of tides had been keenly discussed since Greek philosophers first learned of their existence (because the Mediterranean is so enclosed, its tides are generally small and unnoticed.) The much larger rise and fall of the tides on ocean coasts (twice daily in most places) and the monthly springs and neaps (high and low tidal ranges) seemed to be linked to the Moon and its phases, but how did this come about? Throughout the Middle Ages and the Renaissance numerous explanations were made, usually linked to contemporary thinking on cosmology[8] but without, apparently, any attempt to obtain more accurate knowledge of tides themselves. However, these were clearly well-known to seafarers and during the sixteenth century we find details, usually of the 'establishment' of ports (i.e., time of high water relative to the Moon's passage overhead), appearing in printed works on navigation.

The Scientific Revolution of the Seventeenth Century

The Treasure for Travellers by William Bourne[2], which contains an interesting account of the geography of the sea as seen by a representative of the newly educated professional classes, is a good example of a change in attitude. Bourne takes his own observations as a starting point to suggest how common coastal features, such as cliffs, beaches, and stacks, might have come into existence instead of, as conventional scholars would have done, concentrating on discussion of previous ideas, which could often be traced back to Aristotle. The achievement of the scientific revolution of the seventeenth century was to highlight the need to advance science by experiment and observation, in conjunction with theory.

This was the philosophy which lay behind the foundation of the Royal Society in 1660, but the Fellows' interest in the science of the sea[5] was also influenced by the growing importance of maritime affairs in English national life during the sixteenth and seventeenth centuries. The projects they undertook, therefore, had a dual purpose. The wish to further knowledge of the natural world, in collaboration with scientists in other countries, was combined with the hope of information which would be of practical benefit to seafarers, an intention that was partly humanitarian, but which also had its roots in the desire for national economic and strategic advantage.

Much of the work done by individual Fellows of the Society was directed toward devising apparatus that sailors could use on voyages. They particularly hoped to obtain information about the depth of the sea, but were worried that soundings made in the ordinary way, with lead and line, might be inaccurate because of subsurface water movements bending the line out of the vertical. To overcome this a device suggested in earlier literature was adopted, consisting of a weight with a float attached to it in such a way that when the weight hit the sea bed the float disengaged and rose to the surface through its own buoyancy. The depth of water was calculated from the time this took. Robert Hooke improved the basic design of the apparatus (*Figure 1.2*); during the next 150 years much energy was devoted to improving this method. Although Hooke's device performed well when tested in shallow water, unfortunately both it and its successors suffered from an unsuspected design fault that made it useless in deep water. Pressure of water increases in proportion to its depth, so that in the open sea the float, which was made of wood, became waterlogged and could not rise to the surface. Hooke also designed a sampler to collect sea water from different depths. This was intended to find out if the sea was only salty at the surface, as Aristotle was supposed to have said. If this idea was correct (it was not) then fresh water would be brought up from the depths.

By the provision of instructions and encouragement, the Royal Society[31] had some success in persuading its followers to collect information about the sea, in spite of difficulties experienced with the apparatus. Notably, some of the tidal observations thus obtained were used by Sir Isaac Newton to illustrate his theory of the Universe. He was able to show that tides are due to the gravitational attraction of the Moon and, to a lesser extent, the Sun. This claim proved highly controversial, since the idea of gravity operating through empty space had been rejected by thinkers earlier in the seventeenth century. They were trying to show that nature is governed by physical laws and were therefore reluctant to employ a concept that appeared no more soundly based than discredited ideas about astrological 'influence'. The fact that Newton was able to express the effect of gravity in mathematically demonstrable laws led to the gradual acceptance of his views, but his tidal theory was also crucial in this process. His friend Edmond Halley[11] felt so strongly about it that he wrote an article summarising Newton's arguments in language and terms that could be understood by the layman (the *Principia* had been published in Latin). Information on tides and currents (see *Figure 1.3*) also came from British travellers abroad. Another famous scientist, Robert Boyle, relied on such observations when he wrote three essays on the salinity, depth, and temperature of the sea in the 1670s, still worth

reading by anyone interested in the science of the sea[1]. However, Newton percipiently remarked in one of his letters that, rather than sailors sending information to mathematicians at home, it would be more fruitful to send the mathematicians to sea.

Early scientific interest in the sea was not, of course, confined to Britain. Perhaps the most interesting contribution at this time was made by L.F. Marsigli[33]. A native of Bologna, his university studies seem to have given direction and method to a boundless curiosity about the world in general. As a

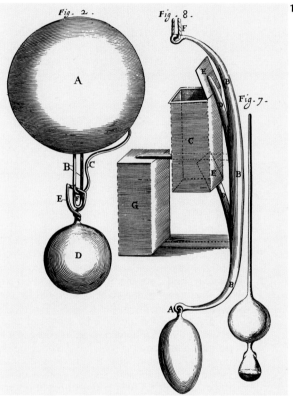

1.2

Figure 1.2 Sounding machine (Fig 2), water sampler (Fig. 8), and hydrometer (Fig. 7); this plate was published in 1667 by The Royal Society[31], founded in 1660, which wanted to collect information on scientific topics as widely as possible, so these instructions were drawn up to show overseas travellers the kind of observations that were wanted, and how to make them. Among other things, they hoped to find how the depth of the sea altered from place to place, and whether the sea was salt throughout, or only at the surface. The Society experimented with apparatus for measuring depth and bringing up water from the lower layers of the sea, but had variable success. Robert Hooke, the society's curator, produced designs which were an improvement on earlier models. Anita McConnell[16] has pointed out, however, that these woodcuts, which appeared in the journal, were not faithful copies of his drawings and would not actually work! (Courtesy of The Royal Society, London, England.)

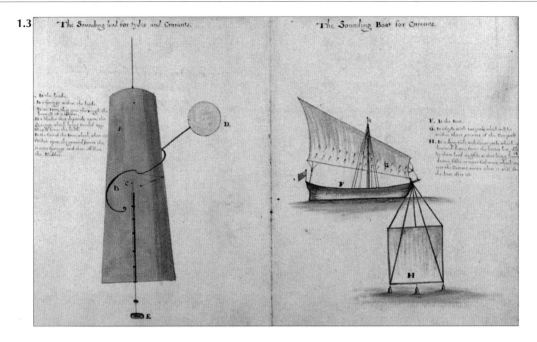

Figure 1.3 Richard Bolland's sounding lead for tides and currents, and sounding boat for currents. These are reproduced from Bolland's manuscript *Mediterranean Journall* of 1675 (in the Pepys Library at Magdalene College, Cambridge). Bolland was stationed at Tangier during its brief period as a British possession (it was part of the dowry of Charles II's queen, Catherine of Braganza). During the British occupation, extensive works were carried out to provide a safe anchorage for shipping. Bolland took part in this work and made a chart of the Strait of Gibraltar, showing its tides and surface-water movements[5]. The existence of a current flowing into the Mediterranean through the Strait had long been known. Sailors believed that there was a compensatory outflow into the Atlantic below. Bolland wanted to obtain proof of the existence of the undercurrent and devised a method of doing this. The sounding lead had a small float (D) attached and a mechanism to release it on striking the bottom. He hoped that by seeing where the float came to the surface, it would be possible to work out the speed of the undercurrent. However, he realised that it would be necessary to allow for the effect of the surface current throughout its depth (which he supposed might be as much as 100 fathoms – about 180 m). To establish this he intended using the drogue attached to the boat, which could be lowered to the desired depth, and the speed of the boat measured relative to that of the surface water. These are quite sophisticated ideas and nothing comparable was achieved until the nineteenth century. However, Bolland, and other supporters of the undercurrent, could offer no explanation of how it was generated. Other people thought it was only a seaman's yarn. A colleague of Bolland's at Tangier, Sir Henry Sheeres, wrote an essay to prove that the inflow from the Atlantic into the Mediterranean was maintained by the climate of the area. Low rainfall and hot sun meant that evaporation exceeded the input of water from rain and rivers (see Chapter 2), and the sea's level would otherwise have fallen below that of the ocean outside. Edmond Halley used this explanation, widely accepted during the next two centuries. However, it did not account for what happened to the salt that is left behind when sea or other salt water evaporates. A German scientist, J.S. von Waitz, who was connected with the salt industry (which relies on this principle), pointed out in the mid-eighteenth century that, as water became more salty, and therefore heavier, it would sink and that, as the depths of the Mediterranean filled with this more saline water, it would flow out into the Atlantic (see text). When more detailed physical surveys of the Mediterranean and the Atlantic came to be made in the late nineteenth and twentieth centuries, these showed that saline water does, indeed, spill out over the lip of the Strait of Gibraltar and spread out into the Atlantic, where it has a significant effect on the wider oceanic circulation (see Chapter 11). Currents can now be measured by moored instruments or acoustic remote sensing (see Chapter 19). (Reproduced by kind permission of the Master and Fellows, Magdalene College, Cambridge, England.)

young man, he accompanied a diplomatic mission to Constantinople and, while there, investigated reports of a counter-current in the Bosphorus, beneath the surface current flowing out of the Black Sea. He showed that the depths of the strait were occupied by more saline, and therefore heavier, water of Mediterranean origin, and that this water must reach the Black Sea, which would otherwise be entirely fresh because of the rivers flowing into it (*Figure 1.3*). He demonstrated the way this could happen with an experiment showing how, when two liquids of differing specific gravity were introduced into a container, they would form layers with the heavier liquid below and the lighter one above.

Marsigli spent many years on military service in eastern Europe before returning to the science of the sea in later life. He wrote about his researches off the southern coast of France in his book, *Histoire Physique de la Mer*[15]. Earlier works on geography and navigation had contained sections on tides and currents, and occasionally other aspects of the sea, but this was the first book on a truly oceanographic theme. However, its title is misleading as it is largely about marine invertebrates. These interested Marsigli because they had so far attracted little attention – there were earlier works on fish. He himself was particularly interested in coral, valued for use in jewellery and decorative objects, but wrongly classed it as plant rather than an animal, because of what he concluded were its 'flowers'[17].

The Eighteenth and Early Nineteenth Centuries

In spite of this promising beginning, marine science did not develop as rapidly as one might have expected during the eighteenth century. An indication of why this was so is found in Marsigli's book[15]. where he points out that science at sea is beyond the resources available to individuals and that further progress would only be made with government aid. This is because oceanographic research demands expensive items, like ships, peo-ple, and apparatus, but state funding of science is a comparatively recent innovation. There were also technical obstacles, especially to the exploration of the deep ocean. Above all, in spite of the work that had already been done, there was at this time no recognised 'science of the sea'. Nevertheless, important advances in understanding were made during the eighteenth century. At the same time, a number of related developments contributed to laying the foundation of oceanography as we know it today. These included improvements in navigation and marine surveying, in particular the discovery of methods to measure longitude at sea, which made it possible for the first time to fix a ship's position accurately when out of sight of land – an essential prerequisite for studying the ocean. These improvements were exploited first on official voyages of exploration, expeditions despatched by governments with political and economic objectives in mind, but from the time of Captain James Cook onward (the circumnavigations he commanded spanned the years 1768–1780) they became increasingly scientific in nature. A considerable amount of oceanographic work was done, especially by French and Russian expeditions, in the early nineteenth century. Their observations of surface temperature and salinity were used by geographers like Humboldt in studies of the world climate. However, as deep-water observations were more difficult, few were made. Much depended on individual scientists being fortunate enough to have the opportunity of observing for themselves, or through interested laymen, especially naval officers.

Even so, sufficient new information was obtained to encourage the development of ideas about the interior of the ocean. Toward the middle of the eighteenth century Stephen Hales[10], who had already tried to improve Hooke's sounding machine (*Figure 1.4*), also produced an apparatus designed to measure temperature by raising water from ocean depths. Such attempts were not entirely new. Hooke had proposed a design for a deep-sea

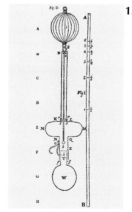

1.4

Figure 1.4 Deep-sea sounding machine devised by Stephen Hales[10]. A clergyman and scientist, Stephen Hales (1677–1761) befriended the naturalist Gilbert White. In the early years of the eighteenth century, he and J.T. Desaguliers, the Royal Society's curator, tried to develop new ways of measuring the depth of the sea which would be more reliable than those suggested by Hooke half-a-century earlier. They relied on the compression of air inside the apparatus, but most were made of glass and would not have been sufficiently robust for use in the marine environment. An example of this type is held in the George III collection at the Science Museum in London. The more durable apparatus illustrated here employed a rifle barrel (Fig. II, K–Z), with a removable rod (Fig. I, A–B) inside. Hales intended using coloured oil to mark the height reached by the water inside, which would enable the pressure, and therefore the depth of water, to be calculated. He did not appreciate the weakness of this design in that it still relied on a float (Fig. II, i) made of wood to bring it to the surface. In the sea's depths, pressure would force water into the pores of the wood so that it would lose its buoyancy and not bob up again. When this machine was tried in mid-ocean it never returned to the surface.

thermometer and Marsigli[15] had measured sea temperatures – until his only thermometer was broken in a raid by pirates. Hales' apparatus[10] consisted of an enclosed bucket with hinged flaps, opening upward only, in both top and bottom. This allowed water to flow freely through the device on the way down, but trapped a sample inside as soon as the observer began to haul it up. Its temperature was measured at the surface. However, the objection could be made that the water's temperature might have altered on the way up. To overcome this problem, the Swiss scientist H.B. de Saussure insulated the thermometer itself and left it down long enough to take on the temperature of the water at that depth. This took such a long time that the method was not much used, though it proved the most reliable at the time. Self-registering thermometers, pioneered by James Six in the 1780s, were more popular, but had disadvantages that did not become apparent until later.

The reason for this growing interest in temperature measurement was initially connected with geological debate – was the interior of the Earth hot or cold? It was some time before people began to speculate what this work was telling them about the sea. In the mid-eighteenth century a German scientist, J.S. von Waitz[6] pointed out that Marsigli's arguments[15] about the Bosphorus could equally well apply to the Strait of Gibraltar. Beneath the surface current from the Atlantic there must be an outflow of more saline water (see *Figure 1.3*, caption), otherwise the Mediterranean's salinity would be far higher. Waitz then suggested that similar imbalances gave rise to currents in the ocean. These were due to differences in density between equatorial regions, where the Sun's heat caused evaporation and increased salinity, and high latitudes where rainfall would lessen it. Saline water in the tropics sank and spread toward the poles in the ocean depths, while lighter, fresher water at the surface flowed toward the equator to replace it. The situation is more complex than Waitz supposed. The density of sea water depends on temperature as well as salinity and one can counteract the other. Heavy rain falls at the equator, and though melting ice reduces salinity in high latitudes in summer, in winter brine is released as the sea freezes. However, he was the first, as far as we know, to suggest the existence of an internal circulation in the ocean.

Waitz's suggestion had little immediate impact, but toward the close of the eighteenth century we find similar ideas appearing, but with an alteration in emphasis that had temperature rather than salinity differences being responsible for maintaining circulation. More ambitious deep-sea temperature measurements made on voyages of exploration in the early nineteenth century provided supporting evidence for this view, in particular those made by the French scientist François Péron. He used de Saussure's method to reveal the existence of low temperatures in the depths of seas in warm latitudes. Since it could not have formed there, it was argued that this colder water must have originated in polar regions.

This idea was widely accepted on the continent, but, in Britain in particular, the supposition that sea water, like fresh water, expands before freezing led to the widely held belief that water in the depths of the sea could not fall below 4°C, the temperature of maximum density of fresh water. In 1819 Alexander Marcet, a Swiss physician living in London, published an important paper on the salinity of sea water in different parts of the world[5]. In this he showed that sea water of average salinity behaves differently from fresh water, and that its density increases with cold until it freezes. This meant that in a theoretical ocean, where salinity was uniform and density was a function of temperature only, the coldest water would always sink to the bottom. The delay in accepting Marcet's findings, which were later confirmed by other scientists, was due principally to poor communication.

There was no recognised science of oceanography during the first half of the nineteenth century and individuals interested in marine science at that time came from a variety of backgrounds. Another reason for the continuing confusion over ocean circulation was that the Six self-registering thermometers (see earlier), then widely used to measure deep-sea temperatures, were not sufficiently protected and gave readings distorted by the effect of pressure. On his voyage of discovery in the Southern Ocean between 1839 and 1843, Sir James Clark Ross[30] measured deep-sea temperatures assiduously, but was not surprised that they never apparently fell below 4°C. As a naval officer, although one who was active in scientific research, Ross could not be expected to be fully up to date in all the branches of science represented on the expedition (its primary task was observations on terrestrial magnetism in the southern hemisphere). He should have been better advised by scientific colleagues at home, but the information he needed in this instance was possessed by chemists and physicists with whom he had no direct contact. It was not until the events leading up to the *Challenger* expedition (see later) that the misunderstanding was exposed, but such widely held misapprehensions are hard to eradicate and the 4°C error can be found in some twentieth century publications, including *Hansard*[12] (1961), in a reply to a question about the operation of Royal Navy submarines in the Arctic!

The situation just described had partly arisen because naval surveyors, geographers, and some

Figure 1.5 Chart of Atlantic currents by James Rennell, from Rennell's book, *An Investigation of the Currents of the Atlantic Ocean*, 1832[23]. Rennell was the first to produce charts of winds and currents, based on observation, for an entire ocean and to show how the course of ocean currents was largely shaped by prevailing winds. A former naval officer and surveyor for the East India Company, he was already interested in currents before his return to England in 1778. The ship in which he was travelling narrowly escaped loss on the Scilly Islands, notorious for shipwrecks. He suggested that part of the problem might be an unidentified current, since known as Rennell's Current, flowing out of the Bay of Biscay. Later literature has generally discounted this, but some recent models of North Atlantic circulation provide a possible explanation for such a feature. Rennell devoted the last 50 years of his life to geographical research, including work on ocean currents throughout the world. Only part of his work, a volume of charts and accompanying memoir on the Atlantic, was published in 1832, two years after his death. These charts were based either on observations made by seafaring friends or derived from ships' log-books. Chapter 4 describes recent discoveries about ocean currents. (Courtesy of the SOC, Southampton, England.)

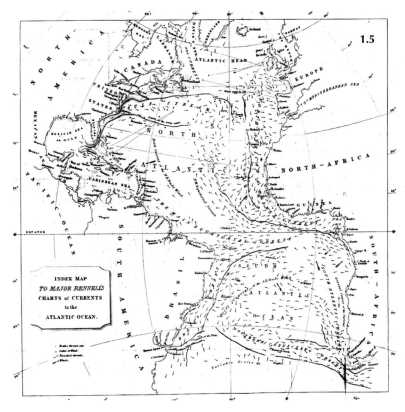

scientists were working in a different tradition, linked to hydrography and the interests and needs of seafarers, rather than to the physical sciences. Though charts of the Gulf Stream had been published somewhat earlier by Benjamin Franklin[27] and his less well-known predecessor, W.G. de Brahm[7], it was the introduction of chronometers toward the end of the eighteenth century which made it possible to collect information on ocean currents on a wider scale. This was because, once ships could fix their position when out of sight of land, the information contained in their log-books enabled the effect of currents upon them to be calculated. The first to take advantage of this was James Rennell[23] (*Figure 1.5*). Other hydrographers followed his example, so mid-nineteenth century sailing directions contained good accounts of the surface currents of the major oceans. Rennell had shown how closely such water movements were allied to the direction of winds blowing over the sea surface, something that had probably always been self-evident to seafarers, but which was slow to take root in the scientific literature. During the 1840s, Matthew Fontaine Maury, Superintendent of the United States Navy's Depot of Charts and Instruments, produced seasonal wind and current charts[20], based on averaging data from log-books and designed to speed the passages of sailing ships. He believed that these charts could be further

improved if ships from all nations systematically collected and recorded details of wind and weather. As a result of his efforts, an international conference was held in Brussels in 1853 at which governments agreed to adopt a standardised scheme of observations. Scientists had always recognised the importance of co-operation and exchange of information and ideas with colleagues in other countries. The Brussels meeting was a milestone in the development of maritime meteorology, which is closely linked to several branches of modern oceanography (in particular the study of air–sea interaction, see Chapter 2). It also introduced the idea of scientific co-operation between governments, which has been of particular significance in the development of modern oceanography.

Maury's initiative came at a time of rapid change – sail was already giving way to steam for naval and commercial purposes. The technological developments of the nineteenth century revolutionised the opportunities for scientific research, and the study of the oceans in particular. It was as a result of this that the establishment of oceanography as a separate discipline took place, and much of the impetus came from increasing maritime activity. The construction of more and larger ships, of ports and harbours to receive them, and of lighthouses to guide their passage made it necessary, for example, to obtain better knowledge of tides

1.6

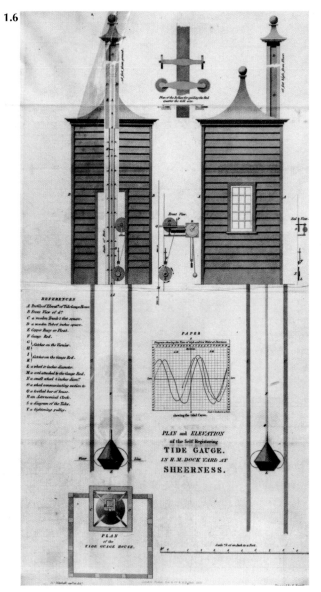

Figure 1.6 A tide gauge (from *Nautical Magazine*, 1832, **1**, 401–404), an apparatus for measuring the rise and fall of the tide devised by an engineer named Mitchell at Sheerness Dockyard. Similar gauges were being installed elsewhere at the time, but Mitchell's gauge incorporated an important innovation – it was self-registering, so did not require the presence of an observer. Tides are caused by the gravitational pull of the Moon (and to a lesser extent of the Sun), which varies with distance. Monthly and annual cycles can be detected, but tidal heights are also affected by weather. By the 1830s the Industrial Revolution was in full swing in Britain, and engineering projects needed precise information of such phenomena. The information was also welcomed by scientists who were trying to work out how the gravitational forces, well-known from the work of Newton and his successors in the eighteenth century, were translated into actual movement in the ocean. Tides can now be measured by instruments set on the sea bed in mid-ocean (Chapter 19).

(*Figure 1.6*) and waves. But perhaps the most important development for marine science came through the technology developed in response to the challenge of laying deep-sea telegraph cables, as this made scientific investigation of the ocean depths possible for the first time.

The Origins of Deep-Sea Exploration

The first functioning submarine telegraph cable was laid across the Straits of Dover in November 1851. From that time the prospect of extending this new means of rapid communication between continents was a powerful incentive to governments and industry alike. New technology had to be developed, not only to protect and lower cables to the sea bed and to raise and repair them if the need arose, but also to find out about the deep-sea environment – knowledge essential for routing and operating the cables[16]. The nature and contours of the sea bed had to be established, and also the temperature of the water. Up to this time, deep soundings had rarely been attempted because of the great effort involved, particularly if line and instruments were to be retrieved (*Figure 1.7*). The introduction of steam power made such operations possible on a more routine basis for the first time, although they were still laborious and time-consuming. By the mid-nineteenth century, hydrographic surveying was already a specialised activity in most navies, so new techniques and apparatus were developed rapidly for use in the deep sea. It was the combination of this new technology, and the accompanying professional expertise, with scientific thought that made possible further advance. Yet there was some delay before this happened – it was marine biologists rather than physical scientists who were the first to make use of the new opportunities that had been created.

Familiarity with the marine life of coastal and surface waters had been greatly extended during the latter part of the eighteenth and early nineteenth centuries, as biologists sought to expand their knowledge of living creatures and establish their affinities through schemes of classification[38]. By the middle of the nineteenth century many European and American zoologists specialised in marine work and their interest was further stimulated by the publication of Darwin's theory of evolution in 1859. Thus, the sea-shore collecting which became a popular craze among Victorian holiday-makers (*Figure 1.8*) served a more serious purpose among the scientific fraternity. They were not only interested in discovering new species, but also in learning more about the physiology and life history of individual organisms. This required working space and equipment, so seasonal laboratories were set up by the sea-shore, from which

Figure 1.7 *Deep Soundings; or, no Bottom with 4600 Fathoms*, a woodcut from Sir James Clark Ross[29]. This shows how laborious a task making deep-sea soundings was in sailing ships. Ross was an experienced polar explorer – before going to the Antarctic he had accompanied his uncle Sir John Ross on naval and private expeditions in search of the North West Passage and the North Magnetic Pole. The primary aim in this expedition, in H.M.S. *Erebus* and H.M.S. *Terror*, was to measure magnetic variation in the southern hemisphere and locate the South Magnetic Pole, in conjunction with a

survey of terrestrial magnetism being made by scientists from many nations. He was also interested in ocean science and made several deep soundings and many sea-temperature measurements. In fact, his soundings greatly exaggerated the depth of water in the areas he covered. During the next few decades surveys of routes for submarine cables using steam vessels resulted in the development of new techniques and greater precision in such observations, and made the scientific study of the deep sea practical, 200 years after it had been proposed by the Royal Society.

more permanent institutions, the marine biological laboratories, emerged. One of the most famous and influential of these, though not the first to be established, was the Stazione Zoologica at Naples, founded by a German zoologist and follower of Darwin, Anton Dorhn, in the early 1870s. Others followed – on both sides of the Atlantic. Both the Woods Hole Biological Laboratory on Cape Cod and the Scripps Oceanographic Institution, part of the University of California, started life in this way.

When these developments took place the leading figures were, increasingly, professional scientists and academics, but in the mid-nineteenth century amateur collectors were still numerous and made an important contribution. Some were keen yachtsmen and began to extend their interests to the waters of the continental shelf. Dredging was also supported by the newly formed British Association for the Advancement of Science[21]. Some of the impetus for this work came from geological discoveries. When modern relatives of fossil remains were discovered in seas further to the north or south, this suggested that climate might have undergone considerable shifts (see Chapter 3). This finding was of much interest in the light of

1.8

Figure 1.8 *Pegwell Bay, Kent – a Recollection of October 5th 1858*, a painting by William Dyce, shows a scene that would have been familiar at the time. The growing interest of nineteenth-century biologists in marine life-forms was for a short while reflected in a more widespread enthusiasm for sea-shore collecting of natural history specimens. This became a feature of seaside holidays, taken increasingly by the expanding middle class as rail transport made travel cheaper. From the 1850s onward many semi-scientific books and guides were written to cater for this market. (Courtesy of the Tate Gallery, London, UK.)

evidence being put forward, particularly by Swiss geologists, for periods of extensive glaciation in the past.

Almost without exception, those engaged in this work failed at first to appreciate the new opportunities which deep-sea surveying work was creating. This was because it was widely believed that life could not exist in the conditions of darkness, cold, and immense pressure that exist in the depths of the sea. There was then no conception of the range of adaptations (which modern biologists are still discovering) that enable creatures to live in such environments (see Chapters 13 and 15). The findings of Edward Forbes, who had worked in the Mediterranean in the 1840s, were often quoted to support the idea of an azoic (without life) zone below 400 fathoms (780 m). There was, it was true, some evidence from elsewhere that seemed contrary to this view, but for some years most people regarded it as unconvincing[25]. However, in the late 1860s accumulating observations from a number of sources suggested to workers in several countries that the supposed limit was erroneous. In 1868, two British biologists, W.B. Carpenter and C. Wyville Thomson, backed by the Royal Society, persuaded the Admiralty to allow them the use of one of its survey ships so that they could dredge in deep water. In three voyages, first in the *Lightning* and then in the *Porcupine*[24], they obtained incontrovertible evidence that the deep sea was populated by a thriving community of creatures previously unknown to science.

The Voyage of H.M.S. Challenger

As the work went on, Carpenter's attention increasingly turned to the physical observations being made by the naval personnel. Improved thermometers gave a more accurate picture of the distribution of temperature with depth, and more was known about the behaviour of sea water at low temperatures. Carpenter adopted the idea that density differences between equatorial and polar regions cause internal circulation in the ocean, with warm, light water moving poleward at the surface to compensate for colder, denser water spreading toward the equator in the depths[5]. Such an idea was largely unfamiliar to a British audience brought up on Rennell and his successors, so it had a mixed reception. Carpenter's fiercest critic was James Croll, who had recently put forward a theory to account for the ice ages which held that shifts in the pattern of trade winds, and the ocean currents which they generate, were responsible for climate change. Carpenter believed that if he could obtain information from the other oceans he would have irrefutable proof of his theory. A respected elder scientist, he used his contacts with other scientists and politicians to win support for a large-scale expedition – the first to have marine science as its primary objective. This resulted in the round-the-world voyage of H.M.S. *Challenger* between 1872 and 1876 (*Figure 1.9*), with a naval crew and a team of civilian scientists led by Wyville Thomson.

The voyage of the *Challenger* (*Figure 1.10*) was a major landmark in the development of oceanography, both in essence and in its findings. The observations of temperature and salinity (*Figure 1.11*) showed hitherto unsuspected features, like the spread of saline water from the Mediterranean into the North Atlantic. Neither instrumentation[16] nor theory were yet good enough to enable a detailed picture of ocean circulation to be made, but its existence in some form could no longer be doubted (*Figure 1.12*).

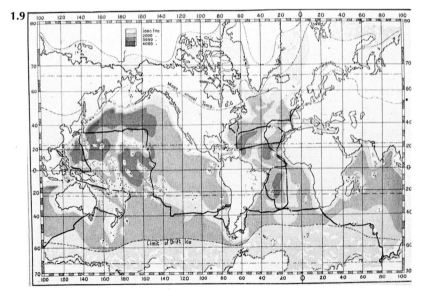

Figure 1.9 The route followed by H.M.S. *Challenger* during her oceanographic voyage round the world of 1872–1876, taken from Wild[36]. Wild was the expedition's artist and secretary to the scientific leader, C. Wyville Thomson. Other circumnavigations, including the voyages of Captain Cook in the eighteenth century, French expeditions, and Russian voyages in the early nineteenth century, had added much to geographical and scientific knowledge, but the *Challenger* expedition was the first large-scale expedition devoted primarily to the science of oceanography. For a non-scientist's view of the voyage, see the recently published letters of Joseph Matkin[22].

Figure 1.10 Challenger in the ice (reproduced from Wild[37]). (Courtesy of the Southampton Oceanography Centre, Southampton, England.)

Figure 1.11 This drawing, by Elizabeth Gulland, was one of a series commissioned as illustrations for the narrative volumes of the *Challenger Report*. It shows a scientist and members of the crew taking readings from deep-sea thermometers. [Courtesy of Edinburgh University Library (Special Collections), Edinburgh, Scotland.]

Figure 1.12 This diagram, showing the deep basins of the Atlantic Ocean, is taken from *H.M.S. Challenger, No. 7. Report on Ocean Soundings and Temperatures, Atlantic Ocean, 1876* (plate VI). This was the last of a series of preliminary reports issued by the Admiralty on the expedition's hydrographic work. It shows how the presence of submarine ridges in the Atlantic influences the distribution of bottom temperatures. The lowest temperature is found in the southwest Atlantic, where cold Antarctic bottom water flows northward. It cannot penetrate the eastern basin of the Atlantic, or the North Atlantic, because its way is blocked by submarine ridges, so the bottom temperature in these basins is slightly warmer. The pattern of topography shown here was partly known from deep-sea soundings, but partly inferred from temperature measurements. They had only a few observations to work from, so the features shown only bear a generalised resemblance to what would be seen on a modern chart – the techniques available to oceanographers today enable the sea bed to be mapped in fine detail. The interesting point to nineteenth-century scientists was that the information contained in this chart could be held to support the idea of internal ocean circulation due to density differences (both the temperature and salinity of sea water affect its specific gravity), as opposed to the pattern of largely wind-driven currents at the surface. This idea was being hotly contested at the time and was one of the reasons for the *Challenger* expedition. This principle has now long been accepted, but the chart also shows a feature which has aroused much excitement within the working lifetime of present-day oceanogra-

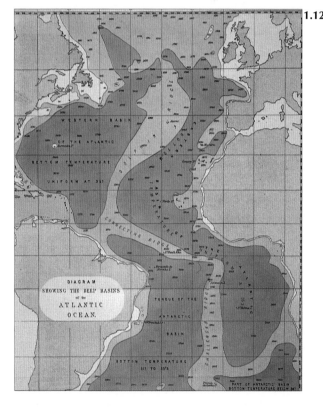

phers. This is what is now known as the Mid-Atlantic Ridge. Nineteenth-century cable surveyors were surprised to find that the greatest depths in the North Atlantic Ocean were not in mid-ocean, but to either side. A combination of soundings and deep-sea temperature observations suggested to the *Challenger* staff that the ridge might continue into the South Atlantic, and on the voyage home in 1876 they carried out soundings that showed this was so. Geological theory at the time could not easily explain such a feature and it was not until the 1960s that it became widely accepted that it is, in fact, a spreading centre at which new ocean floor is being created, and part of a world-wide system whose dynamics are explained by the theory of plate tectonics (see Chapter 8). Connecting ridges, like the Walvis Ridge, whose approximate course is shown here linking to southern Africa, are now thought to be due to hot-spot activity.

1.13a

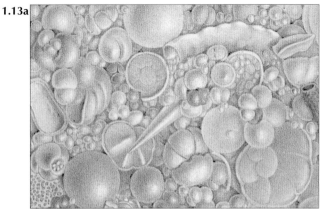

1.13b

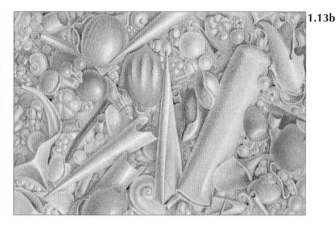

Figure 1.13 (a) Globigerina ooze and (b) Pteropod ooze, from Murray and Renard[19], plate XI, *Figures 5* and *6*, respectively. One of the principal scientific achievements of the *Challenger* expedition was to produce the first global map of what the sea bed is made of. John Murray, who edited the *Challenger Report* after the death of Wyville Thomson in 1882, had made observations during the voyage which enabled him to show that, in most parts of the ocean, sediments reflect the composition of marine life in the surface layers (plankton). Globigerina ooze was named after the remains of microscopic calcareous species of Foraminifera which form it. In the Southern Ocean, siliceous remains of single-celled phytoplankton (plants) predominate in diatom ooze. In deep water, far from land, where calcium carbonate dissolves and there are no terrigenous sediments (derived from land), with the exception of volcanic pumice which can float long distances, the *Challenger* found slowly accumulating 'red clay' and manganese nodules (see Chapters 6 and 7). (Courtesy of the Southampton Oceanography Centre, Southampton, England.)

Sediment samples (*Figure 1.13*) collected during the expedition formed the basis for the first world chart of sea-floor deposits. However, the emphasis both during the voyage and afterward was on marine biology. Of the 50 volumes of the *Challenger Report*, edited by John Murray after Thomson's early death, 33 were reports on the zoological collections, contributed by specialists from both Europe and America. Most of the species described were new to science and the reports are still a basic work of reference for oceanographers. The expedition also aided the growth of marine science in a more general sense. Scientists in other countries used the example of the *Challenger* to obtain government support for their work. As this progressed, the common ground between their researches gave them a sense of identity strong enough to override links with their various parent sciences, such as physics, chemistry, or biology. They called the new science 'Oceanography'.

Oceanography from the 1880s to the 1930s

While single-ship expeditions continued to make important contributions to oceanography[40], unlike the *Challenger*, which had been exploratory in nature, they tended to concentrate more on a particular area or problem. The *Challenger* expedition had been organised without much time to prepare. On the whole, it made use of well-tried techniques which were already a little old-fashioned. It was during the post-*Challenger* period that many basic

types of oceanographic equipment, such as current meters, reversing thermometers, and self-closing nets, became standard[13,16]. There was considerable variation in the provision for oceanographic work from one country to another, depending on local customs and arrangements for supporting science. In some countries, as, for example, Germany, which had only recently been unified, state funding was relatively generous. The German government supported the Stazione Zoologica at Naples (whose founder, Anton Dohrn, was German), as well as oceanographic expeditions. A number of government bodies were involved in various aspects of marine science and a research institute (the Institut für Meereskunde), attached to Berlin University, was set up in 1900. In America, Maury's methods and ideas had been criticised by more orthodox scientists, but when he supported the South in the American Civil War, marine science suffered through the loss of what Schlee[32] has described as 'his stubborn and passionate interest in the sea and his ability to channel funds toward its exploration'. Schlee[32] shows how oceanographic work carried out by US government agencies actually declined in the latter part of the nineteenth century, with the exception of Pillsbury's survey of the Gulf Stream (in the Coast Guard steamer *Blake*) in the early 1880s.

Prince Albert I of Monaco, a wealthy patron of science and himself an active oceanographer[13,24] (*Figures 1.14–1.17*), established an oceanographic institute in Paris and the Musée Océanographique

1.14

1.15

1.16

Figures 1.14–1.17 Prince Albert I of Monaco – his statue, by Françoise Cogné, stands in the gardens adjacent to the Musée Océanographique, Monaco (**1.14**) – used his wealth to further a number of sciences, but oceanography was his principal interest[13]. He was an enthusiastic yachtsman and made almost annual voyages in his own research ships[24] from the 1880s until the outbreak of war in 1914 (he died in 1922). **1.15** shows one of his vessels, the Princesse Alice I (courtesy of Musée Océanographique, Monaco). With the help of specialist assistants, and visiting colleagues from France and other countries, he investigated a wide range of physical and biological problems during these cruises. Much attention was paid to the improvement of existing apparatus and to the development of new methods[16]. One of the Monaco inventions was the 'nasse triédrique'[26], (**1.16**) a baited trap which could be lowered to predetermined depths to catch creatures that evaded traditional nets and trawls. Prince Albert benefited the French oceanographic community by founding an institute for research and teaching in Paris. His superbly situated Musée Océanographique at Monaco (**1.17**), inaugurated in 1910, continues to provide a valuable resource for visitors and students interested in oceanography, and its history (courtesy of Musée Océanographique, Monaco).

1.17

1.18

Figure 1.18 Alexander Agassiz on board the US Fisheries vessel *Albatross*. Agassiz was a mining engineer, and one of the leading marine zoologists of his day (the Swiss zoologist Jean Louis Agassiz was his father[13]). He used his wealth to build up his father's Museum of Comparative Zoology at Harvard and undertake research expeditions, many of them to study coral reefs, in the hope of throwing new light on the origin of atolls. In the late 1870s he made three dredging cruises to the Caribbean in the US Coast Survey Ship *Blake*. The US Fish Commission was founded in the early 1870s, one of a number of such national organisations to come into being at that time (see text). Its ocean-going research vessel, the *Albatross*, built in 1882, was used by Agassiz on several expeditions, for which he paid part of the expenses. The first cruise, in 1891, was undertaken to compare deep-sea fauna on the Pacific side of the Isthmus of Panama with Caribbean forms and perhaps arrive at an approximate date for the closure of the sea-way between North and South America, which his earlier researches on the Atlantic side had suggested must have persisted until comparatively recent geological times. (Courtesy of the Southampton Oceanography Centre, Southampton, England.)

at Monaco in the first decade of the twentieth century. In the US the situation outlined above had led to a greater reliance on private funding. One of the outstanding marine scientists of the late nineteenth century was Alexander Agassiz (*Figure 1.18*), who financed expeditions in the Pacific[13,32]. The Woods Hole Oceanographic Institution was set up in 1930 through the agency of the National Academy of Sciences, with an endowment from the Rockefeller Foundation[32]. Only a few such specialised research institutes for oceanography were founded before World War II, though some others, such as the Geophysics Institute at Bergen in Norway, established in 1917, also did important marine work.

Existing educational and administrative structures did not generally lend themselves to such developments, and private wealth was not usually available on the scale required for such enterprises. In spite of this and the somewhat differing attitudes to science in different countries, marine science developed through a variety of local, national, and international agencies during the period from the 1880s to the 1930s. This happened partly through the expansion of higher education. University departments of oceanography were created – in the UK there was one at Liverpool, established in 1919, and one at Hull, established in 1928. Other scientific departments also did marine work. A considerable number of marine biological laboratories were also attached to universities, although some, like the Marine Biological Association's laboratory at Plymouth, were maintained by private bodies. John

Murray's Scottish Marine Station for Scientific Research in the 1880s had been a short-lived attempt to create a more diversified institution.

In the early twentieth century, with more attention being paid to the importance of physical oceanography than to marine biology, laboratories began to widen their interests. The Scripps Institution of Oceanography in the US took this step to its logical conclusion when it transformed itself from a marine station in 1925[32].

During this period, whatever the prevailing attitude to support for science, most technologically advanced nations began paying more attention to marine research because of a variety of economic and political needs. They also discovered the benefits of international co-operation. By 1900 much of the marine survey work connected with submarine cables was being done by the cable companies themselves, but naval hydrographers also continued their interest in deep-sea work[28]. At the International Geographical Congress held in Berlin in 1899, it was decided that details of all these soundings should be collected onto a continually updated General Bathymetric Chart of the Ocean. This work was undertaken by Prince Albert of Monaco, and after his death, by the International Hydrographic Bureau which had been established in Monaco in 1921.

Owing to the growing economic importance of fisheries, scientific research in this area developed rapidly during the late nineteenth century. National research organisations were set up in many countries[13] to study the life histories of food fish, in the hope of reviving dwindling fisheries and creating new ones. From the start it became apparent that it was not sufficient just to study individual species; more needed to be known about biological diversity in selected regions, and about the physical environment and its influence on populations. This was the starting point for much local activity, such as H.B. Bigelow's biological survey of the Gulf of Maine in the early 1900s[32]. However, the wider questions posed had a considerable impact on the development of oceanography in the broad sense. In the 1890s, Scandinavian oceanographers proposed that there should be a joint programme of observations and this resulted in the setting up of the first intergovernmental body for marine science, the International Council for the Exploration of the Sea (ICES), in 1902. For several years the Council maintained a Central Laboratory in Norway, where important work was done toward improving oceanographic apparatus. Surveys of important fishing grounds were carried out in North America, Australia, and by the European nations, both at home and in their colonies around the world. British scientists working for the Discovery Committee carried out research in the Southern Ocean (*Figure 1.19*) during the 1920s and 1930s, with the aim of putting the whaling industry on a sustainable basis.

Work undertaken by these various organisations, as well as by individual institutions and expeditions, contributed to the growing knowledge of many aspects of the oceans during the late nineteenth and early twentieth centuries. The two most important areas of research developed during these years were ocean circulation studies and biological

oceanography. In the 1870s the American meteorologist, William Ferrel, drew attention to the deflecting effect of the Earth's rotation (the Coriolis force) on ocean currents[32], but it was the circulation theorem of Vilhelm Bjerknes and work by Bjorn Helland-Hansen and Fridtjof Nansen on currents in the Norwegian sea, carried out in the fishery steamer *Michael Sars* in the early 1900s, that formed the basis for modern dynamical oceanography. The behaviour of oceanic winds and currents had been of crucial importance for Nansen's famous attempt to drift to the North Pole in the *Fram* (1893–1896). Observations he made then also formed the basis of two important discoveries by V.W. Ekman[35]. One was the existence of internal waves, now known to occur naturally throughout the ocean, at the interface between layers of water of differing density. In northern seas such waves are generated when a ship moves through a shallow layer of fresh water, originating from rivers or melting ice, that overlays normal sea water and hampers the ship's progress, a phenomenon known to sailors as 'dead water'. The 'Ekman spiral' is the name now given to the discovery that the direction of near-surface currents is increasingly deflected with depth. This results from the Earth's rotation and from frictional forces, and causes a mean current drift to the right of the wind in the northern hemisphere (and to the left in the southern hemisphere).

Important contributions were also made at this time to our knowledge of general oceanic circulation. In his report on the work of the German research ship *Meteor* in the 1920s, Georg Wüst[39] incorporated data from earlier expeditions to show the origin and distribution of the main Atlantic

Figure 1.19 R.R.S. *Discovery II*, at Port Lockroy, Wiencke Island, Palmer Archipelago, off the Antarctic Peninsula, in January 1931. The *Discovery II* was built in 1929 for the Discovery Committee, to replace the sailing vessel *Discovery* (originally built for Captain Scott in 1901) which had been used for initial work in the seas around South Georgia from 1925–1927. *Discovery II* worked throughout the Southern Ocean during the 1930s, contributing to an understanding of ocean circulation and the marine environment. A central theme was the study of the distribution and life history of Antarctic krill, the principal food of southern hemisphere baleen whales. After World War II, *Discovery II* became the research vessel of the newly formed National Institute of Oceanography (later Institute of Oceanographic Sciences), until replaced by the modern R.R.S. *Discovery* (built 1962). (Courtesy of the Southampton Oceanography Centre, Southampton, England.)

1.19

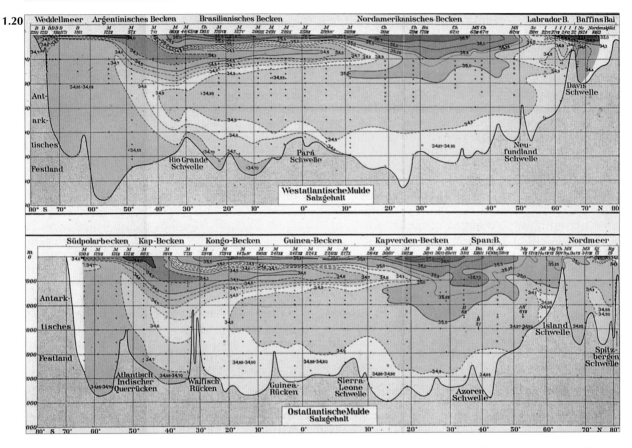

Figure 1.20 Diagram (Plate 33) by Georg Wüst[39] showing the longitudinal distribution of salinity in the Atlantic Ocean. Wüst was one of the leading physical oceanographers of the mid-twentieth century. As a young man in the 1920s he participated in the cruise of the German research vessel *Meteor* and used data collected during the voyage, together with observations made by other expeditions, from the *Challenger* onward, to show how density differences, due to salinity and temperature variations, are responsible for the movement of water masses within the body of the ocean.

water masses (*Figure 1.20*). His picture was extended by Discovery Committee scientists working in the Southern Ocean in the 1920s and 1930s[24].

Fisheries science was also responsible for the growth of interest in marine productivity. This field was developed by German scientists of the 'Kiel School'[18] in the late nineteenth century. Their researches showed that life in the sea depends on primary production – the phytoplankton, single-celled plants grazed by the zooplankton on which all other creatures in the sea depend, either directly or indirectly, for food (modern research also highlights the importance of bacteria in primary production – see Chapter 6). Victor Hensen[18] developed vertical nets for sampling plankton in order to obtain quantitative data on the productivity of the ocean. In the Plankton Expedition of 1889 he extended his work to the North Atlantic in the ship *National*. Karl Brandt[18] showed that the growth of phytoplankton is controlled by the supply of available nutrients.

Modern Oceanography

When Thomas Wayland Vaughan, the Director of the Scripps Institution of Oceanography, surveyed world oceanographic activity on behalf of the US National Academy of Sciences in the 1930s, he found nearly 250 institutions devoted to marine research throughout the world, from Russia to Japan and from Australia to Argentina. He had been asked to undertake this survey because of concern that not enough was being done to encourage the study of the sea, which, as we have seen, requires special conditions to make progress. The trend toward increasing state support for science was soon to be greatly accelerated by the demands of war, and of the Cold War which followed, and by the role of science in post-war economic growth.

Oceanography developed rapidly during and after World War II. It shared in the general expansion of science, as the impact of science and technology on almost every aspect of modern life has led to increased funding for research. National

Figure 1.21 Launching the clover-leaf buoy from R.R.S. *Discovery* in 1967. The 1950s and 1960s saw a dramatic expansion in oceanography, encompassing people, ideas, and methods. This apparatus was one of a number of different types designed and made at the National Institute of Oceanography for studying properties of waves. (Courtesy of Laurence and Pamela Draper, Rossshire, Scotland.)

1.21

defence needs and economic concerns, particularly the search for new energy sources (offshore oil and gas), as well as renewed concern about fisheries stocks, are among the factors that have led to a wider interest in marine research. Such needs have increased rather than diminished with time. While the political international situation has eased overall, environmental worries have come to the fore, in particular the problem of climate change, in which the oceans must play an important part.

However, it is interesting to see how many of the ideas and techniques that are important in present-day oceanography have their roots in the first half of the twentieth century. For example, work on underwater sound was begun before World War I[9], but then developed rapidly in the search to perfect a means of submarine detection by echo location (sonar). The invention of the hydrophone had many other scientific and peace-time applications. Echo-sounding by ships provided more detailed information about the topography of the sea floor. Seismic surveying, to investigate the internal structure of the sea bed, was first employed by Maurice Ewing in the 1930s. After 1945, this work played an important part in obtaining the information on which modern ideas about sea-floor spreading and plate tectonics are based[32]. Magnetic surveys also contributed to these developments, as did gravity measurements, first obtained at sea in submarines in the 1930s by the Dutch scientist, F.A. Vening-Meinesz. Modern oceanographers rely heavily on continuous-recording instruments, which in many areas have replaced the older single-observation measuring devices (see Chapter 19). Such devices were foreshadowed by the bathythermograph, invented by US scientists in the 1930s to measure the temperature of the upper layers of the ocean. This and similar ideas were taken up by scientists during World War II.

Other fields of study originated at that time, including wave research. Until then no way had been found to study sea waves that was not purely descriptive[32]. A major breakthrough occurred when a war-time research team based at the Admiralty Research Laboratory, in Teddington, England, developed a method of analysing wave spectra that enabled their components to be identified. This work continued after the war at the newly established National Institute of Oceanography, where new kinds of wave recorders were developed to measure waves at sea (*Figure 1.21*).

During the past 50 years such developments, and others described in the following chapters, have profoundly transformed our knowledge of the oceans. The main subject areas – marine physics, including ocean-circulation studies, knowledge of the sea floor (which has played a major role in the revolution of the earth sciences leading to modern theories of plate tectonics), marine chemistry, and biological oceanography – have all made important advances. The development and use of more sophisticated apparatus has been assisted by the introduction of computers and satellites, which permit the gathering, transmission, and analysis of data in quantities that would have been inconceivable a generation ago, let alone to the first scientific observers of the sea in the 1660s. Oceanography is still an expensive science, with the ship remaining a fundamental tool, although this too may change in the future. This expense has provided a strong incentive for co-operation and sharing on a more formal basis, so joint expeditions have become an important aspect of modern oceanography, from the International Indian Ocean Expedition of the early 1960s to the World Ocean Circulation Experiment (WOCE), designed to throw new light on the relation between the oceans and climate, in the 1990s.

General References

Deacon, M.B. (1971), *Scientists and the Sea, 1650–1900: A Study of Marine Science*, Academic Press, London and New York, 445 pp.

Herdman, W.A. (1923), *Founders of Oceanography and their Work: An Introduction to the Science of the Sea*, Edward Arnold, London, 340 pp.

McConnell, A. (1982), *No Sea Too Deep: The History of Oceanographic Instruments*, Adam Hilger, Bristol, 162 pp.

Rice, A.L. (1986), *British Oceanographic Vessels, 1800–1950*, The Ray Society and Natural History Museum, London, 193 pp.

Schlee, S. (1973), *The Edge of an Unfamiliar World: A History of Oceanography*, E.P. Dutton, New York, 398 pp.

References

1. Birch, T. (ed.) (1744), *The Works of the Honourable Robert Boyle*, A. Millar, London, Vol. 3, pp 105–113 and 378–388.
2. Bourne, W. (1578), *A Booke Called the Treasure for Traveilers*, Thomas Woodcocke, London, 269 pp.
3. Burstyn, H.L. (1966), Early explanations of the role of the Earth's rotation in the circulation of the atmosphere and the ocean, *Isis*, 57(2), 167–187.
4. Burstyn, H.L. (1971), Theories of winds and ocean currents from the discoveries to the end of the seventeenth century, *Terrae Incognitae*, 3, 7–31.
5. Deacon, M.B. (1971) *Scientists and the Sea, 1650–1900: a Study of Marine Science*, Academic Press, London and New York, 445 pp.
6. Deacon, M.B. (1985), An early theory of ocean circulation: J.S. von Waitz and his explanation of the currents in the Strait of Gibraltar, *Progr. Oceanogr.*, 14, 89–101.
7. De Vorsey, Jr, L. (1976), Pioneer charting of the Gulf Stream: the contributions of Benjamin Franklin and William Gerard de Brahm, *Imago Mundi*, 28, 105–120.
8. Duhem, P. (1913–1959), *Le Système du Monde: Histoire des Doctrines Cosmologiques de Platon à Copernic*, 9 vols, Hermann, Paris.
9. Hackmann, W. (1984), *Seek and Strike: Sonar, Anti-Submarine Warfare and the Royal Navy, 1914–54*, HMSO, London, 487 pp.
10. Hales, S. (1754), A descripton of a sea gage, to measure unfathomable depths, *Gentleman's Magazine*, 24, 215–219.
11. Halley, E. (1697), The true theory of the tides, extracted from that admired treatise of Mr Isaac Newton, entitled *Philosophiae Naturalis Principia Mathematica*; being a discourse presented with that book to the late King James, *Phil. Trans. Roy. Soc. Lond.*, 19, 445–457.
12. Hansard (1961), *Parliamentary Debates*, Fifth Series, 638, 235.
13. Herdman, W.A. (1923), *Founders of Oceanography and their Work: An Introduction to the Science of the Sea*, Edward Arnold, London, 340 pp.
14. Kircher, A. (1678), *Mundus Subterraneus*, Vol 1, 3rd edn, Apud Joannem Janssonium à Waesberge & Filios, Amstelodami, pp 134–135.
15. Marsigli, L.F. (1725), *Histoire Physique de la Mer*, Aux dépens de la Compagnie, Amsterdam, 195 pp.
16. McConnell, A. (1982), *No Sea Too Deep: The History of Oceanographic Instruments*, Adam Hilger, Bristol, 162 pp.
17. McConnell, A. (1990), The flowers of coral – some unpublished conflicts from Montepellier and Paris during the early 18th century, *Hist. Phil. Life Sci.*, 12, 51–66.
18. Mills, E.L. (1989), *Biological Oceanography: An Early History, 1870–1900*, Cornell University Press, Ithaca and London, 378 pp.
19. Murray, J. and Renard, A.F. (1891), Deep-sea deposits, *Report on the Scientific Results of the Voyage of H.M.S. Challenger during the Years 1872–76*, HMSO, London, Plate XI.
20. Pinsel, M.I. (1981), The wind and current chart series produced by Matthew Fontaine Maury, *Navigation*, 28(2), 123–137.
21. Rehbock, P.F. (1979), The early dredgers: 'naturalizing' in British seas, 1830–1850, *J. Hist. Biol.*, 12(2), 293–368.
22. Rehbock, P.F. (ed.) (1992), *At Sea with the Scientifics: The Challenger Letters of Joseph Matkin*, University of Hawaii Press, Honolulu, 415 pp.
23. Rennell, J. (1832), *An Investigation of the Currents of the Atlantic Ocean*, J.G. and F. Rivington, 359 pp.
24. Rice, A.L. (1986), *British Oceanographic Vessels, 1800–1950*, The Ray Society and Natural History Museum, London, 193 pp.
25. Rice, A.L., Burstyn, H.L., and Jones, A.G.E. (1976), G.C. Wallich, M.D. – megalomaniac or mis-used oceanographic genius?' *J. Soc. Bibliogr. Natur. Hist.*, 7, 423–450.
26. Richard, J. (1910), *Les Campagnes Scientifiques de S.A.S. le Prince Albert Ier de Monaco*, Imprimerie de Monaco, p. 33.
27. Richardson, P.F. (1980), The Benjamin Franklin and Timothy Folger charts of the Gulf Stream, in *Oceanography: The Past*, Sears, M. and Merriman, D. (eds), Springer, New York, pp 703–717.
28. Ritchie, G.S. (1967), *The Admiralty Chart. British Naval Hydrography in the Nineteenth Century*, Hollis and Carter, London, 388 pp. (Reprinted 1995 by Pentland Press, Edinburgh.)
29. Ross, Sir J.C. (1847), *A Voyage of Discovery and Research in the Southern and Antarctic Regions, during the years 1839–43*, Vol. 2, John Murray, London, facing p. 354.
30. Ross, M.J. (1982), *Ross in the Antarctic: The Voyages of James Clark Ross in H.M. Ships* Erebus *and* Terror, *1839–43*, Caedmon of Whitby, Whitby, Yorkshire, 276 pp.
31. Royal Society (1667), Directions for observations and experiments to be made by masters of ships, pilots, and other fit persons in their sea-voyages, *Phil. Trans. Roy. Soc., Lond.*, 2, 433–448.
32. Schlee, S. (1973), *The Edge of an Unfamiliar World: A History of Oceanography*, E.P. Dutton, New York, 398 pp.
33. Stoye, J. (1994), *Marsigli's Europe, 1680–1730: The Life and Times of Luigi Ferdinando Marsigli, Soldier and Virtuoso*, Yale University Press, New Haven and London, 356 pp.
34. Vossius, I. (1993), *A Treatise Concerning the Motions of the Seas and the Winds*, together with *De Motu Marium et Ventorum Liber*, Deacon, M.B. (ed.), Scholars' Facsimiles and Reprints, Delmar, New York, for the John Carter Brown Library, 376 pp.
35. Walker, J.M. (1991), Farthest North: Dead water and the Ekman spiral, *Weather*, 46(4), 103–107; 46(6), 158–164.
36. Wild, J.J. (1877), *Thalassa: An Essay on the Depth, Temperature, and Currents of the Sea*, Marcus Ward, London, facing p. 16.
37. Wild, J.J. (1878), *At Anchor. A Narrative of Experiences Afloat and Ashore during the Voyage of H.M.S. Challenger from 1872 to 1876*, Marcus Ward, London, 198 pp.
38. Winsor, M.P. (1976), *Starfish, Jellyfish, and the Order of Life: Issues in Nineteenth Century Science*, Yale University Press, New Haven and London, 288 pp.
39. Wüst, G. (1928), Der Ursprung der Atlantischen Tiefenwässer, *Zeitschrift der Gesellschaft für Erdkunde zu Berlin Sonderband zur Hundertjahrfeier der Gesellschaft*, pp 506–534.
40. Wüst, G. (1964), The major deep-sea expeditions and research vessels, 1873–1960, *Progr. Oceanogr.*, 2, 1–52.

CHAPTER 2:

The Atmosphere and the Ocean

H. Charnock

Introduction

The atmosphere and the ocean are held on the Earth by gravity and irradiated by the Sun. Both are shallow relative to the radius of the Earth and the motions within them are slow relative to that due to the Earth's rotation: they have similar dynamics. As they share a common boundary it is attractive to treat them as a single, coupled, system, but the physical and chemical properties of air and water are so very different that meteorology and oceanography have developed separately and at different rates. Electromagnetic radiation (light, microwaves, radar, radio ...) travels easily through the atmosphere and this, together with the commercial and economic benefit of weather-forecasting, has led to the existence of a global network of meteorological observing stations, making and transmitting regular routine surface and upper-air observations, as well as increasing information from sensors on satellites.

Most meteorological stations are on land, but winds, waves, currents, and weather affect ships, so have been observed by mariners from time immemorial: some selected merchant ships now report their observations as part of the global meteorological system.

Observation of the ocean away from the surface has developed more slowly (Chapter 4): water is almost opaque to electromagnetic radiation so oceanographers are essentially restricted to indirect observation of a fluid through which they cannot see; the period from the *Challenger* Expedition of 1872 has been described as a 'century of undersampling'.

More recent technological development has clarified some processes in restricted areas: it is now accepted that, like the atmosphere, the ocean is a three-dimensional turbulent fluid, with interacting motions and processes on all time- and space-scales. Understanding it requires an observing system which does not yet exist, one that may evolve if national governments perceive a need to forecast conditions in the ocean to the same extent as those in the atmosphere. In the meantime, we can attempt to use our knowledge of the atmosphere and of the underlying ocean surface, together with the limited but increasing observations of the deep ocean, to study the global transfers of heat, of fresh water, and of momentum in the coupled atmosphere–ocean system.

Energy and Water Exchanges in the Atmosphere–Ocean System

The general circulation of the atmosphere and ocean, regarded as a single system, is determined by the distribution of its sources and sinks of energy. Much the dominant external source is the absorption of energy from the Sun (solar radiation of wavelength between 0.2–4 µm). The near constancy of the temperature of the system requires an equal outgoing flux of energy in the form of terrestrial radiation (long-wave radiation of wavelength 4–100 µm).

The distribution and transformation, within the system, of the incoming solar radiation is complicated. Typical global mean values of the major components are shown in *Figure 2.1*, which indicates that nearly half the absorbed solar radiation reaches the Earth's surface, most of which is ocean. This sunlight does not penetrate far into the ocean (even in the clearest water 99% is absorbed in the upper 150 m): heat gains and losses take place at and close to the surface, so the ocean is relatively inefficient thermodynamically. Directly driven motions, due to the cooling of surface water at high latitudes (the thermohaline circulation) are slow. Indirectly driven motions (the wind-driven circulation) arise from the transfer of heat (and water vapour) from the ocean to the atmosphere, where it is converted by complicated processes into depressions and anticyclones, the winds of which provide energy to generate ocean waves and drive ocean currents. Their energy in turn is dissipated into heat by small-scale (viscous) processes and re-radiated. It is a complex, inefficient system with many interlocking components of different scale.

It can be seen from *Figure 2.1* that of the 153 W/m² received at the Earth's surface as solar radiation, 54 W/m² is lost as long-wave radiation

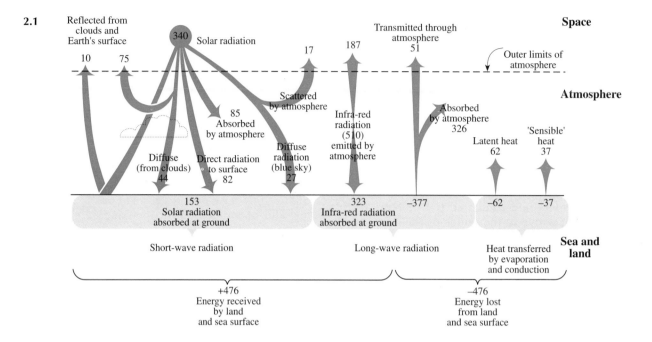

Figure 2.1 The average radiation balance (in W/m²) for the Earth as a whole (based on Nieburger *et al.*[11]).

and 37 W/m² is transferred by the conduction of heat (sensible heat) from a warmer sea to cooler air above. Energy is also used in evaporation of water vapour from the sea surface to the drier air above, heat (latent heat) being required to convert a liquid into a gas. Water has a high latent heat, 62 W/m² being used in evaporation. The latent heat is released to the atmosphere only when the water vapour condenses back into clouds, of liquid water or ice, often far from where it evaporated from the ocean into the atmosphere. Much of the cloud formation happens at relatively low altitudes, justifying the statement that the atmosphere can be regarded as heated from below, making it an active thermodynamic system with important vertical as well as horizontal motion.

Although both atmospheric and oceanic motions are ultimately powered by solar heating, the important working substance of the global heat-engine is water; as vapour, as liquid, and as solid ice. The infra-red characteristics of water vapour make it a major agent of long-wave radiative heat transfers (see *Figure 2.1*) The solar radiation absorbed at the surface is used to evaporate water, the large latent heat of which is released to the atmosphere when and where condensation occurs; the distribution of evaporation, precipitation, continental run-off, and ice are crucial to the determination of the salinity and so to the watermass structure and to the thermohaline circulation of the ocean. For these and

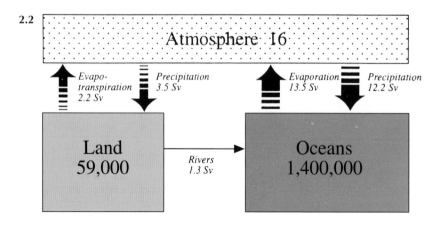

Figure 2.2 Estimates of the amount of water (in 10³ km³) in the atmosphere, the land, and the ocean, and of the fluxes between them (1 Sv = 10⁶ m³/s) (from Schmitt[15]; the figures are derived from Baumgartner and Reichel[2]).

for many other reasons the basic question posed by weather and climate – 'what happens to the sunshine?' – must be supplemented by asking – 'what happens to the water?'

The distribution of evaporation and precipitation over the ocean is clearly vital to understanding climate: unfortunately, this part of the hydrological cycle is not well known. Most treatments of the water cycle concentrate on exchanges over land (they are concerned with man's use of water, for agriculture and industry as well as human consumption), but it is estimated that 78% of the global precipitation goes into and 86% of the global evaporation comes from the ocean (see *Figure 2.2*). The corresponding transfer of (latent) heat represents a major component of the heat balance of the atmosphere and of the ocean.

Meridional Fluxes

The near-constancy of the global mean temperature implies a balance between the absorbed solar radiation and the outgoing long-wave radiation, as indicated in *Figure 2.1*, but their variation with latitude is significantly different. The solar radiation is absorbed mainly in the tropics: the long-wave radiation, determined mainly by the radiative properties of the atmosphere and the underlying surface, is observed to be much less dependent on latitude (*Figure 2.3*). It follows that there is a flux of heat from the tropics to the poles. Measuring this flux is important: it is fundamental to the maintenance of

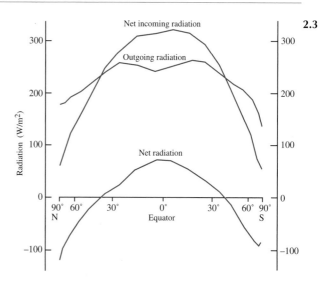

Figure 2.3 Zonal averages of radiation at the top of the atmosphere. Net incoming radiation peaks in the tropical regions; outgoing long-wave radiation varies less with latitude. To maintain a constant temperature, the excess of radiative heating within 35° of the equator is transferred poleward by atmospheric and oceanic motions to compensate for the deficit of radiative heating nearer the poles. A uniform bias of 9 W/m² has been subtracted from the net incoming radiation to ensure a balance between the total incoming and the total outgoing radiation (from Bryden[4]; the figures are derived from satellite observations reported by Stephens *et al.*[18]).

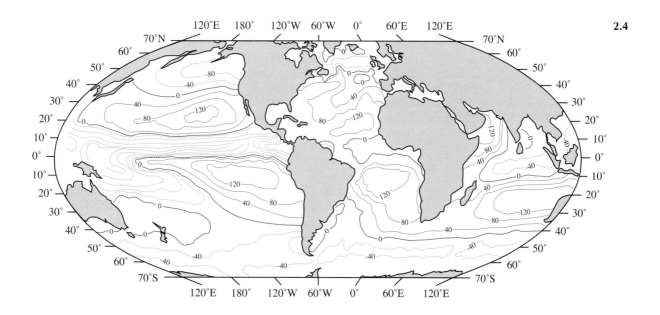

Figure 2.4 A chart of evaporation minus precipitation (*E-P*) over the ocean. Units are cm/yr; solid lines indicate *E>P*, dashed lines *E<P* (from Schmitt and Wijffels[17]; the figures are derived from Schmitt *et al.*[16]).

2.5

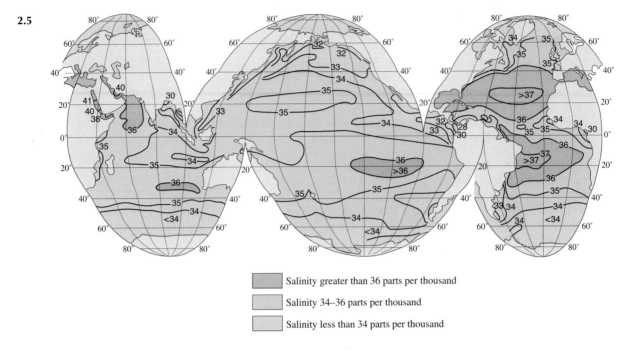

Salinity greater than 36 parts per thousand

Salinity 34–36 parts per thousand

Salinity less than 34 parts per thousand

climate and its magnitude provides a significant constraint on atmospheric, oceanic, and coupled general circulation models. Understanding how the meridional heat flux is maintained requires a detailed knowledge of atmospheric and oceanic structures and processes.

Although the distribution of evaporation and (especially) of precipitation over the ocean is not well known, major features of the net water flux between atmosphere and ocean (evaporation minus precipitation, E–P) can be recognised (*Figure 2.4*). The distribution is roughly zonal (except for the North Indian Ocean); precipitation dominates in subpolar regions and especially in the Intertropical Convergence Zone (the ITCZ) at the thermal equator and in the South Pacific Convergence Zone to the northeast of Australia. Elsewhere, the subtropics have an excess of evaporation. There is a clear association with the surface salinity of the ocean (*Figure 2.5*); the latitudes where (E–P) is high are associated with high salinity at the ocean surface (and with deserts on land). The general pattern is of net annual precipitation at high and at low latitudes, with net annual evaporation between. That the mean structure is not changing implies meridional transports of fresh water by the ocean. As river transports are negligible in comparison, equal and opposite flows of water must occur in the atmosphere. These, in turn, transfer significant quantities of heat.

To assign numerical values to these important meridional fluxes presents difficulties. The radiation balance of the Earth as a whole has for many

Figure 2.5 Salinity of surface waters during the northern summer (from Gross[6]; the figures are derived from Sverdrup *et al.*[20] and from later sources).

years been measured as part of a major programme (the Earth Radiation Budget Experiment, ERBE) using orbiting satellites fitted with radiometers sensitive to solar and to long-wave radiation. Although the measurements are technically demanding and present difficult problems of data analysis, recent results imply a near-balance between the global annual incoming solar radiation and the corresponding outgoing long-wave radiation. The imbalance was small (less than 10 W/m²), but its distribution produces some uncertainty in the derived values of the meridional heat transfer by the atmosphere and ocean combined. *Figure 2.6* shows that it peaks near latitudes 30°N and 40°S where the flux amounts to almost 6 PW (1 petawatt, PW = 10^{15} W).

Estimates of the meridional flux of water and heat (and momentum) by the atmosphere can be made using the observations made daily from meteorological upper-air stations, where balloon-borne instruments measure the wind, the temperature, and the humidity in relation to pressure (height). The local meridional flux of, say, water vapour is carried by the northward component of the wind, V, and is measured by ρVq where q is the specific humidity and ρ the air density. The total meridional flux is given by $[\rho Vq]$ where the square brackets represent a mean value over the depth of the atmosphere and around a latitude circle, over a

2.6

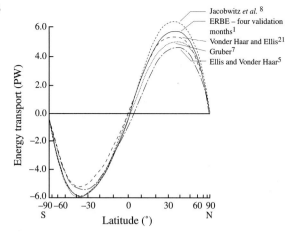

Figure 2.6 The total meridional transport of energy estimated from ERBE data for April, July, and October 1975 and January 1986 (from Barkstrom et al.[1]; other satellite-based estimates are from Vonder Haar and Ellis[21]; Ellis and Vonder Haar[5]; Gruber[7]; Jacobwitz et al.[8]).

time for which [V] vanishes. The calculations can be made by using the upper-air observations directly or by assimilating them together with other observations into a suitable atmospheric computer model and using interpolated values.

The meridional flux of water vapour in the atmosphere (which is discussed later) implies a corresponding flux of latent heat. To estimate the total heat flux this must be supplemented by the sensible heat flux given by $[\rho C_p V q]$, where C_p is the specific heat of air at constant pressure and q the temperature. Some results are shown in *Figure 2.7* for the total heat flux (compare with *Figure 2.2*). The atmospheric fluxes peak at about 40°N and 40°S where the flux amounts to about 4 PW.

There is no oceanic equivalent of the meteorological upper-air observing network, but it is possible to estimate the meridional flux of heat (and that of salt) by an analogous method: the heat transport is estimated by calculating the covariance (Vq) between the temperature and the inferred, as distinct from the directly measured, northward velocities. In the deep ocean away from the surface and the sea floor, the currents can be treated as frictionless and unaccelerated, so can be calculated as if they are in geostrophic balance: that is, the force due to the horizontal variation in pressure is balanced by the force due to the Earth's rotation, which is proportional to the speed of the current. In this case the currents flow along the isobars like winds on a weather map. The observations used are from a high-quality east–west hydrographic section between two continental land masses, giving accurate measurements of temperature and salinity

(and therefore density), at all depths. The difficulty in determining deep-ocean currents geostrophically is that although the density field is known at all depths the pressure field is not known at any one, so the calculation of currents using the geostrophic balance requires a knowledge of the total transport. Given measurements or reliable estimates of the transport of western boundary currents, and making allowance for the near-surface currents due to the frictional drag of the wind, convincing estimates of the meridional heat flux by the ocean can be obtained. Estimates of the heat flux northward across latitude 24°N in the Atlantic have been made three times over the last 35 years and found to be consistently close to 1.2 PW to the north. A transoceanic hydrographic section made in 1985 has provided observations along 24°N in the Pacific from which an oceanic heat flux of 0.8 PW was obtained. Since there is virtually no heat flux across 24°N in the Indian Ocean the total ocean heat-flux across 24°N amounts to some 2.0 PW to the north.

If the atmospheric flux (*Figure 2.7*) across 24°N is taken to be 2.3 PW, the total (atmosphere + ocean) flux is 4.3 PW, significantly less than the 5.7 PW required from the most recent analysis of the ERBE results. Further transoceanic sections are being made as part of the World Ocean Circulation Experiment and should serve to clarify the climatically important meridional heat flux. They will be particularly important in the southern hemisphere,

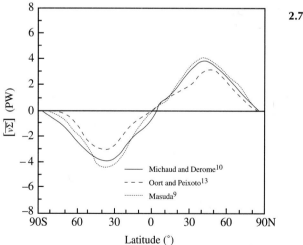

2.7

Figure 2.7 Annual mean northward flux of energy by the atmosphere, December 1985 to November 1986 (from Michaud and Derome[10]). The estimates of Michaud and Derome[10] are compared with those of Oort and Peixoto[13] (observed values from upper-air observations) and of Masuda[9] (ECMWF and GFD model assimilations for 1978–1979).

where the upper-air observing stations are even more sparse than in the northern hemisphere and the estimates of atmospheric heat flux correspondingly less certain.

Information on the meridional transport of heat and water can also be obtained by considering the heat and water balance of the ocean. A particular vertical column of the ocean gains heat from the absorption of solar radiation and loses it by the emission of long-wave radiation and as the latent heat used in evaporation: it gains or loses sensible heat depending on whether the sea is colder or warmer than the air. Empirical formulae are used to estimate the heat gains and losses using long series of surface meteorological observations of cloudiness, windspeed, air temperature, and sea temperature made regularly from merchant ships. In general there is a surplus or a deficit of heat; if the imbalance is averaged over a time such that heat associated with change of the heat content of the ocean can be neglected, the remainder, errors apart, is due to advection of heat by ocean currents. From averages over a latitude circle between continents the meridional oceanic heat flux is obtained by integrating from the North or the South Pole, or from other latitudes where the heat flux is zero or known.

Estimating the meridional flux of fresh water by ocean currents using surface meteorological observations is, in principle, simpler since water is effectively confined to the ocean–ice–atmosphere system. The meridional flux can be estimated by integrating zonal averages of Evaporation minus Precipitation minus Runoff (E–P–R) from some latitude where the flux is known. In practice estimates are uncertain since none of the terms is well known, there being as yet no good way of measuring rainfall at sea. Precipitation estimates are made using statistics of its frequency, type, and intensity from the 'present weather' reports from routine merchant ship weather reports.

The accuracy of meridional flux estimates made from heat and water exchanges at the sea surface appears to be inferior to those made by the 'direct' covariance technique: the surface ships' observations are not well distributed, they have random and probably some systematic errors, and the empirical formulae used are necessarily crude. Also it is difficult, sometimes impossible, properly to allow for the transport brought about by flow from one ocean basin to another, notably the Antarctic Circumpolar Current, and the Indonesian Through Flow from the Pacific to the Indian (see, e.g., *Figure 3.3*). The flow from the North Pacific to the Arctic Basin is small but well established. Some ocean heat budget analyses have

agreed with the 'direct' flux estimates but they are not consistent from ocean to ocean and do not help to resolve the discrepancy between space, atmospheric, and oceanic estimates. Surprisingly, perhaps, recent estimates from the water balance show a gratifying anti-symmetry with the 'direct' atmospheric values (*Figure 2.8*).

The trans-Atlantic and trans-Pacific hydrographic sections used to estimate meridional heat transport have also been used to estimate the corresponding salt transport. Given the transport of water from the North Pacific to the Arctic through the Bering Strait these salt transports can be used to calculate the corresponding freshwater fluxes across 24°N in the Atlantic and the Pacific. The results are 0.9 Sv southward in the Atlantic and 0.6 Sv northward in the Pacific. The former is about 0.1 Sv greater and the latter about 0.1 Sv smaller than that calculated from the surface (E–P–R) distribution, their sum being in reasonable agreement with it and with the global estimates using upper-air observations: the global flux is small, as would be expected for a latitude in the centre of the subtropical gyre.

Meridional flux calculations using climatological data from merchant ships appear to have reached their limits of accuracy. They have produced valuable patterns of heat and water exchange at the sea surface but the extent to which they can be used quantitatively is limited: 'direct' methods in both the atmosphere and the oceans seem inherently more accurate. There is a need for more analysis of meteorological upper-air data and especially for more high-quality trans-oceanic hydrographic sections.

Transfer Mechanisms

The 'direct' methods estimate meridional fluxes by calculating the covariance of temperature and of humidity with observed winds (in the atmosphere) and with inferred currents (in the ocean). They therefore provide information both about the fluxes and also about the mechanisms by which they are brought about.

The structure of the much better observed atmosphere is reasonably well known; it can be thought of as a gigantic – and very inefficient – heat engine, absorbing heat in the hot equatorial belt and losing it nearer the poles. Since much of the heat transfer is due to evaporation and condensation of water vapour, one can think of the mechanism as a steam engine and use such expressive terms as 'the equatorial firebox' and regard the giant cumulo-nimbus cloud towers of the ITCZ as the cylinders in which the steam is converted into liquid water drops and ice crystals. How does the

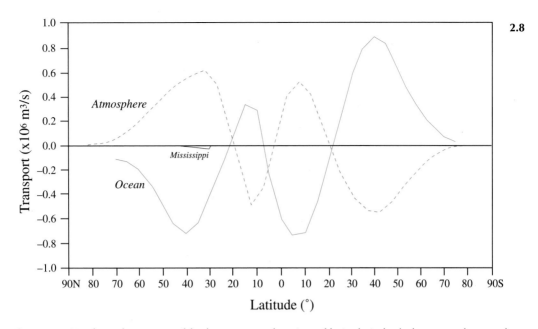

2.8

Figure 2.8 Northward transport of fresh water as a function of latitude in both the atmosphere and the ocean (from Schmitt and Wijffels[17]; the ocean transports are derived from Wijffels *et al.*[22] and the atmospheric transports are derived from Peixoto and Oort[14]). The transport of the River Mississippi is shown for comparison.

atmosphere bring about this huge heat transfer from the tropics to the polar regions? There are two main mechanisms, one which operates between the equator and about 30°N and S and another which operates at higher latitudes.

In the mechanism which operates at lower latitudes the essential feature is that the air rises at the thermal equator and then spreads northward and southward toward the poles. Most of it sinks again at about 30°N or S and returns toward the equator, acquiring an easterly component due to the rotation of the Earth, and forming the Trade Winds. These carry less energy than the poleward winds aloft so there is a net transport of energy from low latitudes to high.

The essential features of this circulation are the two closed wind patterns north and south of the equator; they were described by Hadley as long ago as 1735 and are known as Hadley cells. Their existence provides a rational explanation for the equatorial rain belts (ascending air motion) and the subtropical anticyclones around 30°N or 30°S. The subtropical highs are linked with descending air motions, little rainfall, and hence the desert areas on land and the deep blue and high salinity water at sea. The associated horizontal winds are the Trades, at the surface, and the subtropical jet stream aloft. The subtropical jet stream is a band of strong westerly winds in latitudes 30° to 40° at a height of about 12 km. It arises because the air

which has risen from the equator tends to retain its angular velocity as it moves poleward. When it reaches 30° or so it is moving much faster toward the east than the earth beneath it (*Figures 2.9 and 2.10*).

The Hadley cells provide a good explanation for many observed phenomena, and the mechanism of the general atmospheric circulation in low latitudes is relatively well understood. In both hemispheres, poleward of 30°N and S, the heat transfer is not brought about by motions of the Hadley cell type but by disturbances on a smaller scale. Here the prevailing winds are westerly at all levels, unlike those of the tropical regions where the Trade Winds at the surface have an easterly component. Embedded in the westerlies of middle latitudes are the travelling depressions which provide typically unsettled weather. They have a complicated wind and temperature structure, such that air which is warmer than average tends to be going poleward and air which is colder than average tends to be going toward the equator. On average the heat is transferred in the required sense – from equator to pole. The disturbed westerly regime and the complex travelling depressions are essential features of the heat transfer. The situation is obviously complicated but the overall features of the general circulation of the atmosphere are clear. Dynamical meteorologists are using computers to simulate the behaviour of the atmosphere, with results

2.9

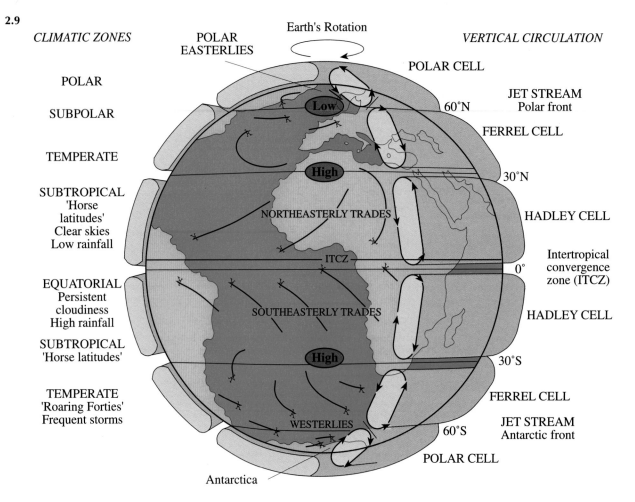

CLIMATIC ZONES   POLAR   Earth's Rotation   VERTICAL CIRCULATION
EASTERLIES

POLAR CELL

POLAR

SUBPOLAR   60°N   JET STREAM
Polar front

FERREL CELL

TEMPERATE

30°N

SUBTROPICAL
'Horse
latitudes'   HADLEY CELL
Clear skies
Low rainfall   NORTHEASTERLY TRADES

Intertropical
ITCZ   0°   convergence
zone (ITCZ)

EQUATORIAL
Persistent
cloudiness
High rainfall   SOUTHEASTERLY TRADES   HADLEY CELL

SUBTROPICAL
'Horse latitudes'   30°S

TEMPERATE   FERREL CELL
'Roaring Forties'
Frequent storms   WESTERLIES   JET STREAM
Antarctic front
60°S

POLAR CELL

Antarctica

Figure 2.9 Schematic representation of features of the general circulation of the atmosphere (from Gross[6]).

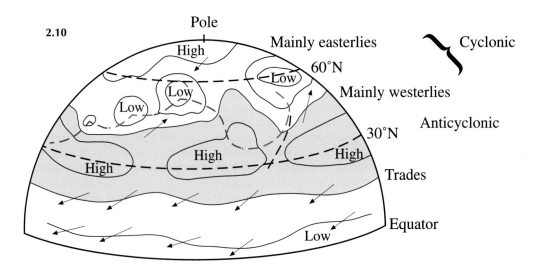

Pole

2.10

Mainly easterlies   Cyclonic

High

60°N

Low

Low   Low   Mainly westerlies

Low   Anticyclonic

30°N

High   High   High   Trades

Equator

Low

Figure 2.10 Atmospheric pressure at the surface of an idealised Earth as it might be on a particular day in comparison with the long-period average of *Figure 2.9* (from Sutcliffe[19]).

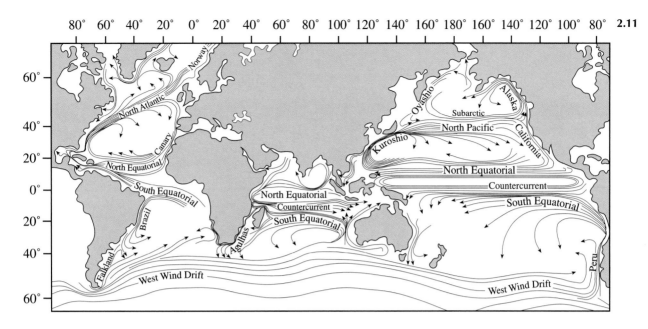

Figure 2.11 Schematic chart of the climatological average of the surface currents of the ocean (based on NRDC[12]).

sufficiently life-like to make one feel that the essential physics of the situation is correct.

The winds that bring about transfer of heat and water in the atmosphere have to be such as to conserve angular momentum about the Earth's axis, and their frictional drag at the Earth's surface such as to maintain the near-constancy of the rate of rotation of the Earth. These considerations limit the latitudinal extent of the Hadley circulation, from its ascent at the ITCZ to latitudes between 30°N and 30°S (*Figure 2.9*). Poleward of this tradewind belt the transfer of momentum, as well as that of heat and water, is brought about by smaller scale features such as depressions and anticyclones.

Calculations of momentum transfer using upper air observations are analogous to those of heat and water vapour transfer: they involve the covariance [ρVq] of the northerly and easterly components of the wind velocity. They demonstrate that the large-scale eddies bring about most of the required transports of momentum, except at low latitudes, where meridional cells play an important role, especially in the vertical exchange of momentum. The middle latitude westerlies and the low latitude North-East and South-East Trades are a consequence of the conservation of angular momentum (*Figure 2.10*): the westerlies have to be the stronger to keep the rotation of the Earth constant.

Although the atmosphere and the ocean have certain basic similarities – both are vast bodies of fluid on a rotating Earth – their differences must be recognised. They have marked differences in physical properties, especially those controlling the transmission of radiant energy, and it must be recognised that their geometry also plays an important role. There are no barriers in the atmosphere which correspond to the continental barriers to the oceans.

Momentum transfer in the ocean is less well known but again it is found that large-scale eddies in the ocean are important in transferring angular momentum from strong surface currents into ocean depths. The distribution of surface currents, thought to be mainly wind-driven, has been compiled from ships' reports of their drift from their calculated course. The general features are shown in *Figure 2.11*: although the North Pacific, and the North Atlantic are quite different in shape they have a rather similar current pattern, or general circulation. There is an anticlockwise circulation (or gyre) in their northern parts and a huge clockwise one in the south. This is conspicuously asymmetric, the currents being much stronger in a narrow region near the western boundary of the North Pacific and the North Atlantic (the situation in the Indian Ocean is complicated by the seasonal variation due to the monsoon). These strong boundary currents (the Atlantic Gulf Stream and the Pacific Kuroshio) are the best known currents of the ocean.

Near the equator in all three oceans there are two west-flowing Equatorial Currents. The South Equatorial Current lies at or south of the equator and the North Equatorial Current to the north of it. In the Pacific and Indian Oceans, and in part of the Atlantic, the two west-flowing Equatorial

35

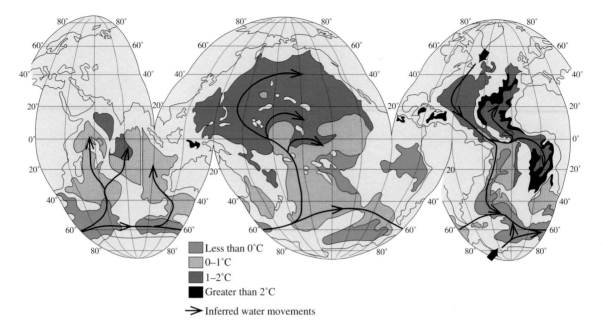

Less than 0°C

0–1°C

1–2°C

Greater than 2°C

→ Inferred water movements

Figure 2.12 An impression of the flow pattern at 4000 m in the ocean. The major inputs are the North Atlantic Deep Water (NADW), which enters at the northern end of the Western Basin of the Atlantic, and the Weddell Sea Bottom Water, which enters from the margin of the Antarctic continent adjacent to the South Atlantic (based on Gross[6] and Broecker and Peng[3]).

Currents are separated by an Equatorial Countercurrent flowing toward the east.

In the Southern Ocean, surrounding the Antarctic, there is no continental barrier (though the relatively narrow Drake Passage may have a similar effect) and the main surface current flows round the Earth as an east-going flow referred to as the Circumpolar Current or as the West Wind Drift.

It must be emphasised that the charts of ocean currents are climatological; that is, they are based on averages of observations made over a long time. On any particular occasion a ship may find a current very different from the average current portrayed on the Pilot Chart. This is especially noticeable in the region of a fast western boundary current, such as the Gulf Stream, which meanders and changes the position of its axis in an unpredictable way (see *Figure 4.8*). In such a case the climatological chart can be misleading, for the observations at a particular place are averaged over a long period, irrespective of whether the current is present or not. It can be seen that a strong narrow current which varies in position is represented on a climatological chart as a broader but slower current. In this way, what may be called the 'climatological Gulf Stream' (as represented on a time-averaged climatological chart) is perhaps ten times wider, and considerably weaker, than the Gulf Stream on any particular occasion.

Currents at greater depth are much less well known; the mean currents are small and are obscured by the variability of the large-scale eddies that have been found to be ubiquitous in the deep ocean. A rough impression of the currents at 4000 m depth is given in *Figure 2.12*.

The estimation of meridional fluxes from transoceanic hydrographic sections has provided information on major differences between the mechanisms in the North Atlantic and the North Pacific. Study of the hydrographic section at 24°N in the North Atlantic shows that the heat flux is mainly due to a deep vertical-meridional cell. Warm and relatively saline water flows north near the surface, ultimately losing enough heat in winter to sink to great depth and return southward. The warm Gulf Stream water flows north near the surface but does not return south at similar depth; only after a high-latitude cooling process does it return to the south as deep water. The smaller northward heat transfer at 24°N in the Pacific is due to a nearly horizontal circulation: relatively warm water flows north in the Kuroshio on the western side and in the near surface layer, loses heat in the subtropical and subpolar North Pacific and returns southward in the central and eastern Pacific, at colder temperatures but still at depths less than 800 m. Unlike the North Atlantic, the North Pacific has no source of deep water and its deep circulation is correspondingly slower. These ocean circulation differences

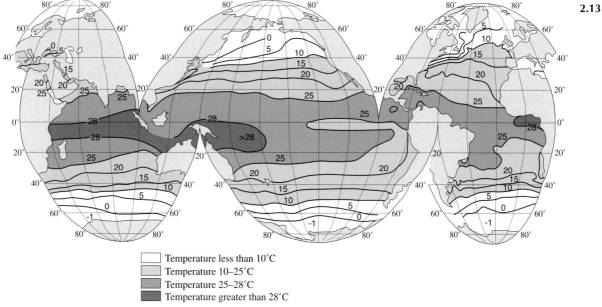

Temperature less than 10°C
Temperature 10–25°C
Temperature 25–28°C
Temperature greater than 28°C

Figure 2.13 Average temperature at the sea surface in February (from Gross[6]; the values are derived from Sverdrup *et al.*[20] and from later sources).

are consistent with the marked climatic differences in the climate of the North Atlantic and the North Pacific – the Atlantic is much warmer, especially at subpolar and polar latitudes (*Figure 2.13*).

The climate of the ocean–atmosphere system depends on complicated interactions between the meridional fluxes, brought about by motions of relatively large scale, and the near-surface vertical fluxes, brought about by motions on a much smaller scale in the complicated turbulent interacting air–sea boundary layer. Processes that determine the properties of the ocean take place at its surface: the resulting horizontal transfers within it are such as to maintain the surface sources and sinks. How this is done is the central problem for observational and theoretical oceanographers alike. It is the target for those making computer models of the ocean and especially for those attempting to model the coupled ocean and atmosphere: they have great difficulty in matching the horizontal motions in the two media to the flux of heat and water between them.

The Coupled Atmosphere–Ocean Boundary Layer

Most of the atmosphere and most of the ocean, most of the time, can be treated as frictionless and adiabatic. But in some places there are vital processes that are more complicated: clouds, fronts in the atmosphere and the ocean, and especially the turbu-

lent boundary layers which exist near the Earth's surface and at the sea floor. Most important is the coupled boundary layer of the atmosphere and ocean which occupies a layer typically 1 km in height above and 100 m in depth below the sea surface. This is a region in which many energy exchanges and transformations take place, processes which determine the properties of the ocean and many of those in the atmosphere. Two processes of the air–sea boundary layer are of particular importance: the production of vertical velocities and the transfer of boundary layer air to the less turbulent free atmosphere above, and of boundary water to the less turbulent deep ocean below.

The frictional stress of the wind on the sea surface, on a rotating Earth, drives a mass transport to the right (left) of stress direction in the northern (southern) hemisphere. If the stress varies from place to place it produces convergences or divergences that lead to vertical motion (*Figure 2.14*); a similar effect happens in the atmosphere. The resulting vertical motions are fundamental to the general circulation of the atmosphere and ocean through their effect on the vorticity balance (see *Box 2.1*). In the atmosphere they lead to the formation of cloud and rain, complications that are not present in the ocean. In the ocean, however, vertical motions (usually smaller than those due to wind stress) are also produced by the difference between Evaporation and Precipitation (*E–P*). Evaporation and precipitation are also fundamental to the near-surface energy exchanges and so to establishing the properties of the lower atmosphere and the upper ocean. These properties are communicated, by complicated and little understood processes, to the

Box 2.1. The Vorticity Balance of the Ocean

All the large-scale flow in the atmosphere and ocean is affected by the rotation of the Earth. The relatively small-scale flows encountered in bathroom, kitchen, or laboratory are dominated by other forces, so the effect of the rotation of the Earth is beyond our usual experience. The flow patterns it produces in the atmosphere and ocean sometimes seem bizarre.

The basic notion is one of the vorticity, or spin, which anything on a rotating globe must have. If one imagines a man standing astride the North Pole, for example, it is obvious that he will be rotating, or spinning, about his own axis, at the same rate as the Earth – once per day. If the same man now stands astride the equator, the Earth continues to rotate about its axis but he no longer rotates about his. His local rate of rotation or spin is zero. So the spin which affects anything on the Earth is zero at the equator and increases to one revolution per day at the poles.

In these examples, our hypothetical man – we could equally have considered a parcel of fluid – was at rest relative to the Earth. He – or the fluid – could also have been rotating on his own axis relative to the Earth. The total spin is obviously made up of two components – the spin relative to the Earth and that due to the rotation of the Earth beneath it. It is a difficult but important concept – important because, in relatively shallow fluids like the atmosphere and the ocean, a body of water moves in such a way that its total spin, its vorticity, stays constant.

The vorticity relative to the Earth of a column of fluid increases if it is stretched: the stretching decreases the diameter and the rotation increases to conserve the column's angular momentum. Conversely shrinking the column increases the diameter and the vorticity decreases. The spinning of an ice skater provides a familiar example. To maintain its total vorticity constant, on the rotating Earth, a shrinking column must move equatorward and a stretched column poleward.

In the upper ocean shrinking and stretching are brought about by a vertical gradient of vertical velocity, produced near the surface by wind-stress convergence (*Figure 2.14*), and by *E–P*. Over much of the North Atlantic the wind distribution is such as to produce shrinking, so motion toward the equator. The *E–P* distribution is such as to produce stretching, so poleward motion of smaller magnitude. The resulting southward transport leads to the westward intensification of wind-driven currents, the necessary poleward return flow being accomplished in narrow western boundary regions – like the Gulf Stream – whose dynamics are more complicated.

In the deep ocean there is a very slow upward velocity to compensate for the sinking of deep water at high latitudes. This stretches the water column, and would be expected to produce generally northward transport with equatorial return flow being confined to a narrow region on the western boundary. Such western boundary currents are observed but the northward transport is obscured by the large-scale eddies and by the effects due to sea floor topography.

2.14

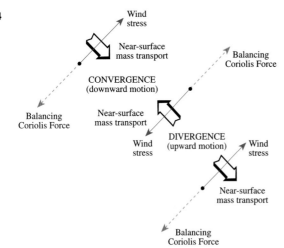

Figure 2.14 The balance of forces in the upper 50 m of the ocean produces a depth mean transport at right angles to the wind stress (to the right in the northern hemisphere). Variations of wind stress from place to place produce convergence or divergence in the surface layers and corresponding vertical velocities below the surface.

air above and to the water below the coupled boundary layer: they determine the properties of the atmospheric airmasses and the oceanic watermasses.

Unfortunately knowledge and understanding of the complicated boundary layer processes is very poor. There is some empirical information about the marine atmospheric boundary layer, especially its lowest 100 m or so (which allows the climatological estimates of heat transfer and evaporation) but uncertainty remains about the interaction of wind and waves. Processes near the top of the marine atmospheric boundary layer, where there is frequently cloud, are much less understood. In the upper layers of the ocean surface waves make observations difficult: there is a need for more information on the effect of spray and of bubbles, as well as on the living and non-living particulate matter that determines the transparency and the absorption of solar radiation. The transfer properties of the helical (Langmuir) circulation (*Figure 2.15*) are still uncertain. In both atmosphere and

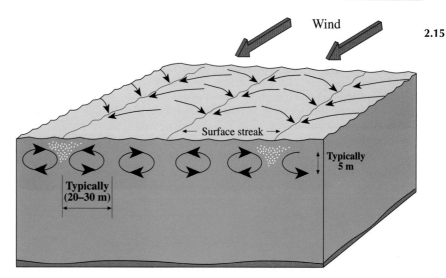

2.15

Figure 2.15 Helical (Langmuir) circulations in the upper layers of the ocean producing wind-rows of floating material on the surface.

ocean we need a reliable way of predicting the vertical extent of the boundary layer and a better understanding of how fluid is transferred from the boundary layer to the free atmosphere and the deep ocean. The problems are technically demanding, and the theory underlying them notoriously difficult, but an improved knowledge of the coupled air–sea boundary layer is vital to our need to understand and to model the atmosphere and the ocean.

Climate Studies

Numerical modelling of the atmosphere is already well advanced, both for weather forecasting and for simulating climate and climatic change. Ocean modelling is advancing rapidly as increasing computer power allows finer resolution to represent ocean eddies, which are smaller than atmospheric disturbances (see Chapter 4). The discrepancies between meridional flux estimates using different methods should soon be clarified: the ERBE project continues, more trans-oceanic hydrographic sections are being made, and there is the prospect of improved atmospheric flux estimates. The results will provide valuable constraints and tests of climate models simulating a coupled atmosphere and ocean.

Near-surface meteorological observations are vital as input to weather forecast models and for the verification of climate models. Their calculation relies on realistic simulation of the atmosphere–ocean boundary layer, but rapid progress in our knowledge of this complicated region is not to be expected. There is also a great need for better observation (preferably from space) and improved simulation of precipitation.

The difference between evaporation and precipitation (E–P) is important as providing a vertical velocity and affecting the vorticity balance, and it provides an input (a haline buoyancy flux) to the surface buoyancy flux which is sometimes comparable to that of the thermal buoyancy flux. It has been suggested that climatic freshening of high-latitude surface water could stop the formation of deep water in the North Atlantic: the overturning meridional cell is thought to be very sensitive to the freshwater flux. Recent estimates show that the thermal buoyancy flux dominates the haline buoyancy flux at high latitudes, suggesting that large changes in (E–P) would be needed to bring about what has been called the 'haline catastrophe'. However the effect of continental run-off, and especially of the freezing and melting of ice (*Figure 2.16*) could be significant, especially in the areas

2.16

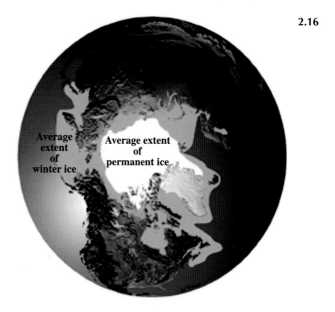

Figure 2.16 Chart to indicate the extent of permanent and winter ice in the Arctic (*from Oceanus,* **37**, 2, 1994).

where deep water formation now occurs. There is increasing evidence that both (*E–P*) and thermal changes are closely connected to changes in ocean circulation, and so to climate on decadal and longer time-scales. The problem requires an improved understanding of how the motions of the atmosphere and the ocean maintain the sources and sinks of heat and water, and of the coupling between the large scale horizontal winds and currents and the small scale turbulent transfer processes of the coupled air–sea boundary layer.

General References

Gill, A.E. (1982), *Atmosphere–Ocean Dynamics*, Academic Press Inc., Orlando.

Gross, M. Grant (1992), *Oceanography, a View of Earth*, Simon and Schuster, Englewood Cliffs, New Jersey.

Peixoto, J.P. and Oort, A.H. (1992), *Physics of Climate*, American Institute of Physics (AID Press), Woodbury, New York.

References

1. Barkstrom, B.R., Harrison, E.F., and Lee, R.B. (1990), Earth Radiation Budget Experiment: preliminary seasonal results, *EOS*, **71**, 297–305.

2. Baumgartner, A. and Reichel, E. (1975), *The World Water Balance*, Elsevier, New York.

3. Broecker, W.S. and Peng, T.-H. (1982), *Tracers in the Sea*, Lamont–Doherty Geological Observatory, Columbia University, New York.

4. Bryden, H. (1993), Ocean heat transport across 24°N latitude, in *Interactions Between Global Climate Subsystems: the Legacy of Hann*, McBean, G.A. and Hantel, M. (eds), Geophysical Monographs, **75**, 65–75.

5. Ellis, J. and Vonder Haar, T.H. (1976), *Zonal Average Earth Radiation Budget Measurement from Satellites*, Atmos. Sci. Papers 240, Colorado State University, Fort Collins, Colorado.

6. Gross, M. Grant (1992), *Oceanography, a View of Earth*, Simon and Schuster, Englewood Cliffs, New Jersey.

7. Gruber, A. (1978), *Determination of the Earth–Atmosphere Radiation Budget from NOAA Satellite Data*, NOAA Tech. Rep. NESS 76, Washington DC.

8. Jacobwitz, H., Smith, W.L., Howell, H.B., and Hagle, F.W. (1979), The first 18 months of planetary radiation budget measurement from the Nimbus-6 ERB experiment, *J. Atmos. Sci.*, **36**, 501–507.

9. Masuda, K. (1988), Meridional heat transport by the atmosphere and the ocean: analysis of FGGE data, *Tellus*, **40A**, 285–302.

10. Michaud, R. and Derome, J. (1991), On the mean meridional transport of energy in the atmosphere and oceans as derived from six years of ECMWF analyses, *Tellus*, **43A**, 1–14.

11. Nieburger, M., Edinger, J.G., and Bonner, W.D. (1982), *Understanding our Atmospheric Environment*, W.H. Freeman and Company, San Francisco.

12. NRDC (1946), *Summary Technical Report*, Division 6, Office of Naval Research, Washington DC.

13. Oort, A.H. and Peixoto, J.P. (1983), Global angular momentum and energy balance requirement from observations, *Advances in Geophysics*, **25**, 355–490.

14. Peixoto, J.P. and Oort, A.H. (1983), The atmospheric branch of the hydrological cycle and climate, in *Variations in the Global Water Budget*, Street Perrott, A. (ed.), pp 5–65, Reidel, Dordrecht.

15. Schmitt, R.W. (1994), *The Ocean Freshwater Cycle*, JSC Ocean Observing System Development Panel, Texas A&M University, College Station, Texas.

16. Schmitt, R.W., Bogden, P.S., and Dorman, C.E. (1989), Evaporation minus precipitation and density fluxes for the North Atlantic, *J. Phys. Oceanogr.*, **19**, 1208–1221.

17. Schmitt, R.W. and Wijffels, S.E. (1993), The role of the ocean in the global water cycle, in *Interactions Between Global Climate Sub-systems: the Legacy of Hann*, McBean, G.A. and Hantel, M. (eds), Geophysical Monographs, **75**, 77–84.

18. Stephens, G.L., Campbell, G.C., and Vonder Haar, T.H. (1981), Earth radiation budgets, *J. Geophys. Res.*, **86**, 9739–9760.

19. Sutcliffe, R.C. (1966), *Weather and Climate*, Wiedenfield and Nicolson, London.

20. Sverdrup, H.U., Johnson, M.W., and Fleming, R.H. (1942), *The Oceans, their Physics, Chemistry and Biology*, Prentice Hall, New York.

21. Vonder Haar, T.H. and Ellis, J. (1974), *Atlas of Radiation Budget Measurements from Satellites*, Atmos. Sci. Papers 231, Colorado State University, Fort Collins, Colorado.

22. Wijffels, S.E., Schmitt, R.W., Bryden, H.L., and Stigebrandt, A. (1992), Transport of fresh water by the oceans, *J. Phys. Oceanogr.*, **22**, 155–162.

CHAPTER 3:

The Role of Ocean Circulation in the Changing Climate

N.C. Wells, W.J. Gould, and A.E.S. Kemp

Overview of Climate System

The relatively warm and stable climate which we have had for the past 10,000 years, since the end of the most recent glaciation, has been essential for the evolution of cultivated crops, which led to the development of human settlements rather than to a nomadic existence, and hence to the development of civilisation.

Recent research has shown that changes in ocean circulation played a key role in controlling climate change and regulating the glacial–interglacial cycles that have been a hallmark of the northern hemisphere climate for the past 2 million years. Understanding the nature of this link between ocean circulation and climate change is now a key goal of research in this area. In particular, the identification of periods of very rapid climate change in the recent past has given a new urgency to these studies.

The measurement of climate change by scientific instruments dates back to the invention of the thermometer by Galileo in the sixteenth century. It was, however, not until considerably later that systematic methods were applied to the measurement of temperature and rainfall, which allowed the concept of climatology to develop. These measurements have been and are unevenly distributed over the globe, with the majority covering the landmasses and, particularly, the well-populated areas of the northern hemisphere. The oceans, covering nearly 71% of the Earth's surface, have not been neglected; indeed, one of the first responsibilities of the UK Meteorological Office, when it was formed in 1854, was to implement the systematic recording of surface observations from commercial ships plying the trade routes of the world. Despite such efforts, measurements in the southern hemisphere, particularly over the Southern and Indian Oceans, remain rather sparse. We shall see demonstrated that these poorly observed oceans are a key element in the Earth's climate.

Recent interpretation of ice cores from Greenland[15] has shown that the climate system is not as stable as was once thought and may undergo extremely rapid changes (e.g., 5–7°C in a decade). As we concern ourselves with possible man-made influences on climate, one of our great challenges in the closing decade of the twentieth century is to measure and understand natural climate variability across the globe, together with the role played by the oceans.

Given the limitations of the observing network, where does our scientific evidence for climate change over longer periods come from?

Indirect evidence in ice cores from Antarctica and Greenland, and from sediment cores studied during the CLIMAP (Climate: Long Range Interpretation, Mapping, and Prediction) initiative and, more recently, from the Ocean Drilling Programme, have provided reliable indicators of past climate variations. The measurement of the air trapped in the Vostok ice core, from Antarctica, has revealed the levels of carbon dioxide from the most recent interglacial period (circa 125,000 years BP) to the present day[16], and thus has given a benchmark from which the more recent anthropogenic contribution can be determined. Measurements of oxygen isotope ratios, contained in bubbles trapped in the ice, are now known to be a proxy for air temperatures (see Chapter 8, *Box 8.3*). Recent analyses of ice cores from Greenland have thus provided time series of temperature from the most recent interglacial to the present day. This has shown that the climate over the last 5,000 years has been remarkable for its stability. The ice cores also provide evidence for rapid changes in climate during the previous interglacial, from temperatures similar to those of today to glacial conditions on the remarkably short time-scales of decades to centuries (see later).

Evidence of longer-term changes, those over millions of years, are found in marine sediments (see Chapter 8). The glacial and interglacial cycles that occurred during the last 2 million years have been linked to variations in solar radiation, associated with variations of the orbital parameters of the Earth around the Sun, known as Milankovitch cycles. Longer-term changes may be associated with the different configurations of the oceans and continents, due to continental drift, and with variations in continental uplift and mountain building.

In addition to these interpretations of past climate we now have new powerful methods which

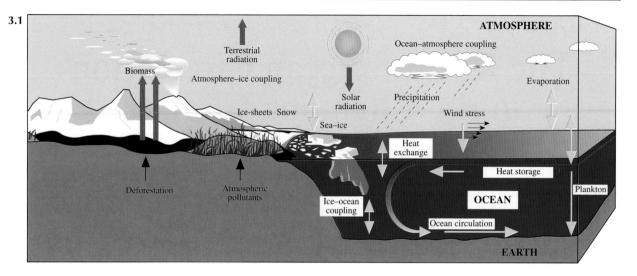

3.1

Figure 3.1 A schematic view of the Earth's climate system showing the roles of land, atmosphere, oceans, and sea ice.

are being applied to understanding how the Earth's climate works. These methods are based on an understanding of the different 'components' of the climate system (air, water, and ice), and the interactions between them (*Figure 3.1*).

Each component responds to change over different periods of time. The atmosphere takes the shortest time (of the order of a week to a month) to communicate changes throughout its mass, because winds are very much faster than ocean currents and mixing through turbulence is efficient. The dominant time-scale of the ocean is of the order of weeks to seasons in the surface layers and decades to centuries at abyssal depths. The ice caps of Antarctica and Greenland are excellent measures of longer-term change as they grow slowly on time-scales of centuries to millennia.

Mathematical computer models of each one of the components, bounded, constrained, or driven by interactions with others, allow a deeper understanding of both the components themselves and their interactions with one another. For example, mathematical models of the ocean (see *Boxes 3.1* and *3.2*) are based on the dynamical laws which govern the behaviour of the ocean. These ocean models can reproduce the large-scale, wind-driven circulations of the ocean basin gyres, and the deep vertical transports, or overturning, of the thermohaline circulation of the oceans driven by cooling at polar latitudes. The advent of faster, more powerful computers has allowed models to reproduce some of the smaller (*ca* 100 km), energetic ocean eddies (see Chapter 4), which are important for both the transport of heat and fresh water in some regions of the world's oceans and for the maintenance of the large ocean circulations. Observations are always needed to initialise and test the models, but while relatively

abundant observations are available for the atmosphere, comprehensive, synoptic ocean measurements are few and far between. The World Ocean Circulation Experiment (WOCE; see later) will provide critical ocean measurements, such as heat and fresh water transports, on a global scale.

Improvements in our ability to model the various components of the climate system will allow the development of more comprehensive climate models in which the components are coupled together. These models will aid our interpretation and understanding of the complex climate system, as well as providing predictions of socioeconomic importance.

The Role of the Ocean in Climate

The ocean and atmosphere together transfer heat from the tropics to the polar regions, at a rate of the order of 5 PW at 30°N, (equivalent to the output of 5 million large power stations) in order to balance the deficit of incoming radiation in the mid-latitudes and polar regions with the excess of radiation in the tropics. The partition of this transfer between atmosphere and ocean is not well-determined, but recent ocean measurements at 24°N have shown transports within the ocean of the order of 2 PW; a value which is comparable with the atmospheric flux at this latitude (*Figure 3.2*)[7]; see Chapter 2 for further discussion.

The oceanic component of the heat flux is provided by ocean currents, which have much longer time-scales (of the order of 10 years for the wind-driven subtropical gyres and of many decades to centuries for the vertical overturning thermohaline circulation) than the atmosphere. When one considers that a 2.5 m deep layer of sea water covering the globe has the same thermal capacity as the entire atmosphere, it can be appreciated that the

Figure 3.2 Poleward transfer of heat by: (a) ocean and atmosphere together (T_A+T_O), (b) atmosphere alone (T_A), and (c) ocean alone (T_O). The total heat transfer (a) is derived from satellite measurements at the top of the atmosphere, that of the atmosphere alone (b) is obtained from measurements of the atmosphere, and (c) is calculated as the difference between (a) and (b) (1 PW = 10^{15} W). (Based on Carrissimo *et al.*[7]; results from other investigations are added for comparison.)

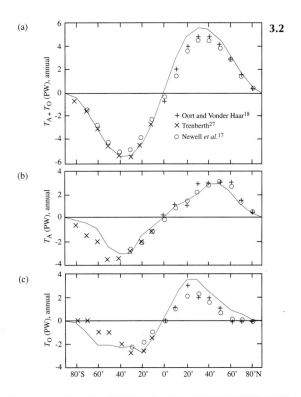

3.2

oceans make a very important contribution to the stabilisation of our climate system.

While, overall, the oceans transfer heat poleward, they also exchange heat between ocean basins in a more complex fashion. For instance, the North Atlantic Ocean loses more heat to the atmosphere than it gains from incoming radiation, so there has to be a net heat transfer from the Pacific and Indian Oceans into the South Atlantic Ocean and thence to the North Atlantic to compensate for this deficit. The thermohaline circulation, driven by the production of denser water at polar latitudes, is a mechanism by which heat is transferred within and between ocean basins. The oceans are interconnected by what has come to be known as The Global Thermohaline Conveyor Belt (*Figure 3.3*)[13].

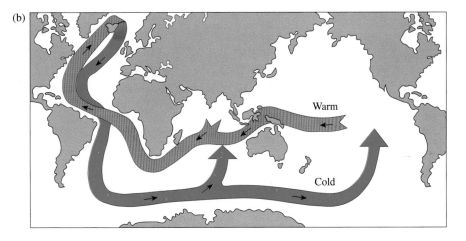

3.3

Figure 3.3 Schematic views of two versions of the 'conveyor' circulation of the oceans. Warm water associated with the surface and intermediate waters of the oceans (upper 1000 m) follows a pathway toward the northern North Atlantic Ocean, where it is subjected to intense winter cooling. This leads to the formation of cold North Atlantic deep water, which spreads southward into the Southern Ocean and returns to the Pacific Ocean. The conveyor is responsible for a northward transfer of heat throughout the whole of the Atlantic Ocean. (a) The upper water moving eastward from the Pacific into the Atlantic. (b) The upper warm water moving through the Indonesian Archipelago into the Indian Ocean and thence into the Atlantic Ocean.

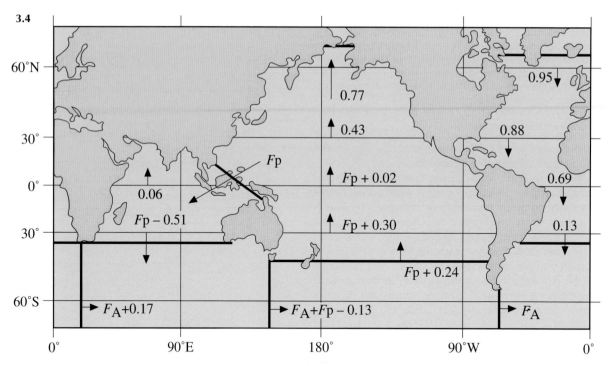

Figure 3.4 An estimate of the transfer of fresh water (x 10^9 kg/s) in the world oceans. In general, in polar and equatorial regions precipitation and river run-off exceed evaporation and hence there is an excess of fresh water, while in the subtropical regions there is a water deficit. A transfer of fresh water is required from the regions of surplus to the regions of deficit. For example, in the North Atlantic, there is southward flow of 950,000 t/s (t = metric tonne) of fresh water at 60°N, while at the equator the southward flow is 690,000 t/s – hence 26,000 tonnes of fresh water per second are evaporated in the North Atlantic. It can be seen that fresh water is exported from the North Pacific to the North Atlantic, through the Arctic Ocean. F_P and F_A refer to the fresh water fluxes of the Pacific–Indian throughflow and of the Antarctic Circumpolar Current in the Drake Passage, respectively.

The ocean also transports fresh water around the globe, another key element of the global conveyor (*Figure 3.4*). The total mass of salts in the ocean remains unchanged on time-scales shorter than geological time. In contrast, the fresh water content of the oceans changes in response to precipitation, evaporation, freezing and melting of ice, and run-off from the land. All these factors influence the dilution or concentration of ocean salt, and hence the salinity. Differences in the input of fresh water into the ocean, from one region to another, have to be balanced by a horizontal transport of fresh water by ocean currents and by sea ice in polar regions. Generally, in the subtropics there is a deficit of fresh water which is reflected in higher surface salinity, while in the higher latitudes there is an excess of fresh water and a lower surface salinity. The North Pacific Ocean is less saline than the North Atlantic, because of lower evaporation, and therefore is a source of fresh water. This water, in turn, is transported into the North Atlantic by the conveyor circulation, to make up for the deficit there.

The density of sea water depends on temperature and salinity. Heating or a decrease in salinity will decrease the density, while cooling or an increase in salinity will increase the density. The low salinity layer at the surface of the North Pacific maintains a stable stratification (low density at the surface), which cannot be destabilised (i.e., made to have a higher density at the surface than deeper in the water column) by surface cooling in the present climate state. Hence, there is no significant deep convection in the North Pacific and, consequently, little formation of cold deep water. By contrast, the northern North Atlantic, with a higher surface salinity, is destabilised by winter cooling, and produces deep cold watermasses by vertical convection. Polar watermasses, formed in the Greenland/Norwegian sea and, to a lesser extent, in the Arctic Ocean, also enter the North Atlantic and provide additional cold deep water. The transport of ice from the Arctic Ocean provides an additional source of fresh water to the North Atlantic, which through its variability can modulate the surface salinity on decadal time-scales[28].

Evidence of Climate Change in the Ocean

Despite its high thermal capacity, the ocean responds to exchanges of heat, fresh water, and momentum with the atmosphere on a range of time-scales from

3.5

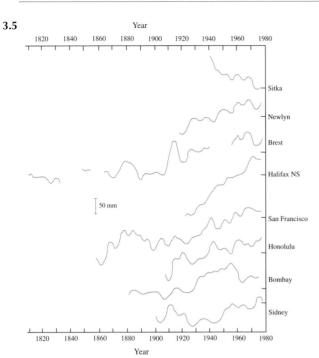

Figure 3.5 Long-term changes in annual mean sea level at selected ports. The general upward trend is seen at all but one station. The downward trend in sea level at Sitka, Alaska, is due to the vertical movements of the land at which sea level is measured (see text). The global mean sea level rise is estimated to be about 1–3 mm/year.

Measurements of temperature and salinity with useful accuracy have only been available for the past 50 years or so, and there are few places where high-quality measurements have been made over several decades. This severely limits our ability to directly measure climate-scale change in the ocean.

At the surface of the ocean there is a great deal of temperature variability caused by daily and seasonal heating and cooling, so here salinity is a better indicator of decadal change. Salinity records made in the area west of Scotland show a remarkable decrease that lasted from 1973–1979, with the lowest values in 1975. The lowest value deviated from the average by almost four times the typical variability.

Subsequent analysis of other salinity data from around the Atlantic showed that this salinity change, now referred to as the 'Great Salinity Anomaly', was not confined to Scottish waters, but was a phenomenon that took over 10 years to propagate around the North Atlantic (*Figure 3.6*)[10,12]. There is still much conjecture about the cause of the 'Great Salinity Anomaly'. Recent evidence has suggested that in the 1960s an unusually large export of ice into the North Atlantic from the Arctic Ocean may have caused the low salinity

a day and shorter right up to geological scales (glacial and interglacial periods of 10,000–100,000 years; see Chapter 8). The thermal inertia of the ocean, however, means that oceanic changes are of much smaller amplitude than those seen in the atmosphere, but nevertheless they are easily measurable on daily to seasonal scales and, with care, on decadal scales. The state of the ocean at any time is predominantly a result of exchanges with the atmosphere over the previous 100 years or so.

By far the longest time series of 'oceanographic' measurement is that for sea level (*Figure 3.5*). Measurements made for the prediction of tides go back over 100 years and, after the tidal signal is removed, show long-term trends, rising in some places, in others falling. It has to be remembered that sea level is measured relative to a fixed point on land, so the changes seen are a summation of changes in true sea level (due primarily to thermal expansion of the water, and melting of glaciers and ice caps) and the not insignificant vertical movement of the land (isostasy). This movement can be caused by 'rebound' after release from the covering of ice during the most recent glacial period or by tectonic activity. Unravelling the changes due to each individual factor is difficult[21] and yet the prediction of sea level change is of immediate and practical importance to low-lying areas and, in particular, to the inhabited coral atolls of the Pacific and Indian Oceans.

Figure 3.6 Timing of the propagation of the 'Great Salinity Anomaly' around the North Atlantic.

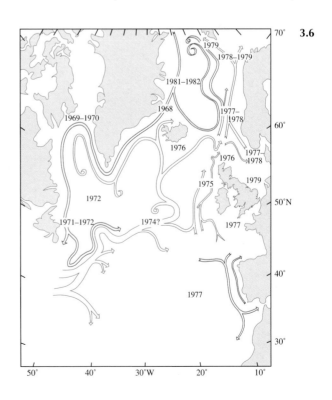

3.6

3.7 (a)

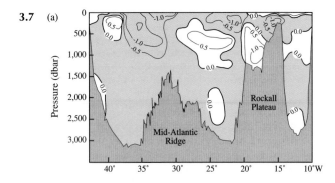

(b)

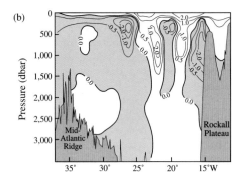

Figure 3.7 Cooling (blue areas, °C) along the section of the subpolar North Atlantic between northwest Europe and Greenland: (a) 1981–1991, (b) 1962–1991.

3.8

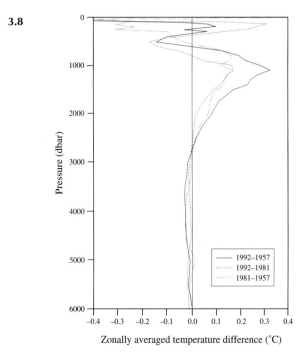

Figure 3.8 Temperature changes (°C) in the subtropical North Atlantic (24°N), 1957–1992. The measurements have been averaged across 24°N between North Africa and Florida.

anomaly, which suppressed deep convection for more than a decade[1].

Whatever the cause, the significance of such changes cannot be over-emphasised. The global thermohaline conveyor belt is driven by deep convection in high latitudes and salinity is, in many cases, the controlling factor that decides whether such deep convection is possible. In simplified terms, fresher water at the surface results in the formation of an insulating layer of ice in winter, rather than plumes of dense cold saline water. So it is possible that extreme salinity anomalies could be responsible for the transitions between glacial and interglacial episodes. We give an example of this in the next section.

In the interior of the ocean, patterns of temperature and salinity change have been determined in three ways, from time series at fixed stations, by the subtraction of averages of values accumulated, say, over one decade from those in another decade, and from the comparison of repeated sections (lines of measurements across an ocean). Each of these techniques has its drawbacks. There are very few time series stations (Bermuda, Hawaii, and off the West Coast of Canada), the decadal averages are subject to errors of non-uniform spatial sampling, and the sections are instantaneous pictures that are contaminated by the presence of energetic transient eddies (see Chapter 4).

Despite these limitations, results from each of these lead to the conclusion that temperatures in the ocean change by a few tenths of a degree over a typical 10-year time-scale. The changes are largest in the upper part of the water column and decrease with depth (see *Figures 3.7* and *3.8*). The horizontal areas of such changes are large, comparable to the width of an ocean basin[19,23].

Thermohaline Catastrophes and the Younger Dryas

During the most recent ice age, the North Atlantic component of the Global Thermohaline Conveyor was partially shut down and the ocean is thought to have operated in a different mode to that of the present day. The northern North Atlantic was considerably cooler and the transport of the North Atlantic current (the northward extension of the Gulf Stream) much reduced (*Figure 3.9*). The North Atlantic component of the conveyor was reactivated at the end of the most recent glaciation, at about 14,000 years BP.

Recent research on deep-sea sediment cores has shown that this reactivation of the conveyor was not without hiccups! Part of this conveyor stopped abruptly at about 11,000 BP – a period known as the Younger Dryas[29]. This led to a catastrophic cooling of the North Atlantic region and caused the build up of small glaciers in the British mountains in what geographers call the Loch Lomond glacial

3.9

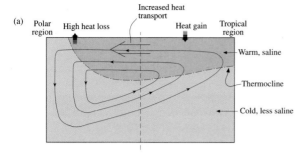

(a)

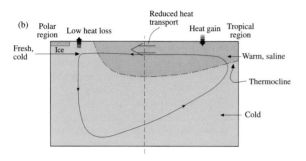

(b)

Figure 3.9 (a) The 'present-day' North Atlantic with a vigorous conveyor (thermohaline) circulation, and a large northward transfer of heat. (b) The 'ice-age' North Atlantic with a weak conveyor (thermohaline) circulation, and a reduced northward heat transfer. The northern North Atlantic was therefore cooler and fresher than the 'present-day' ocean. The extended sea-ice cover is associated with a fresher surface layer, which stabilises the water column and inhibits deep convection.

on the thermohaline system is an example of positive feedback. The fresh water input weakens the thermohaline circulation, which makes the circulation more susceptible to further weakening. This process has been investigated in recent years using mathematical models of ocean circulation. The models show that there are a number of different states of the thermohaline circulation, some of which are stable (*Figure 3.10*). There are, however, transitions between stable states, which occur over periods as short as 40 years[9]. It has been speculated that the present North Atlantic Ocean may be close to one of these transitional states, of which the Younger Dryas is an example. One study has suggested that the transition between states is not necessarily symmetrical. The change from strong to weak thermohaline circulation may be more rapid (40 years) than the re-establishment of the strong circulation (500 years). Further study of the Younger Dryas and similar events in the palaeoclimate record may give us important clues to the

re-advance. This cooling only appears to have lasted a few centuries, but it developed very rapidly over decades. There are several theories about what exactly led to the shut down of the conveyer. One view is that a sudden influx of fresh melt-water from the Laurentide ice sheet into the North Atlantic could have stabilised the vertical stratification and reduced the rate of formation of North Atlantic deep cold water. This, in turn, could have shut down the North Atlantic conveyor circulation, resulting in a cooling of the surface waters of the northern North Atlantic[4].

The effect of changes in fresh water fluctuations

Figure 3.10 A theoretical model for oscillations of the ocean conveyor belt, continental ice volume, surface water flux, and Atlantic Ocean salinity as a function of time (1 unit = 1000 years). The oscillator model is based on bi-stable states for the conveyor circulation [(see *Figure 3.9(a)*]. When the conveyor circulation is turned off, northward heat transfer is reduced and cools (a), which results in a growth in ice volume (b). The reduced export of salt from the Atlantic, because of the 'turning off' of the conveyor circulation, causes an increase of salinity (c). The salinity increases to a threshold value, at which point the deep overturning of the northern ocean occurs, and the conveyor circulation is 'turned on' again. The continental ice volume (b) decreases, because of the increase in melting associated with the increased northward heat flux into the North Atlantic. With more melting and a greater export of salt (due to the 'turning on' of the thermohaline circulation), the Atlantic salinity decreases. These oscillations have a period of the order of 1000 years – the thermohaline circulation time-scale. (Based on Birchfield and Broecker[3].)

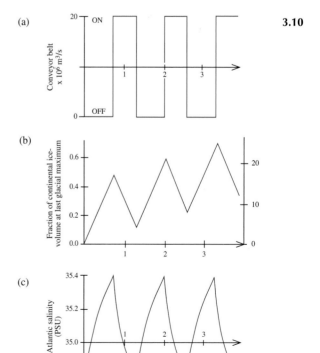

3.10

likely response of the present-day thermohaline circulation in the North Atlantic to global warming (remaining stable or 'flipping' to another state).

Monitoring Climate Change

So we know there are changes in the ocean that occur over periods of tens of years. Similarly, we suspect that effects of this type may be implicated in longer-period and more extreme (glacial and interglacial) changes. We have now to ask the question whether anthropogenic activities can be detected in the ocean and whether new technologies can help us to observe the ocean better.

There is at least one case in which repeated observations in the ocean seem to point to a man-made effect. The western Mediterranean Sea is an interesting 'laboratory' in which the process of deep water convection can be observed in conditions that are less hostile than those found in the high Arctic and Antarctic. The product of western Mediterranean convection is a homogeneous, warm, saline watermass that ultimately leaves the Mediterranean and enters the North Atlantic through the Straits of Gibraltar. Measurements of the properties of this water have been made since early in the twentieth century, since when they show that the temperature, salinity, and density of the watermass have increased slowly up to the mid-1950s and more quickly thereafter, *Figure 3.11*. This change has been attributed to a reduction of fresh water inflow into the Mediterranean caused by the damming of the Nile and those rivers flowing to the Black Sea[25].

The detection of anthropogenic effects in the open ocean is much more difficult. We know little about the inherent ocean variability, so detection requires a long-term commitment to systematic and careful monitoring of the state of the oceans. This has been embarked upon already in the WOCE. Over the period from 1990–1997, this experiment will provide a 'snapshot' of the state of the oceans (*Figure 3.12*), including the distribution of physical and chemical properties and an assessment of the role of ocean circulation in the transport of heat and water. The WOCE measurements, *Figure 3.13*, are being used to test and improve ocean models running on some of the largest computers now available[14].

WOCE will produce a baseline picture of the oceans against which future measurements can be compared. Such future measurements will be made by the Global Ocean Observing System (GOOS), the ocean element of a Global Climate Observing System (GCOS). WOCE observations are based on the use of expensive and sophisticated research ships, and these cannot be expected to provide much longer-term routine monitoring; to do this we must look to other techniques.

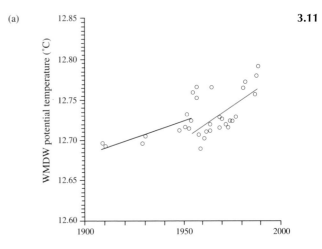

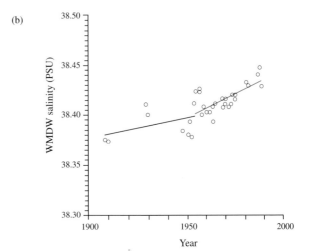

Figure 3.11 Changes in (a) temperature and (b) salinity of the western Mediterranean deep water (WMDW) during the twentieth century.

We already have some elements of a monitoring system in place. Sea level from coastal stations can be monitored centrally in real time and is used to analyse and predict the progress of El Niño events in the Pacific (see later). Satellite-tracked drifters as indicators of surface currents are becoming more and more reliable and can measure not just water movement, but also temperature, salinity (being developed), and meteorological parameters (see Chapter 19). Many merchant ships are equipped to measure temperature using expendable bathythermographs to provide a set of observations of subsurface temperature; but these cover only the major trade routes and contribute almost nothing over the remainder of the globe.

At present, satellites provide the only global rou-

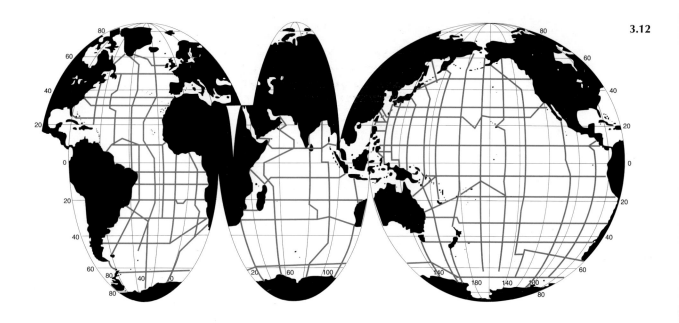

3.12

Figure 3.12 The WOCE hydrographic survey grid. Physical and chemical properties will be measured from surface to sea bed every 50 km along the red lines between 1990–1997.

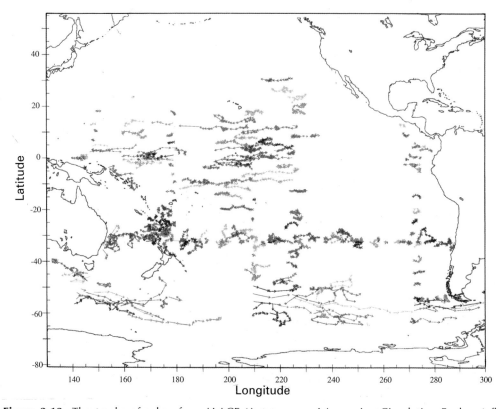

3.13

Figure 3.13 The tracks of subsurface ALACE (Autonomous LAgrangian Circulation Explorer) floats deployed in the Pacific Ocean during WOCE. The floats drift with the currents at a predetermined depth, surface every 20 days to transmit their position and data to a satellite, and then return to their programmed depth. This illustrates data points over 2.5 years.

3.14

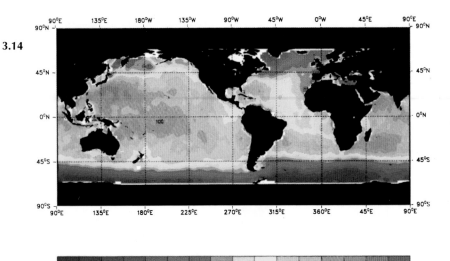

Figure 3.14 Sea surface topography from the Topex–Poseidon satellite altimeter. The range is from –180 cm to +140 cm. Most prominent is the large surface height change across the Antarctic Circumpolar Current and the clear shape of the North Atlantic and North Pacific ocean gyres.

tine monitoring capability (e.g., *Figure 3.14*), but what they can observe about the ocean is limited to surface temperatures and sea surface elevations. However, they do this well (to an accuracy of 0.5°C and less than 5 cm) and will continue to be part of any future ocean monitoring scheme.

For measurements of the interior of the ocean, new techniques will need to be developed (see Chapter 19). The UK is developing unmanned, autonomous submersibles, capable of carrying out many of the observations presently made from research vessels. Such vehicles would be capable of traversing entire ocean basins and making measurements from surface to sea bed on a regular basis (*Figure 3.15*).

Another novel technique is even now being deployed in the Pacific with a view to measuring

ocean temperatures routinely. ATOC (Acoustic Thermometry of Ocean Climate; see *Figure 19.30*) relies on the fact that sound can be transmitted over vast distances in the ocean, and that the speed of sound in sea water is, at any given pressure, predominantly dependent on temperature. An experiment in 1991 transmitted sound from Heard Island in the Southern Ocean to receivers as far away as the east and west coasts of North America (16000 km). Based on these initial encouraging results, low frequency (70 Hz) sound sources off Hawaii and California and receivers around the rim of the Pacific were deployed in early 1994. This ATOC array, *Figure 3.16*, will show whether the kind of temperature anomalies that have been seen from repeated hydrographic station measurements can be reproduced and their evolution monitored.

3.15

Figure 3.15 An artist's impression of the AUTOSUB vehicle carrying out a transoceanic hydrographic survey mission. The project involves the development of a very low drag body, an efficient propulsion system, sensors that can retain their accuracy over long missions, and deployment and recovery systems, as well as navigation and data telemetry schemes.

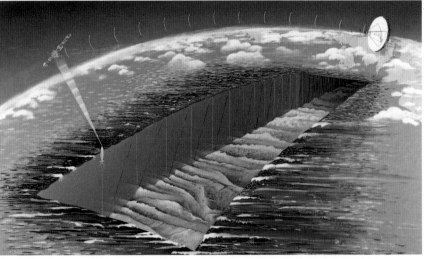

Figure 3.16 A proposed array of sources (north and south of Hawaii, off the Californian coast, and off Kamchatka, in the N.W. Pacific) and receivers (small circles, triangles and stars), to be deployed in the the ATOC project. The sound paths, for each source and receiver, are great circles.

Model studies based on coupled ocean–atmosphere models and a scenario of atmospheric increases in greenhouse gases suggest that ATOC measurements might even be able to detect signals due to global warming.

Modelling of Climate Change

The realisation that many of the large-scale processes involved in the ocean, atmosphere, and ice sheets may be described by a consistent set of dynamical equations has been with us since the early twentieth century. L.F. Richardson[24] described his experiment to predict the weather by numerical iteration of the equations of motion in 1922. He also discussed how the method could be extended to short-term climate prediction. Though his experiment was not a success, the methods were later successfully developed for numerical weather prediction in the 1950s. These methods were also used by N. Phillips[20], who developed the first general circulation model of the global atmosphere. The past 40 years have seen the further improvement of atmospheric general circulation models (AGCM) to include not only the dynamical processes, but also details of the radiations and their interaction with cloud, surface processes associated with hydrology and vegetation, and the ocean. Indeed, the atmospheric models have been at the forefront of estimating the response to the predicted changes in greenhouse gases.

In the late 1960s K. Bryan[5] published the first model of the general circulation of the ocean (see *Box 3.1*). This was developed further by M. Cox[8], and has been used extensively by research groups around the world. The Fine Resolution Antarctic Model (FRAM) project used this general circulation model as its basis (see *Box 3.2*), and it is now being used for other global ocean modelling projects. The development of the AGCM went hand-in-hand with improvements in the power of the computer – climate perturbation experiments for a few decades can be now run in a matter of a few weeks on a supercomputer.

Ocean circulation models, however, have been severely hampered by computer resources. First, the response time-scales in the ocean vary from days and weeks in the surface layers to the order of centuries in the deep ocean. This means that a global ocean model may have to be integrated for at least a 100-year period to reach equilibrium after a climate perturbation, compared with a period of a few years for an AGCM. Second, in some cases there is a need to model ocean eddies (scales of 50–200 km; see Chapter 4), which in some parts of the ocean are significant in the transport and mixing of ocean properties. A finer grid resolution of about 10 km is required to fully resolve these oceanic eddies in ocean models, compared with the lower resolutions needed in AGCMs. It is clear that computers of two orders of magnitude faster than

Box 3.1 An Ocean General Circulation Model

An ocean general circulation model is composed of a set of mathematical equations which describe the time-dependent dynamical flows in an ocean basin. The basin is discretised into a set of boxes of uniform horizontal dimensions, but variable thickness in the vertical dimension. The horizontal flow (northward and eastward components) is predicted by the momentum equation, *Figure 3.17(a)*, at the corners of each box (*Figure 3.18*).

3.17

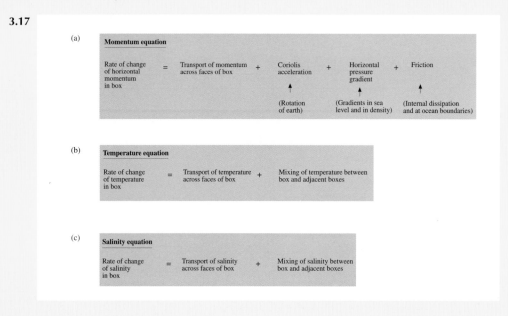

(a)

Momentum equation

Rate of change of horizontal momentum in box	=	Transport of momentum across faces of box	+	Coriolis acceleration	+	Horizontal pressure gradient	+	Friction
				↑		↑		↑
				(Rotation of earth)		(Gradients in sea level and in density)		(Internal dissipation and at ocean boundaries)

(b)

Temperature equation

Rate of change of temperature in box	=	Transport of temperature across faces of box	+	Mixing of temperature between box and adjacent boxes

(c)

Salinity equation

Rate of change of salinity in box	=	Transport of salinity across faces of box	+	Mixing of salinity between box and adjacent boxes

Figure 3.17 The basic equations for an ocean general circulation model.

The forcing for the flow may come from the surface wind stress (the frictional term in the momentum equation) or from surface buoyancy fluxes, arising from heat and fresh water (precipitation–evaporation) exchange with the atmosphere. These buoyancy fluxes change the temperature and salinity in the surface layer of the ocean. However, the horizontal and vertical flow carry these properties far into the interior of the ocean, where they tend to mix with other water masses.

3.18

This process of transport and mixing is described by the temperature and salinity equations, *Figures 3.17(b)* and *3.17(c)*, at the centre of each ocean box (*Figure 3.18*). From these two equations the sea water density and thence the pressure can be obtained for each box. The horizontal pressure gradient is then determined for the momentum equation, while the vertical velocity is calculated from the horizontal divergence of the flow. This set of time-dependent equations can then be used to describe all the dynamical components of the flow field, provided that suitable initial and boundary conditions are specified.

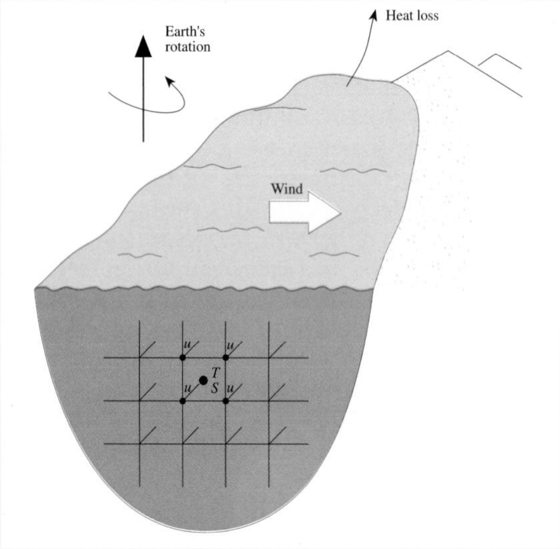

Figure 3.18 A schematic of the model boxes in an ocean general circulation model. The equations for momentum are solved at the corners of the boxes (*u*), while the temperature (*T*), and salinity (*S*) equations are solved at the centres of the boxes. The model is forced by climatological wind stress, surface heat, and fresh water fluxes.

Box 3.2   The Fine Resolution Antarctic Model (FRAM)

FRAM was developed[26] to investigate the role of eddy processes in the circulation of the Southern Ocean, in particular the Antarctic Circumpolar Current. The Southern Ocean comprises 30% of the global ocean and is an important region for the transfer of heat and fresh water between the Antarctic Ice Sheet and the northern land masses. The dynamics of the Antarctic Circumpolar Current are not well understood, although there is strong evidence to suggest that energetic ocean eddies, on horizontal scales of the order of 100 km, play an important role in the dynamics of the current.

The FRAM model forms the basis for the development of a global ocean model, which will be used for climate change experiments. This global model will resolve these energetic ocean eddies and the major frontal zones of the ocean.

The FRAM model subdivides the ocean, south of 22°S to the Antarctic continent, into a regular set of boxes. Each box has a horizontal length of 0.25° latitude by 0.5° longitude (approximately 27 km x 27 km at 60°S). Beneath each surface box a string of boxes reaches to the ocean floor. The thickness of each box varies from 20 m in the surface layer to over 200 m in the deepest parts of the ocean. Within each box, equations for the northward and eastward horizontal components of momentum, temperature, and salinity are specified. There are 5 million boxes which represent the southern ocean, and therefore 20 million prognostic variables to calculate. These variables are calculated by integration in time of the equations from an initial cold, saline motionless ocean. In the first 6 years of the integration the model was forced by the annual mean wind stress and by the observed temperature and salinity. After this period the model was free to run for a further 6 years, subject to seasonal wind forcing, annual mean temperature and salinity at the ocean surface, and an open northern boundary. An example of the model 'output' is shown in Figure 3.19 and a comparison with satellite data is shown in Figure 3.20. Estimates of the meridional heat flux in the model are shown in Figure 3.21.

3.19

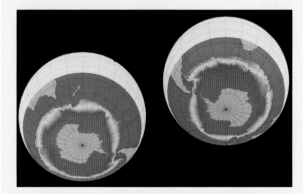

Figure 3.19 Contours of the instantaneous stream function in FRAM. The stream function shows the depth-averaged flow circulating clockwise around the Antarctic continent. The flow is most intense in the Antarctic Circumpolar Current (in the yellow and neighbouring green regions). The flow is unsteady due to the presence of 'eddies', mainly in the Antarctic Current and Agulhas Current south of South Africa.

3.21

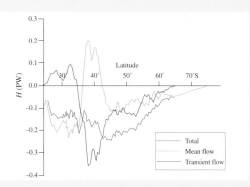

Figure 3.21 The latitudinal variation in meridional-heat transfer (H) by the FRAM model. Northward heat fluxes are positive. The total heat transport is directed southward (negative) toward the pole, although the time-mean circulation drives a heat flow toward the north, between 37°S and 43°S, in the region of the Antarctic Circumpolar Current. The 'eddies' in the flow, however, drive a stronger heat flux toward the south, and thus result in a total heat transport toward the pole. The eddies play an important climatological role in the model.

3.20

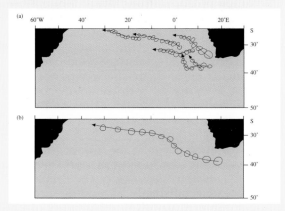

Figure 3.20 (a) Tracks of eddies detected by the Geosat satellite in the South Atlantic. (b) Tracks of eddies in the FRAM. Note that the model produces a more regular eddy track than the observations show.

the present machines (which are capable of 10^9 floating point instructions per second) will be required to manage this task. The recent development of parallel computer systems is expected to deliver this power by end of the twentieth century. Indeed, one of the important tasks of ocean modellers is to develop methods for the analysis and interpretation of the huge quantity of data that these models will produce. Because of their complexity, there will be also be a need to develop simpler models to investigate interactions more completely. To understand and predict the behaviour of sea ice, for example, the sea-ice models will have to be coupled to models of the upper ocean and overlying atmosphere.

A third problem is the importance of ocean chemistry and biology to climate change. The ocean is a depository for the greater part of the Earth's exchangeable fraction of carbon. It is not known how carbon is regulated by the ocean, though it is clear that phytoplankton blooms produce a lowering of the partial pressure of carbon dioxide at the ocean surface and thereby have the ability to alter the flux between ocean and atmosphere (see Chapters 6 and 12). Modelling of the ocean basin ecology and chemistry has commenced in recent years and it is expected that these processes, as they become understood, will be incorporated into the general climate models.

The El Niño–Southern Oscillation (ENSO) Phenomenon – An Example of Ocean Prediction?

The ENSO phenomenon is now recognised as the largest contributor to the perturbation of the climate on a global scale over a period of a few years, and is known to be a natural oscillation of the atmosphere–ocean system. It is a coupled interaction between the atmosphere and the upper layers of the tropical Pacific Ocean, which can result in changes of global surface temperature of a few tenths of a degree Celsius on a time-scale of one year. This is a similar change in temperature to that attributed to the 25% increase in atmospheric carbon dioxide in the past 100 years.

The El Niño is the ocean component of the interaction; it is a general warming of the upper layer of the eastern and central Equatorial Pacific Ocean. [The name El Niño comes from the fact that the impact is felt on the coast of South America around Christmas time and hence it is referred to in Spanish as the (Christ) child.] A major consequence of the oceanic warming is the decline of biological productivity and hence of fish stocks, which are a major source of livelihood for the local population.

It is associated with a reduction in the strength

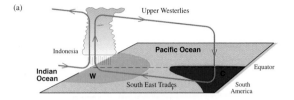

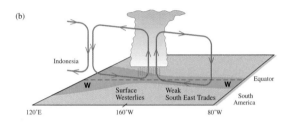

Figure 3.22 The tropical atmospheric circulation over the Pacific Ocean during (a) normal conditions and (b) El Niño conditions. During normal conditions the surface pressure is low over Australia and Indonesia (high rainfall) and high over the southeast Pacific, so the surface trade wind circulation is strong and the Southern Oscillation index ($P_{DARWIN} - P_{TAHITI}$, where P is the seasonal surface pressure) is high. During El Niño conditions, the pressure is higher over Australia and Indonesia (low rainfall) and lower in the southeast Pacific; consequently the trade wind circulation is weaker and the Southern Oscillation index is low (W, warm; C, cold).

of the tropical trade wind system, in particular the southeast trades which occur on the eastern flank of the South Pacific anticyclone. The normal wind circulation produces strong winds, which drive an upwelling of cooler, nutrient-rich, and highly productive thermocline waters along the tropical coast of South America and on the equatorial band of the Eastern Pacific. When the trade winds weaken, the upwelling is reduced and the waters warm by as much as 5°C, due to both the southward movement of warmer equatorial waters and the high solar radiation at the surface.

At first sight this appears to be a one-way forcing of the atmosphere by the surface wind on the ocean, as is the case in most of world's upwelling regions. However, this area of the Pacific Ocean behaves rather differently because the rise in sea-surface temperature over a vast area of ocean changes the distribution of atmospheric heat sources and sinks, which in turn drives the trade wind circulations. The trade winds carry water vapour, evaporated from the ocean, into areas of tropical atmospheric convergence where high rainfall occurs. These convergence zones tend to be

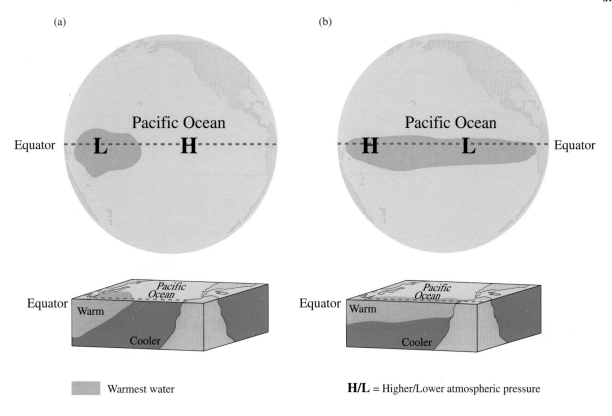

(a)   (b)

Warmest water

H/L = Higher/Lower atmospheric pressure

Figure 3.23 The tropical Pacific Ocean during (a) normal and (b) El Niño years. During normal years the strong trade winds drive the warm water westward and intensify upwelling of cooler subsurface waters in the east Pacific. During El Niño years the weaker atmospheric circulation allows the warmer lighter water to flow eastward, replacing the cooler upwelling waters.

located in regions of maximum surface temperature, that is to the north of the equator and in the tropical southwest Pacific. During El Niño events these convergence zones tend to move southward across the equator and eastward (see *Figures 3.22* and *3.23*). The South Pacific anticyclone becomes weaker and the trade winds weaken. Simultaneously, the surface atmospheric pressure over Indonesia and Australia tends to rise and rainfall decreases.

This see-saw of surface pressure between the southeast Pacific and Indonesia is known as the *Southern Oscillation*. Although its major influence is in the tropical Pacific, its effect is felt throughout the world. For example, in Zimbabwe both rainfall and maize yields are highest during El Niño years[6]. The ENSO phenomenon is a coupled interaction between the surface layers of the ocean and the world wind systems, which occurs two or three times a decade.

The monitoring of ENSO in recent decades, in particular during the exceptional episode of 1982–1983[22], has provided a stimulus to atmospheric and ocean modellers. These efforts have provided good simulations of both the individual components of the system (the ocean response to observed winds and the atmospheric response to observed sea surface temperature).

Coupled models of the tropical ocean and atmosphere are showing promise for the simulation of the ENSO cycle and a number of Climate Centres in the world are now producing experimental forecasts for ENSO with some degree of success.

These models will be improved when the results of the Tropical Ocean Global Atmosphere experiment (an experiment to measure and understand some of the complex processes in the tropical atmosphere and ocean) in the western equatorial Pacific Ocean are analysed. The proposed mooring

3.24

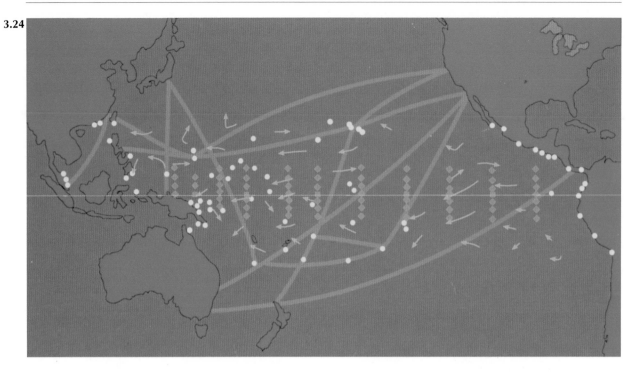

Figure 3.24 The tropical Pacific Ocean observing system. The orange diamonds are ATLAS surface buoys from which continuous temperature measurements are obtained from the surface to a depth of 500 m. These observations together with surface meteorological measurements (e.g., wind, temperature and humidity) are transmitted by satellite to a receiving station. Currents (orange squares) are measured routinely along the equator. The Tide gauge network (yellow circles) provides sea-level height observations, which can be used to calculate horizontal pressure gradients in the oceans. These observing stations are supplemented by routine measurements from satellite-tracked drifting buoys (arrows) and voluntary observing ships (light blue lines).

array (*Figure 3.24*) will provide routine measurements of the upper 500 m of the tropical Pacific Ocean and of surface winds to be used in the ENSO forecast models.

There is also evidence that the intensity of ENSO varies on the longer decadal to centennial time-scales. Longer-term variations in the intensity of El Niño events affecting primary production off California have been related to solar cycles (variations in the intensity of 11–22 year sunspot cycles that cause small changes in incident radiation). Within Californian continental margin sediments, decadal to millennial cycles of preservation of laminated sediments, driven by variation in the intensity of the oxygen minimum zone, have been ascribed to solar cycles affecting the longer-term alternation of El Niño and anti-El Niño[2].

Data on sea surface temperature variations from oxygen isotope studies of Galapagos corals showing 11 and 22 year periods[11] lend support to suggestions of solar cycle modulation of ENSO activity.

The Importance of Climate Change

The most comprehensive study of the scientific evidence for climate change, its potential impact, and the strategies needed to ameliorate its impact has been made by the Intergovernmental Panel on Climate Change (IPCC). While climate change is, in general, regarded as an atmospheric phenomenon, the IPCC reports make it clear that there are direct impacts on and by the ocean and, furthermore, that knowledge of the behaviour of the oceans is central to any climate prediction capability. The potential socioeconomic costs of climate change far outweigh the resources needed to make the measurements and run the models required to improve our ability to predict climate change. The key areas identified in the IPCC report, together with recommendations on improvements in observations and modelling, are given in *Box 3.3*.

Clearly, in all these areas the oceans are important and understanding them represents one of the greatest challenges in the area of climate change prediction.

Box 3.3   The Scientific Uncertainties of Climate Change

IPCC identifies the key areas of scientific uncertainty as:

• *Clouds:* primarily cloud formulation, dissipation, and radiative properties, which influence the response of the atmosphere to greenhouse forcing.

• *Oceans:* the exchange of energy between the oceans and the atmosphere, between the upper layers of the ocean and the deep ocean, and transport within the ocean, all of which control the rate of global climate change and the patterns of regional change.

• *Greenhouse gases:* quantification of the uptake and release of the greenhouse gases, their chemical reactions in the atmosphere, and how these may be influenced by climate change.

• *Polar ice sheets:* affect predictions of sea level rise.

The main observational requirements are:

• The maintenance and improvement of observations (such as those from satellites) provided by the World Weather Watch.

• The maintenance and enhancement of a programme of monitoring, both from satellite-based and surface-based instruments, of key climate elements for which accurate measurements on a continuous basis are required. These include the distribution of important atmospheric constituents, clouds, the Earth's radiation budget, precipitation, winds, sea surface temperatures, and the terrestrial ecosystem extent, type, and productivity.

• The establishment of a Global Ocean Observing System to measure changes in such variables as ocean surface topography, circulation, transport of heat and chemicals, and sea ice extent and thickness.

• The development of new systems to obtain data on the oceans, atmosphere, and terrestrial ecosystem using both satellite-based instruments and instruments based on the surface, on automated vehicles in the ocean, on floating and deep sea buoys, and on aircraft and balloons.

• The use of palaeoclimatological and historical instrumental records to document natural variability and changes in the climate system, and subsequent environmental response.

In the area of modelling the report concludes that any reduction in the uncertainties of climate prediction will be dictated by progress in the areas of:

• Use of the fastest possible computers to take into account coupling of the atmosphere and the oceans in models, and to provide sufficient resolution for regional predictions.

• Development of improved representation of small-scale processes within climate models, as a result of the analysis of data from observational programmes to be conducted on a continuing basis well into the twenty-first century.

General Reference

Wuethrich, B. (1995), El Niño goes critical, *New Scientist*, **145**, 32–35.

References

1. Aagaard, K. and Carmack, E.C. (1989), The role of sea ice and other fresh water in the Arctic Circulation, *J. Geophys. Res.*, **94**(C5), 14485–14498.

2. Anderson, R.Y., Linsley, B.K., and Gardner, J.V. (1990), Expression of seasonal and ENSO forcing in climatic variability at lower than ENSO frequencies: evidence from Pleistocene marine varves off California, *Palaeogeogr., Palaeoclimatol., Palaeoecol.*, **78**, 287–300.

3. Birchfield, G.E. and Broecker, W.S. (1990), A salt oscillator in the glacial Atlantic? A 'scale analysis' model, *Paleoceanogr.*, **5**, 835–843.

4. Broecker, W.S, Kennett, J.P., Flower, B.P., Teller, J.T., Trunbone, S., Bonani, G., and Wolfi, W. (1989), Routing of meltwater from the Laurentide Ice Sheet during the Younger Dryas cold episode, *Nature*, **341**, 318–321.

5. Bryan, K. (1969), A numerical model for the study of the world ocean, *J. Computat. Phys.*, **4**, 347–376.

6. Cane, M.A., Eshel, G., and Buckland, R.W. (1994) Forecasting Zimbabwean maize yield using east equatorial Pacific sea surface temperatures, *Nature*, **370**, 204–205.

7. Carrissimo, B.C., Oort, A.H., and Van de Harr, T.H.V. (1985), Estimating the meridional energy transports in the atmosphere and ocean, *J. Phys. Oceanogr.*, **15**, 52–91.

8. Cox, M.D. (1984), *A Primitive Equation: 3-Dimensional Model of the Ocean*, GFDL Ocean Group Technical Report No.1, GFDL/NOAA, Princeton University.

9. Delworth, T., Manabe, S., and Stouffer, R.J. (1993), Interdecadal variations of the thermohaline circulation, *J. Climate*, **6**, 1993–2011.

10. Dickson, R., Meinke, J., Malmberg, S., and Lee, A. (1988), The Great Salinity Anomaly in the northern North Atlantic 1968–1982, *Progr. Oceanogr.*, **20**, 103–151.

11. Dunbar, R.B., Wellington, G.M., Colgan, M.W., and Glynn, P.W. (1994) Eastern Pacific sea surface temperature since 1600 AD: the Δ¹⁸O record of climate variability in Galapagos corals, *Paleoceanogr.*, **9**, 291–315.

12. Ellett, D.J. and Blindheim, J. (1992), Climate and hydrographic variability in the ICES area during the 1980s, *ICES Mar. Sci. Symp.*, **195**, 11–31.

13. Gordon, A.L. (1986), Interocean exchange of thermocline water, *J. Geophys. Res.*, **91**, 5037–5046.

14. Gould, W.J. (1994), Update: World Ocean Circulation Experiment, *Sea Technology*, **35**(2), 25–32.

15. GRIP Project Members (1993), Climate instability during the last interglacial period recorded in the GRIP ice core, *Nature*, **364**, 203–207.

16. Lorius, C., Jouzel, J., and Reynaud, D. (1993), Glacials–interglacials in Vostok: climate and greenhouse gases, *Global Planet. Change*, **7**, 131–143.

17. Newell, R.E., Kidson, J.W., Vincent, D.G., and Boar, G.J. (1972), *The General Circulation of the Tropical Atmosphere*, Vol. 1, The MIT Press, 258 pp.

18. Oort, A.H. and Vonder Harr, T.H. (1976), On the observed annual cycle in the ocean–atmosphere heat balance over the Northern Hemisphere, *J. Physical. Oceanogr.*, **6**, 781–800.

19. Parrilla, G., Lavin, A., Bryden, H., Garcia, M., and Millard, R. (1994), Rising temperatures in the Subtropical North Atlantic Ocean, *Nature*, **369**, 48–51.

20. Phillips, N.A. (1956), The general circulation of the atmosphere: a numerical experiment, *Quart. J. Roy. Meteor. Soc.*, **82**, 124–164.

21. Pugh, D. (1987), *Tides, Surges and Mean Sea Level: a handbook for engineers and scientists*, John Wiley and Sons, Chichester, 472 pp.

22. Rasmusson, E.M. (1985), *The 1982/83 El Niño Event*, World Meteorological Organisation, Marine Meteorology and Related Oceanographic Activities Report, No. 14, 11–22.

23. Read, J.F. and Gould, W.J. (1992), Cooling and freshening of the subpolar North Atlantic Ocean since the 1960s, *Nature*, **360**, 55–57.

24. Richardson, L.F. (1922), *Weather Prediction by Numerical Process*, Cambridge University Press, Cambridge.

25. Rohling, E.J. and Bryden, H. (1992), Man-induced salinity and temperature increases in western Mediterranean deep water, *J. Geophys. Res.*, **97**(C7), 11191–11198.

26. The FRAM Group (1991), An eddy-resolving model of the Southern Ocean, *EOS Trans. AGU*, **72**(15), 169, 174–175.

27. Trenberth, K.E. (1979), Mean annual poleward energy transports by the oceans in the Southern Hemisphere, *Dynam. Atmos. Ocean*, **4**, 57–64.

28. Wijffels, S., Schmitt, R., Bryden, H., and Stigebrandt, A. (1992), Transport of fresh water by the oceans, *J. Phys. Oceanogr.*, **22**, 155–162.

29. Zahn, R. (1992), Deep ocean circulation puzzle, *Nature*, **356**, 746.

CHAPTER 4:

Ocean Weather –
Eddies in the Sea

K.J. Richards and W.J. Gould

Introduction

Weather maps brought to us daily via television and newspapers have made us all aware of the changing state of the atmosphere (*Figure 4.1*). Cyclones and anticyclones (low and high pressure systems) evolve and interact, produce severe winds, fronts with associated rain, and vary the weather we experience on time-scales from a few hours to several days. Less well-known is that the ocean is populated by very similar systems. The oceanic equivalent of atmospheric highs and lows, the ocean weather is again an ever-changing pattern. As in the atmosphere, intense storms can develop in the ocean to produce strong currents. Fronts separating warm and cold water masses are a common occurrence. The oceanic systems, however, have a much smaller horizontal scale than do their atmos-pheric counterparts and evolve on a much longer time-scale. It is only recently that we have been able to sample the richness in structure of the oceanic eddy field, and we have only just begun to assess the importance of these oceanic eddies in shaping the large-scale ocean circulations and their impact on climate and the biology of the oceans.

Our View of the Ocean

Our understanding of the nature of the ocean circulation is determined by the tools at our disposal to observe it. Just as the development of astronomical telescopes (the most recent of which is the Hubble Space Telescope) has given us successively deeper insights into the structure of the Universe, so has our view of the oceans altered as new measuring techniques have become available.

Figure 4.1 A familiar sight for those in the UK, a low-pressure system and associated rain-bearing fronts sweeping across the country. This illustrates the surface pressure field for a day in January. The low-pressure system, or cyclone, in the centre of the picture is approximately 2000 km across. The wind circles the low pressure in an anticlockwise sense. Interaction between high and low pressure systems dictates the weather we experience on a daily basis. The insert, expanded by a factor of 10, shows a detail from an infra-red image of the sea surface taken from a satellite. Light areas correspond to cool water and dark areas to warm water. Using the sea surface temperature as a tracer, we can clearly see the imprint of an ocean eddy, 100 km across, where the ocean currents associated with the eddy have caused the warm waters to the south and the cool waters to the north to spiral around each other; an oceanic cyclone (in the insert note also the sharp transition from warm to cool water, an ocean front, and the smaller scale struc-tures, 10 km, along the front). Oceanographers call the cyclones and anticyclones of the ocean meso-scale eddies. These ocean eddies are dynamically equivalent to the weather systems in the atmosphere. However, ocean eddies have very different space- and time-scales; typically, horizontal scales of 10–200 km, current speeds of a few tens of centimetres per second, and circulation times of tens of days. In the vertical, some are restricted to the upper levels of the ocean, while others extend to the bottom.

4.1

4.2

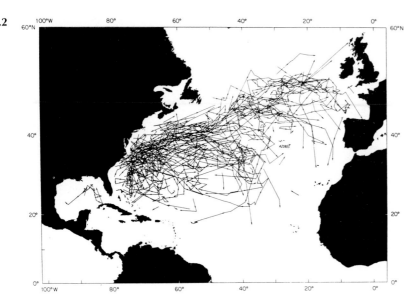

Figure 4.2 The earliest ideas of ocean currents came from the long-term drift of debris from shipwrecks. Some hulks remained water-logged and drifted with the surface currents for years on end. This compilation of such trajectories from the North Atlantic presents a picture not greatly different from our present best estimates of mean ocean currents[9].

Until well into the twentieth century, measurements of ocean surface currents were made by compiling observations of ship drift, made during the course of routine navigation, and of floating objects (*Figure 4.2*). These served to delineate the major currents of the upper ocean with which we are familiar; the intense currents on the west side of ocean basins (Gulf Stream, Kuroshio, and Somali Current), the Antarctic Circumpolar Current system, and the equatorial currents. Maps drawn by Rennell in the eighteenth century (see *Figure 1.5*) and by Maury in the nineteenth century are not very different from those that appear in today's atlases.

Until 40 years ago the measurement of subsurface currents was well-nigh impossible. The only direct method relied on the tracking of surface buoys attached to subsurface drogues from attendant ships, and was limited to short (no longer than two-week) measurements at shallow depths, which were also distorted by wind and wave forces on the surface buoy. The development of the dynamical method by Helland, Hansen, and Nansen, published early in the twentieth century, enabled the vertical structure of currents to be inferred from the vertical profiles of density, but the method was not capable of measuring the changes in currents over long periods, their absolute values, or detailed horizontal spatial structures.

It was not until the 1950s and 1960s that attempts were made to make reliable, absolute measurements of the deep subsurface currents of the ocean. Bill Richardson in the US had limited success in developing and using recording current meters on moorings. The electronics were cumbersome and unreliable and the deployment methods (on moorings) were equally vulnerable.

Simultaneously, in the UK John Swallow[18] (*Figure 4.3*) had the idea of using floats that would drift with the subsurface currents. The floats he used were sealed aluminium tubes designed to be less compressible than sea water and contained what would now be regarded as a very primitive acoustic beacon. By virtue of their compressibility being lower than that of sea water, they could be ballasted to sink at the ocean surface and would

4.3

Figure 4.3 The invention of the neutrally buoyant Swallow float in the 1950s gave the first indication that deep currents in the interior of the ocean were not the then-predicted sluggish drift, but were dominated by energetic (a few centimetres per second) currents associated with meso-scale eddies. Here John Swallow is on board the RRS Discovery II preparing an early float (made from scaffold tube) and watched intently by the ship's cat and two of the crew.

gain buoyancy relative to the water in which they were located as they sank. Then, each individual float would reach neutral buoyancy (its weight equalling the weight of the displaced water; Archimedes principle) at some depth determined by its initial weight. The float would remain at this depth and drift with the currents (since if it moved upward from its balance depth the float would expand less than the water around it, become heavier than the water, and sink back to the original depth, and vice versa for downward movements). The acoustic beacon enabled bearings to be taken from the attendant ship and hence the position and depth of the float could be estimated (see Chapter 19). These floats were designed to investigate the predicted sluggish (of the order a few millimetres per second) deep currents, but were also used to confirm the existence of a more energetic counter-current deep beneath the Gulf Stream, postulated by Stommel and Arons[17].

The measurements of the deep flow south of Bermuda made by Crease and Swallow[19] in 1960 using Swallow's float showed the surprising result that the currents were far from sluggish. Indeed, they were sufficiently energetic that floats rapidly escaped from the acoustic tracking range of the ketch Aries, from which the measurements had been made. The experiment was modified and revealed for the first time the totally unforeseen energetic deep currents.

These measurements made in the late 1950s and early 1960s gave tantalising glimpses of the large-scale complexity of the deep flow, but the techniques were not adequate to fully investigate the nature the currents. This had to await the development, by the early 1970s, of reliable current-recording meters, the moorings to support them, and the design and construction of neutrally buoyant floats that could be tracked for hundreds rather than tens of days and over thousands rather than tens of kilometres.

Gradually, the evidence for the nature of the deep flow accumulated. Long-duration current measurements made with current meters at places like Woods Hole Oceanographic Institution Site D on the continental slope north of the Gulf Stream showed a 'spectral gap' of low energy between the relatively energetic tidal–inertial periods (of the order of 1 day) and the much longer periods of tens and hundreds of days[2] that became known as the meso-scale. In the Soviet Union during the 1960s a number of long-term current measurements made in the Indian Ocean under the direction of Shtockman showed the same time- and space-

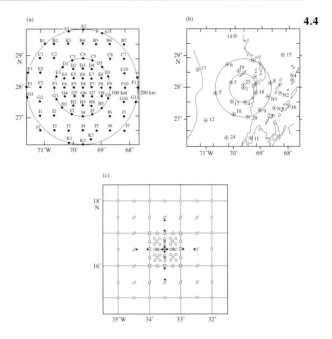

Figure 4.4 In the 1970s arrays of moorings were set in the US MODE and the USSR Polygon experiments; they were used to map the current fields at a number of discrete depths. These were the largest experiments ever undertaken and in the case of the Polygon array involved recovering and replacing each mooring eight times[5,15]. (a) MODE hydrographic grid; (b) MODE current meter array; (c) Polygon-70 hydrographic grid (open circles) and current meter array (filled circles).

scales, and became known to the Soviet scientists as the synoptic scale.

By the early 1970s both the Soviet and western scientists were able to mount concerted investigations of the spatial and temporal characteristics of what we call here meso-scale variability. In 1970 Soviet scientists carried out the Polygon-70 experiment in the Atlantic North Equatorial Current and in 1972 US and UK scientists conducted a 4-month experiment on a similar scale in the area south of Bermuda [the Mid-Ocean Dynamics Experiment, MODE; *Figures 4.4(a) – 4.4(c)*].

The MODE experiment was made possible by the use of neutrally buoyant floats [some from the UK with ranges of up to 100 km and, more importantly, long-range SOFAR (SOund Fixing And Ranging) floats from the US that could be tracked at distances of several hundred kilometres; see Chapter 19). The analyses of these experiments using objective mapping techniques revealed the existence of features in both the flow and density fields with typically a 100 km horizontal scale

4.5

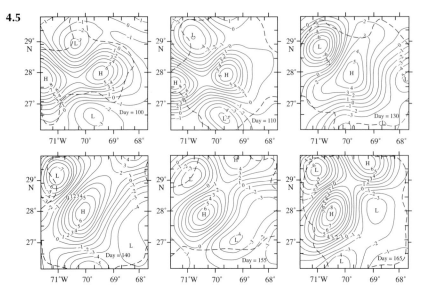

Figure 4.5 The measurements from MODE current meters and neutrally buoyant SOFAR floats were mapped to produce a coherent picture of the flow field, using a technique called objective analysis in which a flow field was devised that had the same statistical characteristics as the data themselves and fitted the observations with minimum error. There is little confidence in the predicted flow field at points outside the observational array marked by the dashed lines[14].

(*Figure 4.5*) and with a complex vertical structure. Features were seen to propagate westward at speeds of about 5 km/day. The complexity of the current field was amply illustrated by the compilation of the SOFAR float tracks into so-called spaghetti plots (*Figure 4.6*), which show the how the eddy-like meso-scale field acted to disperse the floats[10].

Even though the areas of ocean covered by both of these experiments were 200 km across, this is still a relatively small area compared to the size of ocean basins. The experiments, therefore, shed little light on the geographical variability of the characteristics of the meso-scale currents. In order to explore this a joint US–USSR experiment, carried out between August 1974 and April 1975 and called Polymode (a coming together of names as well as scientists), explored the energetics and scales of meso-scale motions over a much wider range of latitudes in the region of anticyclonic circulating water southeast of the Gulf Stream, called the western subtropical Atlantic gyre. In parallel, in the east Atlantic a more limited exploration of eddy variability using long-term moorings (the North East Atlantic Dynamics Study, NEADS) was started by European scientists (one of these current meter sites between the Azores and Madeira is still being

maintained by German scientists – the longest direct current meter measurement series[20]).

The Polygon, MODE, and Polymode experiments studied meso-scale features in detail, but from the early 1970s onward evidence accumulated (from the newly developed observational techniques of SOFAR floats, reliable current meters, satellite infra-red images of the ocean surface, and satellite altimetry data; see Chapter 5, *Box 5.4*) for the ubiquitous nature of the meso-scale eddy field (see, for example, *Figure 4.7*). All these observations confirmed the concentration of high kinetic energy of the time-varying currents (the eddy kinetic energy, EKE) in regions near the major current systems.

So a view was formed, which still holds good

4.6

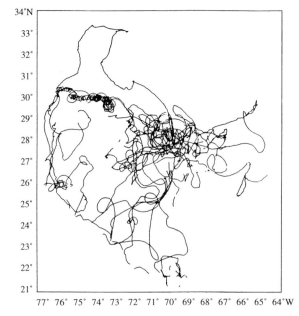

Figure 4.6 A compilation of the tracks (9/1972–6/1976) of all the SOFAR floats launched in the MODE array area near 28°N 70°W (1° of latitude is approximately 111 km). The tracks last several years and show the dispersive nature of currents from a small area to eventually fill much of the western subtropical gyre. In some areas, for instance off the Bahamas and near the Gulf Stream, eddy activity is high. The site chosen for the MODE experiment turned out to one of very low eddy energy[15].

Figure 4.7 The statistics of eddy energy derived from the variability of sea surface slope obtained from the Topex/ Poseidon satellite. High eddy variability is seen in the regions of strong meandering currents, such as the Gulf Stream in the Atlantic, the Kuroshio in the North Pacific, the Agulhas Retroflection off South Africa, and the Antarctic Circumpolar Current in the Southern Ocean. (Courtesy of Prof C. Wunsch, Massachusetts Institute of Technology, Boston, USA.)

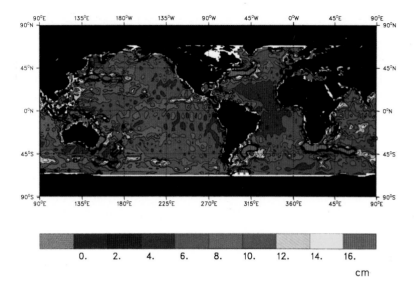

4.7

today, of the ocean populated with eddies, just as the atmosphere is full of cyclones, anticyclones, and frontal systems.

Beasts in the Eddy Zoo

Meso-scale features in the ocean take on a number of guises. Indeed, in the mid-1970s an article was published entitled *New Animals for the Eddy Zoo*, since at that time almost every eddy studied appeared to have different characteristics. For some, there was clear evidence of their presence at the sea surface, while others were confined within the water column; their diameters ranged from over 200 km to about 10 km, and they appeared to have different formation mechanisms. In a review such as this we can only touch on the characteristics of some of the more abundant animals in the eddy zoo. (The reader is referred to Robinson[13] for a collection of papers giving an overview of our knowledge at that time.)

Gulf Stream rings

Some of the best-documented eddies are those that are formed by 'pinching off' meanders from energetic current systems. The area that has been most intensively studied is the Gulf Stream – its variability downstream of Cape Hatteras is well-known. Even as long ago as 1793 evidence for an isolated body of warm water to the north of the stream was noticed, and in the 1930s analysis of ships' thermograph records started to provide evidence of the population of eddies associated with the Gulf Stream, known as Gulf Stream rings.

Undoubtedly, the detailed study of these features and their formation was strongly influenced by the advent of infra-red sensors flown on satellites that could easily identify both warm and cold core rings north and south of the Gulf Stream on cloud-free

images (*Figure 4.8*). The spatial distribution of Gulf Stream rings has been delineated by several censuses (see, for example, *Figure 4.9*), but none is truly comprehensive.

By virtue of being water masses enclosed within water with very different properties the rings can be regarded as isolated ecosystems and the evolution of their physical, chemical, and biological properties over lifetimes of several seasons can be

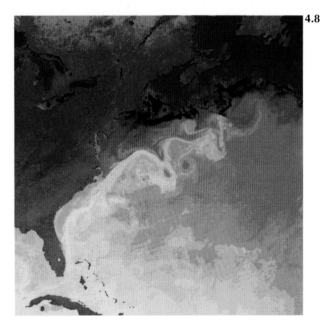

4.8

Figure 4.8 The meandering of a current and the location of eddies detached from the current can be clearly seen in satellite infra-red images of sea surface temperature. Warmer hues denote warmer temperatures. (Courtesy of O. Brown, R. Evans, and M. Carle, University of Miami Rosenstiel School of Marine and Atmosphere Science, Miami, USA.)

4.9 (a)

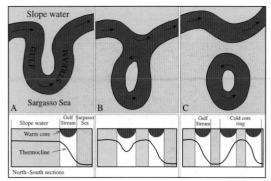

(b)

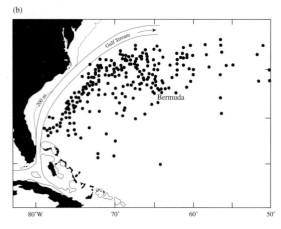

Figure 4.9 (a) A visualistion of the process of 'pinching off' a Gulf Stream ring. (b) The distribution of the centres of Gulf Stream cold core rings[8].

studied. For example, the thermal structure at the centre of a Gulf Stream ring (*Figure 4.9*) is a result of the heating and cooling cycles to which it has been subjected[8].

Mid-ocean eddies

Away from the areas of high currents and strong fronts, such as the Gulf Stream, things are no less chaotic. In these quieter regions, meso-scale eddies dominate the flow. The ratio of the kinetic energy of the eddies to the kinetic energy of the mean flow is typically of the order of 10, and often higher[1]. A typical vertical section of density taken from the northeast Atlantic using SeaSoar (*Figure 19.6*) is shown in *Figure 4.10(b)*. The tell-tale sign of the presence of eddies is the undulating depth of surfaces of constant density, implying variations in the flow speed and direction on a horizontal scale of a few tens of kilometres or less. Upwelling and downwelling motions associated with the eddies, *Figure 4.10(a)*, can be as high as several tens of metres a day, producing an enhanced exchange of water and its properties between the surface and the deep. It is only recently that the technology was developed to map out the density and velocity structure on a fine-enough scale to resolve the eddies at depth[7], and this is only to a depth of around 500 m.

The surface signature of eddies can be viewed from satellites using the sea surface height, sea surface temperature, and ocean colour (*Figure 4.11*

4.10 (a) Density given as σ_0 in kg/m^3

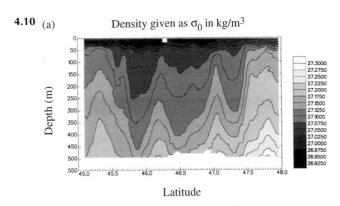

(b) Chlorophyll concentration in mg/m^3

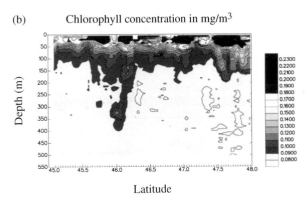

Figure 4.10 A section along 20°W in the North Atlantic. The plots show the vertical structure of both (a) water density as a function of latitude and (b) chlorophyll concentration (a proxy for phytoplankton; see Chapter 6). Large undulations are seen in the depth of surfaces of constant density of several hundreds of metres. The strong density gradient at 47.5°N marks the position of the polar front, the boundary between the warmer subtropical waters to the south and the colder polar waters to the north. South of this are a number of eddy features with a scale of tens of kilometres. The flow at these scales is very much influenced by the earth's rotation, so that in the north hemisphere a density surface shallowing to the north implies a flow to the east and likewise a deepening implies a westward flow (relative to the flow deeper down). Concurrent velocity measurements confirm this. There are important large vertical movements of water induced by the action of the eddies, as evidenced by the filament of high chlorophyll reaching down to 400 m. (This vertical movement must be have been relatively swift as chlorophyll quickly degrades away from the sunlit surface waters.) (Courtesy of Mr G. Griffiths, SOC.)

4.11a 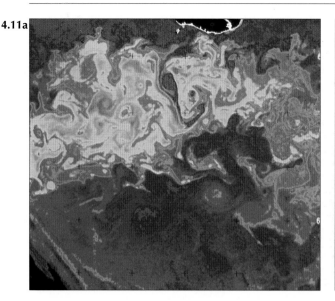 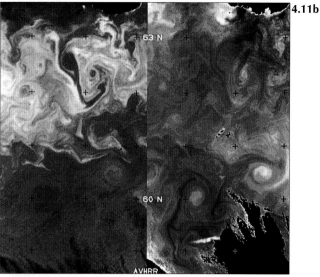 4.11b

Figure 4.11 The use of biological tracers to examine eddying motions. (a) The false colour picture shows the level of visible reflectance as viewed by a satellite using AVHRR (see Chapter 5). Orange and red show areas of high reflectance, while blue and violet indicate low levels. The high reflectance levels are caused by the great abundance of the phytoplankton species, *Emiliania huxleyi*, which produces highly reflective calcite coccoliths (observations from ships report the sea becoming milky white during intense blooms of this species). (b) A smaller area of the same bloom in more detail. Here the visible reflectance is shown using a grey scale and compared with the sea surface temperature (the infra-red band), with a close correspondence between the two. The swirling patterns are very suggestive of eddy motion. An animated sequence of satellite pictures has shown the eddies pulling out filaments of high and low concentration to produce the streaky appearance. The impression is that ocean eddies are efficient stirrers and mixers of properties within the ocean. However, the sharp distinction between the various water masses is maintained for several days, and even enhanced by the eddies. This suggests that the final mixing of properties may take some time. (It must also be remembered that the property being measured is related to a living organism that is growing and dying. It is not simply a passive tracer.) The study of the effect of eddies on biological production in the ocean is an on-going research issue. This particular bloom was found to have a significant impact on the levels of CO_2 in the ocean and on the production of the gas dimethyl sulphide, a possible agent in the making of rain[4]. (Courtesy of Mr S. Groom, Plymouth Marine Laboratory, Plymouth, UK.)

and *Box 5.1*, *Figures 5.12* and *5.13*, for further examples), giving us a large spacial coverage, very impressive pictures of the eddying patterns and their statistical properties, and certainly more detail than can be gained using ships. However, the story does not end there. The relationship between the sea surface fields and the eddy structure at depth is a subject of on-going research. Through the use of mathematical models we know that the behaviour of the eddies is very dependent on the way the flow and density vary in the vertical. We know through experiments such as Polymode that interactions between the eddies can produce rapid changes to the flow conditions, to form intense jets or more quiescent plumes. However, subsurface measurements are limited in time and space. We are at an early stage in characterizing the eddy structures in the open ocean.

And at smaller scales ... ?

And at smaller scales of around 10 km or less there are a host of structures. 'Meddies', 'smeddies', and 'Beaufort eddies' are a few of the names given to a class of animals in the eddy zoo known as sub-meso-scale coherent vortices, or SCVs.

They are lens-like structures that are small in horizontal extent (a few kilometres), have limited vertical extent (typically a few hundred metres), and, in the vast majority of cases, have a circulation in an anticyclonic (clockwise in the northern hemisphere) sense. SCVs are identified by a bulging of density surfaces and a core of water with anomalous properties compared with its surroundings, which may have originated several thousand kilometres away.

The SCVs that have been studied the most are

4.12

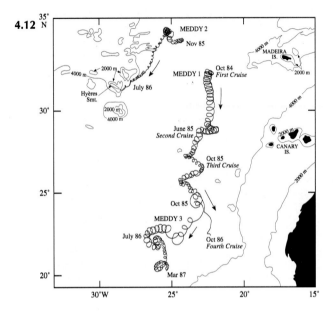

4.13

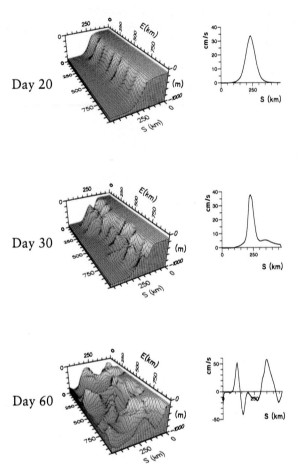

Figure 4.12 Intense 'blobs' of Mediterranean water are formed near Cape St Vincent and keep their rapid rotation and distinct temperature and salinity characteristics for years as they propagate into the ocean interior[11]. The tracks of SOFAR floats that were seeded into 'meddies' demonstrate both the rapid rotation and the southward drift over 2.5 years.

the meddies (*Figure 4.12*). These are lenses of water that originate in the outflow of the Mediterranean and propagate into the North Atlantic, carrying the relatively warm and salty Mediterranean water with them. They can last for a long time. One has been tracked[11] for a period of 2 years.

SCVs are abundant in the ocean. Long-lived lenses of water transporting their chemical composition over large distances may be responsible for a significant flux of water properties.

How do Eddies Originate?

A wide range of fluid flows cannot remain smooth or unchanging in form and are intrinsically 'unstable'; witness the break-up into drops of a jet of water from a tap or the turbulent, chaotic nature of the flow in a stream or river. Small disturbances to the flow are amplified, interact, and produce very irregular motions which totally change the nature of the flow. The flows at larger scales in the atmosphere and ocean are no exception. Currents that are sheared in either the vertical or horizontal may be susceptible to small disturbances that trigger an explosive growth of eddying motions. An example of such an instability is shown in *Figure 4.13*. For a full description of flow instability, and a good guide to ocean dynamics in general, see Gill[3] and Pedlosky[6].

Figure 4.13 Example of an unstable flow. The three panels show the evolution of a jet set-up in a channel. The channel is 500 km wide, 1000 km long, and 2000 m deep and oriented in an east–west direction. The results shown are from a numerical model which simulates flow in the ocean. On the right is a typical cross-section of the surface flow down the channel. On the left is the depth of a density surface. Initially, the jet is in the centre of the channel. Because the flow is strongly constrained by the earth's rotation, there is a corresponding slope in the surfaces of constant density, as seen by the shallowing of the particular density surface shown from 900 m on the southern side of the channel to 100 m on the northern side. At day 0 the jet is perturbed by a random disturbance with a wide range of horizontal scales. These perturbations grow with a growth rate dependent on the scale of the perturbation. By day 20 the fastest growing disturbance, which has a wavelength of 125 km, becomes apparent in the depth of the density surface. By day 30 there has been significant growth. The velocity cross-section shows a tightening up of the core of the jet. By day 60 the density surface has become very distorted and the jet has been replaced by a chaotic eddy field. As the instability grows there is a general slumping of the density surface. The potential energy associated with the initial displacement of the density surface has been converted into the kinetic energy of the eddies.

In an unstable flow, the energy for the eddying motions can be extracted from either the potential or kinetic energy in the system, or from both. In the case of the large-scale flows of the atmosphere and ocean, if the source of the eddy energy is from the (time) mean potential energy the instability is referred to as a baroclinic instability; if the eddy energy comes from the mean kinetic energy the instability is known as a barotropic instability. Although there are some well-studied particular examples of flows that are baroclinically or barotropically unstable, and some known necessary conditions for instability, the stability of a given flow regime, in general, is not known and we have to resort to experiments with numerical models (such as that shown in *Figure 4.13*) to investigate the evolution of the flow.

The growth rate of an unstable perturbation is dependent on the horizontal and vertical structure of the current, as well as on the vertical gradient of density. In regions of highly sheared flow, such as the Gulf Stream, the time-scale for the development of meanders and the pinching off of warm and cold core rings is a few days. In the centre of ocean basins, away from swift currents and fronts, the time-scale for the development of eddies may be around 100 days. The disturbance with the fastest growth rate has a wavelength of around 4–6 times a length-scale called the Rossby radius of deformation (named after the famous Swedish oceanographer, Carl-Gustaf Rossby). From numerical experiments it is found that this length-scale dominates in a fully developed eddying flow. The Rossby radius is a fundamental length-scale of flows on our rotating planet, and is the scale at which rotational forces acting on the fluid motion become comparable to buoyancy forces. It depends on the rotation rate of the earth, the latitude of the motion, the vertical density gradient, and the depth of the fluid. The latitudinal dependence is such that the length-scale decreases with distance away from the equator. In the ocean, the equatorial Rossby radius is around 200 km. At mid-latitudes the Rossby radius is around 30 km, decreasing to less than 10 km in Arctic and Antarctic waters. Meso-scale eddies in the ocean are found to have a similar scale to the Rossby radius, with the same latitudinal dependence. In the atmosphere, the Rossby radius is closer to 1000 km, the scale of the atmospheric weather systems. The eddies in the ocean and atmosphere, with their very different scales of motion, not only have a similar appearance, but also are dynamically equivalent. Thus, the term 'ocean weather' is a particularly apt description of ocean eddies.

Not all meso-scale eddies in the ocean originate from unstable currents. Other generating mechanisms include shedding from topographic features and variable forcing by fluctuating winds. Eddy energy may be exported from a region where it is generated by transport, by the flow, or by radiative mechanisms. Eddies interact with each other, waves, and the mean flow. There are several proposed generation mechanisms for sub-meso-scale coherent vortices – current instability, flow over topographic features, vortex concentration in geostrophic turbulence, and mixing and adjustment of the density field.

The mechanism for the ultimate demise of eddies is not well-understood, but is probably a consequence of bottom friction (if the eddy extends to the bottom), loss of energy by the radiation of large-scale waves, or dissipation by smaller scale motions. Isolated eddies may decay through horizontal interleaving with the surrounding water or may be scattered and broken up over topographic features. We still have much to learn about eddies in the ocean, their dynamics, and the controlling factors for their growth and decay.

Are Eddies Important?

At any one location in the ocean the day-to-day variation in current speed, temperature, and salinity is very much dictated by the evolving eddy field. Predicting the state of the ocean weather is useful for some activities, such as commercial fishing where particular fish stocks are known to favour certain types of eddy and their associated fronts. However, to an oceanographer trying to measure large-scale gradients in the ocean, or the flow averaged over a long period, eddies are a nuisance. The overwhelming effect of the eddy field masks the weaker signal of the larger-scale circulation of the ocean. Time series of currents at many locations in the ocean have to be of several years duration in order to obtain a reliable measure of the mean flow.

But are eddies more than just a nuisance? Are they important in shaping the circulation of the ocean or in the oceanic transport of heat? The answer to many such questions is that we simply do not know. For the atmosphere, the intimate links between weather and longer-term changes in the state of the atmosphere, or climate, are well-established. We have yet to establish the link between ocean weather and the climate of the ocean.

The major difficulty in assessing the role of eddies is the problem of sampling the eddy field. We do know that eddies are effective at dispersing and mixing properties in the ocean (for instance, see *Figures 4.6* and *4.11*), or that they do affect the distributions of plants and animals in the ocean (*Figure 4.10*) and the exchanges between the atmosphere and the ocean. An important aspect of eddy mixing is the homogenisation of potential vorticity, a dynamical property which dictates the behaviour of ocean currents[12]. Eddies also transport heat (a striking example is the Agulhas Current eddies which transport heat from the Indian Ocean to the Atlantic around the tip of South Africa; see Chapter 3, *Figures 3.19* and *3.20*).

Quantifying the effect of eddies is very difficult. Often we have to resort to mathematical models of the ocean to discover what eddy processes may be important. From these models we have learnt much about how eddies interact and affect the mean flow. For instance, eddy–eddy interactions can produce intense vortices; eddy–mean flow interactions can tighten a jet and split a jet into a number of multiple jets. It is only very recently that computer technology has advanced sufficiently that meso-scale eddies can be represented explicitly in models of the circulation in ocean basins or in the entire world's ocean (see, for example, Semtner and Chervin[16]). Whether or not eddies play a major role in the oceanic conveyor belt (Chapter 3) is still unclear.

Epilogue

The future will bring an increase in the power of computers, allowing more detailed and more realistic models of the world's oceans to be developed. We are becoming more skilled in the use of measurements from space to tell us about the dynamics of the oceans. New instruments are being developed to measure not only the physical properties of the ocean, but also biological and chemical properties in greater detail. With new technology and insight we will learn more about ocean eddies and their effect on the ocean's climate, biology, and chemistry. The study of the weather of the oceans has brought together theoreticians, satellite oceanography, and those who make measurements at sea. This collaboration will continue well into the future.

General References

Gill, A.E. (1982), *Atmosphere–Ocean Dynamics*, Academic Press, New York, 662 pp.

Pedlosky, J. (1987), *Geophysical Fluid Dynamics*, Springer-Verlag, New York, 710 pp.

Robinson, A.R. (ed.) (1983), *Eddies in Marine Science*, Springer Verlag, Berlin, 609 pp.

References

1. Dickson, R.R. (1983), Global summaries and intercomparisons: flow statistics from long-term current meter moorings, in *Eddies in Marine Science*, Robinson, A.R. (ed.), Springer Verlag, Berlin.
2. Fofonoff, N.P. and Webster, F. (1971), Current measurements in the Western Atlantic, *Phil. Trans. R. Soc. A*, **270**, 423–436.
3. Gill, A.E. (1982), *Atmosphere–Ocean Dynamics*, Academic Press, New York, 662 pp.
4. Holligan, P.M., *et al.* (1993), A biogeochemical study of the coccolithophore, *Emiliania huxleyi*, in the North Atlantic, *Global Biogeochem. Cycles*, **7**, 879–900.
5. Kort, V.G. and Samoilenko, V.S. (1983), *Atlantic Hydrophysical Polygon-70: Meteorological and Hydrophysical Investigations*, Amerind Publishing Company, New Delhi, for NSF/IDOE, 398 pp.
6. Pedlosky, J. (1987), *Geophysical Fluid Dynamics*, Springer-Verlag, New York, 710 pp.
7. Pollard, R.T. and Reiger, L. (1990), Large potential vorticity variations at small scales in the upper ocean, *Nature*, **348**, 227–229.
8. Richardson, P.L. (1983), Gulf stream rings, in *Eddies in Marine Science*, Robinson, A.R. (ed.), Springer Verlag, Berlin, pp 19–45.
9. Richardson, P.L. (1985), Drifting derelicts in the North Atlantic 1883–1902, *Prog. Oceanogr.*, **14**, 463–483.
10. Richardson, P.L. (1993), A census of eddies observed in the North Atlantic SOFAR float data, *Prog. Oceanogr.*, **31**(1), 1–50.
11. Richardson, P.L., Armi, L., Price, J.F., Walsh, D., and Schroter, M. (1989), Tracking 3 meddies with SOFAR floats, *J. Phys. Oceanogr.*, **19**(3), 371–383.
12. Rhines, P.B. and Young, W.R. (1982), Homogenization of potential vorticity in planetary gyres, *J. Fluid Mechan.*, **122**, 347–367.
13. Robinson, A.R. (ed.) (1983), *Eddies in Marine Science*, Springer Verlag, Berlin, 609 pp.
14. The MODE-1 Atlas Group (1977), *Atlas of the Mid-Ocean Dynamics Experiment (MODE-1)*, MIT, Boston, 274 pp.
15. The MODE Group (1978), The Mid-Ocean Dynamics Experiment, *Deep-Sea Res.*, **25**, 859–910.
16. Semtner, A.J. and Chervin, R.M. (1988), A simulation of the global ocean circulation with resolved eddies, *J. Geophys. Res.*, **93**, 15502–15522.
17. Stommel, H. and Arons, A.B. (1960), On the abyssal circulation of the world ocean, *Deep-Sea Res.*, **7**, 140–154; 217–233.
18. Swallow, J.C. (1955), A neutral-buoyancy float for measuring deep currents, *Deep-Sea Res.*, **3**, 74–81.
19. Swallow, J.C. (1971), The Aries current measurements in the Western North Atlantic, *Phil. Trans. R. Soc. A*, **270**, 460–470.
20. Zenk, W. and Muller, T.J. (1988), Seven-year current meter record in the eastern North Atlantic, *Deep-Sea Res.*, **35A**(8), 1259–1268.

CHAPTER 5:

Observing Oceans from Space

I.S. Robinson and T. Guymer

Introduction – A New Way of Viewing the Ocean

At first consideration it may seem surprising to find a chapter on space in a book about oceanography. But it is not so surprising when we recall that it was the lunar astronauts' first pictures of Earth, the beautiful blue and green planet seen from a barren, dead moonscape, which awakened a new awareness of the fragility of the Earth's natural environment. It took a visit to our nearest neighbour in space to emphasise how important it is to understand the global oceanic and atmospheric environment on which we depend for our continued existence. That global view and wide spatial perspective eventually led to a new branch of oceanography. In this chapter we examine the basic principles of how the oceans can be measured remotely from satellites (and also from aircraft), explain some of the techniques which are used by particular sensors, and present some illustrative examples of image data and their oceanographic interpretation.

The scientific understanding of ocean processes depends upon being able to measure the wide variety of variable parameters which describe the sea. Until recently, oceanographers could make measurements only from ships or buoys (see Chapter 19), and were consequently limited by the sampling constraints of an Earth-bound perspective. The vantage point from aloft, which Earth-orbiting satellites provide, has enabled spatially detailed measurements (using, e.g., the electro-magnetic spectrum, *Figure 5.1*) to be made almost instantaneously over wide areas and provide a novel perspective of the ocean. It has also made possible the regular, repeated monitoring of the ocean on a global scale, and the detection of some of the changes of climate discussed in Chapter 3.

'Satellite oceanography' had its birth as a scientific endeavour in the 1960s, when the first astronauts in orbit around the Earth noticed features in the sea which were so much easier to discern from above than from ground level. Soon satellites were

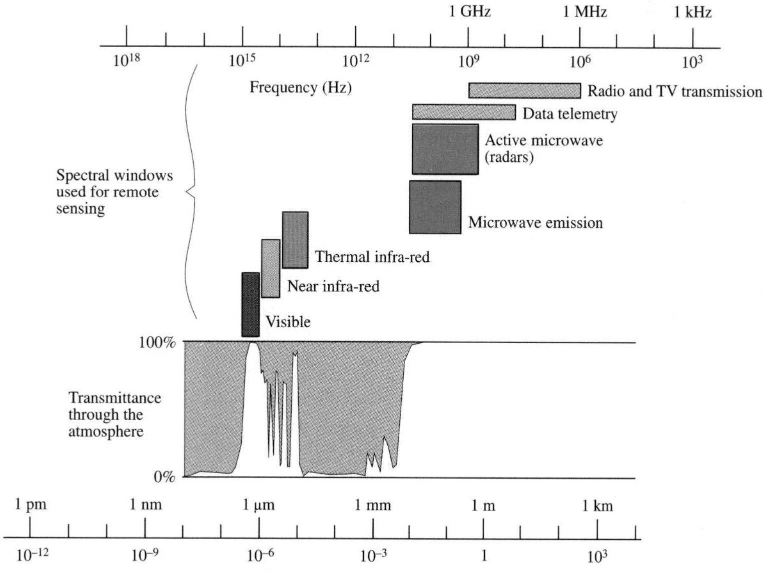

Figure 5.1 The electromagnetic spectrum showing the variation with wavelength of atmospheric transmission and the spectral windows used for remote sensing. Note that microwaves are usually defined in terms of frequency (upper scale), whereas visible and infra-red radiation are generally referred to by wavelength (lower scale).

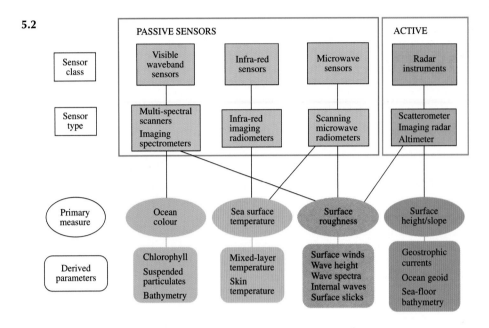

Figure 5.2 Schematic showing the different classes of sensors used in satellite oceanography, the different types of instrument, the primary measurement which each class can make, and examples of the ocean parameters which can be derived from these measurements.

flown specifically for the purpose of Earth observation, and by 1978 sensor technology had developed to the stage where oceanographers could begin to make scientific measurements of the sea. In that year three satellites (Seasat, Tiros-N, and Nimbus-7) were launched with sensors capable of recording a variety of ocean processes, and since then the use of satellite data has become widely accepted in marine science.

Remote Sensing Methods

How can we observe the ocean from space?

Remote sensing relies on electromagnetic radiation to convey information about the sea to a sensor on a satellite or aircraft. Most electromagnetic wavelengths are absorbed or scattered by the atmosphere, and so are of no use for this purpose, but there are three distinct wavebands, 'spectral windows', at which rays can penetrate the atmosphere with less interference (see *Figure 5.1*). These are the **visible** waveband (400–700 nm), parts of the **infrared** (the regions around 3.7 mm and 10–13 mm are used), and **microwaves**, which include radar, longer than about 10 mm. Visible and infra-red light cannot penetrate clouds without being scattered or absorbed, but microwaves are much less affected and permit all-weather monitoring of the sea.

As well as selecting the spectral window, the remote sensing scientist can use a number of sensor types designed with the object of measuring one or more particular ocean parameter (see *Figure 5.2*). Sensors may be either **passive** or **active**. Passive sensors measure naturally occurring radiation, using either the Sun's rays scattered from below the sea surface, or energy emitted by the sea itself in the infra-red or microwave parts of the spectrum. Active sensors produce their own source of electromagnetic energy which is emitted toward the sea and measured after reflection. Sensors that rely on solar illumination, like the human eye, can operate only in the daytime, when the Sun is not too close to the horizon, but the other sensors can operate both day and night. Active sensors deployed on satellites are restricted to microwaves, although visible-wavelength lasers have been used from aircraft.

The instrument designer can choose the particular electromagnetic property which a sensor will measure. This may simply be the magnitude of radiation at a particular wavelength, or the relative magnitude at several wavebands. It may also relate to a particular polarisation state of the radiation. Active sensors provide more options because they can measure not only the magnitude of the returned pulse relative to that emitted, but also its timing, the shape of the pulse, and any frequency shift. For a sensor to be useful for a particular application requires that the measured radiation be influenced in some way by the ocean parameter

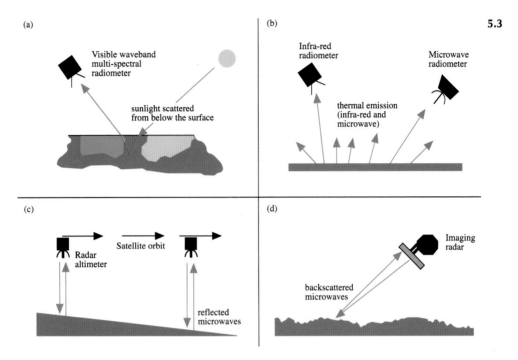

Figure 5.3 Depiction of the four primary measurements of the sea which can be made in satellite oceanography, and an indication of how they are made. (a) Ocean colour; (b) sea surface temperature; (c) sea surface slope (length scale of tens of kilometres); (d) sea surface roughness (length scale of 10 mm to 10 m).

being monitored. The influence can be direct, as with infra-red radiation emitted by the sea surface according to its temperature. It can be more complex, as in the way the spectral composition, or colour, of sunlight scattered from below the surface is influenced by both the chlorophyll and particulate content of the water (see Chapter 14). It may be more subtle, as in the way the sea state changes the shape of the reflected pulse of a nadir-viewing radar. After two decades of experience a number of useful oceanographic instruments have emerged from the wide range of potential design options.

What can we measure from space?

Despite the variety of instruments, remote sensing is able to measure just four basic properties of the sea; the surface **temperature,** the **colour** of the near-surface waters, the surface **roughness** (at short length-scales), and the **slope** of the surface averaged over tens of kilometres (*Figure 5.3*). It is from these primary measurements that a range of other properties can be derived, depending on the detailed design of the sensor. For example, given measurements of the appropriate wavebands, colour can yield an estimate of the chlorophyll content and hence the primary production in the upper ocean (see Chapter 6). From the detection of surface slope can be derived a measure of ocean currents, while

surface roughness can be interpreted in terms of wind speed. Some more details about how oceanographic information is extracted from these primary measurements are given in the examples in *Boxes 5.1–5.6* which describe different applications of remote sensing.

Remote sensing has many limitations. There are many important aspects of the ocean that cannot be detected remotely, such as salinity, at least with the present technology and knowledge. Remote sensing cannot penetrate far below the sea surface and cannot tell us much about how the ocean properties and composition change with depth. Like any measurement tool, we should not misapply it, but use it for what it can do well, and be aware of its particular characteristics.

Sampling Capabilities of Remote Sensing

The greatest advantage of remote sensing lies in its ability to provide measurements of ocean parameters over a wide sea area. The size of the area that can be viewed almost simultaneously, the spatial resolution of the resulting image data, and the frequency with which it can be revisited depend on the type of sensor and the platform carrying the instrument.

Sensors record individual point measurements of a property of the ocean corresponding to an aver-

5.4

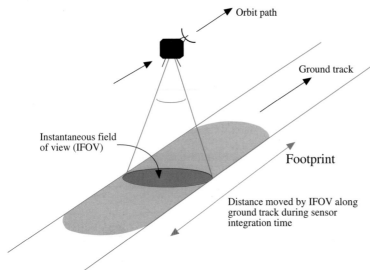

Orbit path

Ground track

Instantaneous field
of view (IFOV)

Footprint

Distance moved by IFOV along
ground track during sensor
integration time

Figure 5.4 Earth viewing by a non-scanning sensor. The IFOV is the area of ground observed at any instant by the sensor, but the recorded measurement is representative of the footprint covered by the IFOV during the instrument integration time.

age over the instantaneous field of view (IFOV), which may be as small as a few metres or as large as hundreds of kilometres, depending on the sensor. Some sensors, such as an altimeter, simply view in a single direction, following the ground-track beneath the moving satellite or aircraft (*Figure 5.4*). Mounted on a polar orbiting satellite such sensors gradually build up coverage of the Earth's surface. There are typically 15 orbits per day during which time the Earth rotates underneath the sensor, which therefore samples along the track illustrated in *Figure 5.5*. Over many days, given a suitable orbit, the gaps between the single-day paths can be filled, providing a fully global picture of the oceans.

If the sensor is one which scans several times per second across the direction of travel, however, an array of data is collected almost simultaneously which corresponds to an image (*Figure 5.6*) of the ocean surface, composed of individual picture elements, or pixels, each of which is an independent measurement. Careful specification of the sampling

5.5

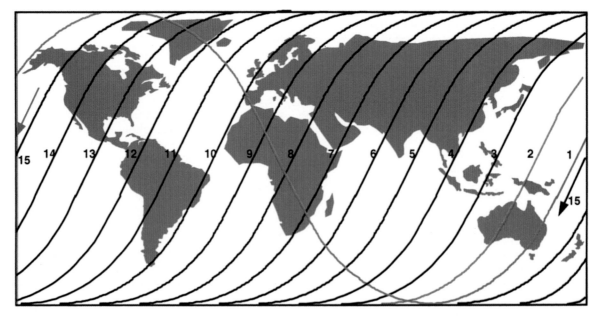

Figure 5.5 The Earth coverage of a typical near-polar satellite orbit during one day. Orbit one is shown in blue, descending in local daytime and ascending in local night. This leads on to orbit two (shown in red). For this and the rest of the orbits (in black) that day, only the daytime pass is shown. Note how the spacing between ground tracks is closer toward the poles.

rate and scan rate relative to the satellite speed can ensure that each IFOV is contiguous with its neighbour, and then sampling is at its most efficient. The swath width depends on the spatial resolution required. For a small IFOV to give high spatial resolution, the swath must be narrower than if the resolution is to be coarser. If the swath is wide enough, the coverage of successive orbits overlaps and coverage of the whole Earth can be achieved daily. Thus, there is a trade-off between frequency of coverage and spatial resolution. Colour scanners and imaging radars with a resolution of about 30 m may take 15–25 days to revisit every point on the Earth, whereas the medium resolution sensors of about 1 km resolution can view the Earth once every one or two days. The sampling capability of the sensor–satellite combination must be matched as carefully as possible to the characteristics of the oceanographic phenomenon to be observed.

Earth observation satellites in near-polar orbit fly at an altitude of about 700–1000 km above the Earth. The orbit is normally arranged to precess slowly so that the orbit plane remains fixed relative to the Sun, despite the seasonal changes of the Sun's position relative to the fixed stars. Such a 'Sun-synchronous' orbit ensures that the local (solar) time is always the same whenever the satellite crosses a given latitude, no matter what the longitude. This ensures a uniformity of solar illumination for ocean colour scanners, and helps to eliminate problems of the diurnal heating cycle. The drawback of the low polar orbit is the difficulty of covering the Earth more frequently than once per day (or twice if night-time images can be used), and the necessity of viewing the sea surface very obliquely at the swath extremities if daily coverage is required.

The only other type of satellite used for remote sensing of the sea is that in geostationary orbit. Parked at a height of about 36000 km above the Earth, this orbits once per day, so that if it is placed over the equator it remains stationary relative to an observer rotating with the Earth. From this vantage point a sensor can scan the whole of the visible disc

5.6

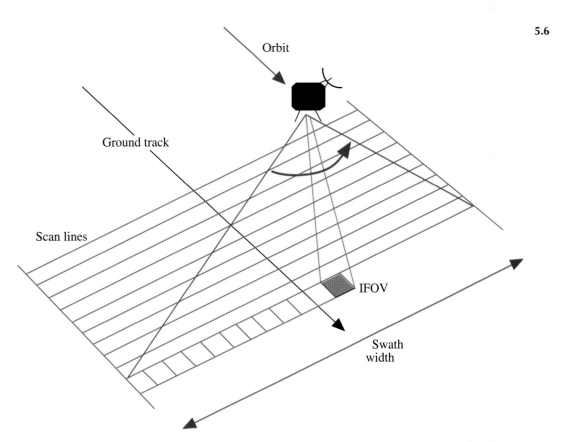

Figure 5.6 An imaging sensor scans across the direction of travel. If the scan rate is matched to the IFOV and synchronised with the speed of the satellite over the ground, the IFOVs are contiguous and the ground area is covered efficiently.

5.7

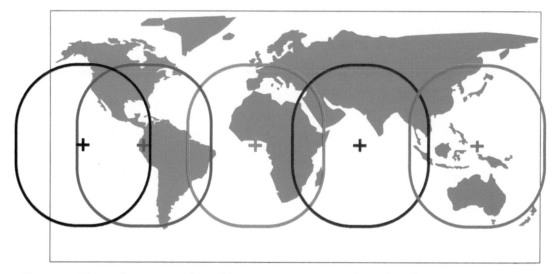

Figure 5.7 The Earth coverage achieved by geostationary meteorological satellites. Locations of the five satellites which span the equator are shown as plus signs. Although the horizon of the Earth-view from geostationary satellites is about 60°, at latitudes greater than 50° the viewing angle is too oblique to be useful, as indicated by the limit lines.

Figure 5.8 Space–time sampling characteristics of major classes of sensor. The capability of each sensor is defined as a box on the diagram. The left boundary represents the sampling frequency; it defines the shortest revisit period (assuming no cloud cover) and depends on a combination of the orbit path and swath width. The bottom boundary represents the spatial resolution of the sensor and defines the length (left scale) or area (right scale) of a picture element (pixel). The top boundary indicates the largest length-scale that can be observed near-instantaneously, as controlled by the swath width. The right boundary depicts the span of time over which image sequences are obtained, limited only by the lifetime of the sensor or series of sensors. **ATSR**, The

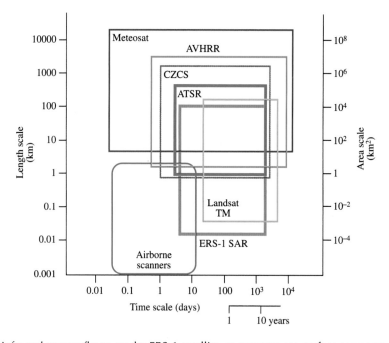

5.8

Along-Track Scanning Radiometer, an infra-red sensor, flown on the ERS-1 satellite, to measure sea surface temperature. **AVHRR**, The Advanced Very High Resolution Radiometer, flown on the NOAA meteorological satellite series, to observe clouds and land surfaces (visible channels), and the temperature of the sea and cloud tops (infra-red channels). **CZCS**, The Coastal Zone Colour Scanner, flown in 1978–1986 by NASA to measure ocean colour in four narrow-band spectral channels. **ERS-1 SAR**, The Synthetic Aperture Radar on the European Space Agency's first remote sensing satellite (ERS-1), launched in 1991. **Landsat TM**, The Thematic Mapper, a high-resolution multichannel visible wavelength sensor mainly for land applications, flown on the US Landsat satellite series. **Meteosat**, The European geostationary meteorological satellite (one of five which by international agreement should be in place around the equator) observes clouds and land, sea, and cloud temperatures using visible and infra-red radiometers.

Figure 5.9 Outline of the procedures required for processing data received from a satellite into oceanographically useful parameters and image arrays.

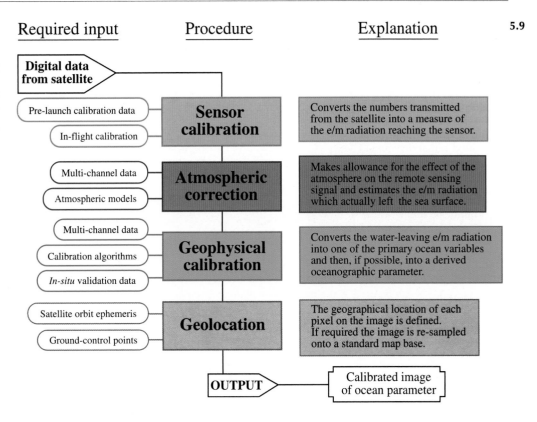

of the Earth (*Figure 5.7*), and can sample at any frequency. Present meteorological sensors scan the full view once in 30 minutes, but spatial resolution is no better than 5 km, and generally worse because of the oblique view. Such sensors cannot view high latitudes above 55° because of the curvature of the Earth.

The sampling characteristics of a number of oceanographically useful sensors are depicted in *Figure 5.8*.

Procedures for Analysing Satellite Data

Before the raw data transmitted from the satellite to the ground station can be of use to the marine scientist, a number of operations have to be performed. *Figure 5.9* outlines the various procedures involved in acquiring satellite data and processing them to yield oceanographic information. Data are recovered from the satellite as a sequence of numbers corresponding to the digitized signal every time it is sampled. Given a knowledge of the detector characteristics, it is possible to apply a **sensor calibration** to the digital signal in terms of the property of radiation being measured at the satellite. The digitization capacity, i.e., whether the values are stored as 6-, 8-, 10-, or 16-bit numbers, determines the resolution and therefore can strongly affect the usefulness of the data.

An **atmospheric correction** also has to be applied in most cases. This attempts to remove the effect of

the atmosphere on the transmission of the ocean information from sea to satellite. The atmospherically corrected value is an estimate of the radiation which the sensor would have detected leaving the surface of the sea if there were no intervening atmosphere. A variety of approaches are used for this, including modelling of the atmospheric effects, and using multispectral methods to correct for variable factors which cannot easily be modelled. For data derived in infra-red and visible wavebands, screening for cloud detection is applied, in which all pixels corrupted by cloud are rejected, leaving blanks in the final image.

While for some applications the value obtained after atmospheric correction may be a useful parameter, e.g., the sea surface temperature (SST), in many cases it is appropriate to apply what is termed a **geophysical calibration** to convert the measured properties into an oceanographically useful parameter, e.g., converting the radiance measured in several bands within the visible spectrum into an estimate of chlorophyll concentration, or converting radar back-scatter into wind speed and direction. Calibration algorithms are generally derived empirically using *in situ* data and calibration or validation experiments.

Another essential requirement is to identify the geographical position of every data value in a **geolocation** procedure. If image data are to be pre-

Box 5.1   Sea Surface Temperature Maps from Infra-Red Sensors

Figure 5.10 Infra-red sensors measure the radiation emitted from the sea surface in the wavebands 3.5–3.9 µm, 10.3–11.3 µm, and 11.5–12.5 µm. The emitted radiation increases with the sea surface temperature (SST), but is reduced by passing through the atmosphere, so the detected (brightness) temperature is not the true value. The atmospheric effect varies with wavelength, and so the difference between brightness temperatures at differ-

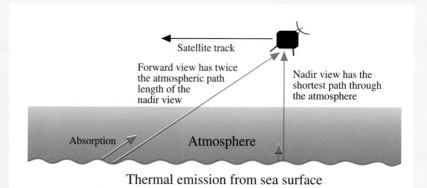

5.10

ent wavebands can be used as the basis for an atmospheric correction. The ATSR sensor also observes each sea area twice, once from above and once obliquely through twice the atmospheric path length. As the difference in temperature between the two views gives a direct measure of the atmospheric effect this approach gives a high absolute accuracy of SST recovery.

Figure 5.11 This image depicts SST in the Bay of Biscay and English Channel, derived from measurements made by the AVHRR sensor on the NOAA-9 satellite on 6 August, 1988. The temperature has been colour-coded according to the scale bar displayed. Black corresponds to the land and white indicates that cloud has been detected. There is a wealth of oceanographic information to be found in this image, including the overall distribution of temperature, showing a drop of 9°C between northern Spain and the English south coast, the discontinuous nature of the thermal structure rather than smooth gradients, and the evidence of meso-scale eddies in the patterns drawn out in the temperature field. There appears to be cool coastal upwelling along the French west coast. Off the Brittany peninsula, the dis-

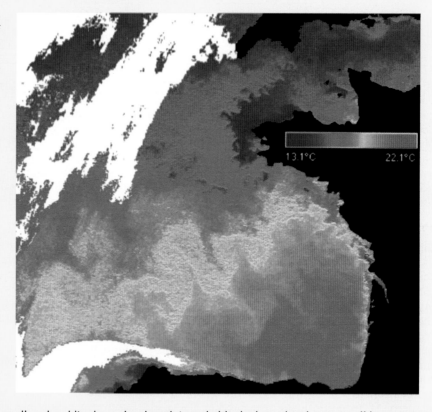

5.11

tinctive cold front, on which small-scale eddies have developed, is probably the boundary between offshore stratified water with a warmer surface layer and the tidally well-mixed water inshore. Further offshore from Brittany, and disappearing under the cloud, is evidence of cool surface temperatures above the shelf edge, caused by breaking internal waves stirring cooler water into the mixed layer.

Figure 5.12 This brightness temperature image of a region 500 km² in the South Atlantic, just north of the Falkland Islands, was obtained from the ATSR 11 µm channel on 13 October, 1992. It illustrates the richness of the dynamical structures which can be found in the ocean and demonstrates the unique capability of remote sensing to capture them. The image is from a region centred about 55°W 45°S. The coldest water comes from the Falkland current, which flows from the southwest. The warmest water derives from the Brazil current flowing from the north. As the two currents interact, large eddying motions occur, entraining warm water into the cold and vice versa. What is striking from the resolution achieved with this sensor is the narrowness of some of the warm and cool filaments drawn out in a complex interleaving, leading in some places to the growth of small eddies. Eventually, these will diffuse into a uniform body of intermediate temperature, but this particular instant within a rapidly evolving pattern has been captured by the satellite in a way that could not have been possible using ships. Remote sensing is providing new challenges for the study of the dynamical processes which control these temperature patterns.

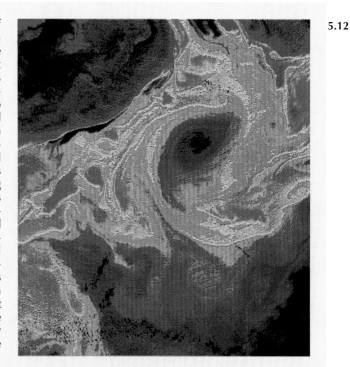

5.12

Figure 5.13 Individual high-resolution images are used to contribute to global compilations of sea surface temperature. (a) The contribution of just one day's overpasses of the ATSR. The regular gaps show the limits of daily coverage; the gaps within the swath are due to cloud being encountered. Over a month the swath covers most of the globe, and very few locations remain persistently cloud-covered (black on the image), leading to the monthly average SST map (b). This is based on a resolution cell of 0.5° latitude and longitude. It reveals the global structure of SST, warmest at low latitudes and coldest toward the poles. However, the east–west asymmetry in the oceans is apparent, particularly in the Atlantic. Also evident is the characteristic cool zone along the equator in the Pacific, and the cool upwelling off the east coasts of Africa and South America. Despite its coarse resolution, an image like this still shows eddy activity (see Chapter 4), particularly in the Atlantic sector of the Antarctic Circumpolar Current.

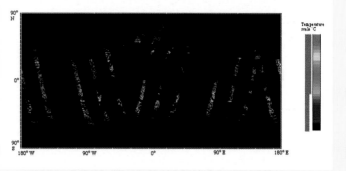

5.13a

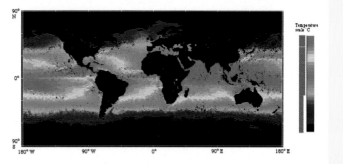

5.13b

sented in the form of a regular map projection, **geometric correction** has to be applied. This may require **resampling** of the image onto a different grid base; since this degrades the data slightly it is usually performed only when all the other corrections have been applied. For many purposes it is convenient to combine the data from several different satellite overpasses into a single map representing average conditions over a given time span, a procedure known as compositing.

For example, a global map of sea surface temperature is built up from the cloud-free pixels derived from all the overpasses on several days (see *Box 5.1*). This enables gaps due to cloud on some days to be filled from clear overpasses on others, although care has to be taken if the ocean variable is likely to vary rapidly within the compositing period.

Ocean Applications of Satellite Data

The routine processing outlined above generally leads to images or maps of a particular ocean variable. Although much challenging scientific work has been done developing these procedures to the point where satellite measurements are scientifically reliable and have a specified accuracy, this aspect of satellite oceanography is just the prelude to the main task which is to **analyse** and **interpret** the data. Here, the exciting scientific opportunity is to use the unique capabilities of remote sensing to extend our understanding of ocean processes. Satellite data potentially have a contribution to make to all aspects of marine science, but here we have room only to look at a few examples.

In *Box 5.1* the measurement of SST using infrared sensors is described. One of the primary objectives in developing these sensors has been to acquire regular global maps of SST distribution and its seasonal variation. Once a long-enough time sequence has been established (useful records commenced in 1980), it will be possible to identify anomalous trends which may herald changes in the patterns of global climate. Already such data provide early indicators of the onset of irregular climate phenomena, such as the El Niño–Southern Oscillation (ENSO) events (see Chapter 3). From the same sensors, but analysing data at higher spatial resolution, maps of SST provide synoptic views of the ocean 'weather', the meso-scale eddies (Chapter 4) which help to distribute heat and ensure that the ocean is far from homogeneous. Such images can be used by fishing fleets to locate regions, such as frontal zones, which are favourable for finding certain species of fish.

A completely different technique, using the measurement of ocean colour, is illustrated in *Box 5.2*. Interestingly, this method can also reveal similar meso-scale dynamical processes to those imaged by their temperature signature. It can also be applied on the scale of ocean basins to reveal clearly how the primary production in the upper ocean varies spatially and seasonally. At the much higher spatial resolutions achievable from aircraft, it is possible to detect small-scale dynamical processes occurring in coastal seas and estuaries, and to use this to understand the processes which control coastal pollution and the dispersion of river discharges.

Images derived from synthetic aperture radar or SAR (*Box 5.3*) represent the variation of radar back-scatter over the sea. They are not so readily interpreted as colour or thermal images, but the spatial patterns often provide clues for interpretation. On some images, swell waves are clearly shown by the way in which back-scatter varies between trough, crest, and leading and trailing face. Sometimes the refraction of wave patterns is imaged. Surprisingly, since the radar is influenced only by the surface roughness, patterns appear which are associated with phenomena beneath the water surface, such as sea-bed bathymetry, or the presence of undulations of density layers known as internal waves. In other images the radar is strongly influenced by the presence of surface films which generate slicks of smoother water having low back-scatter. The spatial coverage of the satellite SAR over 100 km can reveal the coupling between local roughness phenomena and meso-scale dynamical features.

Altimeters do not generate images like scanning sensors, although over time they can build up evidence of sea surface topography which can be interpreted in terms of ocean currents (*Box 5.4*). Until there is an independent measure of the shape of the geoid due to the variation of gravity, only deviations of slope from the mean can be obtained, but these can reveal the variability in ocean currents and map the global distribution of eddy kinetic energy (EKE, see Chapter 4). The altimeter's ability to estimate wave height from the shape of the reflected radar pulse (*Box 5.5*) gives it an important operational role in sea state forecasting.

The scatterometer (*Box 5.6*) measures the average roughness over large areas of the sea surface and enables the local wind speed and direction to be derived. While the direct application of this is in meteorology, it is valuable for indicating the way in which the dynamical forcing of the ocean by the atmosphere (Chapter 2) can vary spatially. It can also be used operationally in improving wave forecasting models.

Although microwave radiometry cannot match the fine spatial and radiometric resolution of infrared sensors, its reduced sensitivity to cloud cover enables it to play a valuable role in cloudy regions. Microwave radiometers can yield not only a measure of SST, but also can measure wind stress and detect the presence of sea ice. They therefore have a valuable operational role to play in polar seas.

Box 5.2   Ocean Colour Images

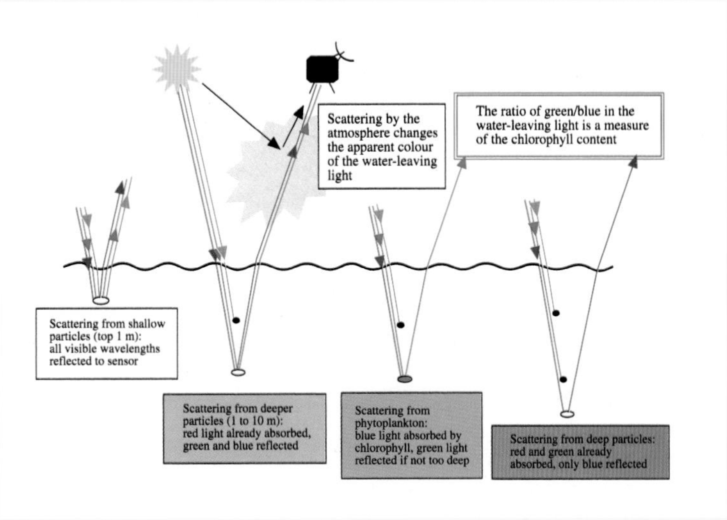

Figure 5.14 The apparent colour of the water is affected by the combination of scattering and absorption of the Sun's light, as illustrated. The red end of the spectrum is preferentially absorbed by sea water, and light scattered from deep in the sea consequently appears blue. However, chlorophyll present in the water absorbs blue light, so that the water-leaving light appears greener in the presence of phytoplankton, provided there is enough particulate material to scatter the green light before it is absorbed by the water itself. Other substances can also influence the water colour, such as dissolved organic material (known as yellow substance) and exceptional 'red tide' plankton blooms (see Chapter 6). The brightness of scattered light also gives a qualitative indication of the amount of suspended particulates in the water. Consequently, the remote sensing of ocean colour can be used to measure the concentration of phytoplankton in the upper layers and to monitor the distribution of suspended particulates, as well as to act as a tracer for the patterns of dynamical features.

5.14

Figure 5.15 An image defining the distribution of chlorophyll pigment concentration in the North Atlantic off the Iberian peninsula on 7th April, 1980, derived from CZCS observations of ocean colour. Images such as this illustrate how variable the ocean can be, and demonstrate the value of remote sensing methods for instantaneously capturing the spatial distribution of phytoplankton. The heterogeneity derives from the patchiness of primary production in spring, which depends on the distribution of nutrients and of overwintering seed populations. Once 'hot spots' of production have commenced, the variable currents associated with meso-scale dynamical features move the patches by advection, stretching and twisting them into complex patterns. One of the most interesting features in this image is the strong filament which extends several hundred kilometres offshore from Lisbon. This, and other less distinct filaments further north, are important for the transport of continental shelf water into the deep ocean. The image suggests, however, that between the offshore flowing filaments are zones of less productive water flowing toward the coast to form an onshore–offshore circulation cell.

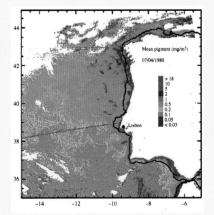

5.15

Figure 5.16 An enhanced colour image of part of Southampton Water, UK, generated from Airborne Thematic Mapper data. The image has been constructed from measurements in the blue, green, and red parts of the spectrum to give a near-real colour image, but because the contrast is enhanced the colours appear brighter than they would to the naked eye. The most apparent feature on this image is the plume in the wake of the large container vessel. Whereas the surrounding water appears dark blue–green, the plume appears yellower, indicating that red as well as green and blue light is being back-scattered. This occurs when there are high suspended-sediment concentrations close to the surface and the image indicates that the passage of the ship has stirred up material from the sea bed. Brighter patches elsewhere also indicate increased suspended sediment, possibly caused by the earlier passage of other ships.

5.16

Box 5.3   Synthetic Aperture Radar Imaging of Small-Scale Ocean Processes

Figure 5.17 It is not possible with a satellite sensor to focus microwaves in the same way as visible or infra-red radiation. To achieve a high-resolution image in the range direction (the radar pointing direction), the radar uses the timing of the return pulse to determine the precise distance to the patch of sea surface being viewed. To achieve comparably high resolution in the azimuth (along track) direction would require a very large aperture antenna. This cannot be constructed physically, but its effect is synthesized digitally using the recorded pulse returns from many positions of the satellite along its orbit, hence the name Synthetic Aperture Radar. SARs on satellites are capable of resolving down to 20 m. A typical SAR image consists of up to

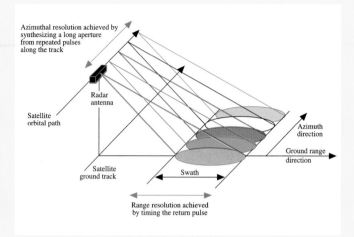

5.17

4000 x 4000 picture elements (pixels). A SAR produces an image corresponding to the magnitude of the radar energy returned from the sea surface. Since the radar views the sea at an oblique angle, the signal relies on detecting back-scattered radiation caused by the interaction between the incident radar waves and the roughness of the sea surface. The microwaves do not penetrate the surface at all, and so the patterns on the radar image are due to variations of the sea surface roughness.

Figure 5.18 An image of the southern North Sea adjacent to the Dover Straits acquired by SAR on Seasat, 19th August, 1978. The original data have been averaged to a resolution cell of 168 m, eliminating the speckle effect and resulting in a very smooth image. The width of the image represents about 90 km and the direction of North is 26.1° clockwise from the top of the image. The radar back-scatter patterns appear to reveal the bathymetry of the sea bed – the image looks rather like a photograph of the sea bed laid bare and illuminated from an oblique angle. Images like this took many oceanographers by surprise when they were first obtained. It must be emphasised that the radar signal does not penetrate through the sea to the sea bed. The imaging mechanism is this: the tidal currents fluctuate in magnitude as they flow over shallow or deep regions, causing horizontal convergence and divergence which concentrates or reduces the surface wave energy, and hence the roughness controlling the radar return. Although such a

5.18

mechanism seems rather complex, the clarity of bathymetric features in the image, which correlate very well with bathymetric charts, demonstrates that this is an effective method for mapping sandbanks and sandwaves. Current research is determining whether quantitative as well as qualitative information can be recovered, and examining the conditions of tide and wind which are necessary to produce such clear images of the sea bed.

Figure 5.19 This ERS-1 SAR image of the English Channel off the Isle of Wight was recorded on 2 July, 1993, and the width of the image represents 80 km. Slicks, regions in which the sea surface is smoother than normal, appear as dark regions on the image and may be caused either by the presence at the surface of material such as organic films, or by dynamical features which produce local divergence of the surface current. Either process reduces the amplitude of the short surface waves which influence the radar back-scatter. In this image the slicks provide a way of detecting other dynamical processes. In some of the coastal embayments narrow slicks appear to be aligned with the local tidal circulation. In other areas the slicks relate to old ship wakes which have left a trail of smoother water behind. Further offshore the larger slicks are probably patches of surface film discharged from ships. The corrugated shape of these patches is due to the shear associated with the strong tidal streams which flow parallel to the coast. (The data from which this image was derived were supplied by the European Space Agency.)

5.19

Figure 5.20 This ERS-1 SAR image of the Atlantic coast off Portugal contains a lot of wave-like phenomena consisting of clusters of between four and eight slightly curved, concentric crests. Since the size of the whole image represents 100 km across, the wavelengths of these phenomena, greater than 1 km, imply that they are the surface manifestation of internal waves. These are subsurface waves producing undulations of the interface between the upper mixed layer of the sea and deeper, cooler, denser layers, and are generated by tidal flow encountering the shelf edge. Packets of several internal waves are produced by each tide and the surface currents associated with them act to compress or stretch the wind waves and thus generate zones of rougher and smoother sea surface. Observations at sea coincident with SAR overpasses confirm that the zones of rough and smooth sea are in phase with the internal waves, which thus acquire a signature in the SAR image. Because these waves propagate quickly, it is not possible to use ship measurements to map their spatial extent. SAR images provide the only systematic way of detecting their occurrence and defining their spatial distribution. By making assumptions about their tidal origin, it is possible to estimate their propagation speed from the spacing between distinct wave packets. (The data from which this image was derived were supplied by the European Space Agency.)

5.20

Box 5.4   Sea Surface Topography

As implied by its name, the altimeter primarily measures height – in this case, that of sea level. The method is conceptually very simple. A sharp pulse is transmitted vertically toward the Earth's surface from an antenna on the satellite. Some of the signal is reflected back to the same antenna from those parts of the surface within the 7 km footprint which are aligned perpendicular to the beam. The key to the technique is the accurate measurement of the time delay between transmission and reception of the pulse by the antenna. If the signal's propagation speed is known then the distance between the satellite and the surface can be obtained. The height of the satellite orbit can be determined by tracking stations and the difference between the two provides an estimate of sea level. As the satellite circles the Earth, so sea level changes in space and time can be monitored.

In practice, it is very difficult to achieve the accuracy required for oceanography (better than 10 cm). Corrections must be made for the effects of the atmosphere on the propagation speed and, because the reflected pulses are very noisy, sophisticated processing techniques are needed to time the arrival of such pulses with high precision. Even the task of precisely determining the position of the satellite presses orbital dynamics computations to their limit, because large satellites in low altitude orbits are influenced significantly by such processes as air drag and solar radiation pressure. It is, indeed, remarkable that the present generation of altimeters flying at altitudes of 1000 km can measure sea level to 4 cm – a precision of 1 in 25 million!

What can we learn from the resulting sea level data? The largest variations by far, amounting to more than 200 m, are those of the geoid, a surface connecting points of equal gravitational potential. The Earth's gravity field is rather uneven because of the way in which mass is distributed, and this is manifested in sea level. Although of great interest to solid Earth geophysicists – the influence of ocean trenches and seamounts is often readily discernible – it is a nuisance to oceanographers wanting to use altimetry because it prevents the calculation of mean currents. What is required is global mapping of the geoid, independently of altimetry. This surface must be subtracted from the altimeter data to yield the oceanographic signal, referred to as the dynamic topography. An example is shown in *Figure 5.21*. Dominant large-scale oceanographic features, such as the Antarctic Circumpolar Current and the Subtropical Gyre, can be identified, but most of the structure represents uncertainties in the geoid which mask the relatively weak signals produced by ocean dynamics. We must await the launch of satellites dedicated to mapping the Earth's gravity field at high spatial resolution before we can significantly improve our ability to separate the ocean's dynamic topography from the geoid.

Happily, the need to know the geoid can be eliminated by studying time-varying altimeter signals of the ocean, since the geoid is constant on the time-scales of interest to us. One of the most obvious changes in sea level with time is that due to tides. Altimeters now provide the most accurate means of mapping open-ocean tides on a global basis. However, after allowing for these regular rises and falls there remain the irregular changes due to ocean currents. Their speed and direction are related to sea level in the same way as winds in the atmosphere are associated with the horizontal distribution of air pressure. Thus, as an oceanic eddy or meander passes through an observation point there is a change in sea level (depressed for a cyclonic eddy and raised for an anticyclonic one). These sea level changes can be expressed as departures from a long-term mean – we call them anomalies. *Figure 5.22* shows sea level anomalies in the South Atlantic obtained from ten days of altimeter data. Over most of the region the signal is less than 5 cm, but two areas (off South Africa and in the Southwest Atlantic) show positive and negative anomalies of 40 cm magnitude. The former corresponds to the Agulhas Retroflection (where the Agulhas Current turns and flows toward the east, south of South Africa) and the latter to the exit region of the Brazil–Falklands Confluence – see also *Figure 5.12*; both are known to be highly energetic and to spawn meso-scale eddies (Chapter 4). Eddies are important in contributing to horizontal heat transports and, through upwelling of nutrients, help to determine biological variability. Altimeter data offers the possibility of monitoring the movement and development of such features and, by assimilation into models, should provide a key element in a future ocean-forecasting system.

5.21

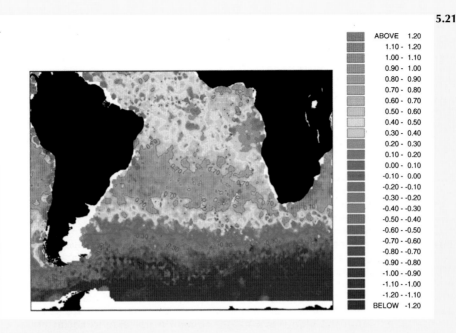

Figure 5.21 Mean sea surface height of the South Atlantic in metres above the geoid. The high in the centre corresponds to the Subtropical Gyre; to the south the Antarctic Circumpolar Current shows up as a region of increased north–south gradient. (Courtesy of Matthew Jones, James Rennell Division for Ocean Circulation, SOC, and Mullard Space Science Laboratory, UCL.)

5.22

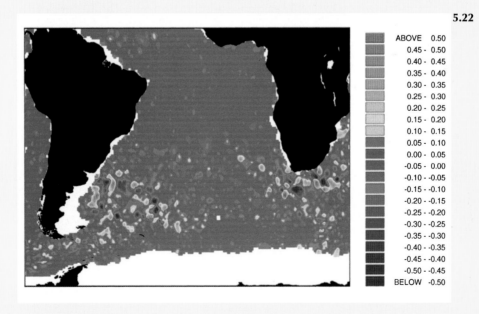

Figure 5.22 Sea surface height anomaly field derived from ten days of TOPEX/POSEI-DON data, calculated with respect to a 2-year mean. The colour scale shows heights of the anomalies in metres. (Courtesy of Matthew Jones, James Rennell Division for Ocean Circulation, SOC, and Mullard Space Science Laboratory, UCL.)

Box 5.5   Wave Height

Contained within the shape of the return pulse of the radar altimeter is information on significant wave height. This quantity corresponds closely to the mean of the highest one-third of the waves being observed at any one time and is similar to the wave height estimated from visual observations. In contrast to the calm sea case, reflection takes place earlier from the wave crests and later from their troughs, which leads to a smearing of the leading edge of the received pulse. The higher the waves the greater is the slope of this leading edge; it is this behaviour that is used to calculate values of wave height. Comparisons with buoys have shown that accuracy of wave height estimates may be better than 10%.

Our knowledge of global wave climate from conventional data is rather poor: buoys equipped with wave sensors are confined to coastal regions and most ship values come from visual observations, which can be very subjective and are biased to fair-weather conditions. Altimeters can provide uniform coverage regardless of weather conditions (except in those areas where sea ice is present). *Figure 5.23* portrays the seasonal variation in significant wave height. Significant differences between winter and summer occur, especially in the northern hemisphere. The high wave heights in the Indian Ocean in June are a consequence of the southwest monsoon.

Less obvious, but more intriguing, is the variation from year-to-year [a selection of cases is shown in *Figures 5.23(b)* and *5.24*] compiled from three altimeter missions – Geosat, ERS-1, and TOPEX/POSEIDON. The plots are for winter in the northern hemisphere. Some features are common to all years – highest waves occur in the mid-latitudes of both hemispheres, corresponding to the stormy westerly winds. The maxima are displaced to the eastern side of ocean basins, where the fetch is greatest. Despite it being summer, wave heights in the Southern Ocean are as large as their northern counterparts, although the high wave height zones do not extend so far equatorward. Wave heights in the tropics are always small when averaged over a month or so. However, some differences can be detected. In some years (e.g., 1992–1993) the North Atlantic and Norwegian Sea are rougher; when this happens the North Pacific is calmer. Likewise, the South Atlantic shows large changes with a minimum in 1987–1988 and a maximum in 1993–1994. Why does this interannual variability occur? Since it presumably reflects changes in the wind forcing, there are important implications for ocean circulation. We are exploring these links as part of the contribution to the World Ocean Circulation Experiment (see Chapter 3).

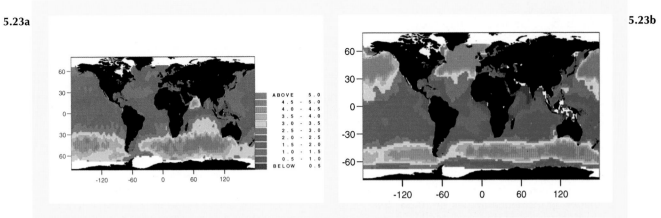

Figure 5.23 Mean significant wave height (m) obtained from TOPEX/POSEIDON altimeter data: (a) June 1994; (b) December 1993–January 1994. (Courtesy of David Cotton, James Rennell Division for Ocean Circulation, SOC.)

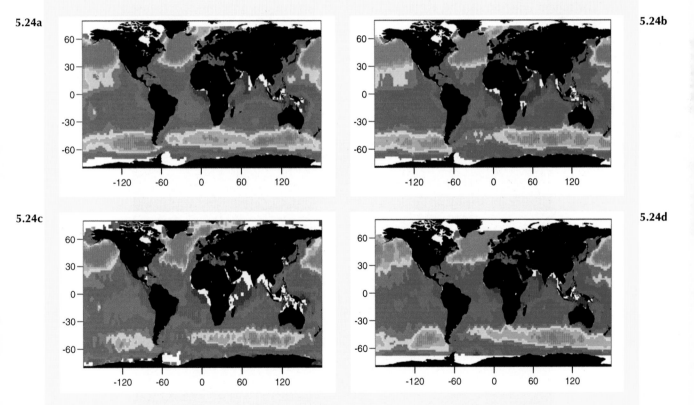

Figure 5.24 Mean significant wave height (m) for four northern hemisphere winters: (a) 1986–1987 from Geosat; (b) 1987–1988 from Geosat; (c) 1991–1992 from ERS-1; (d) 1992–1993 from TOPEX/POSEIDON. Wave heights vary from less than 0.5 m to greater than 5 m as shown in the colour scale. (Courtesy of David Cotton, James Rennell Division for Ocean Circulation, SOC.)

Box 5.6   Scatterometer

Like the altimeter, the scatterometer is also a radar sensor, but instead of measuring time delay or the distortion of pulse shape, it relies on using the amount of back-scattered power to infer an oceanographic quantity. From the early days of radar it has been known that when the sea surface is viewed obliquely 'ground clutter' exists, which increases with sea state. In glassy calms there is no detectable return because the incident radiation is reflected away by the mirror-like surface, with no return along the beam. As the wind speed increases small waves are set up which scatter some energy back toward the antenna. The 'cat's paws' often seen on the sea surface when a breeze springs up are an example. Microwave radars operating at wavelengths corresponding to this ripple-scale (ca 2 cm) are found to be most sensitive to wind variations. Furthermore, the back-scatter is not the same in all directions, being a maximum in the direction of the wind and a minimum across it. This is due to the effect of longer wind waves which are aligned more or less across the local wind.

Such knowledge is exploited in scatterometry in that broad swaths are illuminated with microwave radiation at incidence angles of 20–55º and with spatial resolution cells of side 50 km. The effect of individual waves on the back-scatter is averaged out on this scale. Several antennae are deployed at various azimuth angles so that the same area of ocean can be viewed from different directions over a period of a few minutes, allowing wind direction to be inferred. There is no accepted model of the wind dependence of the measured back-scatter and so the retrieval algorithm is empirically determined from simultaneous radar back-scatter and *in situ* wind measurements (see *Figure 19.13*). In theory this model function can be achieved by comprehensive airborne measurements prior to the launch of the satellite, but in practice it has proved necessary to carry out considerable modifications using the satellite data themselves.

Despite the lack of fundamental understanding, comparisons of scatterometer wind velocities have shown agreement to 2 m/s and 20º with independent *in situ* data. The accuracy, spatial resolution, and all-weather coverage of scatterometer data make them very useful for studying individual meteorological events and for obtaining the time-averaged wind fields needed to calculate surface momentum and heat and water vapour fluxes for large-scale ocean circulation studies. *Figure 5.25* is an example of the detailed winds in a hurricane, generated from scatterometer data and superimposed

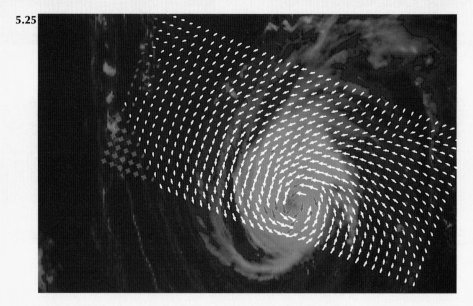

5.25

Figure 5.25 ERS-1 scatterometer surface wind vectors across Hurricane Emily on 30 August, 1993, superimposed on a coincident Meteosat cloud image. The length of the vectors is proportional to wind speed and those in red near the eye of the hurricane correspond to winds greater than 15 m/s. (Courtesy of the European Space Agency.)

on an infra-red cloud image obtained by a geostationary weather satellite. The realistic distribution of wind speed relative to the hurricane and the counterclockwise flow centred on the eye cloud add credibility to the scatterometer data. In some cases tropical storms have been detected in ERS-1 scatterometer data before they have appeared on analysis charts, presumably because of the sparseness of the routine weather observations in relation to the relatively small scale of these tropical systems. For some oceanographic purposes, e.g., study of localised coastal upwelling or deepening of the surface mixed layer, the scatterometer wind data should prove extremely valuable, especially in remote ocean areas where other forms of data may not be available. Global wind fields can be produced by combining data from consecutive passes of the satellite, as in *Figure 5.26*. The Trade Winds can be observed converging on the Equator. At higher latitudes the winds are stronger and more variable in direction, being influenced by the passage of depressions. The largest areas of strong westerlies are found in the Southern Ocean.

5.26

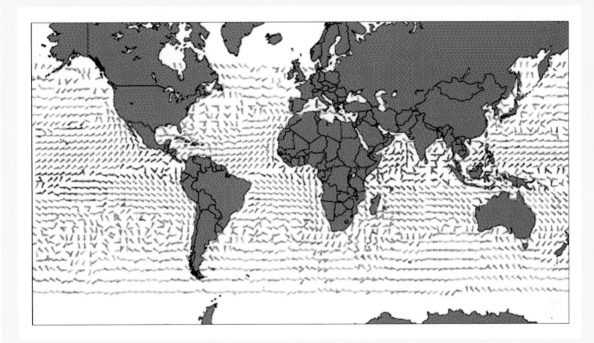

Figure 5.26 Average global wind field for April 1993 derived from ERS-1 scatterometer data. The strength of the wind is indicated by the length of the arrows (lightest winds are in blue, strongest are in red). (Courtesy of the European Space Agency.)

The Future for Satellite Oceanography

From these few examples of the applications of satellite data, the versatility and wide scope of marine remote sensing should be apparent. New sensors with higher performance are being designed, and new methods for processing the data are being developed to improve the accuracy with which ocean parameters can be measured. The space agencies of Europe, North America, and Japan have a continuing programme of Earth Observation Satellites planned, with a number of sensors dedicated to oceanographic applications, and there is scope for new ideas to be injected into the developing technology.

An even more challenging task is to improve the way in which remotely sensed data are integrated with more conventional measurements from ships and buoy-mounted instruments. The goal should be to enable the different types of measurements to complement each other in global ocean-monitoring programmes. An important integrating factor will be the use of computer models. Because of their pixellated nature, satellite data are well adapted for assimilation into numerical models or for validation of model predictions. The rapidly improving performance of global telecommunications will ensure that the vast quantities of numbers being transmitted from Earth Observation Satellites can readily be used to improve operational forecasts, as well as contributing to an improved understanding of oceanic processes.

However, the increased use of satellites is unlikely to reduce the requirement for *in situ* measurements. In fact, the reverse is true. The calibration and validation of satellite sensors requires regular comparisons with data acquired at sea, so there is a demand for new ocean instrumentation to measure parameters, such as surface roughness and surface radiation temperature, which have not been readily measured from ships before.

Remote sensing methods, by offering a new perspective on the ocean, are stimulating new oceanographic endeavours which will embrace all types of oceanographic techniques. The challenge for the next generation of oceanographers is to combine satellite data with more traditional methods, and to use an integrated approach to answer questions about global ocean processes that could otherwise not even be asked.

General References

Maull, G. (1985), *Introduction to Satellite Oceanography*, Martinus Nijhoff, Dordrecht, 606 pp.

Robinson, I.S. (1985), *Satellite Oceanography*, Ellis Horwood, Chichester, 460 pp.

Saltzman, B. (ed.) (1985), Satellite Oceanic Remote Sensing, Advances in Geophysics, Volume 27, Academic Press, London, 511 pp.

Stewart, R.W. (1985), Methods of Satellite Oceanography, University of California Press, Berkeley.

CHAPTER 6:

Marine Phytoplankton Blooms

D.A. Purdie

Introduction

Marine phytoplankton are often referred to as the 'grass of the ocean', since they form the productive basis on which most animal life in the oceans is ultimately dependent, be it a herbivorous zooplankton consumer, a top carnivore, or a benthic deposit feeder. 'Phytoplankton' is a collective term to describe the single-celled microscopic algae (they may also form chains or clumps of cells), that float in the surface, well-illuminated waters of the sea. When conditions are optimum for their rapid growth, or if more dispersed cells are aggregated by some mechanism, such as the fronts described later, planktonic micro-algae become sufficiently abundant to colour the sea; this phenomenon is often referred to as a plankton 'bloom'.

The term bloom was originally used by analogy with terrestrial plants, to describe the spring flowering of the diatoms which characterises most temperate waters. Blooms normally consist of one particular species that dominates the plankton in the surface waters affected. The term 'Red Tide' is also often used to describe blooms, as many of the organisms cause a red colouration to the sea – although green, brown, and yellow colourations are also known. These characteristic colourations of the water are a direct result of high concentrations of photosynthetic compounds (i.e., chlorophylls) and accessory pigments which are found in the micro-algae. The pigments absorb certain wavelengths of visible light, and algal cells scatter a substantial fraction of the submarine light, causing the apparent change in the colour of the water (see also Box 5.2 and Chapter 14).

Blooms of micro-algae represent a large increase in biomass and generally last for relatively short periods, a week or so. They are not exclusive to the sea; they also occur in freshwater lakes and rivers. Blooms become visible as a discolouration of sea water when chlorophyll concentrations exceed 10 mg/m^3 and cell concentrations in the nanoplankton size-range exceed 0.5×10^9 cells/m^3. Typically, however, chlorophyll levels and cell densities may exceed 100 mg/m^3 and 30×10^9 cells/m^3, respectively, and may range up to 2000 mg/m^3 and 100×10^9 cells/m^3 in some localised coastal regions. These very dense blooms are often termed 'exceptional blooms'; they are recurring features in some sea areas and lead to a variety of notable consequences, as described below.

In estuaries or coastal waters, phytoplankton blooms may be small localised features, a square kilometre or so in extent, but they are often much more extensive in deep water, sometimes covering hundreds of square kilometres. The discolouration of the sea that they cause can be detected by aircraft or satellites, and photographed[1] or measured by specialised narrow-range spectral sensors mounted on airborne[5] or space platforms[6,7] (see Chapter 5). An upwelling bloom event along a 60 km front in the central mid Pacific has been photographed from satellites, ships, and the space shuttle *Atlantis*[16]. An extensive bloom of the coccolithophore *Emiliania huxleyi*, detected in the North Atlantic from a series of satellite images[7] (see *Figure 4.11*), covered an area of ca 250,000 km^2.

The mechanisms responsible for changes in the micro-algal carbon biomass with time in the near-surface waters of the sea, known as the euphotic zone (i.e., the zone in which there is sufficient light for growth and cell division of phytoplankton), are controlled by a balance between the primary (algal) production rate and losses from the plankton community through respiration, grazing by herbivores, and sedimentation of intact cells. During non-bloom periods, micro-algal biomass in the sea remains quite constant – the production of biomass is approximately balanced by respiration, grazing, and sedimentation losses. Phytoplankton blooms result when production exceeds loss. The environmental conditions required for a phytoplankton bloom to occur are:

- Availability of the inorganic plant nutrients nitrate and/or ammonium, phosphate, and silicate (in the case of diatoms).

- Sufficient irradiance levels within the surface waters.

- An imbalance or lag between the production of micro-algae and their grazing by herbivores, advection, or sedimentation out of the production zone.

6.1

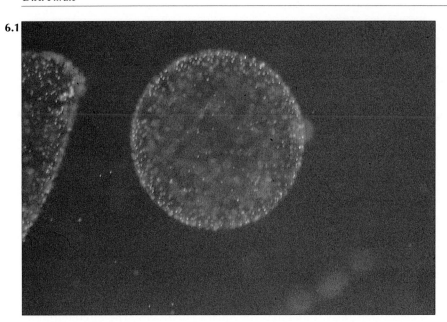

Figure 6.1 Colonies of *Phaeocystis* sp showing small dot-like cells embedded in a mucous colony. The spherical colony is about 1 mm in diameter.

Once nutrients become exhausted in surface waters, cells begin to die and the phytoplankton sediment out of the water column. This may result in production of rapidly sinking amorphous aggregates of dead and decaying phytodetritus, known as 'marine snow'. This material may descend great depths through the water column in a few days, and result in a thick flocculant covering of the surface sediment (see Chapter 7). These processes of bloom growth and decay can be expressed in mathematical terms and represented in biophysical models[15].

Marine phytoplankton blooms are natural phenomena. There is, however, increasing evidence and consequent concern that anthropogenic activity (e.g., the supply of nitrates by rivers) is fertilising coastal waters, resulting in an increase in the intensity and frequency of bloom events, as well as affecting the natural balance of phytoplankton species diversity (see also Chapter 15). These 'unnatural' blooms have a number of both direct and indirect affects on the marine biota, many of which are deleterious and result in severe environmental problems.

Common Examples of Bloom-Forming Marine Phytoplankton

Diatoms are the dominant marine phytoplankton in most marine waters, particularly during the spring bloom. The cells are enclosed in a microscopic porous shell or frustule constructed from silica assimilated as dissolved silicate from sea water. Dinoflagellates are spirally swimming organisms propelled by flagella; they are the major primary producers in many marine regions during summer. *Phaeocystis* sp is a prymnesiophyte that forms large

colonies of cells embedded in a mucous envelope (*Figure 6.1*). Dense blooms of *Phaeocystis* occur in many coastal regions, including the southern North Sea. A related group of marine phytoplankton, the coccolithophores, produce calcite plates that cover the outside of the cell [*Figure 6.2(a)*] and, during bloom conditions, cause the water to take on a milky turquoise colour [due to the scattering of light by the cells; *Figure 6.2(b)*].

The two most frequently observed 'protozoa' that cause discolouration to the sea when present in high abundance are *Noctiluca* and *Mesodinium rubrum*, neither of which are known to posses any toxic effects.

Noctiluca is a heterotrophic (i.e., organism deriving nutrition from assimilation of organic matter) dinoflagellate and a voracious phagotrophic (i.e., particle ingesting) feeder. It is pigmented by orange carotenoids (i.e., non-photosynthetic pigments), possesses the ability to control its buoyancy in sea water, and is often noticed as a tomato-red colouration on the surface during calm conditions. In enclosed shallow harbours blooms of this organism can cause severe oxygen depletion, leading to mortality of marine animals.

The highly motile ciliate *Mesodinium rubrum* (*Figure 6.3*) contains a photosynthetic cryptomonad symbiont (i.e., an algal cell living within the protozoan host) and can be functionally considered a member of the phytoplankton[3]. It is extremely productive in some regions, reaching levels of chlorophyll concentration of 1000 mg/m^3, with a highest recorded production rate of >2000 mgC/m^3/h. *M. rubrum* causes red-water events in coastal, estuarine, and upwelling ecosystems in many parts of the world[3]. It has been intensively studied in the

6.2a  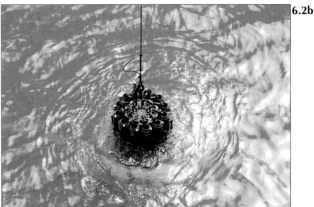 6.2b

Figure 6.2 The coccolithophore *Emiliania huxleyi*: (a) an electron micrograph of cells (about 6 μm in diameter) showing calcite plates covering the cell (courtesy of D. Harbour, Plymouth Marine Laboratory, Plymouth, England); (b) milky turquoise colouration to surface waters (picture taken from RV *Charles Darwin* during June 1991 in the northern North Atlantic).

Southampton Water estuary in southern England, where it forms red tides almost annually (*Figure 6.4*). During the development of these blooms, the inorganic nutrient levels (ammonia and nitrate) in the estuary are considerably reduced; also, bottom-water oxygen levels are depleted by the organism due to its downward displacement at night[4]. The unusually high swimming speed of the organism reduces its chances of being flushed from the estuary, since it avoids increased surface turbulence caused by the ebb tide by swimming to deeper depths.

The Spring Bloom

The best-known type of phytoplankton bloom occurs in spring in temperate and subpolar waters. It was once thought that onset of the spring bloom was triggered by the spring increase in solar irradiance. Subsequently, it was realised that the depth of mixing also plays an important role. A Norwegian scientist, H.H. Gran, first established that, in winter, phytoplankton are mixed vertically to the bottom of the mixed layer and spend considerable periods at depths where there is insufficient light for photosynthesis; with the onset of the spring warming of surface waters and a decrease in storm mixing, the mixed-layer depth and therefore the depth to which phytoplankton cells are regularly mixed, decreases; they are consequently exposed to increasingly higher average light levels, allowing blooms to occur. The classic quantitative explanation of the spring bloom, known as the 'critical depth theory', was first proposed by Sverdrup[14]. We now know that the spring bloom is initiated when the carbon fixed by phytoplankton photosynthesis in the mixed layer exceeds the respiration

6.3a 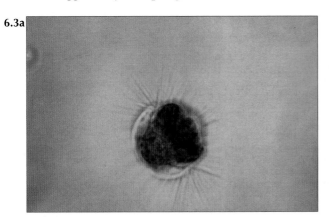  6.3b

Figure 6.3 (a) Photomicrograph of the photosynthetic ciliate *Mesodinium rubrum* (about 100 μm diameter) (Courtesy of Dr D. Crawford, Department of Oceanography, SOC, Southampton, England.) (b) Red tides caused by *Mesodinium rubrum* in Southampton Water estuary, UK.

losses per unit area, or when phytoplankton are confined above a 'critical depth'. Riley[11] showed from empirical evidence that phytoplankton cannot bloom before the average irradiance in the mixed layer exceeds 3.25 E/m²/day. This critical irradiance is reached as a consequence of seasonal increases in both density stratification (shallowing of the mixed layer) and increasing solar irradiance. In some well-mixed coastal regions and estuaries, a decrease in turbidity may also result in a deepening of the critical depth[9].

Diatoms are almost without exception the group of micro-algae that dominate the spring bloom in temperate waters. This is ascribed to their rapid growth rates, which are sustained by the high concentrations of available nutrients in early spring and the tolerance of these organisms to the turbulent conditions which prevail during this period. An example of a spring bloom in an estuary is shown in *Figure 6.4*.

Whereas many studies have been reported of the temporal development of phytoplankton biomass and the species succession associated with the development of spring blooms in coastal waters, few detailed studies of the spring bloom are available for offshore regions. One approach to studying the development of the phytoplankton spring

bloom in the North Atlantic[12] is to follow the water movement rather than to stay in a fixed (e.g., mooring) position; this has highlighted the complexity of spatial and temporal variability in an offshore region, where mesoscale eddies complicate the picture of temporal chlorophyll changes (see also *Figure 4.10*).

The poleward migration of the spring bloom in the North Atlantic has been reported by Strass and Woods[13] from data collected using a towed vehicle, the SeaSoar (see *Figure 19.6*). Chlorophyll distribution throughout the surface 200 m between the Azores (38°N) and Greenland at 58°N was measured from chlorophyll fluorescence using a submersible fluorometer attached to SeaSoar. The near-surface chlorophyll fluorescence showed a patchy bloom development between 40–49°N in April, with a bloom centred on 49°N in June–July and low chlorophyll levels throughout most of the transection in August–September, with bloom conditions existing only north of the polar front at 52°N.

At a particular location, the spring bloom can be a very transient feature and is easily missed if water samples are not collected frequently, a particular problem in offshore waters. This is well-demonstrated in the data obtained during a study in the

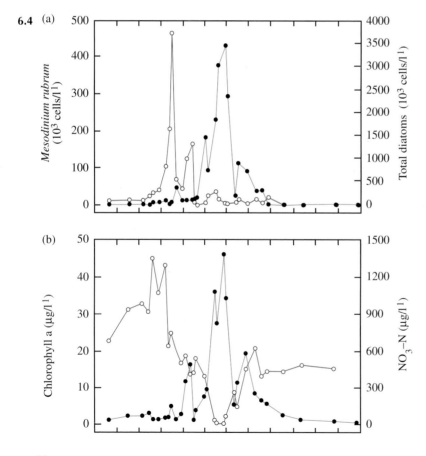

6.4 (a)

Figure 6.4 (a) Seasonal changes in the abundance of diatoms (blue) and the photosynthetic ciliate *Mesodinium rubrum* (red) in the mid-estuary region of Southampton Water, UK, during 1988. The initial peak in diatoms during April was dominated by chains of the small-celled *Skeletonema costatum*, with the second peak dominated by the larger celled *Rhizosolenia delicatula*. Diatom numbers remain relatively low during the rest of the year due to limited silicate availability in the estuary. Note the spring bloom development of diatoms and the summer bloom of the photosynthetic ciliate. (b) These result in large changes in chlorophyll concentration (red) and reduced nitrate levels (blue).

Figure 6.5 Seasonal changes in depth (m) distribution of (a) water temperature, (b) dissolved oxygen saturation, and (c) chlorophyll (derived from fluorescence) at a station in the central North Sea, sampled once a month. These waters become thermally stratified in the spring due to surface heating; in autumn, storms cause complete mixing of the water column. Phytoplankton blooms were detected in late summer in surface waters; however, the transient surface spring bloom chlorophyll peak in April was not detected from the research ship. Settled bloom material in near-bottom waters was observed in late April. Dissolved oxygen is supersaturated during surface blooms and the development of undersaturated oxygen conditions is evident in bottom waters in late summer and autumn.

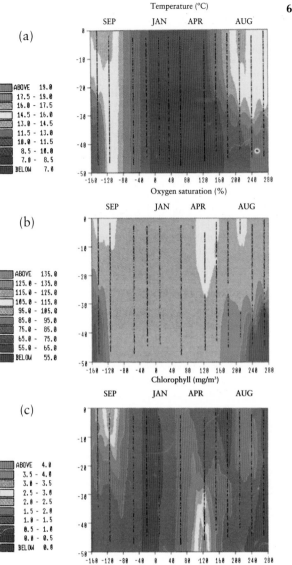

North Sea, in which water samples and water column data were collected by a research ship that visited a particular site in the central North Sea every four weeks and failed to detect the peak chlorophyll levels. Innovative new autonomous instrumentation to monitor chlorophyll fluorescence at intervals of up to four times per hour over a month's deployment on a mooring at this site did detect a transient peak in chlorophyll levels between ship surveys[15]. The levels were up to three times higher than those measured from the ship.

Deleterious Effects of Blooms

Toxic effects and 'Red Tides'

The term 'Red Tide' is often used in an emotive way to describe a poisoning of water. Only a few dozen of the thousands of marine phytoplankton species are known to be toxic, and most of these are dinoflagellates, prymnesiophytes, or chloromonads. Human life can be affected if shellfish, such as clams, mussels, oysters, or scallops, ingest the algae as food and retain and concentrate the toxins in their tissues. Although the shellfish are not greatly affected themselves, a single clam can accumulate sufficient toxin to kill a human being. Shellfish-poisoning syndrome is known as paralytic, diarrhoetic, and neurotoxic shellfish poisoning, or PSP, DSP, and NSP, respectively. A new type of shellfish poisoning was reported in 1990 on the eastern seaboard of Canada, termed amnesiac shellfish poisoning or ASP, in which infected individuals suffered short-term memory loss. The toxins responsible for these poisonings are from families of compounds such as the saxitoxins, which cause PSP by disrupting nerve impulses to the muscles and nervous tissue. It is unclear why certain algae produce toxins, but it is possible they serve as a deterrent to grazers and may give the algae an unpleasant taste.

Other nuisance effects

Many phytoplankton bloom-forming species have no apparent directly toxic effect on organisms in the marine environment. However, the immense increase in organic matter associated with blooms can cause other indirect deleterious effects on local waters and their adjacent environment.

During the day micro-algae photosynthesise and produce new organic matter and oxygen. Rapidly growing phytoplankton populations cause oxygen supersaturation of surface waters, with values of greater than 160% having been recorded; this may cause some motile macrofauna to avoid such regions, since oxygen bubbles may form in their gills. A more frequent problem is that caused by the severe undersaturation of oxygen, or hypoxic conditions, that occur during the latter stages of phytoplankton blooms (*Figure 6.5*). Respiration by the algae themselves, as well as by animals and bacteria, consumes oxygen. In some estuaries, such as

Chesapeake Bay, the bottom waters may become deficient in oxygen after a phytoplankton bloom and, in extreme cases, anoxic (i.e., totally devoid of oxygen), causing mass mortalities of benthic fauna. This process is exacerbated if the water column becomes stratified, either by freshwater inputs to estuaries or by seasonal heating of surface waters (causing a surface mixed-layer above a thermocline). Such conditions occur in the German Bight region of the North Sea; during the early 1980s massive fish kills occurred, which were thought to be in part caused by severe undersaturation of near-bottom waters. Subsequent data, however, showed that the waters below the thermocline become reduced in oxygen following the spring bloom; this is caused by the algal material sedimenting through the thermocline and decomposing in deeper waters[8] (*Figure 6.5*).

Phytoplankton blooms may influence recreational use of coastal regions by causing foam and noxious slime deposits on the beach. Many beaches in Belgium and the Netherlands become covered in thick foam, washed up following the breakdown of the *Phaeocystis* blooms in adjacent coastal waters. The occurrence of *Phaeocystis* in coastal regions of the North Sea has been linked to enhanced inorganic nutrient enrichment (eutrophication) of these waters. A rare time-series of measurements was conducted over a period of 14 years, from 1976–1988, by Cadee[2] at one point in the Dutch Wadden Sea. Water samples were collected weekly in spring and summer, and less frequently, during the rest of the year from the pier at the Netherlands Institute for Sea Research on the island of Texel. Results showed that both the intensity and duration of *Phaeocystis* blooms had significantly increased over the 14-year period, which is probably related to the eutrophication of this region.

Many coastal regions of the Adriatic Sea have been affected in recent years by the production of an extracelluar excretion of mucus or slime caused, apparently, by dense, mixed phytoplankton populations. It appears that no single organism is responsible for the production of this mucous material, but a number of diatoms and dinoflagellate species have been collected within the affected areas, on both the Italian and Yugoslavian (as was) sides of the Adriatic. Eutrophication of these coastal waters, particularly caused by large inorganic nutrient inputs from the river Po outflow, has been blamed for these recent events, with chlorophyll levels rising to more than 800 mg/m^3 in some coastal parts of the Emilia Romagna region. The occurrence in recent years of these events each summer has greatly affected the tourist trade in many regions, since the slime occurs on beaches, yields a noxious smell, and is unpleasant to swim in,

although not apparently directly toxic. The extent of bottom-oxygen deficiency in the central northern Adriatic has also been shown to have increased significantly over the past 20 years, as a consequence of this material sinking to deeper depths following its wind-driven advection offshore.

Phytoplankton blooms can cause a reduction in benthic plant communities due to an increased turbidity of the water, thus reducing the penetration of light to bottom-dwelling plants. This has caused a reduction in sea-grass beds in some coastal tropical regions, and has also affected coral-reef development.

Many of the marine bloom-forming phytoplankton are known to produce the volatile sulphur compound dimethyl sulphide (DMS)[10]. DMS undergoes photochemical oxidation into sulphur dioxide in the atmosphere and contributes to the production of acid rain. Bloom-forming phytoplankton species, such as *Phaeocystis* and *Emiliania huxleyi*, are known to be important producers of DMS; in coastal regions they may add significantly to localised atmospheric sulphur levels[10].

General References

Anderson, D.M. (1994), Red tides, *Sci. Amer.*, **271**(2), 52–58.

Cosper, E.M., Bricelj, J.M., and Carpenter, E.J. (eds) (1989), *Novel Phytoplankton Blooms: Causes and Impacts of Recurrent Brown Tides and Other Common Blooms*, Springer Verlag, Berlin.

Hallegraeff, G.M. (1993), A review of harmful algal blooms and their apparent global increase, *Phycologia*, **32**, 79–99.

Lancelot, C., Billen, G., and Barth, H. (eds) (1990), *Eutrophication and Algal Blooms in North Sea Coastal Zones, the Baltic and Adjacent Areas*, Water Pollution Research Reports, Commission of the European Communities, Brussels, 281 pp.

Parker, M. and Tett, L.P. (eds) (1987), Exceptional phytoplankton blooms, *Rapports et proces-verbaux des reunions*, **187**, 9–18.

Smayda, T.J. and Shimizu, Y. (eds) (1993), *Toxic Phytoplankton Blooms in the Sea*, Elsevier, Amsterdam, 952 pp.

Vollenweider, R.A., Marchetti, R., and Vivani, R. (eds) (1992), *Marine Coastal Eutrophication*, Supplement to *Sci. Total Environ.*, 1310 pp.

References

1. Berge, G. (1962), Discolouration of the sea due to *Coccolithus huxleyi* 'bloom', *Sarsia*, **6**, 27–40.

2. Cadee, G.C. (1990), Increase of *Phaeocystis* blooms in the westernmost inlet of the Wadden Sea, The Marsdiep, since 1973, in *Eutrophication and Algal Blooms in North Sea Coastal Zones, the Baltic and Adjacent Areas*, Lancelot, C., Billen, G., and Barth, H. (eds), Water Pollution Research Report 12, Commission of the European Communities, Brussels, pp 105–112.

3. Crawford, D.W. (1989), *Mesodinium rubrum*: the phytoplankton that wasn't, *Marine Ecol. Progr. Ser.*, **58**, 161–174.

4. Daneri, D., Crawford, D.W., and Purdie, D.A. (1992), Algal blooms in coastal waters: a comparison between two adaptable members of the phytoplankton, *Phaeocystis* sp. and *Mesodinium rubrum*, in *Marine Coastal Eutrophication*, Vollenweider, R.A., Marchetti, R., and Vivani, R. (eds), Supplement to *Sci. Total Environ.*, pp 879–890.

5. Garcia, C.A.E., Purdie, D.A., and Robinson, I.S. (1993), Mapping of the photosynthetic ciliate *Mesodinium rubrum* in an estuary from Airborne Thematic Mapper data, *Estuar. Coastal Shelf Sci.*, 37, 287–298.

6. Holligan, P.M., Aarup, T., and Groom, S.B. (1989), The North Sea satellite atlas, *Continent. Shelf Res.*, 9, 667–765.

7. Holligan, P.M., Fernandez, E., Aiken, J., Balch, W.M., Boyd, P., Burkill, P.H., Finch, M., Groom, S.B., Malin, G., Muller, K., Purdie, D.A., Robinson, C., Trees, C.C., Turner, S.M., van der Wal, P. (1993), A biogeochemical study of the coccolithophore, *Emiliania huxleyi*, in the North Atlantic, *Global Biogeochem. Cycl.*, 7(4), 879–900.

8. Howarth, J.M., Dyer, K.R., Joint, I.R., Hydes, D.J., Purdie, D.A., Edmunds, H., Jones, J.E., Lowry, R.K., Moffat, T.J., Pomroy, A.J., and Proctor, R. (1993), Seasonal cycles and their spatial variability, *Phil. Trans. Roy. Soc. Lond., A.*, 343, 383–403.

9. Kifle, D. and Purdie, D.A. (1993), The seasonal abundance of the phototrophic ciliate *Mesodinium rubrum* in Southampton Water, England, *J. Plankt. Res.*, 15, 823–833.

10. Malin, G., Turner, S.M., and Liss, P.S. (1992), Sulfur: The plankton/climate connection, *J. Phycol.*, 28, 590–597.

11. Riley, G.A. (1942), The relationship of vertical turbulence and spring diatom flowering, *J. Marine Res.*, 5, 67–87.

12. Savidge, G., Turner, R.D., Burkill, P.H., Watson, A.J., Angel, M.V., Pingree, R.D., Leach, H., and Richards, K.J. (1992), The BOFS 1990 spring bloom experiment: temporal evolution and spatial variability of the hydrographic field, *Progr Oceanogr.*, 29, 235–281.

13. Strass, V and Woods, J.D. (1988), Horizontal and seasonal variation of density of chlorophyll profiles between the Azores and Greenland, in *Towards a Theory on Biological–Physical Interactions in the World Ocean*, Rothschild, B.J. (ed.), Kluwer, Dordrecht, pp 113–136.

14. Sverdrup, H.U. (1953), On conditions for the vernal blooming of phytoplankton, *J. Conseil Perm. Int. Exp. Mer.*, 18, 287–295.

15. Tett, P.B., Joint, I.R, Purdie, D.A., Baars, M., Oosterhuis, S., Daneri, G., Hannah, F., Mills, D.K., Plummer, D., Pomroy, A.J., Walne, A.W., and Witte, H.J. (1993), Biological consequences of tidal stirring gradients in the North Sea, *Phil. Trans. Roy. Soc. Lond., A.*, 343, 493-508.

16. Yoder, J.A., Ackleson, S.G., Barber, R.T., Flament, P., and Balch, W.M. (1994), A line in the sea, *Nature*, 371, 689–692.

CHAPTER 7:

Snow Falls in the Open Ocean

R.S. Lampitt

When I think of the floor of the deep sea, the single, overwhelming fact that possesses my imagination is the accumulation of sediments. I see always the steady, unremitting, downward drift of materials from above, flake upon flake, layer upon layer – a drift that has continued for hundreds of millions of years, that will go on for as long as there are seas and continents ... For the sediments are the materials of the most stupendous snowfall the earth has ever seen ...
Rachael Carson[5]

Introduction

It is hard to improve on this inspiring prose, describing so graphically as it does a process which continues to excite and tantalise marine scientists. It is widely accepted that the means by which material is transported down to the abyssal depths and the rates at which this occurs present a range of fundamental questions. These are of direct interest to a wide variety of scientists, from those interested in global biogeochemical cycling (see Chapter 11) to, for instance, those keen to understand the population dynamics of the deep sea benthic fauna (see Chapter 13).

In this brief review I acknowledge the technical and philosophical developments which have occurred over the past 120 years, but focus primarily on the progress which has been made during the last 10, progress which has transformed our understanding of the processes involved in material flux. The means by which small particles at the top of the ocean, with almost negligible sinking rates, are transformed into larger, rapidly sinking aggregates which reach the deep-sea floor in only a few weeks continues to provide exciting insights into the complex processes of material cycling in the oceans.

In the language of the travel agent, the open ocean has become a smaller place, and the linkages between the top sunlit zone and the dark, cold, deep sea interior are now thought to be far closer than previously imagined. Rachael Carson's stupendous snowfall[5] is no less awe-inspiring now than it was to her in 1951, and without doubt the questions surrounding its elucidation are as demanding.

Historical Developments of the Concept

We should not think that interest in the way material is transported and modified has only recently been kindled. In the second half of the ninteenth century, there was a vigorous debate among the giants of old (Jeffreys, Thomson, and Lohmann) about the way in which the sediments of the oceans were formed and if, as some believed, there was life in the deep sea, how it was sustained. While Wallich[25] thought that such life was supported by what is currently referred to as chemo-autotrophy, Thomson[22] thought particulate transport of shallow water macrophytes (seaweeds) and terrestrial run-off were the keys to the food supply of the deep-living animals. Jeffreys[10], however, thought that the dead remains of surface dwelling organisms would be an important source of their food. Lohmann[14] made some surprisingly modern calculations about the rate at which the sediments of the ocean were formed by the deposition of planktonic material. He commented on the fact that near-bottom water above the abyssal sea bed sometimes contained a surprising range of thin-shelled phytoplankton species, some still in chains and with their fine spines well preserved. He deduced that they must have been transported there very quickly from their near-surface habitat, and thought that the faecal pellets from some larger members of the plankton (Doliolids, Salps, and Pteropods) were the likely vehicles. His deductions may, in many cases, have been entirely correct and it remains a poor comment on our science that these observations were largely ignored during the next few decades.

The descriptions of Rachael Carson[5] soon found their mark in the minds of a group of Japanese oceanographers using the submersible observation chamber 'Kuroshio', suspended from a fisheries training ship (*Figure 7.1*). The amorphous particles they could see through the portholes were clearly not living and they coined the term 'marine snow' to describe them[21]. The term is still only loosely defined, but is generally recognised to encompass immotile particles of diameter greater than 0.5 mm. In the open ocean these are all biogenic and are thought to be the main vehicles by which material sinks to the sea floor.

The submersible used by the Japanese oceanographers was a cumbersome device and did not permit anything but the simplest of observations to be made. They did, however, manage to collect some of the material and reported that its main components were the remains of diatoms, although with terrestrial material present that provided nuclei for formation.

In spite of these observations and the outstanding questions surrounding material cycles in the oceans, it was, until the late 1970s, a widely held belief that the deep-sea environment received material as a fine 'rain' of small particles. These, it was assumed, would take many months or even years to reach their ultimate destination on the sea floor. The separation of a few kilometres between the top and bottom of the ocean was thought sufficient to decouple the two ecosystems in a substantial way, such that any seasonal variation in particle production at the surface would be lost by the time the settling particles reached the sea bed. This now seems to have been a fundamental misconception. Part of the reason for this is the lack of understanding of the role of marine snow aggregates.

We now ask the most basic of questions about this important class of material: What is marine snow, how is it distributed in time and space, and why is it of such significance?

A Vertical Profile

Marine snow is found throughout the world's oceans in all parts of the water column. It is not uniformly distributed, either in space or time, but is usually found in higher concentrations in the upper water column and in the more productive regions of the oceans. Although it had been suspected since the early observations of Suzuki and Kato[21] that marine snow concentration decreased with increasing depth, this has been confirmed only recently. The profiles now becoming available do not, however, suggest a simple decrease. There is considerable structure, undoubtedly related to the processes of production, destruction, and sinking. These are all related to the physics, biology, and chemistry of the water column and of the particles themselves.

Figure 7.2 shows some examples of profiles from different parts of the world and using a variety of techniques (*Box 7.1*). Bearing in mind the strong seasonal variation which can occur even well-below the upper mixed layer (see below) and the different techniques employed to obtain these profiles, a common story seems to be emerging. Apart from profiles near the continental slope, where snow concentrations tend to increase near the sea bed due to resuspension, there is generally a rapid fall in concentration over the top 100 m. Peak concentrations are not, however, found

7.1

Figure 7.1 The submersible observation chamber 'Kuroshio' as used to make observations on the distribution and characteristics of marine snow aggregates in the 1950s. The chamber was lowered on a cable from the mother ship, providing a rather cramped view of the sea's interior and sea bottom through the small portholes. Two or three investigators were able to fit into the chamber, descending to depths of up to 200 m for several hours. (Courtesy of Dr Masahiro Kajihara, Hokkaido University.)

throughout the upper mixed layer, but are located at its base, a feature which is directly related to the rates of production and loss of the marine snow particles in this highly dynamic part of the water column.

The upper mixed layer (UML) factory

The mixed layer at the top of the oceanic water column varies in thickness from up to a few hundred metres in winter to a few tens of metres in the spring and summer. It is subject to rapid changes in light, heat, turbulence, nutrient concentration, and depth of mixing on the scale of hours, as well as having distinct seasonal variations. This physical forcing creates changes in the biological processes which depend on them. Furthermore, it is highly variable in the spatial sense, producing a rapidly changing mosaic of physical, chemical, and biological properties not found elsewhere in the water column (Chapters 4 and 5).

It is here in the UML that the primary production of material occurs as a result of phytoplankton growth (see Chapter 6). The cells thus produced (mainly in the range 1–50 μm diameter) are immediately subject to attack from many other elements in the plankton community, but principally from the microplankton (20–200 μm diameter) and the mesozooplankton (0.2–20 mm body length). The

7.2

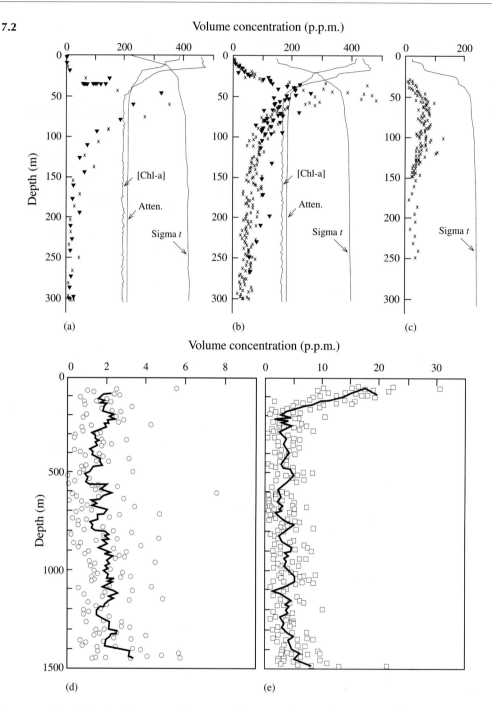

Figure 7.2 Examples of the distribution of marine snow particles greater than 0.6 mm diameter, expressed as volume concentration (p.p.m.). (a) 19 May 1990 (MSP 11), (b) 22 May 1990 (MSP 15), and (c) 26 May 1990 (MSP 19) from the Northeast Atlantic at a water depth of 4800 m (Lampitt et al.[12]). Sigma t is a measurement of water density, [Chl a] (Chlorophyll a) is the concentration of phytoplankton pigment, and Atten (attenuation) is a measure of the concentration of the smaller particles as determined by their effect on light transmission through the water. In terms of abundance, the maximum concentration of particles at around 50 m depth is about 200/l, decreasing to a deep-water minimum of approximately 30/l. (d) 22 November 1987 and (e) 28 January 1988 from Northwestern Gulf of Mexico at a water depth of 1500 m on two occasions; the lines are 9-point running means[26]. In most instances there is a peak in concentration near the surface, but not necessarily at the surface. Near the sea bed there are also elevated levels. There appears to be a very wide range in concentrations and, although some of this may be related to differences in technique, there can be considerable temporal variability at any one site (*Figure 7.6*); there are also likely to be large regional differences reflecting the structure and dynamics of the biological communities in the different environments.

Box 7.1   How to Examine the Distribution in Time and Space of Marine Snow Aggregates

In situ studies: enumeration, measurement, and collection

Marine snow is a highly variable and ephemeral commodity (*Figure 7.2*). It varies dramatically, not only in its abundance, size, and sinking rate, but also in its origins, composition, and value as a food source. One of the greatest and yet apparently simplest current requirements is to obtain good data on its variability in time and space. Most attempts have used *in situ* photographic techniques. A variety of other sensors are now often attached to the cameras in order to measure other environmental variables, such as water density, fluorescence (a measure of phytoplankton concentration), turbidity (determined primarily by the smaller particles), and oxygen. Subaqua divers have also been used to count particles, but because of the difficulty in seeing the smaller classes of marine snow and the biasing influence of different ambient lighting conditions, divers are mainly reserved for collection and *in situ* experimentation. Manned and unmanned submersibles are also used to observe and capture marine snow. *Figure 7.3* shows examples of some of the photographic and diver-operated techniques.

Both video and emulsion-based photography are used; a light source produces a collimated beam which can then be viewed at right angles by the camera. This produces an image of a known water volume with the particles displayed on a dark background. For long-term observations of temporal changes, such devices have been deployed on moorings[11]. Profiling instruments are generally deployed from research ships[20,26].

7.3a

7.3b

7.3c

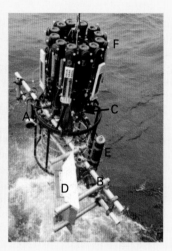

7.3d

Figure 7.3   A variety of photographic devices used to record the spatial and temporal variation in marine snow aggregates *in situ*. (a) The Large Aggregate Profiling System (LAPS)[8] with collimated light beam: 35 mm camera (A), flashlight (B), CTD (C), fresnel lens (D), and transmissometer (E). A volume of about 20 l is photographed and particles binned into six separate size categories from 0.5 mm to >3 mm. (Courtesy of Drs Gardner and Walsh, Texas A & M University.) (b) The Atlantic Geosciences Centre Floc Camera Assembly (FCA)[20]. This carries three cameras (A) and flash (B), and records particles >0.25 mm in a volume of 0.2 l. (Courtesy of Dr Syvitski, Bedford Institute of Oceanography.) (c) In this case the device is associated with a variety of other instruments, such as a fluorometer to measure the concentration of phytoplankton, conductivity and temperature sensors, an echo-sounder (E), and a transmissometer to measure the concentration of the smaller particles. The grey vertical tubes (F) on top of the device are sampling bottles activated by the conducting wire on which the device is suspended. Particles greater than 0.5 mm are recorded in a volume of 40 l using 35 mm film. These are later analysed on an image analyser (see *Figure 7.6*). (d) In contrast to (a)–(c), this instrument is attached to a mooring to record temporal changes (*Figure 7.6*). It has the same arrangement as in *Figure 7.3(c)* and the frames are similarly analysed. Also shown are a buoy (G), vane (H), and battery pack (I).

7.4a 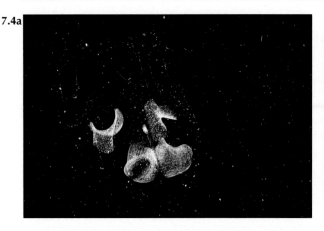 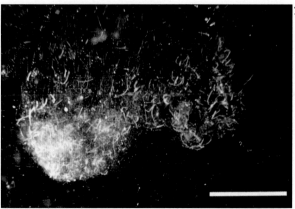 7.4b

Figure 7.4 Examples of Types A and B marine snow aggregates. (a) Type A aggregate comprising an abandoned filter net or 'house' of an appendicularian. It is identifiable by the presence of two in-current filters and a large U-shaped internal filter surrounded by an envelope of particle-studded mucus. These are typically several centimetres in diameter. (Courtesy of Dr Alldredge, University of California.) (b) Type B aggregate comprising living chain-forming diatoms (scale bar = 1 cm). (Courtesy of Dr Gotschalk, University of California.)

complex food web which thus develops is being intensively researched through experimentation and numerical modelling (e.g., Fasham *et al.*[7]). However, the focus of this chapter is the material lost from this UML community, or at least in the process of being lost.

Various parts of the food web produce material which is less attractive to consumers. Some of this material is easily considered as a waste product, such as faeces, senescent phytoplankton cells, gelatinous feeding webs, and zooplankton moults. Marine snow can be divided conveniently into two main classes based on the origin of the components: Type A is derived from the gelatinous houses constructed by some zooplankton species and mucous feeding webs used by others, and Type B is derived from small component particles such as the waste products mentioned above (*Figures 7.4 and 7.5*). The mechanisms by which the snow is formed clearly depend on its class, Type A starting as relatively large particles which remain the dominant component. They are probably not much affected by the physics of the water column in their creation and growth.

Type B snow particles are likely to develop in response to specific biological, chemical, and physical conditions; it is the nature of these conditions which we are only now starting to appreciate. Aggregates can be formed by a variety of processes which cause collisions, such as differential settling during which one particle sinks at a faster rate than the one below and collides with it. Turbulent mixing also increases collision rates between particles, a situation which may be particularly important at

the base of the UML where internal waves, wind driven sheer, and tidal sheer are pronounced[15]. In all cases, significant concentrations of snow can only develop if there are sufficiently high concentrations of the component particles. Until very recently there seemed to be not enough particulate matter in the water column for this to occur. However, the problem may have been related to an underestimate of the abundance of the component particles. Recent observations show that at least one class of particles are naturally transparent unless stained in the laboratory, and yet they are intimately connected to the development of aggregates. These are the transparent exopolymer particles (TEP), which are formed from the polysaccharides excreted by living phytoplankton and bacteria (*Figure 7.5d*)[16]. A further requirement for aggregation is that the particles must be sufficiently sticky, as are the TEPs, to aggregate when they do collide[3].

The material produced in the UML reaches the bottom of this layer both by mixing processes and by gravitation settling; once there, the physical and biological processes conspire to aggregate the material into yet larger particles to produce the characteristic peak at this depth.

In the case of Type A snow particles, the mechanisms of production and, indeed, loss are quite different and depend on the presence of the particular zooplankton species which produce these structures.

As mentioned above, the distributions shown in *Figure 7.2* do not, of course, simply reflect the rates of production, but also the rates of loss either by consumption or sinking. These are discussed below.

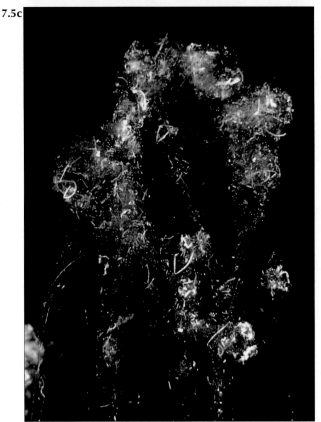

7.5a

7.5b

7.5c

7.5d
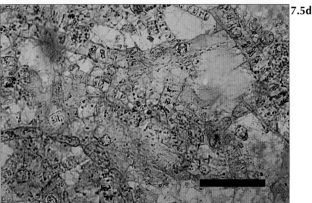

Figure 7.5 (a) Miscellaneous aggregate containing a variety of planktonic particles. (Courtesy of Dr Alldredge, University of California.) (b) Scanning electron micrograph of snow particles dominated by diatoms. In this case they are primarily of the genus *Chaetoceros*. The spherical cell body on the right hand side is about 10 μm in diameter. (Courtesy of Dr Gotschalk, University of California.) (c) *In situ* marine snow aggregate dominated by diatoms; the aggregate is about 1 mm in diameter. (Courtesy of Dr Gotschalk, University of California.) (d) Diatom aggregate after staining with Alcian blue to show transparent exopolymeric material gluing cells together (scale bar = 100 mm). (Courtesy of Dr Alldredge, University of California.)

The long descent

Once it has left the upper reaches of the ocean and, in particular, has passed through the seasonal thermocline at the base of the mixed layer, settling material experiences much shallower gradients in its physical, chemical, and biological environment. Differential settling still occurs and some *de novo* production of Type A snow particles still proceeds. One might think that transformations would take place at a leisurely pace. This is not, however, likely as long as there are significant numbers of zooplankton and nekton migrating from the surface at night and to depths as great as 1000 m by day (see Chapters 14 and 15). This phenomenal diel migration, demanding large energy expenditure by the migrating organisms, may have a significant effect on the vertical transport of sinking material as the migrants consume food primarily in the upper water, where it is most abundant, but excrete, respire, defecate, and die throughout the water column. Furthermore, from the perspective of marine snow, these relatively large organisms are the ones most likely to feed on the sinking material.

Recent observations of long-term changes in marine snow at 270 m in the northeast Atlantic show dramatic changes in their abundance and volume concentration. Peaks correspond to elevated

101

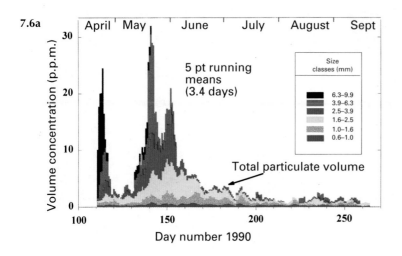

7.6a

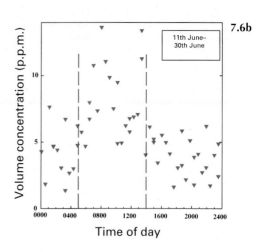

7.6b

Figure 7.6 (a) Temporal variation in marine snow volume concentration (night values only to remove diel periodicity) at 270 m depth in the northeast Atlantic (48°N 20°W), derived from photographs taken using the marine snow camera system, *Figure 7.3(d)*. The photographs obtained with this instrument are analysed by computer to give a maximum and minimum dimension of each particle in the frame. The volume of each particle is calculated and assigned to one of six size categories. Dramatic peaks in concentration are caused primarily by the larger size classes and occur soon after pulses in productivity of the phytoplankton in the surface water layer above[11]. (b) Accumulated data for volume concentration of marine snow at 270 m depth during the period 11–30 June 1990. During this period concentration levels were significantly higher between the hours of 05.00 and 14.00, identified by the dotted vertical lines.

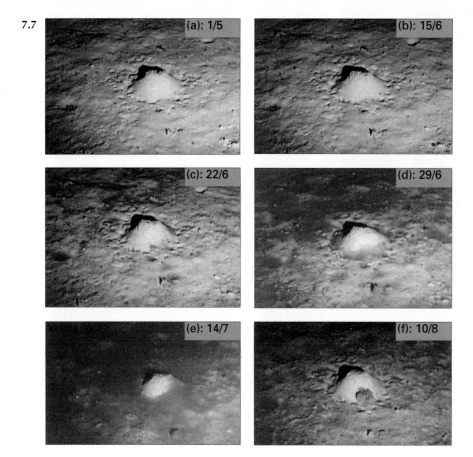

Figure 7.7 Examples of time-lapse photographs of the sea bed at 4025 m depth off the European continental slope, taken using the 'Bathysnap' instrument. Between 1 May and 15 June there is little change in its appearance, but during the rest of the summer there is a progressive increase in the amount of material covering the sea bed, visible as dark patches obscuring the underlying sediment. Between 14 July and 10 August there is a progressive decrease in this covering. The mound in the centre of the frame is 18 cm across.

Figure 7.8 The photographs in *Figure 7.7* were used to derive a semi-quantitative measure of the material lying on the sea bed on each frame (a); the green band is to highlight the trend over the year. Also shown in (a) (vertical bars) is a semi-quantitative estimate of the degree of resuspension in each frame, which reduces visibility of the sea bed. It can be seen that resuspension only occurs after the deposition of phytodetritus and is not a constant feature. The letters A–F indicate the times at which the photographs in *Figure 7.7* were taken. (b) The current speed which shows that it is only when currents exceed about 7.5 cm/s that a significant resuspension occurs. It is thought that the material is not resuspended very high in the water column as, from other Bathysnap deployments, resuspension causes an initial loss of material from the sea bed, which is followed within only a few tens of minutes by its redeposition. The current meter rotor stalls at speeds less than 2 cm/s, as indicated. The tidal cycle is clearly evident, but during late July and early August the minimum of the tidal cycle is above the stall speed of the rotor, indicated by red shading. These are particularly energetic periods.

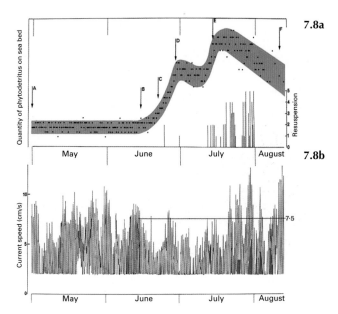

7.8a

7.8b

levels of primary productivity in the overlying water and precede peaks in material flux determined by sediment traps 3000 m below. As can be seen in *Figure 7.6(a)*, the values of volume concentration in the spring are more than ten times those later in the year; this is mainly due to the contribution of the larger-size categories. An additional feature of this data set is that there is a distinct diel variability; in *Figure 7.6(b)* data from one period are presented to demonstrate this signal. It is tempting to relate this signal to the diel migration of most of the larger members of the planktonic biosphere, but at present the mechanism is unknown.

Although much of the material sinks at rates of a few tens to hundreds of metres per day, some of it is modified or remineralised in support of the mid-water community of bacteria and zooplankton. Other components dissolve during the descent, especially calcite particles when below the calcium compensation depth (see Chapter 13), but it seems that in spite of this long descent, some of even the delicate structures of phytoplankton spines are still apparent when the deep-sea floor is reached, just as found by Lohmann[14] back in 1908.

The benthic experience

The ultimate repository for much of the material which sinks into the deeper parts of the water column is the deep-sea floor; it is here we now look for evidence of a close link with the surface of the ocean and indications about the role of marine snow.

Rapid changes in material flux in deep midwater and apparent seasonal reproduction by some of the larger benthic fauna posed a serious threat to the established view of weak linkages between the top and bottom of the oceans. Repeated sampling of the deep-sea sediment, particularly the loose surface layer, was technically very difficult and very demanding on ship time. It was therefore fortunate that in the 1980s, time-lapse photographic techniques using an instrument called the 'Bathysnap' became available for use in an area which was subject to very strong seasonal depositions of phytodetrital material. In *Figure 7.7*, recently deposited material can be readily seen on the sea bed as dark patches of loose fluffy material.

The image of the material above the sediment surface is a reflection of supply and demand. It is the difference between the supply of settling particles from above and the demand by benthic community members, which either ingest or bury these. Under less productive parts of the ocean, such as the Madeira Abyssal Plain, no detrital layer is visible on benthic photos; this not only reflects the lower overall level of productivity, but also the reduced variability in particle flux. Here the supply is more constant and the benthic community consumes or buries it at the same rate as it arrives on the sea floor.

Although difficult to quantify on the photographs, an approximation can be made by measuring the density of the film at one particular location on each frame of the time series (*Figure 7.8*). We can now collect the material using a remarkable coring device, the Scottish Marine Biological Association (SMBA) multiple corer. This has the ability to collect sediment samples with virtually no disturbance to the light interfacial layer, and we can state with certainty that the dark patches seen on the photographs are primarily of phytoplankton

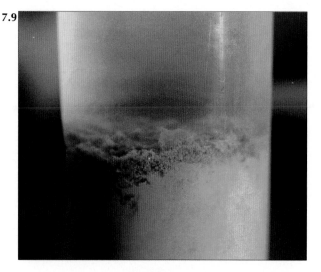

7.9

Figure 7.9 Plastic core tube from the SMBA multiple corer with sediment taken in May 1981. The core tube is 56 cm in diameter and a layer of phytodetritus can be clearly seen lying above the less granular sediment. This core was taken on the European continental slope (Porcupine Seabight) at a depth of 2000 m.

the phytodetrital layer is moved about and resuspended by the near-bottom currents (*Figure 7.8*), fed upon by the benthic biota and buried by some of the larger members of this community. In *Figure 7.11*, visual examples of the benthic fauna's response are shown. Here, at last, was an explanation for the apparent seasonal reproduction of some of the larger benthic species[24], a response that has now been found in the macrofaunal and meiofaunal communities.

Properties and Characteristics of Marine Snow

The physical properties of marine snow particles are as diverse as the origins of the material from which they are derived. Some are sticky, some are fragile, some are porous, and some sink fast. Their biological properties are also diverse, reflecting their origins as mentioned above, but also displaying widely varying rates of biological processes; some have unusually active bacteria within them and others have rapidly growing phytoplankton. In all these cases there are aggregates which display entirely the opposite property, and generalisations are hard to make. The methods used to collect the particles are also far from simple (*Box 7.2*).

Physical and biological properties

As indicated above, the physical characteristics of stickiness and fragility determine whether marine snow aggregates tend to grow in size, and their

origin (*Figure 7.9*). We also know that the phytoplankton species composition is oceanic, demonstrating that it does not slump off the adjacent continental shelf. Finally, in a water depth of 4000 m the peak deposition occurs only 4 weeks after the expected phytoplankton bloom in the overlying water. This all gives strong evidence that this material is transported vertically and rapidly to the deep-sea floor. On closer inspection of some of the benthic time-series photographs, one can even discern the nature of the particles arriving to form this carpet. *Figure 7.10* shows an enlargement of one part of a pair of photographs of a series taken at 2600 m depth on the European continental slope. They are separated in time by only 16 minutes, but during this interval an aggregate of several millimetres diameter arrived on the sea bed. It remained there and, in common with many other similar particles, gradually degraded over the subsequent few days.

In those productive regions, once on the sea bed

7.10a

7.10b

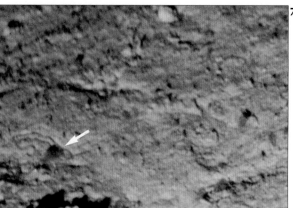

Figure 7.10 Enlargements of consecutive 'Bathysnap' frames of the sea bed in the Porcupine Seabight (northeast Atlantic continental slope) at a depth of 2600 m in May 1981. (a) The arrow shows the spot where 16 minutes later (b) a 10 mm marine snow aggregate (grey) arrived between frames. Over the subsequent few days this particle, and many others like it, degraded and coalesced to form a continuous layer.

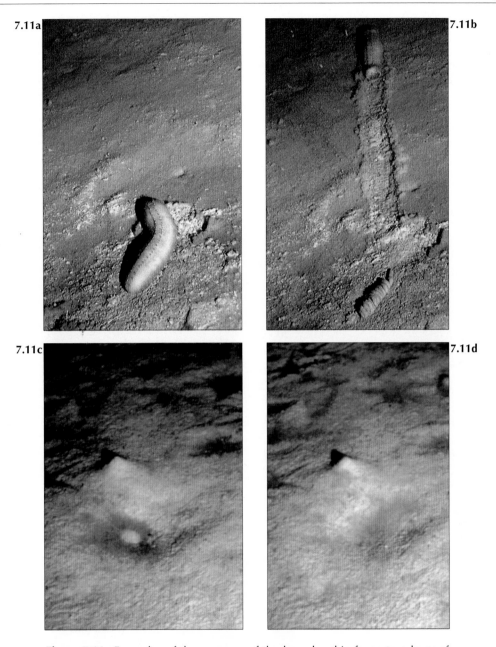

Figure 7.11 Examples of the response of the large benthic fauna to a layer of phytodetritus. (a) The holothurian *Benthogone rosea* of body length 15 cm at a depth of 2008 m in the Porcupine Seabight in June 1982. (b) 30 minutes later the specimen has left a faecal cast to demonstrate its satisfaction with the phytodetrital meal. The echinoid *Echinus affinis* tends to move into depressions in the sediment at times when phytodetritus is present, as in (c), one-third up in the centre, and (d), three-quarters to the right, both found in the Porcupine Seabight at a depth of 2000 m in May 1982.

Box 7.2   How to Collect and Study Marine Snow Aggregates

Sediment traps can be used to collect settling particles, but it is unlikely that the integrity of the individual particles so-collected is maintained, even when they are immersed in a preservative as is usually the case. Because of the fragility of some types of marine snow, and because of their low abundance and high sinking rates, traditional water-bottle techniques are usually of little value. In spite of the inherent difficulties of subaqua diving (limitations of depth, sea state, and personnel), it remains one of the best methods to obtain undamaged snow particles for experimentation (*Figure 7.12*).

Large bottles have been developed by some groups with the specific aim of collecting marine snow, *Figure 7.13(a)*. This 100-litre water bottle is closed at depth by sliding a weight, the 'messenger', down the supporting wire. On recovery it is left on deck for several hours to allow the snow aggregates to settle to the bottom of the vessel. The top 95 litres are drained off and the bottom chamber removed for collection of the particles in the laboratory, *Figure 7.13(b)*. This has proved to be a useful and successful method where diving is not possible.

7.12

Figure 7.12 Subaqua diver behind a very large marine snow aggregate in the northern Adriatic sea. (Courtesy of Dr Stachowitsch, University of Vienna.)

7.13a

7.13b

Figure 7.13 The large-volume marine snow catcher, the 'snatcher', for collecting undamaged samples of the aggregates. (a) The entire device just prior to deployment from RRS Discovery. (b) The lower chamber of the 'snatcher' in the cold laboratory on board ship. Aggregates are being collected from the base of the chamber using a wide-bore pipette to avoid damage. The lights are directed through the transparent sides of the chamber for ease of collection.

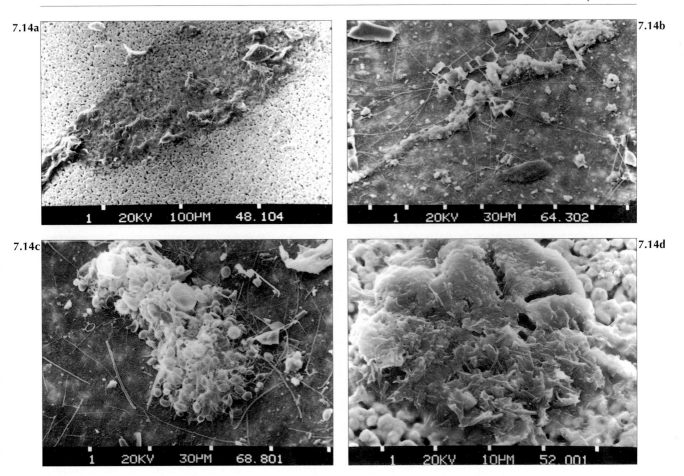

Figure 7.14 Scanning electron micrographs of marine snow particles collected off Baffin Island, showing a wide range of morphologies and composition. (a) Large mucoid aggregate collected at 30 m depth. (b) 'Stringer' collected at 1 m depth. (c) Mixed agglomerate dominated by biogenic material collected at 100 m depth. (d) Aggregate dominated by mineral matter collected at 5 m depth. (Courtesy of Dr Azetzu-Scott, Bedford Institute of Oceanography.)

excess density over that of the surrounding water controls the speed with which they descend through the water column. The proportion of free water in an aggregate, its porosity, determines how fast its internal environment changes in response to varying external conditions; for example, during sinking, porosity controls the rate of exchange of water within the aggregate with that outside. Porosity also influences the rate at which small particles, such as clays, accumulate on the aggregate and the rate at which surface-active elements, such as thorium, are adsorbed onto the snow. Rates of adsorbtion and desorbtion are of considerable relevance to particle-cycling models, which use adsorbtive radioisotopes, such as ^{234}Th, as proxies of solid material (see later).

The physical properties of an aggregate are clearly closely tied to its chemical and biological components. However, as a general pattern, Type B aggregates develop as bloom conditions are reached and nutrients become limiting (see Chapter 6). It is

during this stage that the concentration of TEP increases[16] and the particles probably become more sticky. With increasing size of aggregate, excess density tends to decrease[1], so that the increase in sinking rate is not as pronounced as might be expected. Simultaneously, and when still in the euphotic zone, primary production may be enhanced within the aggregate due to an efficient use of the ammonia released within them[9]. In fact, production may proceed so fast that bubbles of free oxygen are created which, in turn, cause the snow to rise in the water[17]. There is also good evidence that the bacteria[19] and protozoa[13] find the micro-environment within the aggregates attractive, sometimes producing anoxic microzones[18]. This may reduce sinking rates if free gases are produced. The activity of the microbiota may change, however, during the descent of the aggregate; it now seems that this activity may be inhibited by increasing pressure[23].

The composition (*Figure 7.14*) of Type B aggre-

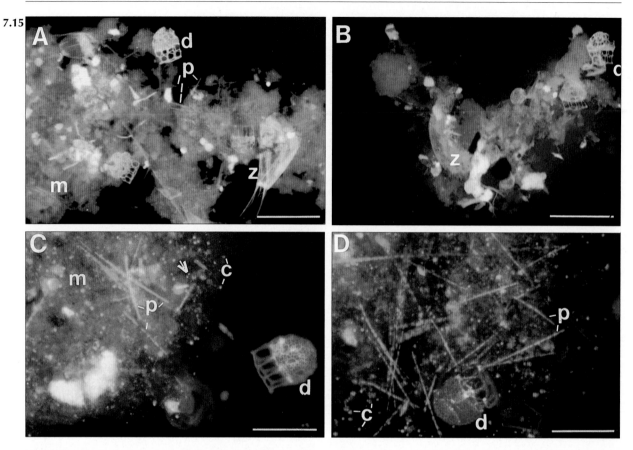

Figure 7.15 Marine snow aggregates collected using the 'Snatcher' (*Figure 7.13*) during May 1990 in the northeast Atlantic over the Porcupine Abyssal Plain from depths of either 45 m (A, C, and D) or 300 m (B). The composition of the aggregates can best be observed under the microscope using ultraviolet excitation, either after staining with acridine orange (A and B) or relying on the autofluorescent properties of the material (C and D). The aggregates contain a wide range of component particles (d, lorica of the tintiniid *Dictyosysta elegans*; p, pennate diatom; z, zooplankton carapace; c, small chlorophyte; m, bacterial matrix). In this instance, the tintiniid lorica tended to be physically damaged when found further down in the water column [scale bars = 100 μm (A and B) and 50 μm (C and D)]. (Courtesy of Dr C Turley, Plymouth Marine Laboratory.)

gates generally reflects that of the suspended particles in the euphotic zone. As such, the aggregates present at any one time at a particular location are of a similar composition[2], whether they be dominated, for instance, by diatoms, faecal material, coccolithophores, or flagellates (*Figure 7.15*).

Whatever the origin of the particles, and however porous the aggregates, it is safe to assume that the environment of a free-living organism or particle changes dramatically when it becomes incorporated into an aggregate. Not only will its chemical environment change, but so too will its sinking rate and, in the case of an organism, its ability to control its depth by, for instance, buoyancy modification. An organism associated with an aggregate will also be subject to quite different predator pressures from its free-living counterparts, as discussed below.

Quantitative Study of Particulate Flux

In order to understand the cycling of biogeochemical components of the oceans, data must be obtained about the time-varying fluxes of the principal compounds, elements, and particle types at different depths in the sea. The concentrations and size distributions of particles can only give a general indication of fluxes, and then only after some major assumptions on sinking rates. There are few ways in which particle flux can be measured; these can be divided broadly into direct measurements using sediment traps and numerical modelling approaches, frequently based on the distributions of various chemicals and particles.

The particle interceptor trap, or sediment trap, is a device akin to the rain gauge, having a funnel into which falling particles are collected. In the case of modern time-series traps (*Figure 7.16*), a rotat-

7.16a

7.16b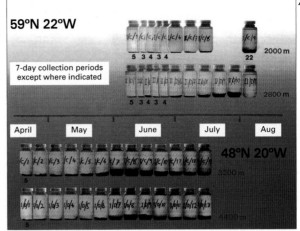

ing carousel moves a fresh collecting jar under the bottom of the trap at predetermined times, such that data may be obtained over several months' duration with a resolution of a week or so. There are several problems related to the accuracy with which sediment traps measure particle flux, but these devices provide the only means of determining directly the flux of material at a particular depth; furthermore, they are the only way in which the material responsible for the downward flux can be collected, described, and analysed.

In 1978, soon after the sediment trap had become a reliable oceanographic technique, a long time-series of measurements was initiated off Bermuda at a depth of 3200 m (*Figure 7.17*). This remarkable series, still continuing, demonstrates a strong seasonal signal of deposition with a peak in the early part of the year. This is certainly not a universal pattern, due to the large physical and biological differences between oceanographic regimes. In this case, the depositional peak lags behind a

Figure 7.16 (a) Time-series sediment trap photographed just after recovery. Settling material enters the yellow cone and from there into the white collecting cups below. The trap had been at a depth of 3200 m for the previous 6 months, in a water depth of 4800 m, and the collected material can be clearly seen in the cups (previously filled with a preservative, formaldehyde). Such devices are suspended on a supporting wire, with a ballast weight on the sea bed attached to the wire by an acoustically operated release mechanism. Buoyancy spheres at the top of the wire carry the entire mooring to the surface after the release has been activated. (b) Sample cups from deep-water sediment traps deployed at two locations in the northeast Atlantic (7 day collection periods except where indicated). The increased flux from the end of May resulted from elevated productivity in the surface waters about 4 weeks previously. (Courtesy of Dr Williamson, University of East Anglia.)

Figure 7.17 Variation in particle flux over an 8-year period at 3200 m in the Sargasso Sea. This is the longest such record from any region and shows the seasonal cycle in particle flux, but with some distinct irregularities during the period 1981–1983. At this location, elevated productivity in the surface is caused by a deepening of the mixed layer, which introduces new nutrients. About 3 weeks later particle flux at 3200 m became enhanced (from Deuser[6]).

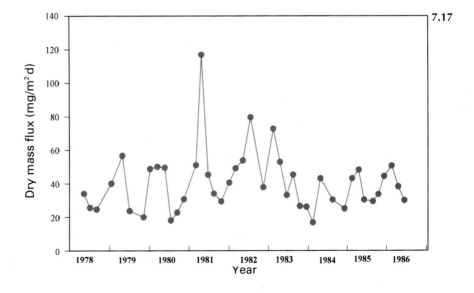

peak in the UML by about 6 weeks; the reason is that the deepening of this layer introduces nutrients into the upper ocean to produce a phytoplankton bloom. The data also demonstrate significant inter-annual variability, a feature which has been found in most studies for which there are sufficient data. Explanations are, however, not always readily available.

In the upper ocean, zooplankton tend to swim into the sampling cups, thus contaminating the material by defecating and/or dying in them. This is particularly unfortunate as it is here where the sharpest gradients in flux occur and where there is the greatest requirement, from the modelling per-spective, for good quality data. Most of the organic carbon lost from the UML in spring and summer is remineralised before it reaches a depth of 1000 m, but at present it is not clear how much is lost from the UML.

Indirect methods of monitoring particle flux usually demand measurements of the vertical distri-bution of the dissolved and particulate phases of certain radionuclides which are particle reactive. Most promising in this regard is ^{234}Th, which is produced from dissolved ^{238}U at a known rate. Using a box model, uptake and removal rates of ^{234}Th can be calculated and carbon fluxes derived[4]. These methods do not, of course, provide material for examination and at present frequently give rather different conclusions from those using the direct methods.

Significance of Marine Snow in Biogeochemical Cycling

There is an ever-increasing interest in the material cycles of the oceans and the relevance of these to global cycles. The significance of marine snow in these cycles may be considered in several ways, depending on the physical dimensions of interest. Marine snow may be considered as an environment in which biogeochemical processes occur (scales ranging from micrometres to centimetres), as a resource on which the plankton and nekton can feed (scales from millimetres to metres), and as a vehicle by which material is transported down through the water column (scales from decimetres to kilometres).

The micro-environment

The micro-environment of a marine snow aggre-gate has already been described from the perspec-tive of its properties. With regard to biogeochemi-cal cycling, the conclusion is that the bulk proper-ties of sea water (nutrient, oxygen, etc.) may have little bearing on the way material is modified if the micro-environment of the marine snow aggregate contains a significant proportion of sea water. It is, however, still an open question as to the size of the material pool which is influenced by this micro-environment, a major challenge to our understand-ing of material cycling in the oceans.

Role as a food source

Considering now the next spatial scale, that of the planktonic feeder, if, as stated above, small parti-cles such as cyanobacteria become incorporated into aggregates, they will be available to quite dif-ferent predators than when they are free living. Observations by divers have, on several occasions, commented on the physical proximity of some zoo-plankton species and marine snow, which suggests that the zooplankton are feeding on the snow parti-cles or a component of them. Preliminary experi-mental evidence is now giving support to this con-clusion, extending the range and diversity of species which feed on snow aggregates (*Figure 7.18*). One of the important conclusions from this with regard to biogeochemical cycling is that it represents a food-chain short-cut in which small particles can be consumed by species normally restricted to a much larger size of food. Once again, if models of bio-geochemical cycling are to be realistic, they must not only consider the bulk properties of the sea water, but also the micro-environments within it.

Role as a transport vehicle

The vast majority of material produced in the UML is recycled there. As described above, there is a sharp reduction in the vertical flux over the top few hundred metres, reflecting the activity of the organ-isms which feed on the sinking material. Particles which sink slowly are particularly susceptible to this recycling, as their residence times are long. Although some marine snow particles have low or even negative sinking rates, those which leave the euphotic zone sink at high rates, between several metres per day and several hundred metres per day. The close temporal coupling between surface processes and the deep sea, the undegraded nature of material collected in deep sediment traps, and the visual observations of snow-sized particles arriving on the deep-sea floor all give strong sup-port to the contention that they are the principle vehicles by which material is transferred to depth. This is important in terms of both the transport of mass and of key compounds, such as those contain-ing organic carbon, trace metals, and radioisotopes adsorbed onto the surfaces of particulate material.

Conclusion

Marine snow is a class of large inanimate particles found throughout the world's oceans. Fragile, and occurring at low abundances, these particles have proved to be difficult to record and collect, but are now thought to be the principal vehicles by which

7.18a 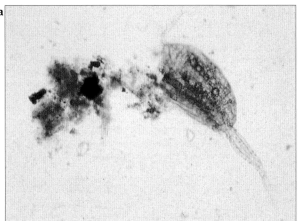 7.18b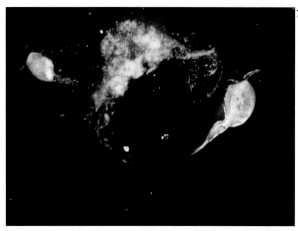

Figure 7.18 Living zooplankton of the genus Oncaea feeding on marine snow aggregates. (a) This specimen, of body length 3 mm, was studied in the laboratory for 3 days during which it made repeated visits to an aggregate comprising cyanobacteria, phytoplankton, and microzooplankton remains (Type B). During this period it produced a large number of its own faecal pellets. (b) *In situ* specimen of about 1.5 mm body length associated with an aggregate derived from a discarded larvacean house (Type A; courtesy of Dr J. King, University of California).

material is transported from the site of primary production in the upper sunlit zone of the water column to the deep sea. Due to their rapid sinking rate, these large aggregated particles link the top and bottom of the ocean in a much closer temporal sense than had been thought the case; they provide an explanation for apparent seasonal changes in the appearance of the deep-sea floor and for the seasonal reproduction of some species of deep-sea benthic animals. It now seems that the rates of production and destruction of marine snow are high so as to facilitate diurnal changes in the particle pool well below the UML of the ocean, a feature which will have far-reaching implications for our understanding of particle cycling.

There are still many unanswered questions surrounding the mechanisms of production and destruction of marine snow particles, but it is expected that, with the techniques now available, major advances will be made over the next decade.

References

1. Alldredge, A.L. and Gotschalk, C. (1988), *In situ* settling behaviour of marine snow, *Limnol. Oceanogr.*, **33**, 339–351.
2. Alldredge, A.L. and Gotschalk, C. (1990), The relative contribution of marine snow of different origins to biological processes in coastal waters, *Contint. Shelf Res.*, **10**(1), 41–58.
3. Alldredge, A.L. and McGillivary, P. (1991), The attachment probabilities of marine snow and their implications for particle coagulation in the ocean, *Deep-Sea Res.*, **38**, 431–443.
4. Buesseler, K.O., Bacon, M.P., Cochran, K., and Livingston, H.D. (1992), Carbon and nitrogen export during the JGOFS North Atlantic Bloom Experiment estimated from ^{234}Th:^{238}U disequilibria, *Deep-Sea Res.*, **39**, 1115–1137.
5. Carson, R. (1951), *The Sea around Us*, Oxford University Press, Oxford, 230 pp.
6. Deuser, W.G. (1987), Variability of hydrography and particle flux: Transient and long term relationships, in *Particle Flux in the Ocean*, Degens *et al.* (eds), Mitt.Geol.-Palaont., Inst.U Hamb., pp 179–193.
7. Fasham, M.J.R., Ducklow, H.W., and McKelvie, S.M. (1990), A nitrogen-based model of plankton dynamics in the oceanic mixed layer, *J. Marine Res.*, **4**, 591–639.
8. Gardner, W.D. and Walsh, I.D. (1990), Distribution of macroaggregates and fine grained particles across a continental margin and their potential role in fluxes, *Deep-Sea Res.*, **37**, 401–411.
9. Gotschalk, C.C. and Alldredge, A.L. (1989), Enhanced primary production and nutrient regeneration within aggregated marine diatoms, *Marine Biol.*, **103**, 119–129.
10. Jeffreys, J.G. (1869), The deep-sea dredging expedition of H.M.S. 'Porcupine', *Nature*, **1**, 135–137.
11. Lampitt, R.S., Hillier, W.R., and Challenor, P.G. (1993a), Seasonal and diel variation in the open ocean concentration of marine snow aggregates, *Nature*, **362**, 737–739.
12. Lampitt, R.S., Wishner, K.F., Turley, C.M., and Angel, M.V. (1993b), Marine snow studies in the Northeast Atlantic: distribution, composition and role as a food source for migrating plankton, *Marine Biol.*, **116**, 689–702.
13. Lochte, K. (1991), Protozoa as makers and breakers of marine aggregates, in *Protozoa and Their Role in Marine Processes*, Reid, P.C., Turley, C.M., and Burkill, P.H. (eds) (NATO ASI Series G: vol 25), Springer-Verlag, Berlin, 506 pp.

14. Lohmann, H. (1908), On the relationship between pelagic deposits and marine plankton, *Int. Rev. Ges. Hydrobiol. Hydrogr.*, **1**(3), 309–323, *in German.*

15. Lueck, R.G. and Osborn, T.R. (1985), Turbulence measurements with a submarine, *J. Phys. Oceanogr.*, **15**, 1502–1520.

16. Passow, U., Alldredge, A.L., and Logan, B.E. (1994), The role of particulate carbohydrate exudates in the flocculation of diatom blooms, *Deep-Sea Res.*, **41**, 335–357.

17. Riebesell, U. (1992), The formation of large marine snow and its sustained residence time in surface waters, *Limnol. Oceanogr.*, **37**(1), 63–76.

18. Shanks, A.L. and Reeder, M.L. (1993), Reducing microzones and sulphide production in marine snow, *Marine Ecol. - Prog. Ser.*, **96**, 43–47.

19. Simon, M., Alldredge, A.L., and Azam, F. (1990), Bacterial carbon dynamics on marine snow, *MEPS*, **65**(3), 205–211.

20. Syvitski, J.P.M., Asprey, K.W., and Heffler, D.E. (1991), The floc camera: A three-dimensional imaging system of suspended particulate matter, in *Microstructure of Fine-Grained Sediments*, Bennet, R.H., Bryant, W.R., and Hulbert, M.H., Springer-Verlag, New York, pp 281–289.

21. Suzuki, N. and Kato, K. (1953), Studies on suspended materials (marine snow) in the sea. Part 1. Sources of marine snow, *Bulletin of the Faculty of Fisheries of Hokkaido Univ.*, **4**, 132–135.

22. Thomson, C.W. (1873), *The Depths of the Sea. An account of the general results of the dredging cruises of H.M.S. Porcupine and Lightning during the summers of 1868, 1869 and 1870 under the scientific direction of Dr Carpenter, F.R.S., J.Gwyn Jeffreys F.R.S. and Dr Wyville Thomson, F.R.S.*, Macmillan, London.

23. Turley, C.M. (1993), The effect of pressure on leucine and thymidine incorporation by free-living bacteria and by bacteria attached to sinking oceanic particles, *Deep-Sea Res.*, **40**, 2193–2206.

24. Tyler, P.A. (1988), Seasonality in the deep sea *Oceanogr. Mar. Biol. Annu. Rev.*, **26**, 227–258.

25. Wallich, G.C. (1862), *The North Atlantic Seabed: comprising a diary of the voyage on board H.M.S. Bulldog, in 1860; and observations on the presence of animal life, and the formation and nature of organic deposits, at great depths in the ocean*, Van Voorst, London.

26. Walsh, I.D. and Gardner, W.D. (1992), A comparison of aggregate profiles with sediment trap fluxes, *Deep-Sea Res.*, **39A**, 1817–1834.

CHAPTER 8:

The Evolution and Structure of Ocean Basins

R.B. Whitmarsh, J.M. Bull, R.G. Rothwell, and J. Thomson

Plate Tectonics

Plate tectonics is the paradigm, developed during the 1960s, by which the majority of large-scale topographic features on the Earth's surface can be explained. Plate tectonics also explains the kinematic and dynamic behaviour of the outer parts of the Earth in terms of rigid lithospheric plates. These plates, which comprise the uppermost mantle and in most cases both continental and oceanic crust, ride on a relatively weak viscous asthenosphere and move relative to each other over the Earth's surface (*Figure 8.1*). The plate motion is driven by mantle convection, the main mechanism by which heat, derived from radioactive decay, is transferred from the Earth's deep interior to the surface.

One of the axioms of plate tectonics is that the plates are strong and rigid, with deformation occurring only at their margins, where the plates interact. This is consistent with the fact that most of the Earth's major topographic features occur at the past or present edges of plates and most seismic (*Figure 8.2*) and volcanic activity is found at active plate boundaries. There are three types of plate boundary (*Figure 8.1*). Accretionary plate boundaries (Chapter 10) are elongate ridges where new oceanic crust is formed; subsequently, the crust is carried away from the ridge axis by sea-floor spreading and subsides as the lithosphere cools.

Convergent plate boundaries are often marked by oceanic trenches, where the oceanic lithosphere is subducted into the asthenosphere (when two continents collide a mountain range results instead). At transform faults, adjacent plates slide past each other and material is neither created nor destroyed.

There are a dozen major plates, most of which contain both oceanic and continental lithosphere, which move relative to each other at 10–200 mm/yr (*Figure 8.3*).

The movement of the lithospheric plates over a spherical Earth can be described by Euler's Theorem. This states that the relative motion between two plates can be defined by a rotation about an axis passing through the centre of the Earth and intersecting the surface at the Euler poles. The most accurate technique for determining the current Euler pole for two adjacent plates uses the locus of the oceanic transform faults between them (*Figure 8.4*). Such faults follow the traces of small circles centred upon the Euler pole; they are mapped using bathymetry, side-scan sonar images, and maps from satellite altimetry (see *Box 8.1*). Thus, the Euler pole can be determined by the intersection of great circles drawn normal to the transform faults. Statistical techniques are used to constrain the likely position of the Euler pole and to give an estimate of its uncertainty.

Figure 8.1 The principal features of plate tectonics, including (a) accretionary (divergent), (b) convergent, and (c) transform plate boundaries. Arrows on the lithosphere represent the relative motion between pairs of plates and arrows in the asthenosphere represent complementary flow in the mantle. The lithosphere is typically 100 km thick; the horizontal extent of the features shown is often at least tens or hundreds of kilometres (after Isacks *et al.*[19]; © American Geophysical Union).

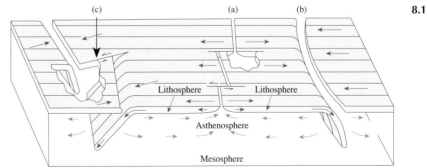

8.1

8.2

Figure 8.2 Global earthquake activity superimposed on a topographic map of the world (based on a cylindrical equidistant map projection). The epicentres of earthquakes with body-wave magnitudes greater than five are shown as coloured dots, colour-coded according to the focal depth of each earthquake (<50 km, red; 50–100 km, yellow; 100–300 km, green; >300 km, blue). Note how most epicentres are concentrated in bands which delineate the plate boundaries (courtesy of the National Oceanographic and Atmospheric Administration/National Geophysical Data Center).

8.3

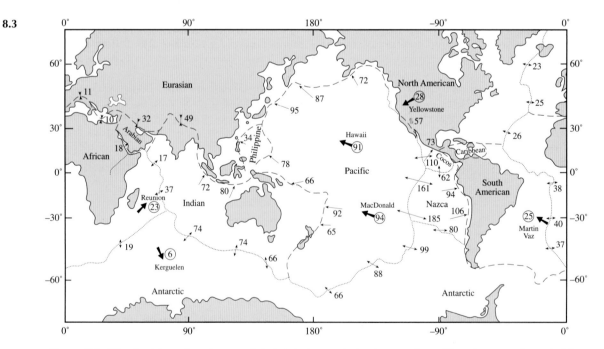

Figure 8.3 The Earth's major plates and their relative and absolute motions. The plates are bounded by oceanic ridges (dotted lines), oceanic trenches, mountain ranges, and transform faults (all as dashed lines). Thin arrows show the directions and rates (in mm/yr) of relative motion at selected points on the plate boundaries [after Bott[6]; reproduced by permission of Edward Arnold (Publishers) Ltd., and based on data in Chase[11]; © Martin H.P. Bott, 1982].

Figure 8.4 Method to determine the Euler pole for a spreading ridge system. Transform faults (thick lines with double arrows), which offset the ridge segments (double lines) describe small circles about the pole. Hence lines drawn normal to the transform faults intersect at the Euler pole.

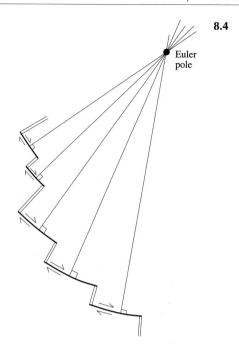

Rates of contemporary sea-floor spreading at accretionary plate boundaries are also used to constrain the angular rotation rate about the Euler pole, which is required to describe current plate movements. However, active convergent plate boundaries are more problematic and relative velocities have to be determined indirectly. For example, if the relative divergent motions between plates *A* and *B* and between plates *A* and *C* are known, the relative convergent motion between plates *B* and *C* can be determined by simple vector algebra. The above approach can be extended to include all known plate boundaries, with the additional constraint that, globally, all the vectors must form a single self-consistent set (*Figure 8.3*). DeMets *et al.*[14] have completed the most recent

Box 8.1   Satellite Gravity Fields

Short-wavelength (<400 km) gravity anomalies are highly correlated with small-scale topography; because of this it is possible to use high-resolution gravity fields computed from satellite data to map the bathymetry and tectonic features of the sea floor. This is particularly useful in parts of the Southern Ocean where ship-collected data are sparse (Figure 8.5).

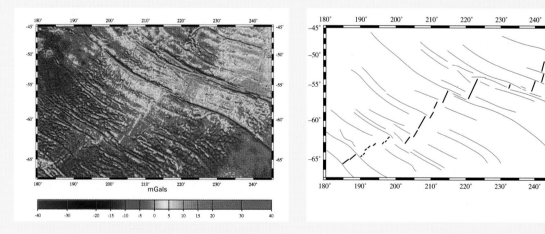

Figure 8.5 (a) An 'illuminated' image of high-resolution gravity anomalies over the Pacific–Antarctic Ridge. This accretionary ridge, outlined in orange and yellow, runs ENE–WSW across the figure, and is offset by numerous transform faults (dark blue and purple; courtesy of W.H.F. Smith, first published in Sandwell and Smith[32]). (b) An interpretation of the gravity image, which reveals features as small as a few tens of kilometres and the complex relationships between spreading segments (thick lines) and small and large offset transform faults and fracture zones (thin lines).

global analysis of current plate motions. The power and use of this technique is that, provided the relevant observations are available, it can be applied to deduce past plate motions; by working backward from the present the relative positions of the plates in the past can be computed.

The definition of the boundary between the lithosphere and asthenosphere depends on the particular physical property under consideration (e.g., temperature, seismic velocity, or flexural rigidity). One of the main factors affecting the strength of sub-surface materials is temperature. Melting occurs where the temperature–depth curve intersects the melting curve or 'solidus' (see *Figure 8.12* later). A thermal definition of lithospheric thickness is given by the depth at which the mantle is closest to melting. In fact, only a small fraction of the asthenosphere is believed to be molten, because it can transmit seismic shear waves which, if it was largely fluid, it could not do. Clearly, the depth of the base of the lithosphere depends on the temperature gradient within the Earth (or geothermal gradient), the melting temperature of mantle minerals, and the relative abundance of such minerals. Therefore, under ocean ridges, where the geothermal gradient is high, the asthenosphere occurs at a

depth of a few kilometres, at the base of the crust. Oceanic lithosphere thickens with time as it cools and reaches about 100 km in the oldest parts of the ocean basins. The continental lithosphere thickness varies between 100–200 km. The thermal definition of the base of the lithosphere is in good general agreement with observations of a zone of low seismic velocity and high shear-wave attenuation at similar depths. These observations, in turn, are consistent with the existence of a weak zone marking the lithosphere–asthenosphere boundary.

Models of Lithospheric Evolution

An important observation first made in the 1960s was that heat flow is highest at mid-ocean ridges and decreases with distance from the ridge-axis, as the mean depth of the ocean increases. These two observations have provided the main constraints on models of the thermal evolution of the oceanic lithosphere. Two models explain these observations. In the first model, the lithosphere behaves as the cold, upper boundary of a cooling half-space, such that the depth varies as $(age)^{0.5}$ and heat flow varies as $(age)^{-0.5}$. In the second model, the lithosphere is treated as a cooling plate with an isothermal lower boundary. The lithosphere thus behaves as a cooling boundary layer until such time as the lithospheric temperature gradient, constrained by the lower boundary condition, causes the curves describing the variation of depth and heat flow with age to flatten and change more slowly with time, as shown in *Figure 8.6*. In this model old lithosphere approaches the asymptotic value of thermal plate thickness. Parsons and Sclater[29] originally found that a 125 km thick lithosphere with a basal temperature of 1350°C fitted the data then available. However, more recently Stein and Stein[35], using improved heat flow data, predicted a hotter and thinner lithosphere (1450°C at the base of a 95 km thick plate; *Figure 8.6*).

Another property of the lithosphere is its response to vertical loading, or flexure. The continental lithosphere is commonly loaded by ice

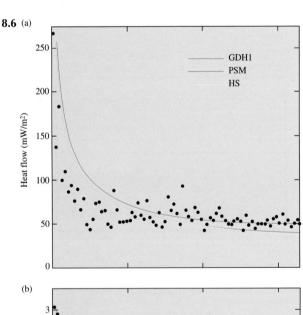

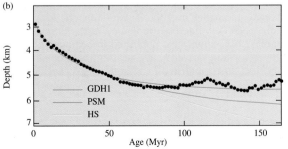

8.6 (a)

Figure 8.6 Observations and models for (a) heat flow and (b) oceanic depth as a function of crustal age. Depths are average values for the North Pacific and Northwest Atlantic Oceans; heat-flow measurements are from sites in these regions. Depths and heat-flow values (dots) have been averaged over 2 Myr intervals. Also shown are functions computed from the plate model of Parsons and Sclater[29] (PSM), a cooling half-space model with the same thermal parameters (HS), and the Stein and Stein[35] (GDH1), plate model. In (a) the HS and PSM curves overlap for ages younger than ca 120 million years (Myr; from Stein and Stein[35]; reprinted with permission from *Nature*, © 1992 Macmillan Magazines Ltd).

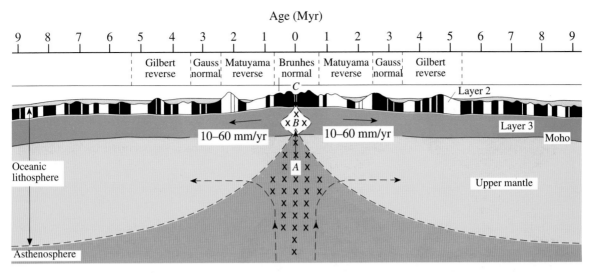

Figure 8.7 Sea-floor spreading and the generation of magnetic lineations by the Vine–Matthews hypothesis. *A* marks the rise of basaltic magma to form the oceanic crust, and the residual material which is left to form the mantle part of the oceanic lithosphere. *B* marks the magma chamber, which eventually cools to form the widespread Layer 3 in the crust. *C* marks the rapid cooling of basaltic magma to form the pillow lavas and dykes of crustal Layer 2. In Layer 2, blocks of normal magnetic polarity are shown black and blocks of reverse polarity are shown white. 'Moho' indicates the seismic discontinuity which marks the base of the crust. Vertical exaggeration is about 10:1 for a half-spreading rate of 30 mm/yr [after Bott[6]; reproduced by permission of Edward Arnold (Publishers) Ltd; © Martin H.P. Bott, 1982].

sheets, sedimentary basins, or mountain ranges, while the oceanic lithosphere is loaded by seamounts, sediments, and aseismic (volcanic) ridges. If the lithosphere behaves perfectly rigidly, then it can be treated as an elastic plate; thus, the amount of flexure depends on its flexural rigidity, the magnitude of the load, and the elastic thickness (defined as the thickness of an equivalent uniform sheet that responds elastically in the same way as the real lithosphere). Within the ocean basins, elastic thickness increases with the age of the lithosphere[39]. However, in general it is found that elastic thickness is much less than thicknesses based on thermal or seismological criteria. This is not surprising, given the simplified rheology inherent in the elastic plate model. The upper part of the lithosphere actually behaves in a brittle manner, while the lower lithosphere behaves plastically. The duration of the loading also needs to be considered. In summary, therefore, the lithosphere can be defined in several ways dependent on the processes being considered.

Sea-Floor Spreading

The magnetic field of the Earth is approximately that of a dipole, with an axis close to, but not normally coincident with, the Earth's axis of rotation. At present, the polarity of the geomagnetic field is such that a compass needle points to the North Magnetic Pole, but sometimes in the past the geomagnetic field had the opposite polarity for 10^5

years or more, and a compass needle would have pointed to the South Magnetic Pole. Vine and Matthews[38] were the first to combine the notion of sea-floor spreading with the phenomenon of polarity reversals of the geomagnetic field. They envisaged that new oceanic crust acquires a stable thermal remanent magnetisation after cooling through the Curie temperature (the temperature at which magnetic minerals acquire the magnetisation vector of the ambient geomagnetic field). They proposed that the formation of oceanic crust is symmetric and continuous, on geological time-scales, as the lithosphere moves away on either side of the axis of an accretionary ridge. Thus, the stripy magnetic anomalies characteristic of the oceanic crust are produced by alternating crustal blocks of opposite polarity that have recorded the polarity reversals (*Figure 8.7*).

Sequences of polarity reversals observed on land enable magnetic anomaly patterns to be predicted at sea. The use of the geomagnetic polarity reversal time-scale to date the oceanic lithosphere is based on the identification of characteristic magnetic anomaly profiles and their relation to the independently established reversal chronology (see *Box 8.2*). The Vine–Matthews hypothesis was verified by confirming the age of the oceanic crust predicted from the magnetic anomaly pattern by drilling, and dating with microfossils, the sediments immediately overlying the oceanic basement. The magnetic anomalies preserved in the oceanic crust have

8.8

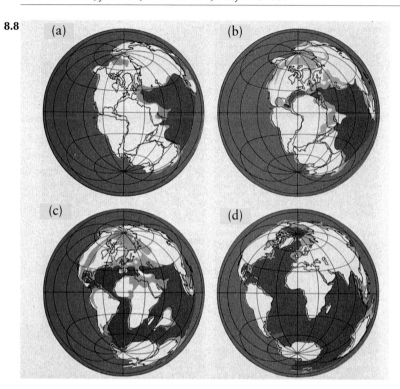

(a) (b) (c) (d)

Figure 8.8 A sequence of maps, reconstructed using sea-floor spreading and palaeomagnetic data, to illustrate the break-up of the Pangaea supercontinent. (a) The configuration of land about 200 Myr ago; the dark blue on the east side of the supercontinent is the Tethys Ocean. The configurations for 160 Myr (b) and 80 Myr ago (c), and the present day (d) show that the dispersal of the continents occurred by the opening of the Atlantic and Indian Oceans, and the resultant closure of the Tethys Ocean to form the Alpine–Himalayan mountains (dark blue, Atlantic-type – interior – ocean; medium blue, Pacific-type – exterior – ocean; light blue, flooded continent; from Nance *et al.*[27]; © 1988 by Scientific American, Inc. All rights reserved).

allowed the sea-floor spreading history of about the past 200 Myr to be reconstructed in detail. For example, *Figure 8.8* shows reconstructed positions of the continents following the break-up of Pangaea, a supercontinent that existed about 200 Myr ago.

The Life Cycle of Ocean Basins

The creation, evolution, and eventual destruction of the ocean basins is cyclical and has been called the Wilson Cycle [*Figure 8.9(a)–(e)*] after the famous Canadian Earth scientist Tuzo Wilson. The cycle starts when a continental area is thrown into tension, either because of a sub-crustal heat source

causing uplift or because of the geometry of the plate boundaries surrounding the continent. This tension causes the development of normal (extensional) faults in the brittle crust and ductile flow at depth, leading to significant crustal thinning. The East African Rift Valley and the Rio Grande Rift in New Mexico are examples of this phase, *Figure 8.9(a)*.

With continued extension the crust is eventually thinned to such an extent that hot mantle material rises to the surface, oceanic crust begins to form, and an ocean basin is born (see later for a fuller explanation). The uplifted continental flanks are gradually eroded (see, e.g., Chapter 9) and sediments are deposited in the new ocean basin. The

---

Box 8.2   Sediment and Rock Dating

Fossiliferous sediments and rocks younger than 600 Myr may be dated by a variety of techniques. These involve the recognition of individual fossil species, or the content of assemblages of such species, within a biostratigraphic framework which describes the sequence of appearances and disappearances of species over geological time. The marine oxygen-isotope record (see *Box 8.3*) also provides a well-constrained means for dating Late Cenozoic biogenic carbonate sections. Physical methods, such as geomagnetic polarity reversal stratigraphy (see text) and radiometric techniques, may also be used for dating; such methods are especially useful when applied to nonfossiliferous sediments. Radiometric dating methods depend on the high-precision measurement of the parent–daughter pairs of isotopes by mass spectrometry; knowing the natural radioactive decay chains and decay constants of the individual isotopes in such chains, it is possible to compute ages from these measurements. Sediment dates tend to be based on the decay of shorter-lived natural or cosmogenic radionuclides; rock dates are generally based on the predictable rates of conversion of long-lived natural parent radionuclides into daughter products by radioactive decay over time.

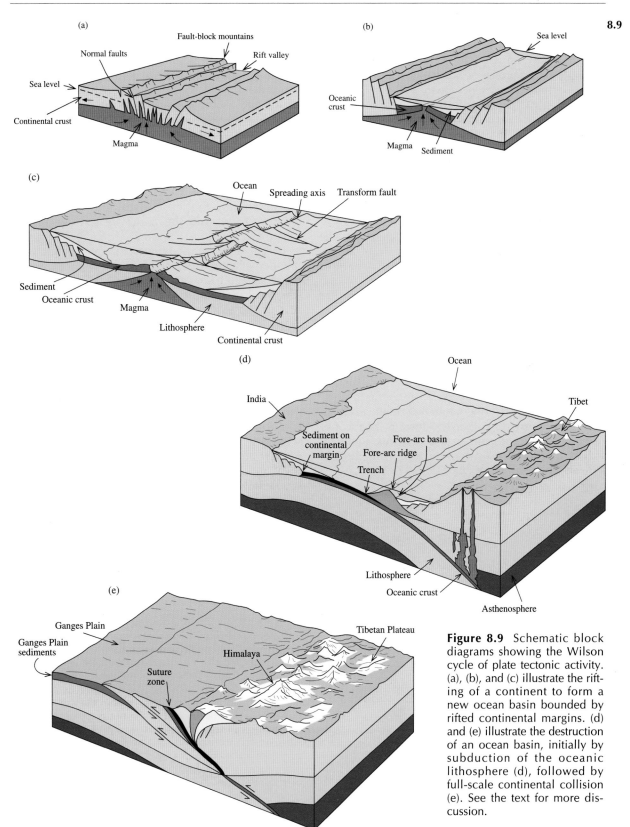

Figure 8.9 Schematic block diagrams showing the Wilson cycle of plate tectonic activity. (a), (b), and (c) illustrate the rifting of a continent to form a new ocean basin bounded by rifted continental margins. (d) and (e) illustrate the destruction of an ocean basin, initially by subduction of the oceanic lithosphere (d), followed by full-scale continental collision (e). See the text for more discussion.

Red Sea is an example of such a young oceanic rift, *Figure 8.9(b)*. The ocean basin then expands by sea-floor spreading. The Atlantic Ocean is an example of a mature ocean basin surrounded by rifted continental margins, *Figure 8.9(c)*.

Cooling of the oceanic lithosphere continues until its density exceeds that of the asthenosphere and a potentially unstable situation arises. A subduction zone may form. The machanism for this is presently controversial; it is the one part of the cycle which is poorly understood. The main characteristics of a subduction zone are shown in *Figure 8.9(d)*, which illustrates the incipient collision of India and Tibet. The subduction of oceanic lithosphere may cause the ocean basin to contract. The Pacific Ocean is a present-day example of a contracting ocean delimited by active plate margins (subduction zones and transform faults).

Eventually, if all the oceanic lithosphere has been subducted, continent–continent collision takes place. The collision of India with Asia, which led to the formation of the Himalayan mountains and the high Tibetan Plateau, is a dramatic example of the closure of an ocean basin (the Tethys Ocean – see *Figure 8.8*) and continent–continent collision, *Figure 8.9(e)*. Present-day Asia represents a complex assembly of a large number of small continental blocks that have collided to form a large continent, or supercontinent. Eventually, this continent will be rifted, a new ocean basin will form, and the Wilson Cycle will begin again.

Intraplate Deformation

Although most volcanic and seismic activity is concentrated at plate boundaries, some activity occurs within plates. For example, the ocean basins are marked by the frequent occurrence of seamount chains and aseismic ridges associated with one or more active volcanoes. Much of this intraplate volcanic activity is related to 'hot spots' caused by thermal plumes rising from the lower mantle. Over time, the motion of the plates over the near-stationary hot spots leads to elongate chains of volcanoes and aseismic ridges. Two conspicuous examples are the Hawaiian–Emperor seamount chain in the Pacific Ocean and the Ninetyeast Ridge in the Indian Ocean.

Bergman[4] recognised that oceanic intraplate seismic activity could be divided into two categories based on the focal mechanisms of the earthquakes (the focal mechanism is the form of rupture at the seat of the earthquake inferred from the distribution of initial compressional or dilatational motions, generated by the earthquake, at the Earth's surface). A large proportion of the seismic activity occurs in oceanic lithosphere less than 35 Myr old; focal mechanisms of these events are a poor indicator of the state of regional stress, because they are dominated by the early thermal evolution of the lithosphere. The second type of seismic activity occurs in lithosphere older than 35 Myr and is characterised by thrust or strike–slip focal mechanisms. These indicate an intraplate stress field dominated by maximum horizontal compression oriented normal to the ridge crest. The focal mechanisms associated with this type accurately reflect the regional state of stress.

Seismicity in old oceanic lithosphere is not entirely randomly distributed throughout the oceans and may be associated with a few areas of intraplate

8.10

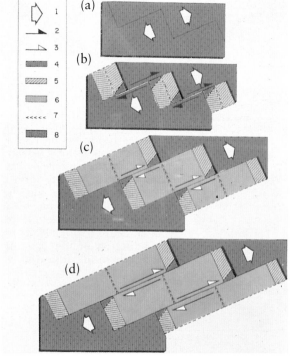

Figure 8.10 A simplified model for the evolution of transform and rifted continental margins based on the example of the Côte d'Ivoire and Ghana margins off West Africa. (a) Onset of continent-to-continent active transform contact. (b) Continent-to-continent and continent-to-rifted-margin contacts. Shearing produces a lateral marginal ridge and associated tectonic deformation. (c) Progressive drift of the end of the hot accretionary ridge along the transform margin. Thermal exchange occurs between continental and oceanic lithosphere, leading to vertical adjustments. (d) Mature stage of rifting; evolution of the margin includes thermal subsidence. 1, Direction of relative plate motion; 2, transform motion between continental crusts; 3, transform motion between oceanic crusts (transform faults); 4, thick continental crust; 5, thinned continental crust; 6, oceanic crust; 7, axis of accretionary plate boundary; 8, marginal ridge and region of related tectonic deformation (courtesy of J. Mascle, with permission).

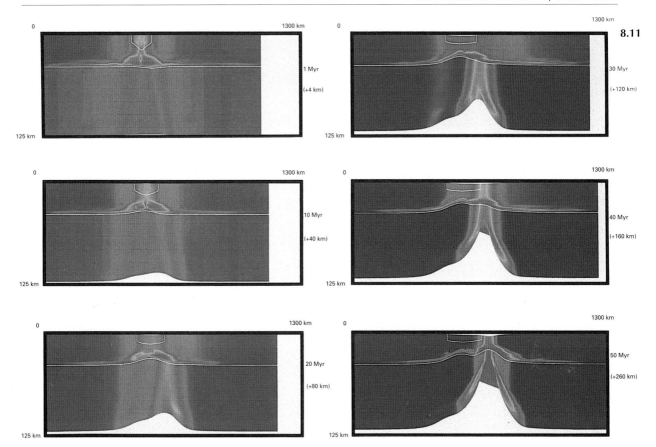

Figure 8.11 Numerical modelling of strain rate within a continental lithosphere which is stretched under tension by 200 km over 50 Myr relative to the fixed left side of the model. The intensity of deformation is shown by the colours; greater redness indicates more intense deformation and emphasises the loci of extension. The colour scale is logarithmic from 10^{-15}/s (darkest blue) to 10^{-11}/s (red). The crust–mantle boundary is shown in white (horizontal line at 1 Myr). From the start the model has an in-built mid-crustal zone of weakness (white lines outlining the enclosed region at centre top) and an upper mantle weakness caused by the thickened crust to the right of the upper crustal weakness. Note how the locus of extension begins in the upper crust, but shifts to the right and into the lower lithosphere after 20 Myr. Models such as this explain why crustal rifting precedes the appearance of volcanic products at the surface and why the rifting and volcanism occur in different places. If the rifting illustrated here continues beyond 50 Myr, eventually new oceanic crust will be formed in the centre of the model (courtesy of D.S. Sawyer, with permission).

deformation. These areas have also been called diffuse plate boundaries[42]. The best-developed example is in the northeastern Indian Ocean, where a 1500-km-wide zone of intense tectonic deformation separates the Indian and Australian plates[9,14].

The Margins of the Ocean Basins

An ocean basin is created by the rifting and breaking apart of a continental lithospheric plate and is bounded by rifted or transform continental margins (*Figure 8.10*). Rifted margins extend from the edge of the continental shelf into oceanic depths; consequently, such margins are major bathymetric features on the Earth's surface. They contain evidence of the processes which accompanied the initiation of sea-floor spreading, as well as the poorly understood transition from continental to oceanic crust.

Rifted margins are more common; for example, they border the Atlantic Ocean from 50°S to the Norwegian Sea. They are of particular interest for economic reasons, since the same processes which lead to rifted margins can also form major sedimentary basins in which oil and gas may accumulate (see Chapter 21). Transform margins are often shorter and have been less widely studied.

Continental rifting, like most other major Earth processes, is driven by heat from within the Earth. The temperature of the asthenosphere largely determines the amount of melt produced during rifting; the temperature of the lithosphere determines when, and how much, uplift and/or subsidence is experienced by the margin. The rifting and breakup of continents is often thought to be a response to tensional forces resulting from the sub-lithos-

8.12

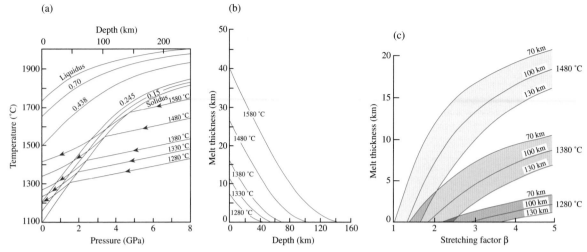

Figure 8.12 (a) Adiabatic decompression paths in temperature–pressure space (asthenospheric potential temperature is given on the right). Melting begins when material rising along an adiabatic decompression curve (arrowed lines) meets the solidus curve. The melt fraction is indicated on the curves between the solidus and the liquidus. (b) The higher the temperature of the adiabatic decompression curve, the greater the thickness of melt produced. The total thickness of melt present below a given depth, calculated by integrating the volume of melt implied by the curves in (a), is plotted as a function of depth. (c) Predicted thicknesses of melt produced as a result of different lithospheric stretching factors ß. The numbers against the curves give the thickness of the mechanical boundary layer and the temperatures on the right are the interior potential temperatures, i.e., the temperatures the rocks would have if brought to the surface without loss of heat (*Figures 7* and *22* in McKenzie and Bickle[24]; by permission of Oxford University Press).

pheric loading associated with regions of reduced density in the mantle. Numerical dynamic modelling of rifting suggests that a continental lithospheric plate subjected to prolonged tension begins to stretch and thin. At the top it behaves brittly, so the upper crust is dissected by normal faults; deeper, the deformation is viscous (*Figure 8.11*). The asthenosphere responds passively to this thinning in a very important way. Material rises to fill the space vacated by the lithosphere and consequently experiences lower pressure. Usually this happens adiabatically (i.e., heat is conserved), the temperature rises, and eventually the more volatile components of the rock begin to melt (*Figure 8.12*). The higher the initial temperature and the greater the ascent of the asthenosphere, the more magma is produced. The magma separates from the host rock and rises by its own buoyancy. Eventually, after several million years, following the surface faulting, magma may reach the surface. Even later, the continental lithosphere ruptures and the asthenosphere itself essentially reaches the surface. This is when

8.13

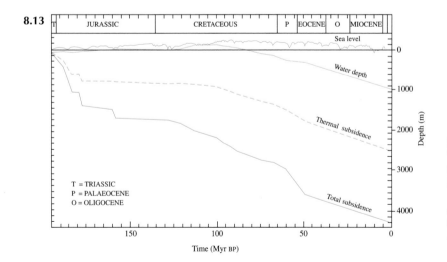

Figure 8.13 Subsidence history of Well 62/7-1 in the Goban Spur Basin, Southwestern Approaches to the English Channel. The total subsidence is the sum of the thermal subsidence and the isostatic subsidence due to sediment loading (*Figure 4* in Colin *et al.*[12]; reproduced by permission of © Butterworth Heinemann Ltd).

Figure 8.14 Models of gravity and surface and deep-towed magnetometer profiles across the West Iberia margin (see inset for location; OC = oceanic crust; TC = transitional crust; CC = continental crust). (a) Deep-tow and (b) surface magnetic profiles – the magnetic models (intensities in A/m, black blocks are normally magnetised oceanic crust) indicate that sea-floor spreading began at about Chron M3 (130 Myr ago) at 9 mm/yr; extrapolating the reversal time-scale further back in time does not yield computed anomalies that fit the observations (crosses). Hence, east of the peridotite ridge normally magnetised blocks with variable intensities of magnetisation were used, such as might be produced by continental crust. Note how the deep-towed magnetometer observations provide better resolution of the sea-floor spreading anomalies. (c) Gravity profile – the model used to fit the observed profile (crosses) is constrained by layer thicknesses and velocities calculated from seismic refraction lines L1 to L3 (aligned normal to the profile) and L4; densities are in Mg/m³; PR indicates a peridotite basement ridge drilled during Ocean Drilling Program Leg 149 and numbered arrows indicate other Leg 149 boreholes (work in progress, R.B. Whitmarsh and P.R. Miles).

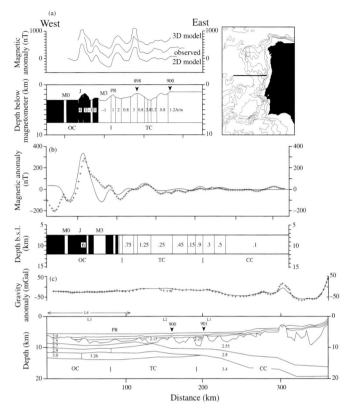

the steady-state process of magma production, and its intrusion and extrusion at the ocean floor, which we know as sea-floor spreading, begins. An ocean basin has been born.

As well as influencing horizontal movements, the above events also determine the vertical motion of the margin. If the asthenosphere is hot enough, then initially the lithosphere expands sufficiently to cause uplift, even well-above sea-level, of the continent adjacent to the rifting. However, in all cases following break-up the lithosphere cools, and therefore subsides, with a time constant of about 60 Myr. Erosion of the adjacent continents frequently leads to substantial offshore sediment accumulations, which load the margin; these can flex the margin and cause further isostatic subsidence. It is very difficult to measure accurately the subsidence history of a margin; palaeodepth estimates based on benthic microfossils from deep wells are often crude. However, given the uncertainties, numerical cooling and isostatic models can usually be fitted to such curves quite well (*Figure 8.13*).

The West Iberia margin is a good example of one which rifted over a cool asthenosphere, as indicated by the apparent absence of syn-rift volcanism (rifting associated with a relatively hot asthenosphere is accompanied by the production of large quantities of volcanic lavas). The crust of the mar-

gin has been studied using seismic refraction and reflection profiles, magnetic and gravity anomalies, heat flow measurements, and cores from scientific drilling[41] (*Figure 8.14*). The first-formed oceanic crust is abnormally thin (only about 3–4 km), indicating a poor magma supply, and is bounded on the landward side by a basement ridge of serpentinized peridotite (peridotite is an iron- and magnesium-rich silicate rock originating in the mantle; this rock can become altered by a process known as serpentinization, which occurs when peridotite comes into contact with water at low temperatures). Between this ridge and the thinned continental crust, typified by gently landward-tilted fault blocks and half graben, there is a recently discovered 130 km wide transitional zone. This zone has linear magnetic anomalies parallel to the isochrons (i.e., lines of constant crustal age) of sea-floor spreading, yet does not have the magnetic or seismic character of oceanic crust. It is probably either a mixture of continental crust and igneous intrusives–extrusives or unroofed mantle. A similar transitional zone may be exposed today on the Arabian margin of the central Red Sea.

The western edge of the Rockall Plateau in the northeast Atlantic is a good example of a margin which rifted over a hot asthenosphere. Here, the uppermost crust overlying the continent–ocean transition is marked by seaward-dipping reflectors

8.15

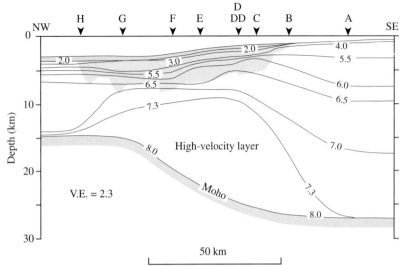

Figure 8.15 Seismic velocity model from NW (left) to SE (right) of the northwest margin of Rockall Plateau, northeast Atlantic Ocean. A–H and DD indicate locations of expanding-spread profiles aligned normal to the northwest–southeast section. VE is the vertical exaggeration. The diagonal shading indicates the estimated extent of the seaward-dipping reflectors and 'late stage volcanics'. Moho represents the Mohorovičic discontinuity, the depth at which seismic velocity exceeds 7.8 km/s and where the mantle begins (*Figure 4* in Morgan and Barton[25]; by permission of Elsevier Science).

(*Figure 8.15*). Drilling of similar reflectors off Norway has shown that they represent a thick sequence of dipping lava flows. Seismic refraction measurements show that the crust off the western edge of the Rockall Plateau is unusually thick, with up to 15 km of material with a velocity over 7.3 km/s at the base of the crust (*Figure 8.15*). Both the dipping volcanic reflectors and the thick crust can be explained by a high asthenospheric temperature and extensive melt production at the time of break-up. The lavas denote magma which reached the surface and the high-velocity lower crust signifies denser magma 'underplated' at the base of the crust. The precise reason for the prodigious melt production at this margin is debatable. It may have been due to the proximity of the Iceland plume or hot spot. An alternative hypothesis is that the melt was produced by vigorous local convection within the asthenosphere.

World-wide, few transform margins have been studied. Such margins are believed to develop very differently from rifted margins. Principally, they experience shear motion, perhaps with a minor component of extension or compression, along one or more deep crustal faults (faults which are very difficult to detect on reflection profiles). Every point on the margin also experiences a temporary rise in temperature, and consequent uplift, as the end of the adjacent spreading ridge traverses along the margin (*Figure 8.10*). *Figure 8.16* is a section across the southern Exmouth Plateau margin off northwest Australia. Here there is a relatively abrupt transition from 26 km thick underplated continental crust to 8 km thick oceanic crust. Between the two there is a 50 km wide zone of complicated structure which may include a zone of intrusion and, seaward, a tilted fault block.

The Structure of the Oceanic Crust

A vertical section through the Earth beneath the ocean basins generally consists of hundreds or thousands of metres of sediment overlying 7 km of igneous crust, formed by sea-floor spreading, which in turn overlies the upper part of the mantle. Our knowledge of the structure and composition of the igneous crust comes from direct sampling and from remote sensing by geophysical measurements.

Sampling by dredging, where rocks are exposed on the sea floor, reveals a wide variety of rock types from basalts and gabbros of the crust to ultramafics (iron- and magnesium-rich rocks) derived from the uppermost mantle. However, it is frequently difficult to relate even well-located dredged rocks to an unambiguous layering of the crust. Drilling is a better way to do this (e.g., see Chapter 19). The international Ocean Drilling Program (ODP) has succeeded in drilling hundreds of metres into the crust at several holes around the world. The deepest is the 2111 m deep Hole 504B in the eastern Pacific Ocean[2] (*Figure 8.17*). In spite of this unique achievement the hole has so far probably penetrated only basaltic pillow lavas and sheeted dikes of the uppermost crust. Down-hole physical measurements demonstrate that the crust is vertically variable on a scale of metres or tens of metres. Lateral variability is also expected from models of crustal creation, which include a varying magma supply and magmatic, as well as purely tectonic, extension. A simpler picture emerges if we use remote geophysical measurements to obtain a more averaged view of the crust.

The best way to investigate remotely the Earth beyond the reach of the drill is through seismic energy. Using the wide-angle refraction technique, where the seismic source (usually towed at the sea

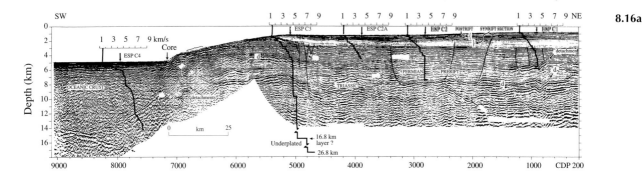

8.16a

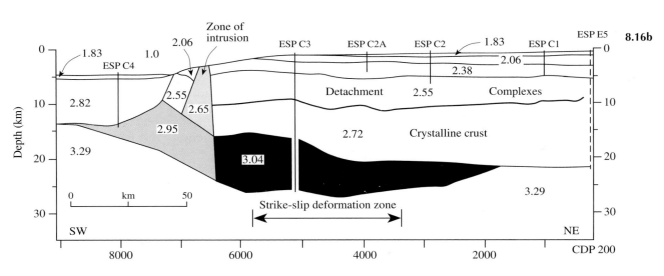

8.16b

Figure 8.16 (a) Multichannel seismic reflection profile after migration and depth conversion, from southwest (left) to northeast (right) across the Exmouth Plateau transform margin (vertical exaggeration = 3). Near vertical lines denote features interpreted as faults; other possible near-horizontal faults are labelled detachment or d, d_1, d_2. The positions of five intersecting expanding-spread profiles (ESPs), labelled C1, C2, etc., and the velocity–depth profiles computed from them, are also shown. (b) A two-dimensional gravity model (densities in Mg/m^3), based on the seismic observations, which incorporates transitional crust (grey) between common depth point (CDP) gathers 6500–8000 (see horizontal axes) and underplated crust (black) landward of this zone. The transitional and underplated crusts may have been intruded as the adjacent rift axis migrated along the transform margin (*Figures 2* and *3* in Lorenzo *et al.*[23]; by permission of the authors).

surface) and receivers (usually placed on the ocean floor) are many kilometres apart, it is possible to compute the increase of seismic velocity with depth and to measure the thickness of the crust (*Table 8.1*; see *Figure 19.16*). Although the crustal structure is often reported in terms of two or more distinct layers, in practice, velocity increases more or less steadily with depth. We know from sampling that the uppermost layers (at least 2.5 km/s) represent basaltic pillow lavas. Although solid basalt has a velocity of about 6.3 km/s, cracks, fissures, and widespread voids in the lavas lead to lower velocities. In time, many of these spaces are filled by hydrothermally deposited minerals; consequently, the upper crustal velocity is greater in older crust. The lava flows are underlain by dikes, subvertical

Table 8.1. Mean Oceanic Crustal Structure (from *Table 8* in White *et al.*[40])

	Velocity (km/s)	Thickness (km)
Layer 2	2.5–6.6	2.11±0.55
Layer 3	6.6–7.6	4.97±0.90
Mantle	>7.6	
Total igneous crust		7.08±0.78

8.17

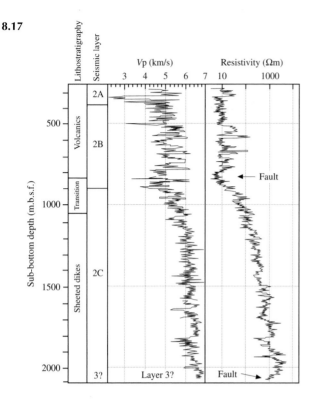

Figure 8.17 Down-hole logs from Ocean Drilling Program Site 504B (as at the end of Leg 148, February, 1993). Note the broad changes in velocity (V_p) and resistivity between the volcanic and sheeted-dike layers, and the short wavelength (ca. 1–10 m) variations in these properties (profiles begin at 250 mbsf; mbsf = metres below sea floor; *Figure 2* in ODP Leg 148 Shipboard Scientific Party[28]; © American Geophysical Union).

density of rocks is strongly correlated with their seismic velocity (*Figure 8.18*); this useful property means that we can infer density from velocity and check our seismic models by computing their gravitational effect and comparing this with independent gravity observations (e.g., *Figure 8.14*).

Another important property of the igneous crust is its magnetisation. When magma, containing a few percent of certain iron oxides, cools it acquires a remanent magnetisation, in the direction of the contemporary Earth's field, which is stable over millions of years (see above). This magnetisation provides the 'memory' in the rocks, whereby sea-floor spreading records reversals of the Earth's magnetic field. The remanent magnetisation slowly decreases with time as some iron oxides undergo further oxidation, a process accelerated by hydrothermal circulation. Older basalts acquire a stronger, secondary, possibly chemical, magnetisation (*Figure 8.19*). Hence a minimum magnetisation occurs in rocks which are 8–20 Myr old. The magnetic susceptibility of the crustal rocks is usually relatively insignificant, so that it does not have to be included when

sheets that intrude parallel to the ridge axis during sea-floor spreading. Below the dikes there is a layer of 6.7–7.2 km/s material which forms the greater part of the crust. This oceanic layer, or Layer 3, has rarely been drilled *in situ*. From samples, and by analogy with ophiolites (sequences of mainly igneous rocks which contain the same rock types, and in the same order, as are found within the oceanic crust) exposed on land, most geologists think it largely consists of gabbro, a coarser grained rock representing the frozen melt which was the source of the basaltic flows and dikes. At the base of the crust, velocity increases, often abruptly, to around 8.0 km/s, comparable to the Mohorovičic discontinuity at the base of the continental crust; this marks the top of the Earth's mantle. The velocity–depth structure in *Table 8.1* is an average for normal crust; in fracture zones the crust is often thinner and may be underlain by velocities thought to represent serpentinized peridotite; and near hot spots, such as Iceland, it is thicker. The

Figure 8.18 Wet-bulk density of samples of oceanic crust and ophiolites plotted against the inverse of compressional-wave velocity. The solid line represents a least squares fit, with the standard error indicated by the dashed lines (*Figure 1* in Carlson and Raskin[10]; reprinted with permission from *Nature*, © 1984 Macmillan Magazines Limited).

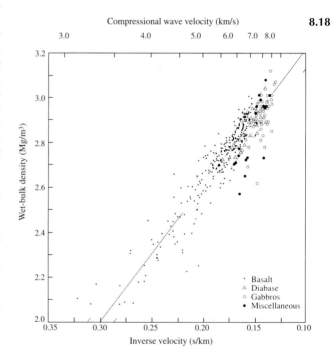

8.18

Figure 8.19 Summaries of crustal magnetisation against crustal age. (a) Obtained from the inversion of marine magnetic anomalies. (b) Derived from measurements of normal remanent magnetisation made on drilled basalt samples. Both curves show the decrease in magnetisation caused by the oxidation of magnetic minerals in the first 20 Myr and the subsequent increase indicating the acquisition of a remanent (chemical?) magnetisation (*Figure 4* in Sayanagi and Tamaki[33]; © the American Geophysical Union).

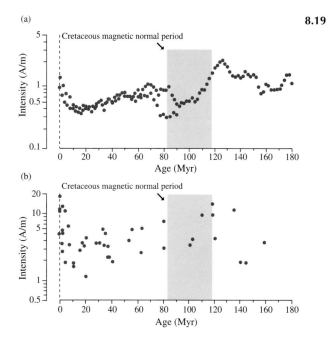

computing the magnetic effect of these rocks.

Other important physical properties of crustal rocks include electrical resistivity, shear wave velocity, permeability, and thermal conductivity; many are anisotropic.

Recently, our perception of the igneous crust has been improved by seismic-reflection profiling[26]. Using the same equipment and ships as in the search for oil and gas, but with specially designed configurations, intriguing reflecting surfaces have been detected below the Atlantic Ocean, deep within the crust (*Figure 8.20*). On profiles acquired along isochrons, these surfaces tend to have low

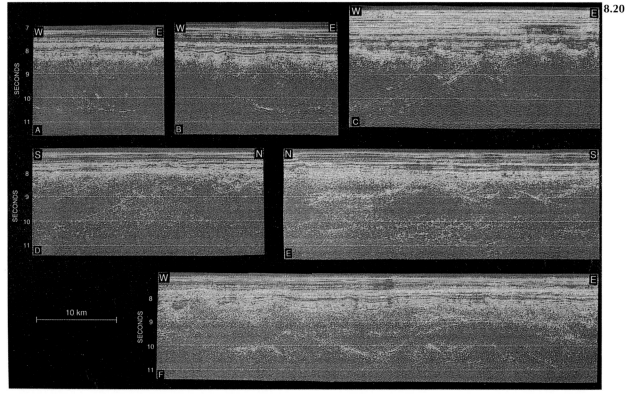

Figure 8.20 Multichannel seismic-reflection profiles of oceanic crust produced at slow-spreading rates in the North Atlantic Ocean, illustrating a variety of forms of reflectivity (the vertical scale in seconds is the time required for sound to be reflected back to the sea surface). A, B, D, E, and F are from the western North Atlantic; C is from the eastern North Atlantic. A and B are from flow-line profiles, C is oblique to the spreading direction, D and E are isochron lines, and F is along the trough of the small-offset Blake Spur fracture zone. Typically, the shallow crust contains distinct sub-horizontal reflections and the middle crust is almost reflection-free. The lower crust exhibits the strongest and most diverse reflectivity, including banded patterns of dipping reflectors. The dipping reflectors may have a tectonic or igneous origin. A distinct Moho reflection is seldom seen; the reflective lower crust typically merges downward to a reflection-free upper mantle (*Figure 4* in Mutter and Karson[26]; © 1992 by the AAAS).

8.21

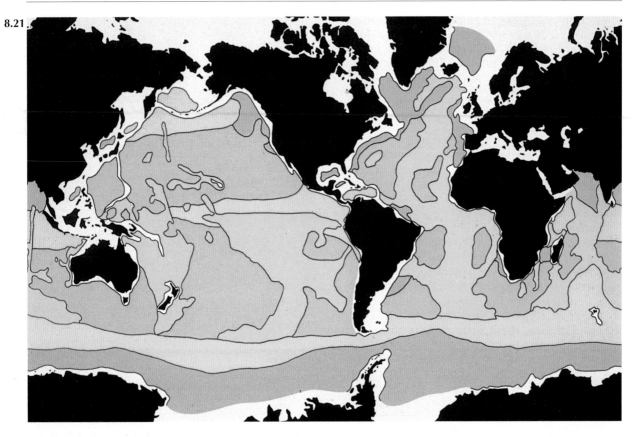

Figure 8.21 Chart of the global distribution of sediment in the ocean (green, calcareous sediments; yellow, siliceous sediments; brown, terrigenous sediments; blue, glaciogenic sediments; pink, deep-sea clay; white, margin sediments; drawn by R.G. Rothwell).

dips, be highly reflective throughout the igneous crust, and dip bi-directionally; on profiles parallel to the spreading direction they are steeper, usually dip toward the spreading centre, and offset the basement surface. The former may represent either contrasts developed during the igneous creation of the crust or faults; the latter are probably ridge-parallel normal faults, mostly active in the early development of the crust.

Sediment Provenance and Transport Processes

Sediments cover most of the ocean floor. Our knowledge of these sediments, and the Earth history they record, has increased markedly in the past three decades, through gravity and piston coring (see *Figure 19.4*), and deep-sea drilling. The sediments comprise, in varying amounts, detrital material derived from the weathering of the continents, biogenic debris derived from planktonic organisms, and clay-size material. Sediment types are distinguished by particular constituents; different sediments show well-defined global distributions (*Figure 8.21*; see also *Figure 11.8*). Sedimentary

material is transported to the ocean floor by a number of mechanisms and processes (*Figure 8.22*; also see Chapter 9).

Rivers form the main pathway of terrigenous sediment to the oceans, although wind transport is particularly important for fine-grained detrital material. Glacier input is important at high latitudes. The main factors controlling the flux of sediment derived from continental erosion are climate, precipitation, type of weathering, character of the coarse-grained material, topography, and land area in the source regions. When sea-level was low, such as during glacial periods, deep-sea terrigenous sedimentation was especially dynamic. At such times, the mechanical erosion of continents and the sediment loads of rivers were much greater. Much of the continental shelves were exposed as coastal plains, resulting in rivers that transported their loads to the outer edge of the continental shelf for more rapid deposition into the deep-sea basins.

Pelagic sediments are typically dominated by biogenic material, but vary considerably with latitude and water depth[20]. They include the deep-sea calcareous and siliceous oozes, composed largely

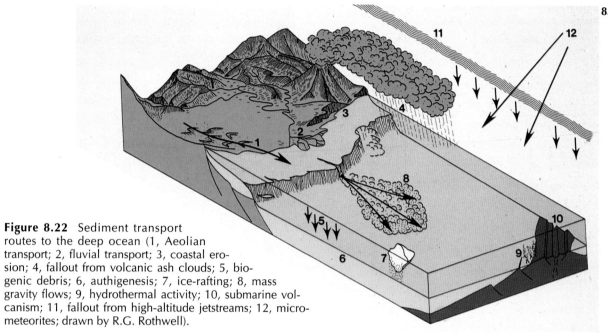

8.22

Figure 8.22 Sediment transport routes to the deep ocean (1, Aeolian transport; 2, fluvial transport; 3, coastal erosion; 4, fallout from volcanic ash clouds; 5, biogenic debris; 6, authigenesis; 7, ice-rafting; 8, mass gravity flows; 9, hydrothermal activity; 10, submarine volcanism; 11, fallout from high-altitude jetstreams; 12, micrometeorites; drawn by R.G. Rothwell).

of the remains of planktonic organisms, and deep-water clays (*Figure 8.23*). Four main processes control the character of biogenic oozes: the supply of biogenic material, its dissolution in the water column, its dilution by nonbiogenic material, and subsequent diagenetic alteration (see later). Pelagic sedimentation can be viewed as a form of interaction between the near-surface ocean and the deep ocean. Locally, these two distinct environments interact through wind-driven upwelling of deep water, and the consequent downwelling of near-surface water, and through the constant 'rain' of skeletons from dead planktonic organisms (*Figure 8.23*). This 'rain' of biogenic particulate matter forms the primary sink in the ocean basins (Chapter 7).

The spatial distribution of calcareous oozes is controlled by depth due to dissolution (Chapter 11). Calcite, which forms the main skeletal material of many planktonic organisms (such as foraminifera and coccolithophores), shows increasing solubility with water depth; this is related to increased hydrostatic pressure, increasing CO_2 content within the water, and decreasing temperature. Therefore, there is a depth, called the lysocline, that separates well-preserved from poorly preserved, solution-etched foraminifera and coccolithophores

8.23

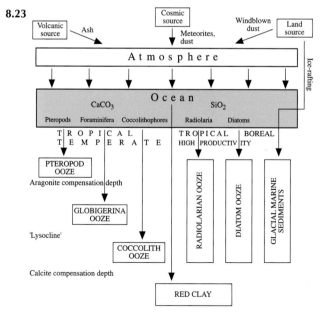

Figure 8.23 Sources and pathways of pelagic sedimentation in the oceans (after Hay[17]).

8.24

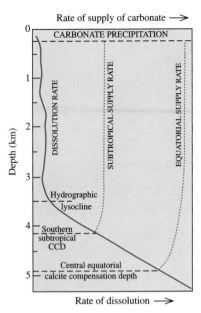

Rate of supply of carbonate →

CARBONATE PRECIPITATION

Depth (km)

DISSOLUTION RATE

SUBTROPICAL SUPPLY RATE

EQUATORIAL SUPPLY RATE

Hydrographic
lysocline

Southern
subtropical
CCD

Central equatorial
calcite compensation depth

Rate of dissolution →

Figure 8.24 Conceptual model for calcite dissolution rate in the ocean, showing the relationship between the calcite compensation depth (CCD) and the lysocline. Increased carbonate supply at the equator depresses the CCD (*Figure 11* from Berger *et al.*[3]; © American Geophysical Union).

Terrigenous sediments are composed largely of detrital material derived from the weathering of continents. They include turbidite muds, which cover the abyssal plains, and glacial material, which covers substantial parts of the sea floor in polar regions. Such sediments are characterised by high rates of deposition, usually contain small quantities of biogenic material, and are usually transported to the deep sea by some form of sediment gravity flow, such as debris flows or turbidity currents[31,36] (*Figure 8.25* and Chapter 9).

Sediment Diagenesis

The constituents which comprise a newly deposited marine sediment include relatively unreactive detrital materials, introduced to the ocean from the continents by winds and rivers (and by ice at high latitudes), and more reactive biogenic materials, such as $CaCO_3$, opal (an amorphous form of silica which contains water) and organic matter supplied by ocean-surface productivity. The concentration of a particular component subsequently observed in

(*Figure 8.24*). The lysocline varies, but generally lies at depths of 3000–5000 m. At some greater depth, called the calcite compensation depth (CCD), the rate of supply of biogenic calcite equals its rate of dissolution. Below the CCD, only carbonate-free sediments accumulate. Regionally, the depth of the CCD is a function of a number of variables which reflect oceanic productivity patterns and the shoaling of the lysocline near continental margins. The CCD varies between 3500 and 5500 m in the Atlantic and Pacific Oceans, but has a mean depth of around 4500 m.

Siliceous oozes are the typical pelagic sediments found beneath regions of high productivity (the equatorial and polar belts and areas of coastal upwelling), especially where the sea floor is deeper than the CCD. The dissolution of opaline skeletons within the surface sediments releases silica to deep waters; its dissolution within the sediments forms silica-rich interstitial solutions which migrate along bedding planes and fractures to precipitate in nearby permeable lenses and layers, as deep-sea cherts. Such deposits seem to occur more frequently at particular times in the geological record (e.g., in the late Eocene), which may reflect changes in the global supply of silica to the ocean or changes in oceanic productivity.

Pelagic clays are generally found only in deep areas far from land. They generally contain less than 10% biogenic material and are mainly composed of clay minerals and fine-grained quartz, the bulk of which have been derived from aeolian fallout. They are commonly reddish brown to chocolate brown and have accumulated slowly, at generally less than 1 mm per 1000 years, compared to the 10–30 mm per 1000 years typical of calcareous and siliceous oozes.

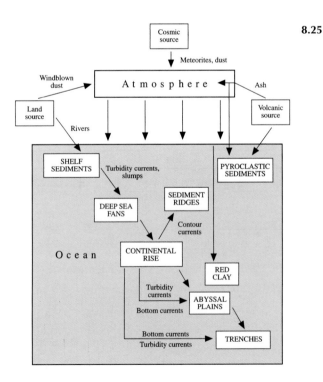

8.25

Figure 8.25 Processes of clastic sedimentation in the oceans (after Hay[17]).

Figure 8.26 Schematic representation of trends of pore-water concentration profiles of oxidants (O_2 and NO_3^-) and oxidation–reduction products (Mn^{2+} and Fe^{2+}) against sub-bottom depth in deep-sea sediments as a result of bacterially mediated early diagenesis. Oxygen and nitrate from sea water are consumed during oxidation of organic matter [formally $(CH_2O)_{106}(NH_3)_{16}(H_3PO_4)$] according to reactions (a) and (b) below. Mn^{2+} and Fe^{2+} do not co-exist with high pore-water oxygen concentrations, but are present in pore waters as a consequence of similar reactions of organic-matter oxidation deeper in the sediments, with Mn and Fe oxyhydroxides initially present in the solid phase. These reactions may take place within a few centimetres or over several metres in deep-sea sediments, dependent mainly on the relative fluxes of organic matter and oxygen (reprinted from Froelich *et al.*[15]; © 1979 Elsevier Science Ltd, with kind permission).

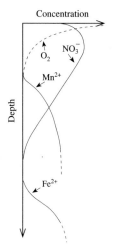

8.26

$$(CH_2O)_{106}(NH_3)_{16}(H_3PO_4) + 138O_2 = 106CO_2 + 16HNO_3 + H_3PO_4 + 122H_2O \qquad (a)$$

$$(CH_2O)_{106}(NH_3)_{16}(H_3PO_4) + 94.4HNO_3 = 106CO_2 + 55.2N_2 + H_3PO_4 + 177.2H_2O \qquad (b)$$

a sediment of a particular age, however, depends not only on the original relative rates of supply of all the sedimentary constituents, but also on the preservation of each constituent after burial. The physical, chemical, and biological processes responsible for converting an original, water-rich, unconsolidated sediment into solid rock are collectively termed *diagenesis*.

Deep-sea sediments contain about 0.2–0.5% organic carbon, while shelf sediments contain up to 5%. Although organic matter is therefore a minor component, its importance is out of all proportion to its abundance. As organic matter is the only reductant supplied to sediments in any quantity, its microbially mediated degradation drives early diagenetic reactions, which progressively consume the oxidants (electron acceptors) available in sediment pore-waters (e.g., oxygen, nitrate, and sulphate) and as coatings on sediment grains (e.g., manganese and iron oxyhydroxides). This gives rise to a vertical succession of geochemical environments or zones in the sediments[15] (*Figure 8.26*) as the electron acceptors are consumed in the order of decreasing thermodynamic advantage (i.e., in the order of decreasing energy produced by each mole of carbon that is oxidised). One classification of these successive early diagenetic environments is in terms of the concentrations of oxygen and total dissolved sulphide of the pore water; it recognises *oxic, post-oxic, sulphidic,* and *methanic* environments with increasing depth. Other redox-sensitive elements in the sediments also respond to the changes in geochemical conditions experienced within these different environments. In particular, the abundant metals iron and manganese form a variety of authigenic minerals (i.e., new minerals formed from constituents pre-existing within the sediment), which can also impart characteristic colours to the sediments[5].

As sediments are progressively buried, they experience both physical changes, which result in a reduction of water content and porosity with burial depth and time, and concurrent chemical changes, which result in cementation and chemical compaction. It is these diagenetic processes of lithification which convert carbonate sediments into limestones, siliceous oozes into cherts, clays into claystones, and so on. Organic-rich marine sediments which have undergone these later diagenetic modifications are also the potential source rocks for most of the world's oil reserves (see Chapter 21). What conditions are required for the initial formation of organic-rich sediments remains controversial. The traditional interpretation that black (i.e., organic-rich) shales develop under anoxic water columns[13] has been challenged by the alternative contention that high primary productivity provides the first-order control[30]. Regardless of the relative importance of preservation and productivity, however, organic matter undergoes progressive complex diagenesis as it is buried and experiences increases in pressure and temperature over a prolonged period (*Figure 8.27*). This diagenesis causes a loss of hydrogen and oxygen relative to carbon in the residual sediment, compared with the organic matter originally deposited.

The low temperature, bacterially mediated, early diagenesis discussed above is referred to as *eogenesis*, while the later reactions deeper in the sedimentary column are referred to as *catagenesis* or *metagenesis*. Catagenesis takes place at moderate temperatures (50–150°C) and pressures (30–150 MPa), and includes the 'oil window', which generates liquid hydrocarbons of medium-to-low molecular weight. Metagenesis occurs at higher temperatures and pressures, and generates methane in the 'dry gas' zone[1,7].

The Oceans in the Past

Deep-sea drilling has shown that conditions in ancient oceans were sometimes different from those of today[16,18,22]. Mid-Cretaceous sediments, cored by

8.27

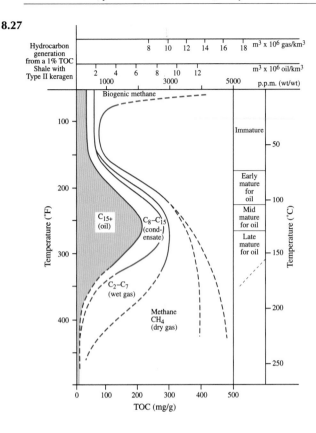

Figure 8.27 Calculating the volume of hydrocarbons generated from a given source-rock unit. The yield of hydrocarbons generated per 1% total organic carbon (TOC) in the source rock is indicated on the horizontal axes in p.p.m. (wt/wt) and other units. As shown, C_8–C_{15} and C_{15+} hydrocarbons (the major components of a typical North Sea oil) are generated in large quantities from 80–130°C, while over this temperature range the light hydrocarbons (C_2–C_7 and methane) are present in relatively small quantities. Once heavy hydrocarbon generation has ceased, at about 130°C, a presumed cracking reaction takes over, increasing the yield of the C_2–C_7 fraction and CH_4 at the expense of heavier hydrocarbons (from Brooks et al.[8], reproduced with permission).

the Deep Sea Drilling Project and the Ocean Drilling Program from the Atlantic and some parts of the Pacific, commonly contain organic-rich black shales, testifying to periods of probably quite brief, but widespread, anoxia. The black shales may have been caused by the lack of a regular supply of cold, dense, well-oxygenated water to the deep oceans, due to the absence of ice-caps, and by abundant biological production encouraged by the warm conditions and extensive continental shelves of the time. Recently, it has been suggested that widespread volcanism during the Cretaceous may have played a role in causing contemporary deep-

Box 8.3   Oxygen Isotope Stratigraphy

Oxygen has three stable isotopes (^{16}O, ^{17}O, and ^{18}O) with atomic mass numbers of 16, 17, and 18:

- ^{16}O makes up 99.763% of natural oxygen.
- ^{17}O makes up 0.033% of natural oxygen.
- ^{18}O makes up 0.204% of natural oxygen.

Oxygen makes up 90% of water by weight; the ^{16}O isotope is lighter than the ^{18}O isotope. Therefore, ^{16}O is preferentially evaporated relative to ^{18}O.

During glacial periods, ^{16}O-enriched water vapour is precipitated as snow which builds up to form glacier ice and ice caps. This ice is relatively depleted in ^{18}O. The oceans, however, become relatively enriched in ^{18}O, because of evaporation of ^{16}O-enriched water vapour. The larger the ice caps, the larger the proportion of ^{16}O removed from seawater, and the more the $^{18}O:^{16}O$ ratio of the sea water increases.

Marine organisms, such as foraminifera, which form skeletons or tests of calcium carbonate, incorporate different proportions of ^{16}O and ^{18}O from the water, according to the temperature; but, more importantly, according to the background ratio of $^{18}O:^{16}O$ in the sea water, which reflects global ice volumes. Measurements of the small differences in the $^{18}O:^{16}O$ ratio in different samples using a mass spectrometer allow the sequence and age of warm and cold conditions to be determined.

The $^{18}O:^{16}O$ ratio of foraminifera, especially benthic species which live in low-temperature bottom water (and hence are not affected by temperature changes), can therefore be taken as a measure of the amount of water held in ice sheets at any given time, and hence also as an indicator of global sea level.

Figure 8.28 Combined plot of global production of oceanic crust, high latitude sea-surface palaeotemperatures, long-term eustatic sea-level, black-shale deposition, and rate of production of the world's oil resources against geological time (from Late Jurassic to Pleistocene; *Figure 1* from Larson[21], by permission of the author).

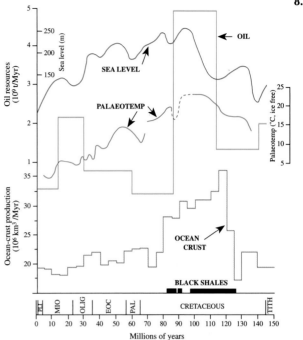

sea anoxia. The resulting increase in outgassing of mantle volatiles and CO_2 and increased ocean crust production (*Figure 8.28*) at this time may have resulted in an enlarged supply of nutrients and carbon to the ocean. The 'greenhouse' effect of increased CO_2 in the atmosphere led to relatively high sea-surface temperatures. The resultant explosion in productivity led to vast increases of organic carbon in the marine system.

During the Cenozoic, the CCD fluctuated widely (*Figure 8.29*), which may have been related partly to changes in sea level. Continental shelves, being shallow, are favourable places for carbonate accumulation in times of high sea-level and may, therefore, act as carbonate traps, thereby removing $CaCO_3$ from the oceanic chemical cycle. However, changes in productivity also lead to changes in the CCD, since biogenic production of $CaCO_3$ skeletons lowers saturation. Therefore, fluctuations in the CCD may possibly reflect productivity fluctuations too.

Oxygen isotope studies (*Box 8.3*) of the skeletons of benthic foraminifera have shown a general cooling trend in the oceans since the Cretaceous. Palaeoceanographic changes, caused by plate motions over the same period, led to greater partitioning of the world ocean system with time and played a major role in causing this trend. Palaeoceanographic studies show that polar cooling

was particularly pronounced toward the end of the Eocene (*Figure 8.28*), presumably due to the thermal isolation of the Arctic and Antarctic oceans from the rest of the world ocean, but also perhaps due to albedo changes resulting from changing vegetation and snow cover. This led to the equatorward shifting of climatic belts and, on high latitude shelves, to the cooling of water, which became cold and dense enough to sink and fill the deep ocean basins. The late Eocene cooling therefore resulted in a new type of world ocean – one characterised by the development of marked contrasts between high and low latitudes, between different oceans, and between the deep sea and the ocean margins. Large amounts of ice-rafted debris in sediments around Antarctica since the middle Miocene testify to the build up of the Antarctic ice cap. The microfossil record, particularly of siliceous types, indicates that fundamental changes in deep-water circulation were occurring in the Miocene, concurrently with the build up of Antarctic ice and the world-wide cooling of abyssal waters. Subsequent changes in palaeogeography, particularly the northward drift of land masses and possibly the closing of the Panama seaway (about the middle Pliocene), reinforced by other mechanisms, such as mountain building, resulted in the onset of northern hemisphere glaciation 2.5–3.5 Myr ago and in the oceans we know today.

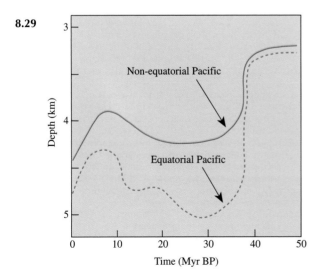

Figure 8.29 Reconstructions of past fluctuations in calcite compensation depth in the Pacific Ocean (reconstruction from Van Andel[37]; after Seibold and Berger[34]).

General References

Barnes, M.A., Barnes, W.C., and Bustin, R.M. (1984), Diagenesis 8: Chemistry and evolution of organic matter, *Geosci. Canada*, **11**, 103–114.

Brooks, J. and Fleet, A.J. (eds) (1987), *Marine Petroleum Source Rocks*, Blackwell Scientific Publications, Oxford, for The Geological Society.

DeMets, C., Gordon, R.G., Argus, D.F., and Stein, S. (1990), Current plate motions, *Geophys. J.*, **10**, 425–478.

Kennett, J.P. (1982), *Marine Geology*, Prentice-Hall, Eaglewood Cliffs, NJ, 813 pp.

McKenzie, D., (1978), Some remarks on the development of sedimentary basins, *Earth Planet. Sci. Lett.*, **40**, 25–32.

Mutter, J.C., Buck, W.R., and Zehnder, C.M. (1988), Convective partial melting: I. A model for the formation of thick basaltic sequences during the initiation of spreading, *J. Geophys. Res.*, **93**, 1031–1048.

Open University Course Team (1989), *Ocean Chemistry and Deep Sea Sediments*, Pergamon Press, Oxford, 134 pp.

Parsons, B. and Sclater, J.G. (1977), An analysis of the variation of ocean floor bathymetry and heat flow with age, *J. Geophys. Res.*, **82**, 803–827.

Seibold, E. and Berger, W. (1993) *The Sea Floor, An Introduction to Marine Geology*, Springer Verlag, Berlin, 356 pp.

Vine, F.J. and Matthews, D.H. (1963), Magnetic anomalies over oceanic ridges, *Nature*, **199**, 947–949.

White, R.S. and McKenzie, D.P. (1989), Magmatism at rift zones: the generation of volcanic continental margins and flood basalts, *J. Geophys. Res.*, **94**, 7685–7730.

White, R.S., McKenzie, D.P., and O'Nions, R.K. (1992), Oceanic crustal thickness from seismic measurements and rare earth element inversions, *J. Geophys. Res.*, **97**, 19683–19715.

References

1. Barnes, M.A., Barnes, W.C., and Bustin, R.M. (1984), Diagenesis 8: Chemistry and evolution of organic matter, *Geosci. Canada*, **11**, 103–114.

2. Becker, K. *et al.* (1989), Drilling deep into young oceanic crust at Hole 504B, Costa Rica Rift, *Rev. Geophys.*, **27**, 79–102.

3. Berger, W.H., Adelseck, C.G., and Mayer, L.A. (1976), Distribution of carbonate in surface sediments of the Pacific Ocean, *J. Geophys. Res.*, **81**, 2617–2627.

4. Bergman, E.A. (1986), Intraplate earthquakes and the state of stress in the oceanic lithosphere, *Tectonophys.*, **132**, 1–35.

5. Berner, R.A. (1981), A new geochemical classification of sedimentary environments, *J. Sediment. Petrol.*, **51**, 359–365.

6. Bott, M.H.P. (1982), *The Interior of the Earth*, 2nd edn, Edward Arnold (Publishers) Ltd, London.

7. Brooks, J. and Fleet, A.J. (eds) (1987), *Marine Petroleum Source Rocks*, Geological Society Special Publication No.26, Blackwell Scientific Publications, Oxford, for The Geological Society.

8. Brooks, J. *et al.*, (1987), The role of hydrocarbon source rocks in petroleum exploration, in *Marine Petroleum Source Rocks*, Brooks, J. and Fleet, A.J. (eds), Geological Society Special Publication No.26, Blackwell Scientific Publications, Oxford, for The Geological Society, pp 17–46.

9. Bull, J.M. and Scrutton, R.A. (1990), Fault reactivation in the Central Indian Ocean Basin, and the rheology of the oceanic lithosphere, *Nature*, **344**, 855–858.

10. Carlson, R.L. and Raskin, G.S. (1984), Density of the ocean crust, *Nature*, **311**, 555–558.

11. Chase, C.G. (1978), Plate kinematics: The Americas, East Africa, and the rest of the world, *Earth Planet. Sci. Lett.*, **37**, 355–368.

12. Colin, J.P., Ionnides, N.S., and Vining, B. (1992), Mesozoic stratigraphy of the Goban Spur, south-west Ireland, *Marine Petrol. Geol.*, **9**, 527–541.

13. Demaison, G.J. and Moore, G.T. (1980), Anoxic environments and oil source bed genesis, *American Association of Petroleum Geologists Bulletin*, **64**, 1179–1180.

14. DeMets, C., Gordon, R.G., Argus, D.F., and Stein, S. (1990), Current plate motions, *Geophys. J.*, **10**, 425–478.

15. Froelich, P.N., Klinkhammer, G.P., Bender, M.L., Luedtke, N.A., Heath, G.R., Cullen, D., Dauphin, P., Hammond, D., and Hartman, B. (1979), Early oxidation of organic matter in pelagic sediments of the eastern equatorial Atlantic: suboxic diagenesis, *Geochim. Cosmochim. Acta*, **43**, 1075–1090.

16. Haq, B.U. (1984), Palaeoceanography: A synoptic overview of 200 million years of ocean history, in *Marine Geology and Oceanography of Arabian Sea and Coastal Pakistan*, Haq, B.U. and Milliman, J.D. (eds), Van Norstrand Reinhold Co., New York, pp 201–231.

17. Hay, W.W. (1974), *Studies in Paleooceanography*, Society of Economic Palaeontologists and Mineralogists, Tulsa, Oklahoma, p. 2.

18. Hsu, K.J. (1986), *Mesozoic and Cenozoic Oceans*, American Geophysical Union, Geological Society of America (Geodynamics Series Volume 15), 153 pp.

19. Isacks, B., Oliver, J, and Sykes, L.R. (1968) Seismology and the new global tectonics, *J. Geophys. Res.*, **73**, 5855–5899.

20. Jenkyns, H.C. (1986), Pelagic environments, in *Sedimentary Environments and Facies*, 2nd edn, Reading, H.G. (ed.), Blackwell Scientific Publications, Oxford, pp 343–397.

21. Larson, R.L. (1991), Geological consequences of superplumes, *Geology*, **19**, 963–966.

22. Leggett, J.K. (1985), Deep-sea pelagic sediments and palaeo-oceanography: a review of recent progress, in *Sedimentology: Recent Developments and Applied Aspects*, Brenchley, P.J. and Williams, B.P.J. (eds), Geological Society Special Publication No. 18, Blackwell Scientific Publications, Oxford, pp 95–121.

23. Lorenzo, J.M. *et al.* (1991), Development of the continent–ocean transform boundary of the southern Exmouth Plateau, *Geology*, **19**, 843–846.

24. McKenzie, D.P. and Bickle, M.J. (1988), The volume and composition of melt generated by extension of the lithosphere, *J. Petrol.*, **29**, 625–679.

25. Morgan, J.V. and Barton, P.J. (1990), A geophysical study of the Hatton Bank volcanic margin: a summary of the results from a combined seismic, gravity and magnetic experiment, *Tectonophys.*, **173**, 517–526.

26. Mutter, J.C. and Karson, J.A. (1992), Structural processes at slow-spreading ridges, *Science*, **257**, 627–634.

27. Nance, R.R., Worsley, T.R., and Moody, J.B. (1988), The supercontinent cycle, in *Shaping the Earth Tectonics of Continents and Oceans*, Moores, E.D. (ed.), Readings from Scientific American, W.H. Freeman and Co., New York, pp 177–188.

28. ODP Leg 148 Shipboard Scientific Party (1993), ODP Leg 148 barely misses deepest layer, *EOS Trans. Am. Geophys. Un.*, **74**(43), 489–494.

29. Parsons, B. and Sclater, J.G. (1977), An analysis of the variation of ocean floor bathymetry and heat flow with age, *J. Geophys. Res.*, **82**, 803–827.

30. Pedersen, T.F. and Calvert, S.E. (1990), Anoxia versus productivity: What controls the formation of organic-carbon-rich sediments and sedimentary rocks?, *Amer. Assoc. Petrol. Geol. Bull.*, **74**, 454–466.

31. Pickering, K.T., Hiscott, R.N., and Hein, F.J. (1989), *Deep Marine Environments, Clastic Sedimentation and Tectonics*, Unwin Hyman, London, 416 pp.

32. Sandwell, D.T. and Smith, W.H.F. (1994), New global marine gravity map/grid based on stacked ERS-1, Geosat and Topex Altimetry, *EOS Trans. Am. Geophys. Un.*, **75**(16), 321.

33. Sayanagi, K. and Tamaki, K. (1992), Long-term variations in magnetization intensity with crustal age in the northeast Pacific, Atlantic and southeast Indian Oceans, *Geophys. Res. Lett.*, **19**, 2369–2372.

34. Seibold, E. and Berger, W. (1993) *The Sea Floor, an introduction to marine geology*, Springer Verlag, Berlin, 356 pp.

35. Stein, C.A. and Stein, S. (1992), A model for the global variation in oceanic depth and heat flow with lithospheric age, *Nature*, **359**, 123–129.

36. Stow, D.A.V. (1986), Deep clastic seas, in *Sedimentary Environments and Facies*, 2nd edn, Reading, H.G. (ed.), Blackwell Scientific Publications, Oxford, pp 399–444.

37. Van Andel, Tj.H., Thiecke, J., Sclater, J.G., and Hay, W.W. (1977), Depositional history of the South Atlantic Ocean during the last 125 million years, *J. Geol.*, **85**, 651–698.

38. Vine, F.J. and Matthews, D.H. (1963), Magnetic anomalies over oceanic ridges, *Nature*, **199**, 947–949.

39. Watts, A.B., Bodine, J.H., and Steckler, M.S. (1980), Observations of flexure and the state of stress in the oceanic lithosphere *J. Geophys. Res.*, **8**, 6369–6376.

40. White, R.S., McKenzie, D.P., and O'Nions, R.K. (1992), Oceanic crustal thickness from seismic measurements and rare earth element inversions, *J. Geophys. Res.*, **97**, 19683–19715.

41. Whitmarsh, R.B., Pinheiro, L.M., Miles, P.R., Recq, M., and Sibuet, J.-C. (1993), Thin crust at the western Iberia ocean–continent transition and ophiolites, *Tectonics*, **12**, 1230–1239.

CHAPTER 9:

Slides, Debris Flows, and Turbidity Currents

D.G. Masson, N.H. Kenyon, and P.P.E. Weaver

Introduction

Gravity-driven flows, in a variety of forms ranging from turbulent suspensions to coherent sliding masses, are the major agents of downslope sediment transport in the deep sea. They sculpt the continental slopes into complex shapes, carry land-derived sediment into the deep ocean basins, and redistribute biogenic sediment on a vast scale. Slope failures and resultant flows are often near-instantaneous events, capable of the destruction of marine installations and submarine telecommunications cables and, in some extreme cases, of generating deadly tsunamis. In ancient rocks, sand bodies once deposited by gravity flows, such as the sands found in submarine sediment fans, are a major reservoir facies for oil and gas, and have considerable economic importance.

The study of gravity flows in the deep ocean has progressed rapidly since the early 1950s. Perhaps the best-known study is the analysis of cable breaks caused by the turbidity current (a sediment-laden flow) associated with the 1929 Grand Banks earthquake, from which the first velocity estimate for a turbidity current was produced[5] (*Figure 9.1*); this is discussed later. Understanding of downslope sediment transport processes has developed, not only through studies in the modern ocean, but also through studies of ancient marine sequences now exposed on land, and through experimental work in the laboratory. Important contributions include the concept of sequential deposition of fining-upward sediment in individual turbidites (the sedimentary layers or deposits laid down by turbidity currents[1]), and the comprehensive theoretical analysis of turbidity current flow and turbidite deposition[12–14]. The development of seismic profiling and side-scan sonar equipment over the past 40 years (see Chapter 20) has revolutionised the way in which we analyse the sea floor, leading to a new appreciation of the extent and importance of gravity flows and their deposits. The discovery of huge sediment slides on the flanks of the Hawaiian

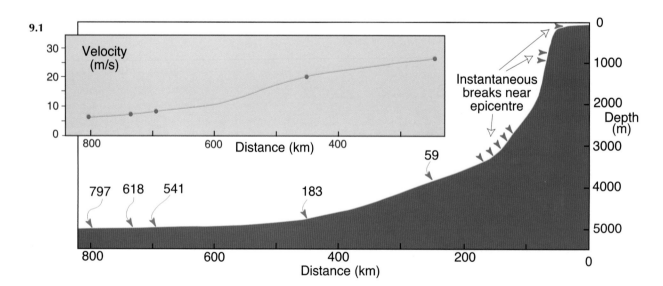

Figure 9.1 Cross-section through the continental slope and rise south of Newfoundland in the area affected by the 1929 Grand Banks earthquake and turbidity current. Green arrows mark the positions of cable breaks, with the time after the earthquake in minutes. Inset shows the turbidity current velocity as deduced from the timing of cable breaks (redrawn from Heezen and Ewing[5]).

Table 9.1. Statistics of some major slides, debris flows, and turbidity currents.[a]

Name/Location	Waterdepth (m)	Area (km²)	Length (km)	Thickness (m)	Volume (km³)	Slope
Nuuanu Slide (Hawaii)[b]	0–4600	23,000	230	up to 2000	5000	?5→–0.1°
Storegga Slide/Debris Flow[c]	150–3000	112,500	850	up to 430	5580	?1.5→0.05°
Saharan Slide/Debris Flow	1700–4800	48,000	700	5–40	600	1.5→0.1°
Canary Debris Flow	4000–5400	40,000	600	up to 20	400	1→0°
f turbidite (Madeira Abyssal Plain)	?–5400	>60,000	1000+	up to 5	190	?0.2° average
1929 Grand Banks Turbidite	600–6000	160,000	1100	? up to 3	185	?→0.01°

[a] Turbidite areas are those covered by deposit only, slide and debris flow areas include scar and deposit.
[b] Nuuanu Slide flowed uphill for final 140 km.
[c] total of three slide events.

Islands using the GLORIA long-range side-scan sonar (see Chapter 19) is one example of the application of these technical advances[15].

Classification of Gravity-Driven Sediment Flows

Gravity-driven sediment transport includes a wide variety of processes, such as slumping, sliding, debris flow, grain flow, and turbidity currents. However, sediment slides, debris flows, and turbidity currents are the three major gravity-driven processes which transport significant volumes of sediment over large distances in the deep ocean.

A slide is defined as the movement of an upper layer on a basal failure surface. It can result in the downslope transport of large coherent blocks of material, with internal deformation ranging from negligible to severe.

Debris flow has been described as the movement of granular solids, sometimes mixed with minor amounts of entrained water (or, on land, air) on a low slope. A common and effective analogue is with the movement of wet concrete.

A turbidity current is a type of gravity or density current driven by gravitational buoyancy forces resulting from the difference in density between two fluids. To the geologist, it is the downslope flow, under the influence of gravity, of a suspension of sediment in water. The sediment particles, kept in suspension primarily by turbulence, provide the excess density which drives the flow.

Size and Scale

The largest slope failures on earth occur around the margins of and in the ocean basins. This is a consequence of the relief and shape of ocean basins (see Chapter 8), as well as the huge quantities of unconsolidated or partially consolidated sediment which occur on smoothly sloping continental margins, often under geotechnical conditions only marginally in favour of slope stability. Should failure occur, the ocean floor offers unimpeded slopes and flat-floored basins hundreds of kilometres in length, allowing flow over enormous distances.

Individual sediment slides and debris flows can involve many thousands of cubic kilometres of material (*Table 9.1*). The largest of the huge slides on the flanks of the Hawaiian Islands is up to 2 km in thickness, with a volume of 5000 km³. A volume as great as 20,000 km³ has been ascribed to the Agulhas Slide, off South Africa, but the available evidence perhaps suggests that this is a complex of failures rather than a single gigantic event. The Storegga Slide[2], off Norway, and the Canary and Saharan Debris Flows[11], off West Africa, all have runout distances of 600–800 km, much of this on slopes less than 0.5°.

The volume of sediment carried by the largest known turbidity currents is an order of magnitude less than that of the largest sediment slides and debris flows. Individual turbidites of 100–200 km³ are known from several abyssal plain basins in the Atlantic. However, transport distances can be

9.2

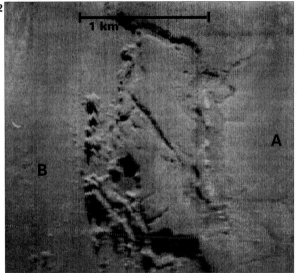

1 km

A

B

Figure 9.2 TOBI 30 kHz side-scan sonar image showing a slope failure in sediments on the flank of an abyssal hill (A), almost certainly caused by undercutting of the slope by erosion of the channel floor (B). Note the progressive disintegration of the large slabs of sediment as they slide toward the channel floor. The topographic relief between the channel floor and the crest of the abyssal hill is about 80 m.

rupted matrix. Deposits of this type are widespread in the geological record (*Figure 9.3*). For example, Macdonald *et al.*[9] have described a sediment slide in Mesozoic sediments in Antarctica which covers an area of at least 20 km x 6 km. This slide is made up of coherent blocks, some in excess of 1 km across, in a mudstone matrix, although the proportion of matrix to blocks is small. Some blocks are completely undeformed, some show minor deformation only at their edges, while a few are folded.

Almost all observations of debris flows have been made in the subaerial environment (*Figure 9.4*), but there is no reason to believe that submarine debris flow processes differ significantly. The classic model of debris flow is of an upper raft of semi-rigid material carried along on a basal layer undergoing intense shearing. The flow is predominantly laminar, although some internal mixing is clearly required to account for the observed chaotic clast structure of most debris-flow deposits. Debris flows differ from simple viscous fluid flows in that they have a finite strength. This manifests itself in

spectacular, frequently exceeding 1000 km. It has been suggested, but not proven, that single turbidity currents may travel up to 4000 km in the Northwest Atlantic Mid-Ocean Channel, which extends from the northern Labrador Sea to the Sohm Abyssal Plain south of Newfoundland.

It is important, however, to realise that the low resolution of the tools commonly used to survey the ocean floor tends to lead to overemphasis of the role of large-scale failures. In addition to most small failures, the role played by slow creep goes unrecognised in the modern submarine environment, even though well-represented in ancient rocks.

Sediment Slides and Debris Flows

Processes

Sediment sliding and debris flow are clearly closely related, with the principal difference being in the degree of fluidity, deformation, and clast (i.e., fragments of broken sediment layers) mixing. It is clear that a slide can be transformed into a debris flow as downslope movement leads to its progressive disintegration. Many of the larger 'sediment slides' described in the literature (e.g., the Saharan Slide) appear to be complex events involving elements of both sliding and debris flow.

Most, if not all, submarine slides in which transport over a significant distance is known to have occurred show evidence for deformation and disruption of the slide material. Although a slide may begin as a single displaced block (*Figure 9.2*), most slides rapidly disintegrate under the stresses imposed during transportation. The typical end-product, when imaged from the sea surface, is an area of hummocky topography indicative of a mass of displaced blocks embedded in a more highly dis-

9.3

Figure 9.3 Section through a debris flow exposed in a road cutting in N.W. Ecuador (photograph courtesy of C.D. Evans, British Geological Survey).

9.4  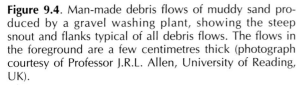  9.5

Figure 9.4. Man-made debris flows of muddy sand produced by a gravel washing plant, showing the steep snout and flanks typical of all debris flows. The flows in the foreground are a few centimetres thick (photograph courtesy of Professor J.R.L. Allen, University of Reading, UK).

Figure 9.5 A large, rafted block in a subaerial debris avalanche on the flanks of Mount Rainier, in the north-western US. The block is 50 m long, 40 m wide, and 18 m high (note the person on top of the block for scale; reprinted from *Rockslides and Avalanches, 1, Natural Phenomena*[20], with permission of Elsevier Scientific Publishing).

the typical steep margins of debris flow deposits and contributes to the ability of debris flows to support large clasts.

Subaerial debris flows are renowned for their ability to carry seemingly impossibly large boulders (*Figure 9.5*). In theory, clasts in debris flows are supported primarily by a combination of clast buoyancy and the cohesive strength of the matrix. However, in many flows the largest clasts may not be totally supported by the matrix, and their trans-port may include a component of sliding or rolling, with the matrix giving some buoyancy and acting as a lubricant. In submarine debris flows, it seems likely that sliding is important in the emplacement of the largest rafted blocks. Off Northwest Africa, for example, rafted blocks in both the Saharan and Canary Debris Flows show evidence of having moved more slowly than the bulk of the flow, suggesting some frictional drag on the underlying sea bed (*Figure 9.6*).

9.6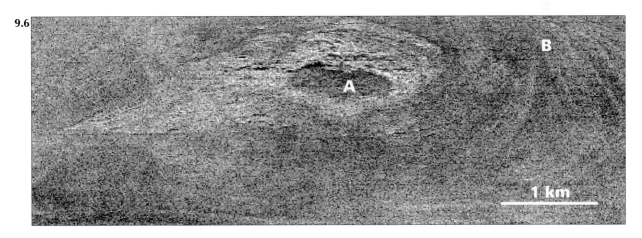

Figure 9.6 TOBI 30 kHz side-scan sonar image of part of the Saharan debris flow deposit (see *Table 9.1*), illustrating a rafted block (A) with a streamlined 'halo' of chaotic debris. The block is about 1 km in length, no more than 50 m thick, and rises only a few metres above the general level of the flow. The block appears as a dark feature, because of its relatively flat and smooth surface which back-scatters little sound. The flow is from right to left. The halo, and pressure ridges (B) upstream of the block, suggest that the block moved more slowly than the bulk of the flow, possibly because it was sliding on the underlying substrate.

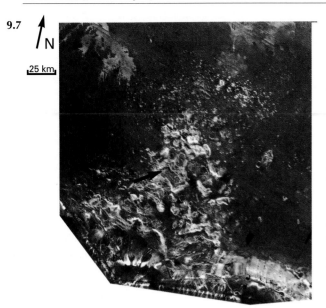

Figure 9.7 Mosaic of GLORIA 6.5 kHz side-scan sonar images showing the blocky surface of the combined Nuuanu and Wailau debris avalanches north of the Hawaiian Islands of Oahu (bottom, left) and Molokai (bottom, right). The largest transported block (arrow, centre) is 20 km long and stands up to 2000 m above the surrounding sea floor.

Hawaiian Island slides occur within poorly bedded volcanic rock sequences. Slide deposits range from relatively coherent masses up to 10 km in thickness (called slumps in the literature) to thinner (up to 2 km thick), more disaggregated masses (called debris avalanches). The former are probably emplaced by slow, intermittent movement, the latter by individual catastrophic failures. Many of the mapped slides may be intermediate between these two end members. The best-known slow moving failure is the Hilnia Slump on the southern flank of Kilauea volcano on the main island of Hawaii. This affects an area of about 5200 km², some three times greater than that of the subaerially exposed volcano. Bathymetric evidence, showing a stepped submarine slope, suggests that the slump consists of two or more enormous rotational slide blocks, each up to 60 km long and 20 km wide. Alternatively, the steeper areas of sea floor may represent the fronts of enormous slump folds. In 1975, a large earthquake located beneath the head of the slump was associated with 3.5 m of subsidence and several metres of seaward movement along much of the south coast of Hawaii. Two similar events occurred in the nineteenth century. The earthquake epicentre, at a depth of around 10 km and approximately coincident with the pre-volcanic sea floor, indicates that the slump probably involves the entire 10 km thick volcanic pile, with lateral movement concentrated within the pre-volcanic pelagic sediment layer which covers the oceanic crust.

Case studies

Hawaii

The idea that submarine landslides were of fundamental importance in shaping the Hawaiian Islands has provoked controversy since the late 1890s, but was only confirmed following extensive GLORIA surveys in the late 1980s; submarine landslides have now been recognised on the flanks of every Hawaiian Island[15] (*Figure 9.7*). Similar failures are now recognised on the flanks of many other oceanic volcanic islands[6] (*Figure 9.8*). Landsliding appears to begin even before a submarine volcano grows enough to reach sea level, peaks in frequency during the shield-building stage of volcanic construction, and continues at a decreasing rate long after volcanic dormancy.

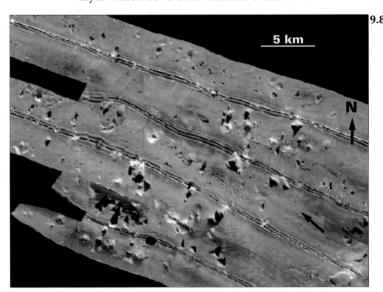

9.8

5 km

N

Figure 9.8 Mosaic of TOBI 30 kHz side-scan sonar images showing part of a debris avalanche off Hierro in the Canary Islands (the arrow shows the flow direction). The largest blocks are up to 1 km across and 200–300 m high. The lack of any obvious flow fabric seems to be typical of this type of flow (see also *Figure 9.7*).

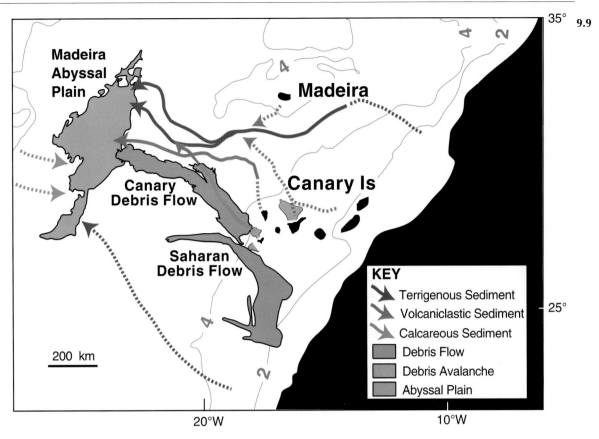

Figure 9.9 Map of the northwest African continental margin around the Canary Islands, showing the locations of known, major, debris flow deposits and turbidity current pathways. Solid arrows show mapped turbidity current channels; dashed arrows are more generalised pathways where detailed information is not available. The debris flows have volumes of several hundred cubic kilometres and transport material for up to 600 km. Individual turbidity currents, carrying up to 200 km³ of sediment, can travel distances in excess of 1000 km. Contours are in kilometres.

The largest Hawaiian slide mapped to date is the Nuuanu Debris avalanche, originating on the north flank of Oahu Island. This avalanche is 230 km long, has a maximum thickness of 2 km, covers 23,000 km², and has a volume of perhaps 5000 km³. The surface of the avalanche has a distinctive blocky texture, with a clear downslope decrease in the size of the slide blocks (i.e., toward the top of *Figure 9.7*). The largest block, originally mapped as a seamount, is 30 km long by 17 km wide, with a flat summit about 1800 m above the adjacent avalanche surface. The Nuuanu Avalanche flowed across the Hawaiian Deep, the moat-like feature surrounding the islands, and climbed at least 300 m up the flank of the Hawaiian Arch to the north. This indicates that the slide had considerable momentum. There is evidence that some debris avalanches give rise to associated turbidity currents which can climb at least 500 m, to overtop the Hawaiian Arch, and which may travel up to 1000 km from the islands. These catastrophic events can also generate huge tsunamis, with one astonishing example, which occurred around 105,000 years ago, reaching over 300 m above sealevel on the island of Lanai. This same tsunami may even have affected the east coast of Australia, some 7000 km distant.

West Africa

The most common submarine slope failures typically involve only the upper few metres to tens of metres of relatively unconsolidated sediment. These failures often produce a thin, narrow tongue of debris extending downslope from a distinct failure scar. They can occur on very low slopes (<1–5°) and have enormous runout distances. The Saharan and Canary Debris Flows (often referred to as slides) on the northwest African margin near the Canary Islands, are large examples of this type of failure[11] (*Figure 9.9*).

The Saharan Debris Flow originated at about 2000 m water depth on the African continental slope south of the Canary Islands, and flowed northwest and then west for 700 km across the

141

9.10

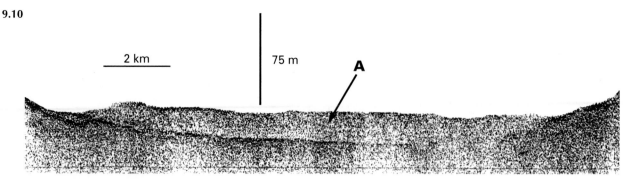

Figure 9.10 3.5 kHz high-resolution profile across the Saharan debris flow deposit (A), immediately southwest of the western Canary Islands (see *Figure 9.9*), showing the typical expression of debris flow deposits (for survey methods see Chapter 20). In this location, the 25 m thick debris flow deposit sits within and partially fills a broad channel across which the profile has been taken.

slope and upper continental rise. The sea-floor gradient decreases downslope from about 1.5° in the source area to as little as 0.1° near the end of the flow. The debris flow incorporated around 600 km³ of sediment. Its failure scar is bounded by a complex scarp 20–80 m in relief. Southwest of the Canaries, on the continental slope below 4000 m water depth, the debris flow deposit forms a narrow tongue about 25 km wide, ranging in thickness from 5–40 m (*Figure 9.10*). In this area, high-resolution side-scan sonar data show spectacular images of flow banding, longitudinal shears, lateral ridges, and transported blocks (*Figures 9.6 and 9.11*).

The Canary Debris Flow originated on the western slopes of the Canary Islands, at about 4000 m water depth. It produced a relatively broad (60–100 km wide), but thin (usually <20 m thick) debris sheet, which extends for 600 km from the source area to the edge of the Madeira Abyssal Plain. This sheet has an average thickness of 10 m, a volume of about 400 km³, and covers an area of 40,000 km². Gradients decrease from 1° in the source area to effectively 0° at the edge of the abyssal plain. The Canary Debris Flow has a complex outline, which appears to have been strongly influenced by even the gentlest topography, particularly at its distal end, where very subtle topographic lows (e.g., pre-existing shallow channels)

9.11

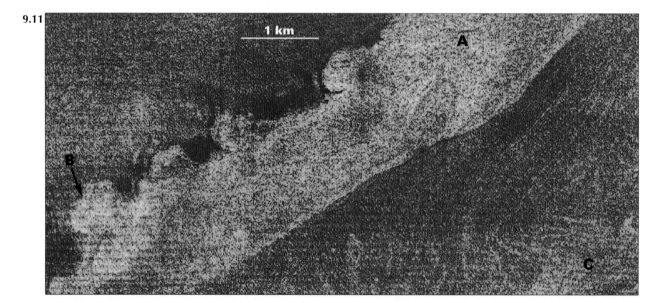

Figure 9.11 False-colour TOBI 30 kHz side-scan sonar image showing the edge of the Saharan debris flow deposit southwest of the western Canary Islands (see *Figure 9.9*). Blue is low back-scatter, and yellow is high back-scatter. The lateral ridge (A), which has a relief of about 5 m, is believed to comprise chaotic rubble deposited along the edge of the flow (B). It is separated from the main part of the flow deposit by a distinct longitudinal shear. The main flow has a characteristic 'woodgrain' fabric (C), which may be evidence for drawing out of the debris into a flow-parallel banding.

clearly control the path of narrow tongues of debris. The head of the Canary Debris Flow is somewhat unusual because no clear headwall scarp is present. Instead, there is a broad zone of apparent shallow rotational faults some 30 km in width. The sediment surface between the faults shows disruption increasing downslope until a featureless, apparently completely homogenised facies is reached. Within this facies, rafted blocks of undisturbed sediments up to 5 km across are seen. Shears within the debris surrounding the blocks, extensional depressions adjacent to their downslope margins, and trails of fragments behind some blocks all suggest that the blocks moved more slowly than the bulk of the flow, presumably because they were in contact with, and dragging on, the underlying sea floor.

Storegga Slides

Some of the world's largest known catastrophic earth movements are found beneath the sea off the heavily populated coasts of northwestern Europe[2]. They are in an area of the mid-Norwegian margin known as the Storegga ('great edge'), because the 290 km wide headwall of a submarine slide forms the top of the continental slope.

The last major slide event occurred about 7000 years ago. It involved erosion of up to 300 m thickness of slope sediments and displacement of about 1700 km³ of material. The deposits include very blocky slide deposits, containing some huge, largely intact slabs up to 10 km x 30 km in plan view and 200 m thick. The blocky nature is due to the failure having cut down into more consolidated sediments. A very thick (up to 20 m) homogeneous fine-grained turbidite covers the entire deep basin of the Norwegian Sea. It is believed to be related to this latest slide event. Probable tsunami deposits on the coasts of Scotland and Norway are also believed to be related to the latest slide. They confirm the age of about 7000 years.

An earlier slide affected an even wider area and displaced about 4000 km³ of sediment. It involved shallower, less consolidated sediments and the resulting deposits are less blocky. This event occurred about 30,000 years ago. Two older slide events are now known to have occurred in the region and there were probably others whose scars have been removed by later erosion, but whose deposits may be present in the deep basin. The slides are believed to have been triggered by earthquakes and the decomposition of gas hydrates (see later). Gas hydrates have been recognised on seismic profiles from the slope just to the north of the slide.

Turbidity Currents

Processes

Turbidity currents may be generated directly from sediment suspensions delivered to the shelf edge by agencies such as tidal currents, rivers, or storms. Many others appear to have their origins in sediment failures on the continental slope, although the actual initiation mechanisms remain poorly understood. One possibility is that they evolve from debris flows, in effect continuing the process of disintegration which first led from slide to debris flow. This evolution requires the dilution of the flow by incorporation of water and a transition from laminar to turbulent flow. One elegant mechanism, which has been demonstrated in flume experiments, is the generation of turbidity currents by erosion of the steep snout of a debris flow as it moves downslope[4]. An alternative mechanism involves mixing of water into the body of the flow, perhaps due to internal flow turbulence. This latter process is attractive because it offers a mechanism for transforming entire debris flows into turbidity currents. However, it has proved difficult to reproduce in experiments, and its occurrence in nature remains hypothetical.

Proving turbidite–debris flow relationships in the modern ocean basins is difficult, the main problem being the collection of appropriate samples from deposits which may be spread over thousands of square kilometres. One set of cores, collected across the snout of the Canary Debris Flow deposit at the edge of the Madeira Abyssal Plain, does, however, conclusively demonstrate such a relationship (*Figures 9.12* and *9.13*). Here, the debris flow actually occurs within the turbidite, interrupting the latter's fining-upward depositional sequence. It appears that the faster moving turbidity current began depositing sediment at the edge of the plain, perhaps a few hours before the arrival of the debris flow, which then buried the lower turbidite layers. Deposition of the finer fraction of the turbidite then continued on top of the debris deposit. A wider study of this debris flow–turbidity current pairing indicates that the two phases have markedly different (although overlapping) depositional patterns, and that they clearly had divergent paths as they crossed the lower continental slope and rise[10]. This strongly suggests that the debris flow did not continuously spawn a turbidity current as it moved downslope, but that the two phases evolved in or near the source area and then travelled independently.

The flow characteristics of large turbidity currents in the modern ocean have not been directly observed. Flow models are based on indirect obser-

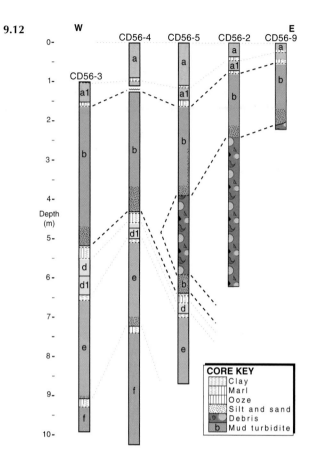

9.12

CORE KEY
- Clay
- Marl
- Ooze
- Silt and sand
- Debris
- b Mud turbidite

Figure 9.12 Diagrammatic logs of a transect of cores (for a description of coring devices, see Chapter 19) across the snout of the Canary debris flow deposit, showing the relationship between the debris flow and a coeval turbidite, identified as 'b' in the Madeira Abyssal Plain turbidite sequence (see *Figure 9.15*). Lettered units are turbidites which can be correlated across the Madeira Abyssal Plain. The feather edge of the debris flow deposit occurs within 'b', indicating that the two flow phases must be part of the same event (see text for a more detailed explanation).

vations of both the flow (e.g., cable breaks, *Figure 9.1*) and the structures it leaves along its path (e.g., channels, levees; see later), geological evidence from ancient rock sequences (*Figure 9.14*), and flume experiments. The available evidence suggests that flows range widely in concentration, thickness, turbulence, and velocity, although some of these parameters are obviously linked. For example, one model for the emplacement of the thick, ungraded mud turbidites found on abyssal plains is based on thin (<20 m thick), high concentration (50–100 kg/m³) and low velocity (<1 m/s) flows, which, as they approach their point of deposition, are almost nonturbulent. In contrast, a model for thin, fine-grained turbidites, such as those which characterise channel-overbank sequences, indicates thick (up to several hundred metres), low density (≈ 2 kg/m³) flows, but with similar velocities. Overall, velocities are known to vary from <1 m/s to at least 25 m/s.

The classic turbidite, described in minute detail from ancient rocks, is a fining-upward sequence with grain size ranging from sand to mud. Turbidites are usually described in terms of the Bouma sequence[1], a five-fold division based on sedimentary structures and grain size. Using flume experiments, the progression of sedimentary structures has been shown to relate to the progression of bedforms seen under a decelerating flow. In nature, few turbidites exhibit a complete Bouma sequence. For example, the thick fine-grained turbidites which characterise many abyssal plains may consist entirely of only the upper two Bouma divisions.

Case studies

The 1929 Grand Banks turbidity current

In November 1929, a magnitude 7.2 earthquake occurred beneath the upper continental slope just south of the Grand Banks of Newfoundland. Six submarine telephone cables in the immediate vicinity of the epicentre were broken instantaneously and a further six, in an orderly downslope sequence, over the next 13 hours 20 minutes (*Figure 9.1*). In attempts to explain these observations, various theories, based on sea-floor faulting or movement of

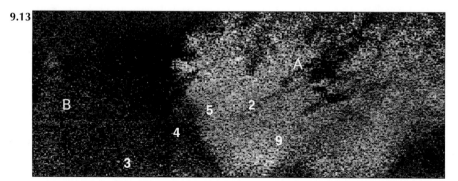

9.13

Figure 9.13 Gloria 6.5 kHz side-scan sonar image of the Canary debris flow snout. The chaotic nature of the debris-flow material (A) generates a high level of back-scatter (light tones) and ensures a strong acoustic contrast with the flat (low back-scatter) abyssal plain (B). Numbers show locations of the CD56 cores shown in *Figure 9.12*.

Figure 9.14 Cyclical turbidite sequences believed to be characteristic of sandy fan lobes (reproduced from E. Mutti's *Turbidite Sandstones*[16], by permission of AGIP, Milan, Italy).

sediment leaving sections of cable unsupported, were proposed during the next 20 years. None, however, satisfactorily explained the orderly sequence of cable breaks, the fact that substantial sections of cables were buried (but only on the deeper, less steep, area of the continental slope), or the lack of damage to cables on the continental shelf. The hypothesis that the cables were broken by a turbidity current is the only explanation for this combination of observations, as was realised by Heezen and Ewing[5] in their classic paper. Subsequent sampling in the Sohm Abyssal Plain to the south of the cable break area proved the existence of a basin-wide turbidite, underlain by Holocene sediments, at the sea floor. This turbidite covers an area of some 160,000 km^2, has a maximum thickness in excess of 3 m, and a volume of about 185 km^3.

From the timing of the cable breaks, Heezen and Ewing realised that it was possible to calculate the velocity of the turbidity current. They estimated a maximum velocity of about 25 m/s on the continental slope, decreasing, over a distance of about 500 km, to about 6 m/s at the edge of the abyssal plain. Lack of precise knowledge of the point of origin of the turbidity current, within the general source area, casts some doubt on their maximum value. Most authorities, however, agree that velocities of at least 15–20 m/s were attained. Similar velocities have since been calculated for turbidity currents associated with both the 1954 Orleansville (Algeria) earthquake and the 1979 slope failure off Nice (southern France).

The Madeira Abyssal Plain

The turbidites of the Madeira Abyssal Plain (for location, see *Figure 9.9*), the deepest part of the Canary Basin off northwest Africa, are probably the best-studied in the modern ocean basins[21]. This turbidite sequence has been created by enormous turbidity currents which carry sediment from the African continental margin, often over distances in excess of 1000 km. On the plain, these currents become 'ponded' by the surrounding higher topography and deposit their sediment load. Indeed, the flat plain results from the stacking of numerous turbidites on top of each other, forming a 350 m thick layer which has levelled off the otherwise irregular topography. The upper 35 m of this sequence, corresponding to the past 750,000 years, has been sampled, allowing its depositional history to be determined.

During that time, a turbidity current reached the Madeira Abyssal Plain, on average, once every 30,000 years. Most occurred during periods of rapidly changing sealevel (both rises and falls) associated with Pleistocene glacial cycles (*Figure 9.15*). The turbidity currents derive from the northwest African continental slope, from the flanks of the Canary Islands, and, occasionally, from seamounts to the west of the plain.

The turbidites deposited on the abyssal plain range in volume from a few cubic kilometres to almost 200 km^3, with those over a few tens of cubic kilometres forming layers over the whole plain. They consist predominantly of fine-grained mud; any sand present tends to be deposited at the break of slope at the edge of the plain (*Figure 9.16*). Much of the coarser sediment is transported across the continental rise through deep-sea channels, while, in contrast, the finer material appears to move as an unconfined sheet flow. The incoming turbidity currents cause no significant erosion of the underlying sediments, and successive turbidites

9.15

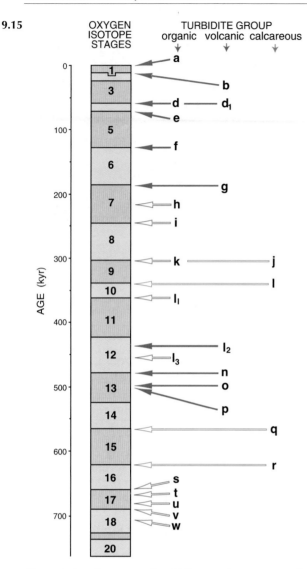

are separated by sediment layers built up by a slow pelagic rain of biogenic carbonate and wind-blown dust from the Sahara Desert. Fossils in this pelagic record allow us to date the turbidite sequence, and even to assign approximate ages to individual turbidity currents.

Turbidites such as those sampled on the Madeira Abyssal Plain (*Figures 9.15* and *9.16*), which originate as failures on sedimented slopes, contain a mixture of sediments (and microfossils) with an age range corresponding to that of the failed sediment mass. In theory, if the pelagic fossil record in the source area is well-known, it should be possible to estimate the age range and thus thickness of the original sediment failure. If, in addition, the volume of the resultant turbidite is known, then the area eroded to form the corresponding turbidity current can also be calculated. In the case of the Madeira Abyssal Plain turbidites, this theory has been put into practice, allowing a typical sediment failure on the northwest African margin, 1000 km away from the study area, to be described – this failure is a few

Figure 9.15 Summary of turbidite emplacement on the Madeira Abyssal Plain, showing the ages of individual turbidites (each identified by a letter) and turbidite groups based on source area (organic turbidites – green arrows – from the northwest African continental margin, volcanic – red arrows – from the Canary Islands and Madeira, and calcareous – blue arrows – from seamounts to the west of the plain). These groups can be further subdivided into turbidites from north (filled arrows) and south (open arrows) of the Canaries. Ages of turbidites have been determined relative to the oxygen isotope time-scale[18] (blue and even numbers are glacial periods, brown and odd numbers are interglacials). Note the strong correlation of turbidite emplacement with oxygen isotope stage boundaries, suggesting a relationship between turbidity currents and changing sea level.

9.16

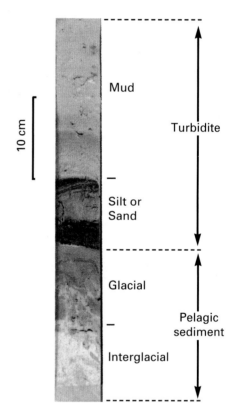

Figure 9.16 Core photograph of some typical Madeira Abyssal Plain sediments. The upper part consists of a turbidite with a black volcanic sand/silt basal unit and a brown mud top. The lower part consists of pelagic sediment, with the brown sediment marking deposition during a cold glacial climate and the white that during a warmer interglacial.

Figure 9.17 (a) The distribution, through time, of five key coccolith species occurring in pelagic sediment in the Madeira Abyssal Plain area. For sediments deposited during the most recent half million years, variation in the ratio of these species gives an age accurate to within one or two oxygen isotope stages, i.e. a few tens of thousands of years. (b) Erosion of sediment representing more than a few tens of thousand years produces coccolith mixtures not seen in the pelagic record, but dependent upon the age range of sediments which were eroded. Using the distribution of coccolith species through time (a), it is relatively simple to calculate what age range any observed coccolith mixture represents. In the example shown, for turbidite 'f' of the Madeira Abyssal Plain sequence (emplaced at the end of oxygen isotope stage 6), synthetic mixtures can be created for the erosion of stage 6 sediments only, 6 + 7, 6–8, and so on. By comparing these with the actual mixture found in 'f' (inset), it can be seen that 'f' contains sediments originally deposited during isotope stages 6–12, i.e. between 130,000 and 480,000 years ago. This corresponds to erosion of about 50 m of sediment from the source region of 'f' on the African margin south of the Canaries.

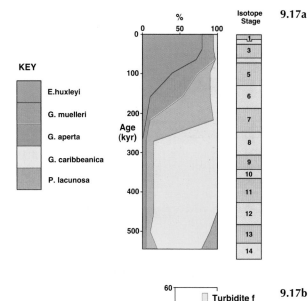
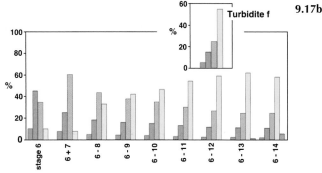

tens of metres thick, covers an area as great as 6000 km², and incorporates sediment with an age range of 50,000–500,000 years (*Figure 9.17*). An important additional observation based on this study is that none of the turbidites examined contains a significant excess of surface sediment, suggesting that, once formed, the turbidity currents which transported them were virtually nonerosional, and that they travelled many hundreds of kilometres in this state.

Turbidites and sediment fans

Turbidites and related deposits, such as debris flows and debris avalanches, have the greatest volume of any types of sediment in basin fills. The processes that form them concentrate the sands into bodies that are potential reservoirs for hydrocarbons. It is necessary for effective hydrocarbon exploration that the size, shape, and relationships of the bodies are known. These deposits usually lie in front of a subaerial and shelf-feeder system, such as a river drainage basin or a glacially carved cross-shelf trough. The largest, high-input, feeder systems, such as the Indus, Amazon, Mississippi, and Nile, can supply enough material to the deep sea to form bodies of sediment (sediment fans) that are 10 km or more thick. Glacial-fed submarine fans, like that in front of the Barents Sea, can be of an equivalent size. However, small fans fed through the almost ubiquitous submarine canyons which dissect the continental slopes are much more common. Submarine canyons can be thousands of metres deep, and are usually fed by submarine tributary systems with a hierarchy of gullies (*Figure 9.18*). The fans at the mouths of these canyons are

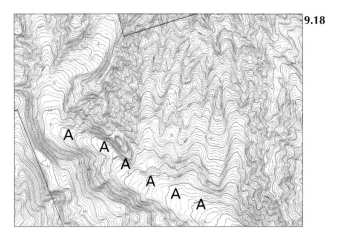

Figure 9.18 A hierarchy of tributary gullies feeding into the main, flat-floored Var Canyon (A), off Nice in the northwest Mediterranean Sea. The main canyon is up to 300 m deep; depth contours are in metres.

9.19

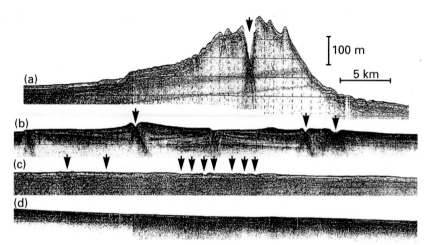

Figure 9.19 High resolution acoustic profiles showing the down-fan changes in the morphology of the Indus Fan. (a) Large channel–levee system about 300 m high; (b) small channel–levee systems nested above each other; (c) and (d) the flat sandy lobe where the sand content is preventing much sound penetration (arrows show locations of channels).

difficult to detect as positive morphologic features, often merging together with the neighbouring systems to form a continental rise or slope apron. Paradoxically, these more common fan types have been relatively little-studied because of the difficulty of detecting them and because they are usually sand rich, which makes them difficult to sample with conventional corers.

The high-input fans have extensive distributary channel–levee systems, only one of which seems to be active at any one time. This implies that there are many instances where there has been a switch to a new channel. The gradients of these large fans are low. The levees (banks on either side of the channel) can be hundreds of metres high, as on the upper Indus Fan (*Figure 9.19*), in contrast to the levees of rivers, which reach a maximum height of only 10 m or so. The submarine chan-

nels, which decrease in depth down fan, are seen on swath mapping systems, such as GLORIA, to have a remarkable resemblance to river channels in their sinuosity. They tend to be highly sinuous or meandering in their middle reaches[8] (*Figure 9.20*).

Migration of meanders can leave riverine features, such as abandoned reaches reminiscent of oxbow lakes. Just as sands in rivers tend to deposit on the inside of meander bends, in what are called point bars, so too do sands in submarine channels. Submarine point bars can be as wide as the meander belts (typically 2–10 km). Sands can also be deposited within channels, in what are called channel lag deposits, causing build-up of the channel floor. This accumulation of sand can be imaged on geophysical records, where it is seen as strong acoustic reflectors beneath the channel floor. Sand

9.20

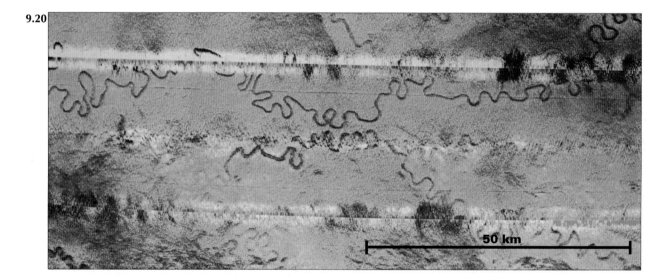

Figure 9.20 A mosaic of GLORIA images of meandering channels on the middle Indus Fan. The channels are about 1 km across; water depth varies from 3550–3650 m, from left to right. The stronger backscatter is purple.

Figure 9.21 Interpretation of the youngest major sequence of the Mississippi Fan based on GLORIA survey data. The sediment slides have a swirly pattern, whereas the extensive sandy lobes, fed by a single channel, have a distributary, frond-like pattern (small grey box at the right shows area of *Figure 9.22*).

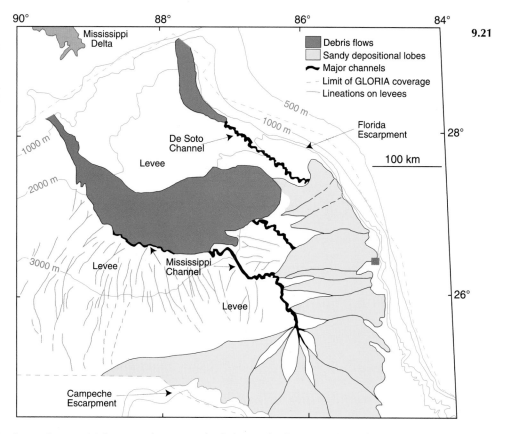

bodies associated with channels are thickest in the upper or proximal fan (typically tens to hundreds of metres thick). Sandy lobes which occur beyond the ends of channels (i.e., on the distal fan) have only recently been detected on large fans. The Mississippi fan (*Figure 9.21*) has a hierarchy of sandy lobes which are only a few metres thick, reaching a maximum of about 20 m. When seen in high resolution, individual sandy flow deposits display a remarkable frond-like pattern[19] (*Figure 9.22*). Each of these deposits was laid down over a relatively short period, probably at an early stage in

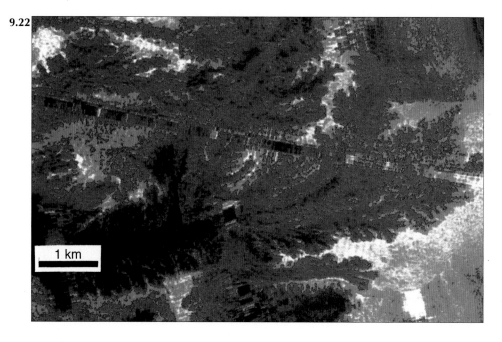

Figure 9.22 Frond-like pattern at the end of one of the sandy lobes of the Mississippi Fan, seen on high-resolution side-scan sonar. The relatively high level of acoustic back-scatter from the sandy lobes is shown by dark colours. The diagonal line (top left to middle right) across the image lies directly beneath the sonar vehicle path (courtesy of D. Twichell, U.S. Geological Survey).

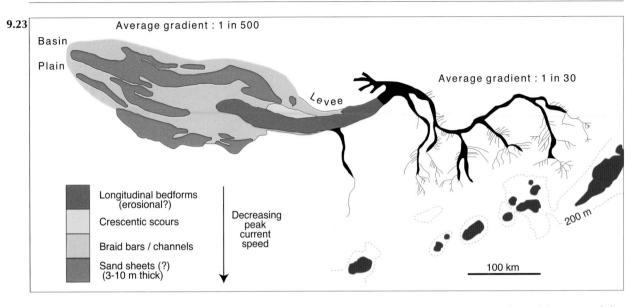

9.23

Average gradient : 1 in 500

Basin Plain

Levee

Average gradient : 1 in 30

Longitudinal bedforms (erosional?)

Crescentic scours

Braid bars / channels

Sand sheets (?) (3-10 m thick)

Decreasing peak current speed

200 m

100 km

Figure 9.23 The Umnak turbidite system in the Bering Sea basin, mapped using GLORIA. The subdivisions of the extensive sandy lobe are shown in blue, green, and yellow; islands of the Aleutian chain are shown in brown (from Kenyon and Millington[7]).

a phase of rising sealevel. Hence the flows that gave rise to them must have been frequent. In recent times, flows have occurred approximately once a year on the active Zaire Fan.

The more common low-input types of fan generally have small subaerial drainage basins and well-developed tributary canyons that feed poorly developed, usually single, channel–levee systems. A good example has been mapped using GLORIA side-scan sonar on the northern flanks of the volcanic islands of the Aleutian Chain[7] (*Figure 9.23*). It is thought that this fan is particularly well-imaged by the sonar because it is young and fresh, being in an area of very active earthquakes. The single, straight leveed channel has a trumpet-shaped mouth fed from an elaborate distributary channel system, beyond which is an enormous spread of flat and strongly patterned ground that is thought to be a sandy lobe. The patterns are arranged in zones that indicate a decreasing energy of flow away from the channel mouth. Maximum channel gradients are usually greater than for the high-input fans and channels are fairly straight. Sands should be found mainly in the channel mouth lobes (typically tens of kilometres across). Flows in low-input fans are believed to be infrequent. For instance, flows on the low-input Var system, off Nice in southern France (*Figure 9.18*), have occurred about once every 1000 years throughout the Holocene.

Triggering Mechanisms

Although individual slope failures may be catastrophic in nature and, cumulatively, their products

may dominate the sediment sequences in deep-sea basins, in historical terms they are infrequent events, except on the largest river-fed submarine fans. Since they also occur beneath an opaque blanket of sea water, it is not surprising that little is known about the circumstances in which they are generated. Among the many factors which may contribute to the triggering of slope failures are:

- Earthquakes.
- Loading and oversteepening of slopes.
- Underconsolidation (i.e., when the fluid pressure within the sediment exceeds hydrostatic - pressure), usually due to rapid sedimentation, to gas build-up, or to gas hydrate decomposition within the sediment.
- Sealevel change.
- The occurrence of slope-parallel weak layers within bedded sequences.

The majority of historically recorded slope failures, recognised because they broke telephone cables or caused tsunamis, have been triggered by earthquakes. Other historical failures, such as the 1979 Nice Slide, may have been caused by man-made slope loading. One example of earthquake-induced failure, from off northern California, occurred during a magnitude 6.5 earthquake in 1980. Parts of the area had fortuitously been surveyed with a high-resolution profiler in the late 1970s, when no deformation was found; repeat surveys immediately after the earthquake found evidence[3] for failure on slopes as low as 0.25°. The common occurrence of failures in areas of very rapid sedimentation, such as major river delta fronts, is persuasive evidence that underconsolidation contributes to the failure process. Such failures have been implicated in the damage, and even loss, of oil-drilling platforms in the Mississippi Delta area. In this area, failures have occurred primarily during hurricanes, suggesting that loading by wave action may be a triggering factor.

Many slope failures, however, have no obvious trigger. This is particularly true on passive continental margins around the North Atlantic, for example off the east coast of the US. In these areas slopes are low, as is seismicity. Sedimentation rates are low-to-moderate, and there is no evidence for gas in the upper sediment column. Failure on slope-parallel bedding planes is the common type of failure in this situation[17]. One suggestion is that these failures are the end product of slow deformation under the action of gravity or under the repeated loading effects of earthquakes, none of which, individually, are capable of causing failure. Eventually, one or more weak layers fail under the cumulative strain effects.

The prediction of submarine slope failures is not a simple task. Nevertheless, the ability to forecast geological hazards, or at least areas prone to such hazards, is of fundamental importance to man's activities, both offshore and in low-lying coastal areas. The rewards for a better understanding of submarine landslides and their causes could be very large indeed.

Acknowledgement

We would like to acknowledge financial support through EC MAST II contracts MAS2-CT94-0083 and MAS2-CT93-0064.

General References

Johnson, A.M. (1970), *Physical Processes in Geology*, Freeman, Cooper and Co., San Francisco. 577 pp.

Pickering, K.T., Hiscott, R.N., and Hein, F.J. (1989), *Deep Marine Environments*, Unwin Hyman, London, 416 pp.

Pickering, K.T., Hiscott, R.N., Kenyon, N.H., Ricci Lucci, F., and Smith, R.D.A. (1995), *Atlas of Deep Water Environments: Architectural style in turbidite systems*, Chapman and Hall, London, 333 pp.

Simpson, J.E. (1987), *Gravity Currents in the Environment and the Laboratory*, Ellis Horwood, Chichester, 244 pp.

References

1. Bouma, A.H. (1962) *Sedimentology of some Flysch Deposits*, Elsevier, Amsterdam, 168 pp.
2. Bugge, T., Belderson, R.H., and Kenyon, N.H. (1988), The Storegga Slide, *Phil. Trans. Roy. Soc. London, Ser. A*, **325**, 357–388.
3. Field, M.E., Gardner, J.V., Jennings, A.E., and Edwards, B.E. (1982), Earthquake induced sediment failures on a 0.25° slope, Klamath River Delta, California, *Geology*, **10**, 542–546.
4. Hampton, M.A. (1972), The role of subaqueous debris flow in generating turbidity currents, *J. Sedimen. Petrol.*, **42**, 775–793.
5. Heezen, B.C. and Ewing, M. (1952), Turbidity currents and submarine slumps, and the 1929 Grand Banks Earthquake, *Am. J. Sci.*, **250**, 849–873.
6. Holcomb, R.T. and Searle, R.C. (1991), Large landslides from oceanic volcanoes, *Marine Geotech.*, **10**, 19–32.
7. Kenyon, N.H. and Millington, J. (1995), Contrasting deep-sea depositional systems in the Bering Sea, in *Atlas of Deep Water Environments: Architectural style in turbidite systems*, Pickering, K.T., Hiscott, R.N., Kenyon, N.H., Ricci Lucci, F., and Smith, R.D.A. (eds), Chapman and Hall, London, pp 196–202.
8. Kenyon, N.H., Amir, A., and Cramp, A. (1995), Geometry of the younger sediment bodies of the Indus Fan, in *Atlas of Deep Water Environments: Architectural style in turbidite systems*, Pickering, K.T., Hiscott, R.N., Kenyon, N.H., Ricci Lucci, F., and Smith, R.D.A. (eds), Chapman and Hall, London, pp 89–93.
9. Macdonald, D.I.M., Moncreiff, A.C.M., and Butterworth, P.J. (1993), Giant slide deposits from a Mesozoic fore-arc basin, Alexander Island, Antarctica, *Geology*, **21**, 1047–1050.
10. Masson, D.G. (1994), Late Quaternary turbidity current pathways to the Madeira Abyssal Plain and some constraints on turbidity current mechanisms, *Basin Res.*, **6**, 17–33.
11. Masson, D.G., Kidd, R.B., Gardner, J.V., Huggett, Q.J., and Weaver, P.P.E. (1992), Saharan continental rise: Facies distribution and sediment slides, in *Geologic Evolution of Atlantic Continental Rises*, C.W. Poag and P.C. de Graciansky (eds), Van Nostrand Reinhold, New York, pp 327–343.
12. Middleton, G.V. (1966a), Experiments on density and turbidity currents: I. Motion of the head, *Can. J. Earth Sci.*, **3**, 523–546.
13. Middleton, G.V. (1966b), Experiments on density and turbidity currents: II. Uniform flow of density currents, *Can. J. Earth Sci.*, **3**, 627-637.
14. Middleton, G.V. 1967. Experiments on density and turbidity currents: III. Deposition of sediment. *Can. J. Earth Sci.*, **4**, 475–505.
15. Moore, J.G., Clague, D.A., Holcomb, R.T., Lipman, P.W., Normark, W.R., and Torresan, M.E. (1989), Prodigious submarine landslides on the Hawaiian Ridge, *J. Geophys. Res.*, **94**, 17465–17484.
16. Muti, E. (1992), *Turbidite Sandstones*, Agip and Istituto di Geologica, Universita di Parma, p. 176.
17. O'Leary, D.W. (1991), Structure and morphology of submarine slab slides: clues to origin and behaviour, *Marine Geotech.*, **10**, 53–69.
18. Shackleton, N.J. and Opdyke, N.D. (1973), Oxygen isotope and palaeomagnetic stratigraphy of equatorial Pacific core V28-238: Oxygen isotope temperatures and ice volumes on a 10^5 year and 10^6 year scale, *Quatern. Res.*, **3**, 39–55.
19. Twichell, D.W., Schwab, W.C., Nelson, C.H., Kenyon, N.H., and Lee, H.J. (1992), Characteristics of a sandy depositional lobe on the outer Mississippi Fan from SeaMARC 1A side-scan sonar images, *Geology*, **20**, 689–692.
20. Voight, B. (1978), *Rockslides and Sandstones, 1, Natural Phenomena*, Elsevier Scientific, Amsterdam, p. 186.
21. Weaver, P.P.E., Rothwell, R.G., Ebbing, J., Gunn, D.E., and Hunter, P.M. (1992), Correlation, frequency of emplacement and source directions of megaturbidites on the Madeira Abyssal Plain, *Marine Geology*, **109**, 1–20.

CHAPTER 10:

Mid-Ocean Ridges and Hydrothermal Activity

C.R. German, L.M. Parson, and R.A. Mills

Mid-Ocean Ridges

The mid-ocean ridge system is the largest continuous topographic feature on the Earth's surface, comprising a mountain range of volcanoes and faulted blocks that, in places, rise several kilometres from the surrounding sea floor. Sections of the ridge system extend throughout all of the world's oceans (*Figure 10.1*) and total more than 50,000 km in length, some four times the diameter of the globe.

The ridge marks the zone along which the tectonic plates that comprise the Earth's surface are separating, allowing new sea floor to be created[15] (see also Chapter 8). The new sea floor is generated at the ridge axes in the form of a dense igneous rock called basalt. The mid-ocean ridge is the most

volcanically active continuous zone on the Earth's surface, where hot, soft mantle rises from a depth of several tens of kilometres in the Earth's interior, to melt and form abundant basalt magmas (semi-molten rock) in ponds or reservoirs beneath the sea floor. The basalt seeps through the overlying rock in complex plumbing systems and is extruded onto the sea-floor surface. Flows become broken and fragmented as they are shunted off-axis by successive additions of new sea floor. The lava flows, volcanoes, and dyke systems combine to create some 20 km^3 of new crust per year in a layer 6 km thick (*Figure 10.2*).

Continuous separation of the two tectonic plates at a spreading centre induces extensional stress on those sections of ridge where no new sea floor is

10.1

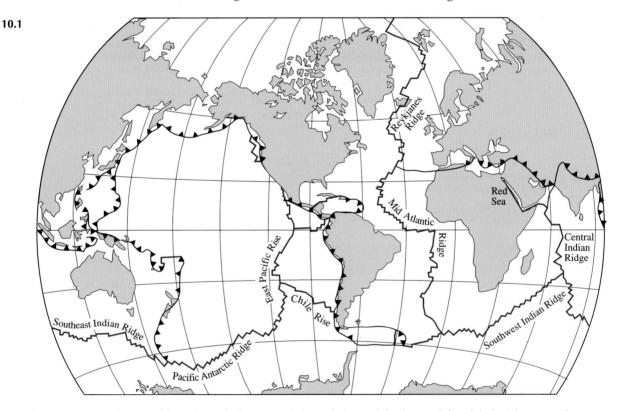

Figure 10.1 Distribution of the principal plate tectonic boundaries and the locus of the global mid-ocean ridge system is shown in blue. Convergent margins are shown in black, the triangles indicating the direction of subduction.

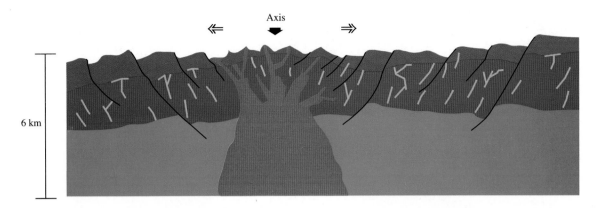

10.2

Figure 10.2 The likely complexities of magmatic plumbing in the shallow crust beneath a slow-spreading ridge system. Basaltic sheet flows and axial volcanoes (brown) are fed by a complex plumbing arrangement (red). Steeply inclined, sheeted intrusives (purple) carry complex sill/dyke fabrics (orange) and overlie gabbros in the lower crust (green). As the new crust moves away from the ridge axis, it cools and large faults and fissures develop (black).

being generated. These stresses cause thinning and eventual cracking of the crust. The degree of fissuring and faulting of the crust varies with the rate of spreading; hence, to a first approximation, the shape of the ridge system can be used as a clue to the rate of spreading activity[26].

The rate at which different tectonic plates separate varies considerably. While the American and European–African plates are moving apart at rates of between 20 and 60 mm/yr (roughly the speed at which fingernails grow), the American and Pacific plates are distancing themselves at nearly ten times that speed. The type of sea floor generated at these different speeds is usually very different, in terms of the topography and the composition of the basaltic rocks generated to make the crust. In general, faster-spreading ridges are characterised by an elevated crestal region, while slower spreading systems have a well-defined axial valley or trough,

marked by inward-facing fault scarps which may have more than 500 m of throw. The faults bound an axial floor, commonly 10–12 km wide, within which a range of neovolcanic and neotectonic activity takes place. Volcanoes, which are commonly distributed across the axial floor, are fed through a complex system of pipes and cracks which connects them to magma chambers. The magma has risen and ponded in shallow levels of the crust, in what are believed to be discrete supply systems following the melting of crustal rock during the uprise of heat from the upper mantle (*Figure 10.3*).

Inevitably, as the plates continue to move apart, sediments accumulate on the surface of the crust, burying the volcanic and tectonic features which characterise the ridge.

The technology which has become available to marine scientists during the past two decades has allowed the mid-ocean ridge system to be studied

10.3

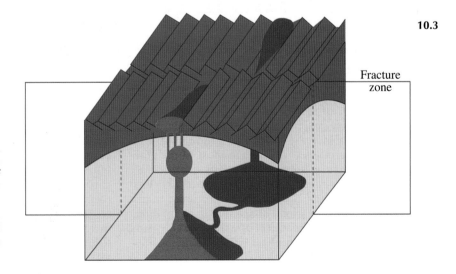

Fracture zone

Figure 10.3 The deeper magma supply system to the mid-ocean ridge. Regularly spaced diapirs of magma rise from the mantle, pond as reservoirs, and feed the central portions of ridge segments. Here, the ridge segments are separated by a fracture-zone offset, which perturbs the magma supply.

10.4

Figure 10.4 Enhanced Lansat TM satellite image of the southern branch of the Afar rift triple junction, northeast Africa, showing an area of 120 km by 120 km. Marked NNE linear fabrics indicate faulting, which locally cuts and is cross-cut by fresh volcanic features, such as volcanoes and basaltic lava flows (© C. Oppenheimer and J.P. Rogers, Open University).

with a degree of detail hitherto impossible. On the broadest scale, satellite imagery (see Chapter 5) has been used to increase the understanding of mid-ocean ridge processes[10]. Although most of the global ridge system is submerged, a few sections are subaerially exposed (e.g., Iceland and the East African Rift). These provide the opportunity to calibrate our interpretations of satellite data by direct examination of the outcrop. The developing rift–ridge systems of the East African Rift, as illustrated in the LANDSAT thematic mapper images in *Figure 10.4*, can be elucidated using computer programmes that generate high-resolution structural maps, which are then interpreted by geologists. From satellite data, faults can be mapped over many hundreds of kilometres; their relative chronology of activity can be estimated, along with their relationship with volcanic activity, from intersecting or cross-cutting features. On the ground, as in Iceland (*Figure 10.5*), where the Mid-Atlantic Ridge is exposed, even finer details are recorded; features of ridge systems can be measured, such as

Figure 10.5 One of the steep-walled rift valleys exposed on land in Iceland, where the Mid-Atlantic Ridge passes from its submarine to a sub-aerial setting. A volcanic cone can be seen in the distance. This section of the rift displays intense hydrothermal venting (hot springs, geysers, etc.).

10.5

valley width, slope angles of faults, and volumes of lava flows.

Satellite altimetry has been used offshore to map the mid-ocean ridge in detail, in places where ships seldom venture – like the Southern Ocean – and to demonstrate the extent to which the ridge system is cut by closely spaced fracture zones. Otherwise, offshore surveying of the ridge relies on the acquisition of remotely sensed marine geophysical data obtained from ships. Methods include conventional seismic-reflection profiling and magnetic and gravity surveying, but also new methods of mapping whole swaths of sea floor in terms of their reflectivity or topography (see Chapter 20); these new methods have provided the greatest recent advances in understanding the origins of the ridge system (*Figure 10.6*).

On the scale of a few metres, molten magma is extruded at the sea floor and rapidly cooled to form sheet-like lava flows, rounded pillow basalts, hummocky pillowed topography, or volcanoes of varied shapes (*Figure 10.7*). Piles of pillows accumulate to form pillow mounds which, with continued supply, build into volcanoes.

The factors which control the shapes of volcanoes, their size, and their number along the mid-

10.6

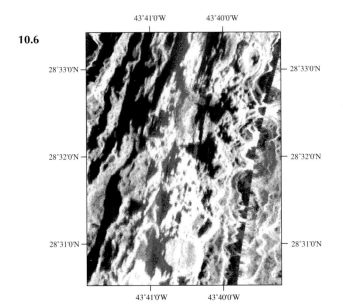

10.7

Figure 10.7 Photograph taken from a submersible at 3080 m water depth, showing pillow basalts outcropping on the floor of the Mid-Atlantic Ridge at 29°N (the view is ca 6 m across; © Deep Submergence Laboratory, Woods Hole Oceanographic Institution).

Figure 10.6 TOBI deeptow side-scan image of the central portion of the Mid-Atlantic Ridge at 29°N (the image width is ca 5 km). The lighter tones represent areas which are rough unsedimented fresh basalt, or are orientated normal to the acoustic beam and have higher levels of back-scattered energy. Darker tones mark surfaces with higher acoustic absorption or shadows. Clear examples of volcanoes and hummocky sea-floor surfaces are recognisable.

ocean ridge system are not yet well-understood[27]. Variations in the composition of volcanic material, speed of plate motion, and rate at which magma is fed to the ridge are certainly contributory. More significantly, we recognise that the localised coalescence of pillow basalt mounds and seamounts into linear ridges several kilometres in length is a fundamental mechanism in the generation of new crust. It is a major thrust of marine geoscientific endeavour in the 1990s to understand the relationship of these volcanic features, and their subsurface plumbing, to the occurrences of hot-fluid circulation systems (hydrothermal activity), and to predict their location along the world's spreading ridges.

Hydrothermal Circulation

One of the most exciting developments since the beginning of the study of oceanography was the discovery, in the late 1970s, that submarine hydrothermal vents were associated with the volcanically active zone at the crest of the mid-ocean ridge system (*Figure 10.1*). The base of the oceanic crust is extremely hot (>1000°C), yet its upper boundary is in contact with sediments and sea water at temperatures that are only a few degrees above 0°C. Since the earliest recognition of mid-ocean ridges and their significance to plate tectonics in the 1960s, geophysicists have known that conductive heat flow measured through young ocean crust could not account for all the heat lost at these tectonic spreading centres. So, they predicted that some alternative *convective* heat-transfer process must also occur near the crests of mid-ocean ridges[33].

It was not until the late 1970s that these predictions were proved correct, with the discovery of, first, low-temperature submarine hydrothermal activity (10–30°C) at the Galapagos Spreading Centre[3] and, second, high-temperature (350°C) hydrothermal activity on the East Pacific Rise[30]. The importance of these discoveries went much further than simply proving the geophysicists' theories correct. To marine geochemists, hydrothermal activity represented a new supply of chemicals to the oceans, comparable in importance to the influence of rivers flowing from the land[4]. For marine biologists, the discoveries were, perhaps, even more outstanding. Here, fed by the chemicals emanating from the vents, living in total darkness, and isolated from almost all other strands of evolution, were extraordinary, undescribed species. These animals were previously unsuspected and certainly not looked for, yet were soon recognised to be living in an entirely novel ecosystem whose origins could be traced back to the earliest life on Earth (see also Chapter 13).

The pattern of hydrothermal circulation is one in which sea water percolates downward through fractured ocean crust toward the base of the ocean-ic crust and, in some cases, close to molten magma. In these hot rocks, the sea water is first heated before it reacts chemically with the surrounding host basalt. As it is heated, the water expands and its viscosity reduces. If these processes occurred on land, at atmospheric pressure, catastrophic explosions would result as temperatures would rise above 100°C and the water would turn into steam. But, because mid-ocean ridges lie under 2000–4000 m of sea water, at pressures 200–400 times greater than atmospheric pressure, the reacting sea water reaches temperatures up to 350–400°C without boiling. At these temperatures, the altered fluids do become extremely buoyant, however, with densities only about two-thirds that of the downwelling sea water; thus, they rise rapidly back to the surface as hydrothermal fluids. The movement of the fluid through the rock is such that, while the downward flow proceeds by gradual percolation over a wide area, the consequent upflow is often much more rapid and tends to be focussed into natural channels emerging at 'vents' on the sea floor.

Beneath the sea floor, the reactions between sea water and fresh basalt remove the dissolved Mg^{2+} and SO_4^{2-} ions that are typically abundant in sea water, resulting in the precipitation of a number of sulphate and clay minerals. As the water seeps lower into the crust and the temperature rises, metals, silica, and sulphide are all leached from the rock to replace the original Mg^{2+} and SO_4^{2-} ions. The hot and, by now, metal-rich and sulphide-bearing fluids then ascend rapidly through the ocean crust to the sea floor[1]. As soon as they begin to mix with the ambient, cold, alkaline, well-oxygenated deep-ocean waters there is an instantaneous precipitation of a cloud of tiny metal-rich sulphide and oxide mineral grains[6]. These rise within the ascending columns of hot water, giving the impression of smoke. Precipitation around the mouths of the vents over time builds chimneys through which the smoke pours, hence the term 'black smokers' (*Figure 10.8*); hot water gushes out of these tall chimney-like sulphide spires at temperatures of ca 350°C and at velocities of 1–5 m/s. Upon eruption, this hydrothermal fluid continues to rise several hundred metres above the sea bed, mixing with ordinary sea water all the time, in a buoyant turbulent plume (see later).

At slightly lower temperatures (below 330°C), the fluid may cool and mix with sea water sufficiently to deposit some metal-rich precipitates in the walls of the channels up which the fluids rise, before reaching the surface. In such cases, the particulate material formed when the ascending hot water finally emerges at the sea bed is made up predominantly of amorphous silica and various sul-

Figure 10.8 Black smokers on the sea floor of the East Pacific Rise at approximately 2600 m depth, near 21°N. Individual chimneys measure approximately 30–40 cm tall and 10 cm across, and are seen venting hydrothermal fluids at temperatures of 350–400°C and velocities of 1–5 m/s. The fluid within the chimneys is extremely clear, but rich in dissolved metals and hydrogen sulphide. As soon as this fluid mixes with cold oxygen-rich sea water, at the very mouths of the vents, precipitation of a range of sulphide and oxide minerals occurs, giving rise to the clouds of tiny black 'smoke' particles, which are seen billowing upward above the chimneys into the overlying sea water. A previously active vent chimney, which has been cemented solid by mineral precipitation, is at the extreme right. The light, angular area to the top of the structure represents the relatively fresh internal composition of the dark grey, almost cylindrical, chimney structure, where it has been broken open using the robot arm of the submersible (photograph courtesy of Woods Hole Oceanographic Institution and the American Geophysical Union[17]).

phate and oxide minerals, yielding a white cloud of mineral precipitates; the common name for these slightly cooler vents is thus 'white smokers'[5,31].

Individual high-temperature vents at mid-ocean ridges may only be ca 10 cm in diameter at their mouth, yet over time, growing like stalagmites from the sea floor, they can form chimneys anywhere from 1 m to 30 m tall. A typical vent field might comprise several such chimney structures spread over a circular area ca 100 m in diameter. Throughout this area there may also be a number of lower temperature vents emitting hot, shimmering water from the sea bed. Even these vents are as hot as 10–30°C, which is notably warmer than typical deep-ocean water (2–3°C)[18,25]. It is in the vicinity of these warm and more diffuse emissions that the majority of vent-specific biota are most abundant.

Life at Hydrothermal Vents

Hydrothermal fluids are enriched in dissolved hydrogen sulphide, a substance which is toxic to most forms of life; elsewhere in the oceans it is found only in lifeless, stagnant anoxic basins such as the Black Sea. It is quite remarkable, therefore, that sites of active hydrothermal venting are not barren wastelands where no life can exist. Instead, scientists diving at the very first vent site to be discovered, on the Galapagos rift in the eastern equatorial Pacific, were surprised to find dense concentrations of benthic (sea-bed dwelling) megafauna (large animals) living within the vent-field area. Since that time a number of new vent sites have been discovered around the world – and most have remarkably high concentrations of animals around them[17,32]. The reason why these areas attract such abundant life is even more remarkable. A common observation at all the vent fields that have been

studied to date is that the dominant species of animals at any site often appear to be extremely large. This has raised the question: "Where do the animals get food to grow at all, let alone enough to reach such a size?" The answer is that hydrothermal vent animals derive their food from a chain that is driven by geothermal (terrestrial) energy, unlike all the other organisms that rely, directly or indirectly, upon sunlight for their survival. In hydrothermal-vent communities, free-living bacteria, which are anchored on the sea bed or float free in the water column, coexist with symbiotic sulphide-oxidising bacteria, which live within the larger vent-specific organisms; these exploit the free energy of reaction released when hydrogen sulphide present in the vent fluids interacts with dissolved carbon dioxide and oxygen in ordinary sea water to form organic matter (equation 10.1), where $(CH_2O)_n$ is a carbohydrate. Because it is a *chemical*, hydrogen sulphide, which plays the role that sunlight plays in the more familiar process of photosynthesis in the warm surface waters of the oceans and on land, this unique deep-ocean, sunlight-starved process has been given the name *chemo*synthesis[2].

$$n(CO_2 + H_2S + O_2 + H_2O) = (CH_2O)_n + n(H_2SO_4) \qquad (10.1)$$

Approximately 95% of all animals discovered at hydrothermal vent sites are previously unknown species. So far, over 300 new species have been identified and, for many of these, the differences from previously known fauna are so great that new taxonomic families have had to be established in order to classify them satisfactorily[32]. Some of the most exciting examples of hydrothermal vent species discovered include the spectacular tubeworms found along the East Pacific Rise vent sites,

10.9

Figure 10.9 Dense populations of tube worms (*Riftia pachyptila*) and brachyuran crabs inhabiting the Genesis hydrothermal vent site, 13°N, East Pacific Rise (depth, 2636 m). The tubeworms, which can grow up to 2–3 m in length and may be 2–5 cm across, derive their nutrition from *symbiotic* sulphide-oxidising bacteria which live within their gut. The crabs, which may only grow to a modest 5–10 cm across, feed by scavenging on the red haemoglobin-filled 'plumes' of the tubeworms; these can extend up to 10 cm beyond the end of their protective chitin tubes to draw in the hydrogen sulphide rich waters necessary for chemosynthesis, but can be withdrawn rapidly back into their tubes for protection (photograph by R.A. Lutz, Institute of Marine and Coastal Studies, Rutgers University, and the American Geophysical Union[17]).

which can measure 2–3 m or more long, and which typically appear in thick clusters as shown in *Figure 10.9*. Also common along the East Pacific Rise are giant clams and mussels (*Figures 10.10* and *10.11*), which can often reach the size of a large dinner plate. The total biomass at any one hydrothermal site is typically very high. Indeed, hydrothermal vent fields have been likened to submarine oases which punctuate the deserted barren plains of the deep-sea floor (Chapter 13). In contrast, biodiversity at individual vent sites (i.e., the total range of different species present; see Chapter 15) is surprisingly low. Not only that, but the species present at vent sites in the different oceans show remarkably little similarity. For example, no giant tube-worms or giant clams have been found at any of the five known hydrothermal fields discovered so far in the North Atlantic Ocean. Instead, for example, the

10.10

10.11

Figure 10.10 Giant white clams (*Calyptogena magnifica*) living in clusters within the crevices between basalt pillows, in an area known as Clam Acres at 21°N on the East Pacific Rise. Other vent fauna include a clump of tube worms (*Riftia pachyptila*), galatheid crabs (*Munidopsis susquamosa*), and limpets. Note the robot arm of the submersible 'Alvin' collecting a clam approximately 25 cm in length (photograph by R.A. Lutz, Institute of Marine and Coastal Studies, Rutgers University, and the American Geophysical Union[17]).

Figure 10.11 Dense beds of mussels (*Bathymodiolus thermophiolus*) and clams (*Calyptogena magnifica*) at the Rose Garden vent site along the Galapagos Rift. Again, individual clams and mussels are typically 20–30 cm in length. Other vent fauna seen include galatheid crabs and snails (photograph by R.A. Lutz, Institute of Marine and Coastal Studies, Rutgers University, and the American Geophysical Union[17]).

Figure 10.12 A schematic cross-section of the TAG hydrothermal mound, 26°N Mid-Atlantic Ridge (the co-ordinates give the dive site). Hot vent-fluid (shown in pale blue) flows up an open fissure in the deeply faulted oceanic crust, and then percolates out through the entire sulphide mound along a tortuous network of interconnected channels, giving rise to the highest-temperature (350–365°C) black smoker fluids at the apex (50 m across and 50 m above the sea floor) and to the lower temperature (270–300°C), partly diluted white smokers around the outer section of the mound (200 m across and 20–30 m above the sea floor). Extinct chimneys are also seen across much of the outer mound, where earlier fluid flow has ceased because subsurface mineral precipitation has choked the flow-channels solid. Toward the flanks of the mounds, rubble deposits occur where oxidised and altered material from the hydrothermal mound has been broken up by mass-wasting ('landslide') events and carried out across the sea floor, to be deposited upon more typical volcanic basement and a thin veneer of more typical pelagic sediments (courtesy of Pierre Minon, © National Geographic Society).

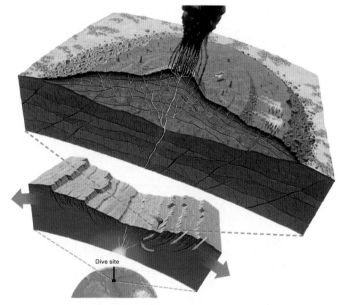

10.12

Dive site

southernmost three of these sites are characterised by abundant small shrimp, which cluster in their millions around the black-smoker chimneys.

Such completely isolated biological communities indicate separate paths of evolution over many generations – perhaps stretching back millions of years. However, we know that individual vent fields and chimneys may only remain active for periods of perhaps 100–1000 years at a time, before the chimneys themselves become choked with minerals and the flow of warm fluids becomes blocked. For any one species of hydrothermal organism to have survived down through the generations, therefore, we know that the ability to migrate from one vent site to another must be vitally important. Thus, an important question – particularly given the globe-encircling nature of mid-ocean ridges – is "How do animals move from one hydrothermal field to colonise another on an intra-basinal scale, yet live within communities which, on an inter-basinal scale, are evidently quite isolated?" One key area of research currently underway involves mapping the entire mid-ocean ridge system in sufficient detail to determine the total number of hydrothermal vent sites world-wide, their average spacing one from another, and how that spacing is controlled by the tectonic and volcanic nature of the mid-ocean ridges which host them. The second key issue to explain the biodiversity and separate evolution problems is to understand how animals reproduce and migrate along the lengths of these ridges. It is proposed that the processes of reproduction and migration must be intimately related to each other, because a large majority of vent-specific organisms exhibit quite sessile lifestyles as adults.

Therefore, only if these species give rise to planktonic larval stages is there any prospect of them being able to migrate along mid-ocean ridge axes and colonise new vent fields as and when they occur[32].

Hydrothermal Deposits

Hydrothermal deposits on the sea floor range from single chimneys to large, sprawling mounds topped by clusters of chimneys[20]. In general, the size of the hydrothermal deposits appears to be related to the length of time for which active venting persisted at those sites. Pacific vent fields, with very localised chimney deposits, have life-spans of just tens of years to perhaps a hundred years or so before mineral precipitation cements their chimneys solid[11]. By contrast, certain Atlantic sites are characterised by mounds stretching in excess of 200 m across and reaching 30–50 m high, and which dating reveals have remained active for thousands and even tens of thousands of years[13].

One such site that has been studied extensively by the international community is called TAG (named after a North Atlantic basin study in the 1970s – the Trans-Atlantic Geotraverse)[22]. The TAG site comprises one large active mound and several mounds that are now extinct. The deposit is one of the largest active sites known and is approximately 200 m in diameter and 50 m high. There is a vast range in the style of fluid venting and type of mineral deposit at different sites on the mound. Fluid flow through the mound is pervasive, fed by a complex network of channels through which high-temperature fluid flows, rising through the mound from the underlying basement (*Figure 10.12*). Hot

10.13

Figure 10.13 A cluster of inactive chimneys from the outer portion of the TAG hydrothermal mound. The chimneys each measure several metres in height and 10–30 cm in width. They have become clogged with minerals, as the supply of hydrothermal fluid from the underlying network of channels has waned – presumably due to further mineral precipitation subsurface, within the mound. Oxidation of the iron-, copper-, and zinc-sulphide minerals gives rise to the alteration of the chimneys' colour from blue–grey to brown. In time, these chimneys will become sufficiently altered and weakened to collapse, adding to the unconsolidated sulphide rubble which makes up much of the TAG hydrothermal mound. Evidence that this particular site has not long been inactive is also provided, however, by the presence of the luminous pale-blue mineral seen plated to the exterior of many of the chimneys. This mineral is anhydrite (a form of calcium sulphate), which only precipitates in the oceans at high temperatures and will continue to dissolve as the chimneys age and cool (courtesy of Woods Hole Oceanographic Institution).

(360°C) acidic fluids gush from the summit at speeds of several metres per second; the whole mound apex is shrouded in black smoke which is rapidly entrained upward into a buoyant hydrothermal plume. Those chimneys that have been sampled are 1–10 cm in diameter and consist predominantly of copper- and iron-sulphides (chalcopyrite, marcasite, bornite, pyrite) and calcium sulphate (anhydrite)[31]. Lower temperature, 'white smokers' vent at rates of centimetres per second to the southeast of the main black smoker complex. Here the temperatures are up to 300°C, and the chimneys are bulbous and zinc sulphide (sphalerite) rich. The white 'smoke' consists of amorphous silica mixed with zinc- and iron-sulphides. Much of the rest of the mound surface is covered with red and orange iron oxides, through which diffuse, low-temperature fluids percolate. Areas of diffuse flow are delineated by clusters of white anemones and shimmering water[20].

The range in fluid-venting styles is evident from the distribution of fauna over the mound surface.

Three different species of shrimp have been discovered at TAG; the most abundant is *Rimicaris exoculata*, which swarms over the black-smoker edifice. The lower-temperature diffuse flow areas host a community of anemones and crabs. The distribution of organisms can give semi-quantitative information as to the fluid flow regime[32].

Hydrothermal activity is intermittent, even at individual sites. Inactive chimneys have been discovered on the outer portions of the main mound and are shown in *Figure 10.13*. The chimneys have been oxidised to orange iron oxides and are beginning to crumble and collapse. Eventually, they will be completely weathered down to the mound surface. The mound itself is unstable and subject to collapse and mass wasting events, in which submarine equivalents of landslides, perhaps triggered by minor earthquakes, sweep altered chimney material down off the slopes of the mound and out onto the surrounding sea-floor sediments. As a result, the sediments immediately adjacent to the flanks of the mound are also full of metal-rich sulphide and oxide minerals, just like the hydrothermal chimneys from which they are derived[7,19]. Warm fluids can continue to percolate up through these flanking sediments, over time altering their mineral assemblage to clays. Despite this mineral alteration, the metal concentrations preserved within these sediments remain extraordinarily high – up to 45% iron and 34% copper, and as much as 10–15% zinc. The potential for mining such deposits in the future is discussed in Chapter 21.

Eventually, hydrothermal sediments are buried beneath the normal background pelagic sediments and transported away from the ridge axis by plate-spreading processes (see earlier). Deep-sea sediments have been drilled by the Ocean Drilling Program at over 700 sites. On those occasions where drilling has reached the basement rock, a metal-rich layer has often been observed at the bottom of the sediment pile, representing ancient metalliferous sediments[14]. Such basal metal enrichments are often the only evidence that hydrothermal activity has been extensive, not only spatially, but also throughout geological time.

Hydrothermal Activity, Ocean Circulation, and Ocean Composition

The effects of hydrothermal activity are not restricted to the immediate vicinity of black-smoker vent sites – although this is where their most visually spectacular impact is best observed. As hydrothermal fluids erupt from the sea floor, they remain buoyant as they mix with sea water, and rise, carrying their mix of particles and fluid upward in a conical expanding plume (*Figure 10.14*). As this turbulent, continuously mixing plume rises it eventually reaches a stage where it is no longer more buoyant than the surrounding water column, and so ceases to rise[28]. This particle-rich fluid is then dispersed as an approximately horizontal layer (often referred to as the neutrally buoyant plume), flowing along isopycnal (constant density) surfaces. Dispersion of this material through the oceans is then driven, primarily, by large-scale ocean circulation patterns[16].

It takes the fluid, solutes, and particles in a buoyant hydrothermal plume less than an hour to rise 100–300 m above the sea bed, before being carried away by the prevailing deep-ocean currents. During this time, mixing is so turbulent that the initial vent fluid is typically diluted approximately 10,000 fold by ordinary sea water[24]. Because the initial vent fluid is enriched by factors of up to a million in certain key chemical tracers (e.g., dissolved methane, dissolved manganese, and total suspended particulate matter), strong enrichments can still be detected after emplacement into and dispersion within neutrally buoyant hydrothermal plumes. Recognition of this characteristic of active

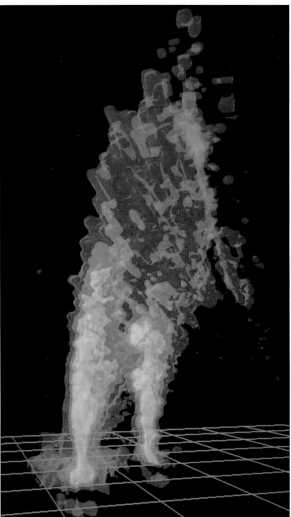

10.14

Figure 10.14 A three-dimensional acoustic image of the lower 40 m of two buoyant hydrothermal plumes discharging from adjacent black smoker vents at 2635 m depth on the East Pacific Rise, near 21°N. This image, obtained using a sonar (echo-sounder) system mounted on the sail of the US submersible 'Alvin', shows coalescence of the two plumes as they rise, as well as a 'bending-over' of the uppermost portion of their merged plume in the prevailing ocean-current direction. (courtesy of P.A. Rona, Institute of Marine and Coastal Studies, Rutgers University, and the American Geophysical Union[23]).

venting has greatly increased the ability of geochemists to locate new sites of hydrothermal activity[8,9,12,21]. Instead of needing to photograph every square metre of sea bed to look for individual vent fields, it has become sufficient simply to sample the overlying water column, perhaps just once every few kilometres, to detect the presence of any chemical or physical anomalies characteristic of hydrothermal discharge in any particular area.

10.15

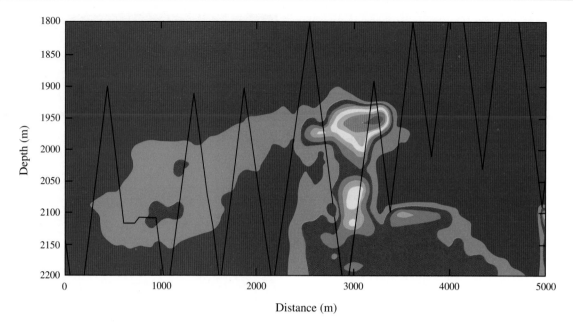

Figure 10.15 A two-dimensional cross-section of suspended particle enrichment along a 5 km segment of the 'Rainbow' hydrothermal plume at 36°15′N, Mid-Atlantic Ridge. Particle enrichment is caused by the precipitation of thousands of tiny (1–10 µm across) metal oxide and sulphide particles from black smoker chimneys, which are then swept up into the water column, up to 300 m above the sea bed, before being dispersed by prevailing ocean currents. Because the deep ocean is typically very clear and contains very little suspended particulate matter, optical devices to measure sea-water transparency (in this case, a nephelometer) can readily detect the presence of hydrothermal plumes and, hence, deduce the location of active hydrothermal vent fields. In this case, suspended particle concentrations increase approximately 10-fold from the background (dark blue; ≤35 µg of suspended particles per litre of sea water) to maximum plume concentrations directly above the vent site (shown in pink and purple) of 300–350 µg/l.

Figure 10.15 shows an example of a section of ridge crest some 5 km long that was surveyed in this way in September 1994. A nephelometer, an optical device which measures relative particle concentrations in sea water, was towed through the water column, and detected and mapped the presence of a particle-rich plume of water approximately 300 m above the sea bed over the Mid-Atlantic Ridge near 36°N – the 'Rainbow' hydrothermal field[9].

A new study of hydrothermal vents at Steinahóll, close to Iceland, has thrown up a new technique for studying hydrothermal activity. At this shallow site, the confining pressures (from only 300 m of water) are insufficient to prevent dissolved gases – including carbon dioxide, methane, and hydrogen – from bubbling out of solution. These bubbles rise all the way up the water column and can be imaged acoustically by an echo-sounder as they travel from the sea bed (*Figure 10.16*). In extremely calm weather, they can also be seen by the naked eye, 'popping' at the sea-surface, but in more typical, rougher, weather the sea surface is too disturbed and only sonar can reveal the hydrothermal activity below.

Although it has long been believed that neutrally buoyant hydrothermal plumes are dispersed passively by the prevailing deep-ocean currents, recent work suggests that this may not be the case. New research by physical oceanographers indicates that a process known as 'vortex shedding' may occur, in which the forces of the Earth's rotation combine with the upward motion of a buoyant hydrothermal plume to break off hydrothermal eddies (essentially, spiralling rings of neutrally buoyant plume material); these then migrate through the water column, away from their source at an active vent site[29]. This idea, however, has only recently been predicted, following laboratory-based tank experiments and theoretical calculations. No such features have been positively identified in the real oceans, yet, but the possibility that they might exist raises some very important questions. If hydrothermal eddies are formed regularly above vent sites, could they represent 'incubating' parcels of chemically enriched sea water, in which young larval stages of vent fauna might survive for hundreds of days as they are carried tens of kilometres up and down the world's mid-ocean ridge axes? Such a mechanism might go a long way toward resolving the key problem of how vent fauna colonise newly formed hydrothermal vent sites. In the extreme

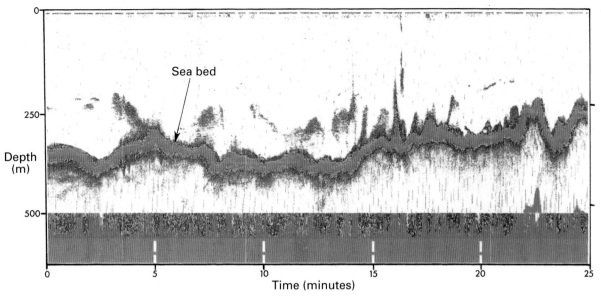

10.16

Figure 10.16 A high-frequency (38 kHz) echo-sounder trace of gas-rich venting along a 1000 m section of the Reykjanes Ridge through the Steinahóll vent field at 63°06'N, immediately southwest of the Reykjanes Peninsula, Iceland. Echoes from the solid sea-bed produce a strong (bright red) acoustic reflection, which is overlain by a more diffuse (blue) absorption–reflection pattern caused by plumes of bubbles rising from the sea bed. (At a ship's speed of 1.0–1.5 knots, 5 minutes of survey time, as indicated along the *x*-axis, is approximately equivalent to 200 m survey coverage over the sea bed.)

case, on a fast-spreading section of mid-ocean ridge crest, where abundant hydrothermal activity occurs at very close spacing, could the cumulative heat flux and consequent numerous hydrothermal eddies be of sufficient magnitude (acting in concert) to exert some control on the general physical circulation of the deep-water column over the length-scale of entire ocean basins?

Perhaps unsurprisingly for a field that is scarcely more than 15 years old, the questions currently arising from hydrothermal study greatly outweigh the answers that have been discovered, so far. Indeed, current lines of research seem to suggest that the full significance of hydrothermal activity and its impact upon numerous other aspects of oceanography have yet to be fully appreciated. It seems certain that hydrothermal circulation will continue to be a focus of research for many years to come.

General References

Humphris, S.E, Zierenberg, R.A., Mullineaux, L.S., and Thomson, R.E. (eds) (1995), *Physical, Chemical, Biological and Geological Interactions within Hydrothermal Systems*, Geophysical Monograph, American Geophysical Union, Washington, DC.

Parson, L.M., Walker, C.L., and Dixon, D. (eds) (1995), *Hydrothermal Vents and Processes*, Special Publication, The Geological Society, London, 396 pp.

Sinton, J.M. (ed.) (1989), *Evolution of Mid-Ocean Ridges*, Geophysical Monograph 57, American Geophysical Union, Washington, DC, 77 pp.

References

1. Bowers, T.S., Von Damm, K.L., and Edmond, J.M. (1985), Chemical evolution of mid-ocean ridge hot springs, *Geochim. Cosmochim. Acta*, **49**, 2239–2252.
2. Childress, J.J. and Fisher, C.R. (1992), The biology of hydrothermal vent animals: physiology, biochemistry and autotrophic symbioses, *Oceanogr. Marine Biol. Annu. Rev.*, **30**, 337–441.
3. Corliss, J.B., Dymond, J., Gordon, L.I., Edmond, J.M., von Herzen, R.P., Ballard, R.D., Green, K., Williams, D., Bainbridge, A., Crane, K. and van Andel, Tj.H. (1979), Submarine thermal springs on the Galapagos Rift, *Science*, **203**, 1073–1083.
4. Edmond, J.M., von Damm, K.L., McDuff, R.E., and Measures, C.I. (1982), Chemistry of hot springs on the East Pacific Rise and their effluent dispersal, *Nature*, **297**, 187–191.

5. Edmond, J.M., Campbell, A.C., Palmer, M.R., Klinkhammer, G.P., German, C.R., Edmonds, H.N., Elderfield, H., Thompson, G., and Rona, P. (1995), Time series studies of vent fluids from the TAG and MARK sites (1986, 1990) Mid-Atlantic Ridge: a new solution chemistry model and a mechanism for Cu/Zn zonation in massive sulphide orebodies, in *Hydrothermal Vents and Processes*, Parson, L.M., Walker, C.L., and Dixon, D. (eds), Special Publication, The Geological Society, London, pp 77–86.

6. Feely, R.A., Lewison, M., Massoth, G.J., Robert-Baldo, G., Lavelle, J.W., Byrne, R.H., von Damm, K.L., and Curl, H.C. J. (1987), Composition and dissolution of black smoker particulates from active vents on the Juan de Fuca Ridge, *J. Geophys. Res.*, **92**, 11347–11363.

7. German, C.R., Higgs, N.C., Thomson, J., Mills, R.A., Elderfield, H., Blustajn, J., Fleer, A.P., and Bacon, M.P. (1993), A geochemical study of metalliferous sediment from the TAG hydrothermal mound, 26°08'N, Mar, *J. Geophys. Res.*, **98**, 9683–9692.

8. German, C.R., Briem, J., Chin, C., Danielsen, M., Holland, S., James, R., Jónsodottir, A., Ludford, E., Moser, C., Ólafsson, J., Palmer, M.R., and Rudnicki, M.D. (1994a), Hydrothermal activity on the Reykjanes Ridge: the Steinahóll vent-field at 63°06'N, *Earth Planet. Sci. Lett.*, **121**, 647–654.

9. German, C.R., Parson, L.M., and Scientific Party of RRS *Charles Darwin* cruise CD89'HEAT' (1994b), Hydrothermal exploration at the Azores triple-junction, *EOS, Trans. Am Geophys. Union.*, **75**, 308.

10. Haxby, W.F. (1987), *Gravity Field of the World's Oceans: A Portrayal of Gridded Geophysical Data Derived from SEASAT Radar Altimeter Measurements of the Shape of the Ocean Surface (Scale 1:40,000 at the Equator)*, World Data Centre for Marine Geology and Geophysics.

11. Haymon, R.M. and Kastner, M. (1981), Hot spring deposits on the East Pacific Rise at 21°N: preliminary description of mineralogy and genesis, *Earth Planet. Sci. Lett.*, **53**, 363–381.

12. Klinkhammer, G., Rona, P., Greaves, M.J., and Elderfield, H. (1985), Hydrothermal manganese plumes over the Mid-Atlantic Ridge rift valley, *Nature*, **314**, 727–731.

13. Lalou, C., Thompson, G., Arnold, M., Brichet, E., Druffel, E., and Rona, P.A. (1990), Geochronology of TAG and Snakepit hydrothermal fields, Mid-Atlantic Ridge: witness to a long and complex hydrothermal history, *Earth Planet. Sci. Lett.*, **97**, 113–128.

14. Leinen, M. (1981), Metal-rich basal sediments from northeastern Pacific Deep Sea Drilling Project sites, in *Initial Reports of the Deep Sea Drilling Programme 63*, Yeats, R.S. and Haq, B.U. (eds), US Govt. Printing Office, Washington, pp 667–676.

15. Le Pichon, X. (1968), Sea floor spreading and continental drift, *J. Geohpys. Res.*, **73**, 3661–3697.

16. Lupton, J.E. and Craig, H. (1981), A major ^{3}He source at 15°S on the East Pacific Rise, *Science*, **214**, 13–18.

17. Lutz, R.A. and Kennish, M.J.(1993), Ecology of deep-sea hydrothermal vent communities: a review, *Rev. Geophys.*, **31**(3), 211–242.

18. Mills, R.A., Thomson, J., Elderfield, H., and Rona, P.A. (1993a), Pore-water geochemistry of metalliferous sediments from the Mid-Atlantic Ridge: diagenesis and low-temperature fluxes, *EOS, Trans. Am. Geophys. Union*, **74**, 101.

19. Mills, R.A., Thomson, J., and Elderfield, H. (1993b), A dual origin for the hydrothermal component in a metalliferous sediment core from the Mid-Atlantic Ridge, *J. Geophys. Res.*, **98**, 9671–9681.

20. Mills, R.A. (1995), Hydrothermal deposits and metalliferous sediments from TAG, 26°N Mid-Atlantic Ridge, in *Hydrothermal Vents and Processes*, Parson, L.M., Walker, C.L., and Dixon, D. (eds), Special Publication, The Geological Society, London, pp. 121–132.

21. Murton, B.J., Klinkhammer, G., Becker, K., Briais, A., Edge, D., Hayward, N., Millard, N., Mitchell, I., Rouse, I., Rudnicki, M., Sayanagi, K., Sloan, H., and Parson, L. (1994), Direct evidence for the distribution and occurrence of hydrothermal activity between 27°N–30°N on the Mid-Atlantic Ridge, *Earth Planet. Sci. Lett.*, **125**, 119–128.

22. Rona, P.A., Klinkhammer, G., Nelsen, T.A., Trefry, J.H., and Elderfield, H. (1986), Black smokers, massive sulphides and vent biota at the Mid-Atlantic Ridge, *Nature*, **321**, 33–37.

23. Rona, P.A., Palmer, D.R., Jones, C., Chayes, D.A., Czarnecki, M., Carey, E.W., and Geurrero, J.C. (1991), Acoustic imaging of hydrothermal plumes, East Pacific Rise, 21°N, 109°W, *Geophys Res. Lett.*, **18**(12), 2233–2236.

24. Rudnicki, M.D. and Elderfield, H. (1993), A chemical model of the buoyant and neutrally buoyant plume above the TAG vent field, 26°N, Mid-Atlantic Ridge, *Geochim. Cosmochim. Acta*, **57**, 2939–2957.

25. Schultz, A., Delaney, J.R., and McDuff, R.E. (1992), On the partitioning of heat flux between diffuse and point-source seafloor venting, *J. Geophys. Res.*, **97**, 12229–12314.

26. Searle, R.C. (1992), The volcano-tectonic setting of oceanic lithosphere generation, in *Ophiolites and their modern oceanic analogues*, Parson, L.M., Murton, B.J., and Browning, P. (eds), Special Publication 60, The Geological Society, London, 65–80.

27. Smith, D.K. and Cann, J.R. (1993), Building the crust at the Mid-Atlantic Ridge, *Nature*, **365**, 707–715.

28. Speer, K.G. and Rona, P.A. (1989), A model of an Atlantic and Pacific hydrothermal plume, *J. Geophys. Res.*, **94**, 6213–6220.

29. Speer, K.G. and Helfrich, K.R. (1995), Hydrothermal plumes: a review of flow and fluxes, in *Hydrothermal Vents and Processes*, Parson, L.M., Walker, C.L., and Dixon, D. (eds), Special Publication, The Geological Society, London, pp 373–385.

30. Spiess, F.N., MacDonald, K.C., Atwater, T., Ballard, R., Carranza, A., Cordoba, D., Cox, C., Diaz Garcia, V.M., Francheteau, J., Guerrero, J., Hawkins, J., Haymon, R., Hessler, R., Juteau, T., Kastner, M., Larson, R., Luyendyke, B., Macdougall, J.D., Miller, S., Normark, W., Orcutt, J., and Rangin, C. (1980), East Pacific Rise; hotsprings and geophysical experiments, *Science*, **207**, 1421–1433.

31. Thomson, G., Humphris, S.E., Schroeder, B., Sulanowska, M., and Rona, P. (1988), Active vents and massive sulfides at 26°N (TAG) and 23°N (Snakepit) on the Mid-Atlantic Ridge, *Canad. Mineralog.*, **26**, 697–711.

32. Tunnicliffe, V. (1991), The biology of hydrothermal vents: ecology and evolution, *Oceanogr. Marine Biol. Ann. Rev. 29*, 319–407.

33. Wolery, T.J. and Sleep, N.H. (1976), Hydrothermal circulation and geochemical flux at mid-ocean ridges, *J. Geophys. Res.*, **84**, 249–276.

CHAPTER 11:

The Ocean: A Global Geochemical System

J.D. Burton

Introduction

When the first systematic measurements of the chemical composition of sea water were made as part of the work of the *Challenger* Expedition of 1872–1876, the constituents which could be analysed with any accuracy were very limited in number. They comprised some of the more abundant dissolved salt components (*Figure 11.1*) and certain dissolved gases. Subsequent advances in analytical chemical techniques, allied with improved methods of sampling, gradually enabled a fuller knowledge of the concentrations of a wider range of constituents to be acquired. These advances accelerated remarkably from about the mid 1970s, so that there is now information not only on the concentrations, but also on the patterns of oceanic distribution for the majority of the elements. Information on the concentrations of some important dissolved elements in sea water is given in *Table 11.1*.

Marine scientists need information on chemical constituents for various purposes. Some constituents are useful as tracers, providing information on the circulation and mixing of water bodies in the ocean. They include the chlorofluorocarbons (freons), added to the atmosphere and thence to the ocean by man's use of them from about 1950 onward, and anthropogenic radioactive materials, some of which entered the ocean from the atmos-

pheric testing of nuclear weapons in the late 1950s and early 1960s, and some of which are due to discharges of low-level waste from nuclear fuel reprocessing. Examples of the kind of information obtained by the use of such tracers are given in *Figure 11.2*. Constituents such as nitrate, phosphate, and dissolved silicon have been studied very intensively because they are important plant nutrients, with an influence on the photosynthetic production of organic matter. Others, including trace metals (such as cadmium, mercury, and lead) and organic micropollutants (such as pesticides and polynuclear aromatic hydrocarbons) are of concern because of their potential impact on marine life in estuaries and coastal waters, which now receive increased inputs of these substances as a result of agricultural uses and the disposal of industrial and domestic wastes.

In a solution with so complicated a composition as sea water, many elements are present in a variety of physicochemical forms. For example, in the presence of the various inorganic, negatively charged ions (anions) in sea water, the dissolved element copper occurs to only a small extent as the positively charged divalent ion, the cupric Cu^{2+} cation, and is largely associated (complexed) with the anions. The distribution of copper between the simple cation and the complexes formed with various

Figure 11.1 Relative concentrations of dissolved major constituents in sea water. The bulk content of dissolved material is expressed by salinity (*S*), a defined dimensionless quantity which approximates to the total concentration (in g/kg) of the substances present in solution. Salinity can be measured at best to 0.001; only the few major constituents in the diagram contribute significantly to it. The actual mean concentrations of the constituents are given in *Table 11.1*. The mean concentration of dissolved silicon exceeds 1 mg/kg (approximately 0.001 in salinity), but its distribution is much more variable than those of the elements shown, and it is in a chemical form largely undetected by the widely employed conductimetric determination of salinity. It is therefore grouped with the minor or trace constituents, many of which occur at concentrations between 1 μg (10^{-6} g) and 1 ng (10^{-9} g)/kg, and others at still lower concentrations.

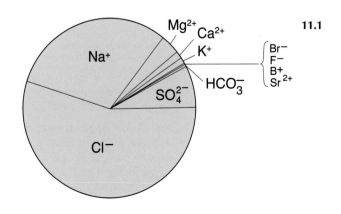

11.1

Table 11.1 Concentrations and most important chemical species of some dissolved elements in sea water.

Element	Average concentration	Range	Units (per kg)	Most abundant chemical species[a]
Lithium	174	b	µg	Li^+
Boron	4.5	b	mg	H_3BO_3
Carbon	27.6	24–30	mg	HCO_3^-, CO_3^{2-}
Nitrogen[c]	420	<1–630	µg	NO_3^-
Fluorine	1.3	b	mg	F^-, MgF^+
Sodium	10.77	b	g	Na^+
Magnesium	1.29	b	g	Mg^{2+}
Aluminium	540	<10–1200	ng	$Al(OH)_4^-$, $Al(OH)_3^0$
Silicon	2.8	<0.02–5	mg	$H_4SiO_4^0$
Phosphorus	70	<0.1–110	µg	HPO_4^{2-}, $NaHPO_4^-$, $MgHPO_4^0$
Sulphur	0.904	b	g	SO_4^{2-}, $NaSO_4^-$, $MgSO_4^0$
Chlorine	19.354	b	g	Cl^-
Potassium	0.399	b	g	K^+
Calcium	0.412	b	g	Ca^{2+}
Manganese	14	5–200	ng	Mn^{2+}, $MnCl^+$
Iron	55	5–140	ng	$Fe(OH)_3^0$
Nickel	0.50	0.10–0.70	µg	Ni^+, $NiCO_3^0$, $NiCl^+$
Copper	0.25	0.03–0.40	µg	$CuCO_3^0$, $CuOH^+$, Cu^{2+}
Zinc	0.40	<0.01–0.60	µg	Zn^{2+}, $ZnOH^+$, $ZnCO_3^0$, $ZnCl^+$
Arsenic	1.7	1.1–1.9	µg	$HAsO_4^{2-}$
Bromine	67	b	mg	Br^-
Rubidium	120	b	µg	Rb^+
Strontium	7.9	b	mg	Sr^{2+}
Cadmium	80	0.1–120	ng	$CdCl_2^0$
Iodine	50	25–65	µg	IO_3^-
Caesium	0.29	b	µg	Cs^+
Barium	14	4–20	µg	Ba^{2+}
Mercury	1	0.4–2	ng	$HgCl_4^{2-}$
Lead	2	1–35[d]	ng	$PbCO_3^0$, $Pb(CO_3)_2^{2-}$, $PbCl^+$
Uranium	3.3	b	µg	$[UO_2(CO_3)_3]^{4-}$

[a] Refers to the inorganic speciation in oxygenated waters.

[b] Variations are determined entirely or largely by those in salinity, i.e., the element is essentially conservative (see text for definition). For these elements, average concentration given is for sea water of salinity 35.

[c] Concentrations refer to combined nitrogen; element occurs also as dissolved nitrogen (N_2) gas. Species other than NO_3^- are often important in the upper ocean (e.g., NO_2^-, NH_4^+).

[d] Concentrations are affected by inputs to surface ocean of atmospherically transported lead from combustion of leaded petroleum.

Note:

Based mainly upon information in Bruland[4]. Usual ranges for oceanic waters are shown; concentrations of certain elements can be higher in some coastal waters.

Chemical oceanographers often employ molar units instead of the mass units shown here. For sodium (atomic weight 22.99) the concentration given above of 10.77 g/kg can alternatively be expressed as 0.449 mol/kg.

anions is shown in *Figure 11.3*. Copper is one of a number of metals which are also strongly complexed by the organic matter present in sea water, and so the speciation shown in *Figure 11.3* actually applies to only a small fraction of the dissolved copper. The dissolved organic material in sea water comprises a great diversity of molecules dominated by compounds which are described as 'marine humic material', because they show some analogies to the humic and fulvic acids produced in soils by the decomposition of terrestrial vegetation. These marine humic substances have not been fully characterised as regards their molecular structures, but show a range of capacities to form complexes with

Figure 11.2 (a) Distribution of tritium (^{3}H) in a longitudinal section in the western North Atlantic Ocean in the 1970s (Broecker and Peng[3]; with permission from the authors). Values are given in tritium units (TU; 1 TU = 1 x 10^{-18} atoms of tritium per atom of hydrogen). Surface concentrations are greatly enhanced above natural background values by the input of tritium from the atmosphere, due to thermonuclear weapon testing, particularly in the late 1950s and early 1960s. The penetration of tritium into deeper waters at high latitudes reflects the formation of North Atlantic Deep Water. Because tritium is an isotope of hydrogen and enters the ocean largely as tritium-labelled water molecules, it is an ideal tracer for the advection and diffusion of water from the surface to the deeper ocean. (b) Distribution of a chlorinated fluorocarbon (freon), CFC 11, in a section across the South Atlantic Ocean in the 1990s. The section runs along 45°S from the western margin to the mid-Atlantic Ridge and then in a northeastern direction to the eastern margin at 30°S. The higher concentrations in surface waters reflect exchange with the atmosphere and transfer downward by advection and diffusion. The increase toward the sea bed reflects the inputs to the deepest ocean by transport of Antarctic Bottom Water of more recent surface origin than the overlying North Atlantic Deep Water. This penetrates the Argentine Basin, west of the mid-Atlantic Ridge, more effectively than the Cape Basin to the east (unpublished data provided by D. Smythe-Wright and S. Boswell).

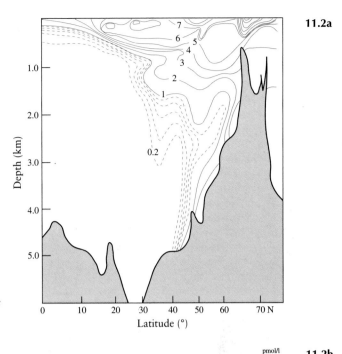

11.2a

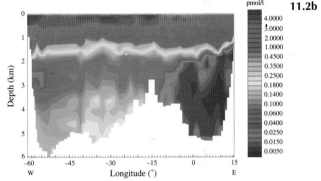

11.2b

Figure 11.3 The distribution of copper between the free cation and the complexes formed with various inorganic anions in sea water (chloride, Cl$^-$; sulphate, SO$_4^{2-}$; carbonate, CO$_3^{2-}$; and hydroxyl, OH$^-$). For trace constituents, information of this kind cannot be obtained directly by measurements on sea water, but is derived from models which assume equilibrium between the different species and use data on the stabilities of the various complexes. The nature of the organic complexes that account for a large part of the actual chemical speciation of copper in sea water cannot be similarly specified (see text).

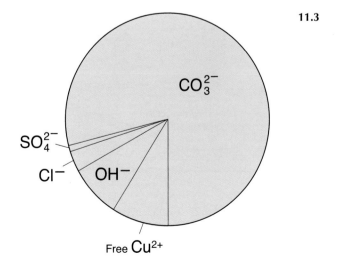

11.3

metals. In contrast with copper, the abundant element sodium has a very simple chemistry; it occurs in sea water almost entirely as the Na$^+$ ion.

The probable main inorganic chemical species for some important elements are shown in *Table 11.1*. Understanding chemical speciation is important because it affects the reactions of the elements in sea water. For example, the toxicity of copper to organisms such as phytoplankton (see Chapter 6), is related to the concentration (or, more strictly, the equivalent thermodynamic quantity, the activity) of the cupric ion rather than that of total copper. In chemical analyses, however, the chemical speciation is rarely resolved, so dissolved concentrations generally refer here to the total element or ion.

Knowledge of the chemistry of the ocean relies heavily on the analytical chemical measurements which reveal the spatial distributions of chemical constituents and their variations in time. Because many constituents of interest are present in sea water at very low concentrations, they can be determined only by sensitive methods; special precautions are necessary to avoid contamination during sampling and analysis. Given the complexity of the mixture and the different levels of concentration at which constituents occur, the potential interference of one in the analysis of another must be avoided. For these reasons, analytical measurements on sea water often involve quite complicated chemical manipulations to concentrate and separate the required chemical forms. Very few chemical constituents can be measured by probing the ocean with a sensor, in the way that salinity can be measured (see Chapter 19). Much of the picture of their oceanic distributions is, therefore, coarsely resolved, but the great progress made in the acquisition of reliable data over the past two decades has, nevertheless, underpinned comparable advances in understanding the chemical processes that determine these distributions.

The Geochemical Context

The oceanographic study of chemical properties and processes is undertaken, as already indicated, from a variety of standpoints. The most central, however, and the one which provides the most valuable unifying insights, is that of geochemistry – the science which addresses questions concerning the occurrence and distribution of the chemical elements within Earth as a whole and within its major components, in space and time. These major components include the crust (lithosphere), the atmos-

11.4

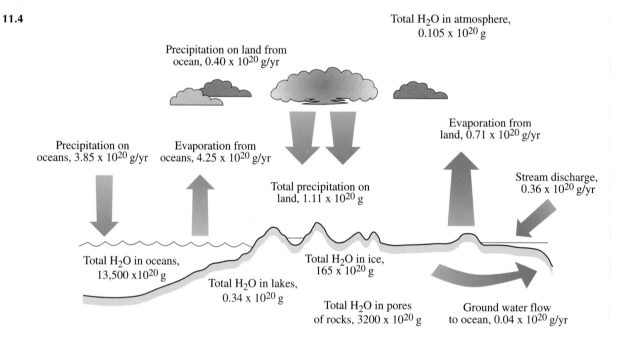

Figure 11.4 The global water cycle – the numbers show the total amounts of water held in the major reservoirs and the fluxes of water between reservoirs. In effect, the cycle acts not only as a transport system for weathered continental material, but also as a liquid extraction system for continental solid phases. Some of the water distilled by evaporation from the ocean, containing only a small proportion of sea-water salts as aerosol (see text), is deposited as precipitation on the continents and returned to the ocean containing dissolved material leached from continental rocks. Repeated cycles create a continuous net transfer of dissolved constituents to the ocean (based on data from Baumgartner and Reichel[2]).

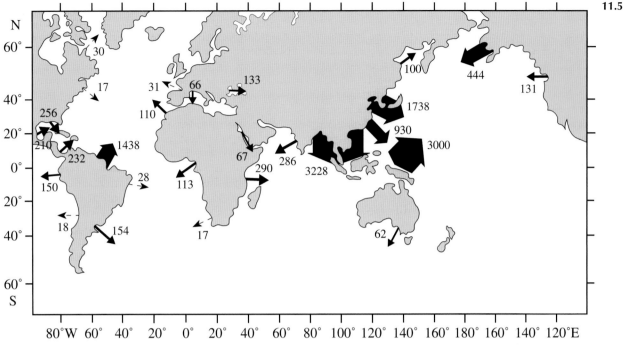

Figure 11.5 Fluxes of river-borne particulate material discharged from the major drainage basins. The widths of the arrows are proportional to the discharge; numerical values are in units of 10^6 tonne/yr. Fluxes to the Arctic Sea, totalling 226×10^6 tonnes/yr, are not shown. The directions of the arrows are arbitrary. Continental erosion in South East Asia and adjacent islands accounts for much of the discharge. The Amazon carries about 20% of the total water discharge to the ocean, but is not proportionately important as a source of either particulate or dissolved material. In some catchments, weathered particulate material is efficiently trapped in lakes or artificial reservoirs so that relatively little reaches the coast, e.g., the St. Lawrence (after Milliman and Meade[15]; with permission from the University of Chicago Press).

phere, and the aquatic environments collectively described as the hydrosphere, of which the ocean comprises the major mass, accounting for more than 98% of the free water at Earth's surface. The geochemical perspective emphasises the fact that the ocean is a dynamic system, chemically as well as physically, continuously exchanging material at its boundaries and also distributing it internally among its major water masses.

The major sedimentary cycle

The role of the ocean is central in the major sedimentary cycle. Continental rocks undergo physical weathering, which produces particles (mineral dust) small enough to be lifted from the ground by wind action and transported, particularly from arid regions, through the atmosphere (aeolian transport); particles are also carried by surface water run-off. Water, containing carbon dioxide dissolved from the atmosphere, is a powerful agent of chemical weathering, reacting with rock minerals to produce dissolved constituents and solid-phase

products, of which the clay minerals are particularly important.

The global water cycle (*Figure 11.4*; see also *Figure 2.2*) thus drives a continuous flux of dissolved and particulate material to the seas, globally amounting to estimated annual inputs of about 3.7×10^{15} g of dissolved material and about 15×10^{15} g of particles. Estimates of global fluxes are essential for understanding the ocean's material budget, but these values are averages of discharges, which show large differences regionally, as illustrated for particulate material in *Figure 11.5*. About 10% of the riverine flux of dissolved material is recycled from the ocean to the continents, mainly through the agency of bubbles of air: these bubbles are produced largely by breaking waves, are then carried below the sea surface, and, on returning to the surface, burst to create sea-salt aerosols, a fraction of which is transported through the atmosphere and deposited on land.

Much of the particulate material discharged from rivers settles out to form bottom sediments in estuaries, deltas, and the deeper-water submarine fans formed by some major tropical rivers. It is estimated that, globally, under 10% of the input of particles from rivers reaches the open ocean. Aeolian transport carries to the ocean an estimated 0.9×10^{15} g of mineral dust annually. As this mechanism is capable of distributing material over great distances before deposition on the ocean surface, mainly within the latitude belt of injection into the atmosphere, it represents a source of particles for the open ocean comparable in magnitude with the rivers. Transport of the products of weathering by water and wind is not the only process by which material enters the ocean, but it is certainly the best understood and is probably the dominant input for most elements. It is therefore a logical starting point for considering the way in which the ocean fits into the larger framework of geological processes.

The *primary* source for weathered material is igneous crustal rock, because the sedimentary rocks, such as limestones, sandstones, and shales, are themselves products of the major sedimentary cycle. The present-day supply of weathered material to the ocean, however, contains a large contribution, approximately 75% of the total, derived from reweathering of sedimentary rocks formed from sediments deposited earlier in geological time and emplaced on the continents tectonically (see Chapter 8). Early geochemical calculations indicated that the total amount of some constituents present in the ocean, sediments, and sedimentary rocks could not be accounted for by the weathering of the same quantity of igneous rock that is sufficient to supply the amounts of the major rock-forming elements, such as sodium and calcium, found in these reservoirs. These 'excess' constituents in the sedimentary and oceanic domains share the property that they are all volatile compounds, or can be derived from volatile compounds; they include the major anions present in sea water (chloride, sulphate, bicarbonate, and bromide), water itself, and nitrogen, which is the dominant constituent of the atmosphere.

The chemical evolution of the ocean

On this basis the evolution of the ocean, atmosphere, sediments, and sedimentary rocks can be considered as a global-scale reaction, over geological time, between the minerals (mainly silicates and oxides) contributed by the igneous rocks and acidic volatiles, particularly hydrochloric acid, sulphur-containing gases, and carbon dioxide. The primary supply of the volatiles is by outgassing of the mantle. There are several models for this process and its time course, and for the rate of deposition of the sedimentary rocks. Mantle outgassing is a continu-

ous process; direct evidence for this is provided by inputs of ^3He of mantle origin at mid-ocean ridge spreading centres. It is reasonably certain, however, that there was substantial degassing in the early stages of the formation of igneous rocks from magma, the released gases subsequently reacting with the crystallised rock minerals at low temperatures.

The major stages in the evolution of the ocean and the atmosphere were largely completed by the beginning of the Phanerozoic Eon (about 6×10^8 years ago); subsequently, the mass of sedimentary material has changed little. During the past 2.5–3×10^9 years, the mass of sedimentary rock that formed is about five times as much as that existing now[7], material having been recycled by reweathering of sedimentary rocks or destroyed by high-temperature metamorphism, after subduction of sediments into the mantle. The present ocean can be regarded broadly as being close to a steady-state system, with the production of sediments by weathering largely balanced by their high-temperature metamorphism, leading to a recycling of the volatiles consumed in weathering. The only one-way process is the continuing release of primary volatiles from the mantle. Carbon dioxide is the main volatile involved in present-day weathering.

While many features of oceanic composition can be understood in terms of the interaction of mantle-derived volatiles and continental rocks, reactions between sea water and the basaltic oceanic crust must also be considered. These reactions occur under a wide range of temperatures and ratios of rock-to-water mass[21]. The most investigated are the hydrothermal reactions involving sea water that has penetrated deep into newly formed basalt at mid-ocean ridge spreading centres, and which is returned to the ocean by convective circulation in the crust (Chapter 8). For some constituents, such as magnesium and sulphate, these reactions lead to removal from sea water; for others, such as calcium and important trace metals, including manganese, they create an additional input to the ocean. The rate at which sea water circulates through hydrothermal reaction zones is a matter of much controversy. This is reflected in uncertainties as to the global-scale impact of the primary hydrothermal reactions and also the reactions which occur when hydrothermal fluids mix with waters of normal composition in the depths of the ocean.

A major aspect of the evolution of the ocean is the establishment of oxygen as a permanent constituent of the atmosphere, since this enabled the change to occur from a chemically reducing Earth-surface environment to one which is oxidising. Oxygen is produced by photosynthesis and by dissociation of water vapour in the upper atmosphere, with loss to space of some of the hydrogen released in the process. While photosynthesis heavily dominates the present-day production of oxygen, the relative importance of the processes in earlier periods of geological time is uncertain. In the initial stages, the oxygen produced would have been consumed in reactions with the abundant reducing materials at the Earth's surface. The build-up of atmospheric oxygen from the photodissociation of water vapour requires that the production rate exceeds the consumption rate in such reactions. For photosynthesis to contribute to the build-up of atmospheric oxygen, some of the organic matter must be buried and not oxidised in respiration, otherwise all the oxygen produced would be consumed. The intervention of man, acting as a geochemical agent on a global scale, has led to the combustion, in less than two centuries, of a significant amount of this buried carbon. This has produced the fossil-fuel effect of significantly rising atmospheric concentrations of carbon dioxide (e.g., see *Figure 12.3*). This combustion has removed some oxygen from the atmosphere, but because the atmosphere contains so much higher a concentration of oxygen (21%) compared with that of carbon dioxide (0.03%), the change in concentration of oxygen is very difficult to detect.

At the present time the atmosphere can be regarded as close to a steady-state, in which oxygen production is balanced by respiration and oxygen participation in weathering reactions with reducing materials, such as sulphides and ferrous iron. Probably, the partial pressure of oxygen was still rising in the early Phanerozoic Eon, but has changed relatively little during the Cenozoic Era. Although ocean surface waters typically contain dissolved oxygen at saturation concentrations, or somewhat above, respiratory processes of organisms, including bacteria, can lead to low concentrations in some subsurface waters. In environments where the deeper waters are isolated from mixing with surface waters, such as the Black Sea, anoxic conditions can develop. The extent of these environments is very limited at the present time, but because of their sensitivity to changes in circulation the extent has varied episodically within the Phanerozoic Eon.

The role of the ocean in relation to Earth surface conditions appears differently from the standpoints of different time-scales. Over the whole period of the evolution of the sedimentary system major shifts have occurred in, for example, the distribution of sulphur between sulphides and sulphates, and that of carbon between carbonates and organic carbon[7,8]. In terms of these transfers, the ocean acts as a medium of transfer rather than of storage. An

extreme difference in perspective occurs when considering those reservoirs of carbon (atmosphere, ocean, biota, humus), which constitute a very short-term exchange system in which carbon is turned over on time-scales up to about 1000 years. In this system, the ocean is a major storage reservoir.

Evidence, particularly from salt deposits formed by crystallisation from sea water under conditions of high evaporation, suggests that over the last 10^8 years, and perhaps for most of the Phanerozoic Eon, the average composition of sea water, with regard to major constituents, has not undergone a major evolutionary change. Some constituents may have varied in concentration by less than a factor of two, but constraints on the likely concentration of calcium do not preclude variations during the Phanerozoic Eon in a range 25–400% of the present value[8]. The present-day river input of dissolved calcium maintained over the period would, however, have introduced about 80 times the amount of calcium now present in the ocean. Such considerations support the assumption made by geochemists that, in many respects, the ocean can be modelled as a steady-state system in which supply of material is balanced by its removal. As discussed later, the turnover time for the major constituents in the ocean is long. The steady-state assumption, therefore, is not incompatible with evidence that supply has exceeded removal, or vice versa, over shorter periods, such as during periods of rising or falling sea level. For constituents with more rapid rates of turnover in the ocean, significant changes in oceanic concentration can occur over shorter periods and there is considerable evidence, discussed below, of changes in the ocean's regulation of atmospheric concentrations of carbon dioxide over the glacial to inter-glacial time-scale (see also Chapter 12).

The marine sedimentary sink

Incorporation of material into bottom sediments is a removal mechanism from the ocean, so that the sediments function dominantly as an ocean sink. The particulate material settling down the water column and depositing as sediments has two dominant sources. The input to the ocean of particles derived from the lithosphere (lithogenous material) has already been described. In the remote ocean, this material accretes slowly, typically of the order of 1 mm in a thousand years. The other major source is the production within the ocean of biogenous particles, mainly from surface-living organisms within the ocean. This includes a small fraction of the organic matter which escapes oxidation by respiration in the water column or at the sediment surface (see Chapter 7 for an account of the deposition of 'marine snow'). The major

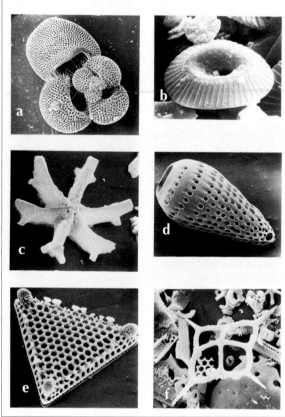

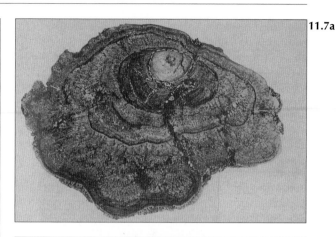

5 cm

Figure 11.6 Skeletons of planktonic organisms constitute a major part of the sediments over extensive areas of the deep sea. The photographs, obtained from oceanic oozes with a scanning electron microscope, show calcareous [(a) foraminiferan, x 53; (b) coccolith, x 4000; (c) discoaster, x 2700] and siliceous [(d) radiolarian, x 440; (e) diatom, x 440; (f) silicoflagellate, x 970] planktonic remains (*Figure 7.7* from Kempe[9], with permission from the Natural History Museum, London).

Figure 11.7 Cross sections of ferromanganese nodular concretions ('manganese nodules') from the sea floor. (a) Polished radial section of a nodule from the equatorial North Pacific Ocean, showing asymmetrical growth of layers around a nucleus (dashed white line) of consolidated sediment (note the discontinuity in the dark layer, indicated by white arrows, which shows that the orientation of the nodule on the sea bed has not remained constant; from von Stackelberg[23]; with permission from the author and D. Reidel Publishing Company). (b) Polished radial section of a nodule from the Blake Plateau (Atlantic Ocean). The light-coloured veins consist largely of calcite and clay, which have accumulated in cracks in the growing nodule. The nodule accreted around a piece of phosphorite, which is no longer clearly distinguishable as a nucleus, having been partly replaced by ferromanganese material (photograph by J. Mallinson).

biogenous components, however, are the minerals produced as skeletal material by organisms (*Figure 11.6*). Where these are important sedimentary components, they accumulate typically at rates of about 1 cm every thousand years.

Of less importance in terms of sediment mass, but of great geochemical interest, are the solid phases formed by chemical reactions in sea water; these sediments are sometimes termed hydrogenous (formed in the hydrosphere), but more commonly are described as authigenic. Important examples

are the sediments deposited by hydrothermal plumes (see Chapter 10) and the ferromanganese concretions (*Figure 11.7*), which are abundant on the sea bed in some regions and have potential economic value (see Chapter 21). Other notable authigenic phases include phosphorite (carbonate fluorapatite), which occurs particularly in eastern boundary regions of the ocean, where coastal upwelling is frequent, and glauconite (a potassium-rich aluminosilicate). Glassy basalts reacting with sea water are so extensively altered that the minerals formed

11.8

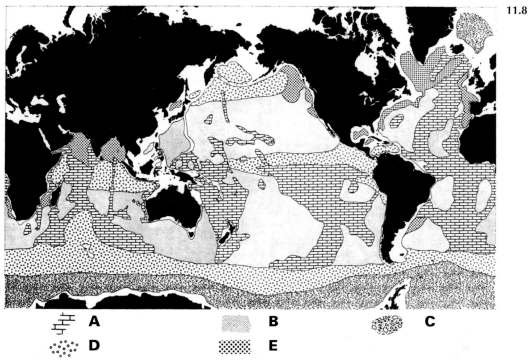

Figure 11.8 Dominant sediment types, classified by origin, in different areas of the deep-sea bed. Low concentrations of calcareous sediments in the North Pacific Ocean reflect a shallower calcite compensation depth (see text) in this region; a general tendency can be seen for high concentrations of calcium carbonate to occur on the mid-ocean ridges, where the depth of water is usually less than the calcite compensation depth (A, calcareous sediment; B, deep-sea clay; C, glacial sediment; D, siliceous sediment; E, terrigenous sediment; ocean margin sediments occur in the areas left blank; from Davies and Gorsline[6], with permission from the authors and Academic Press).

should also be regarded as authigenic; these reactions account for the abundant zeolite mineral, phillipsite, and the clay mineral, montmorillonite, in sediments of the South Pacific basin. During some geological periods, large masses of evaporite sediments have been produced by the crystallisation of salts from sea water that has become concentrated by evaporation in shallow marginal areas. Such deposits are the starting materials for continental salt beds of economic importance. The major features in the distribution of sediment types in the deep sea are shown in *Figure 11.8* (see also *Figure 8.21*).

The Oceanic Particle Conspiracy

The dissolved constituents in the ocean can be removed, apart from those trapped in accumulating sediments as pore waters (see also *Figure 8.26*), by conversion into particles or by binding to particle surfaces. The solution composition is controlled by the balance between the input processes and the particle–solution interactions that occur in the water column. Turekian[22] summed up these interactions as follows: "The great particle conspiracy is active from land to sea to dominate the behaviour of dissolved species ... As these scrubbing agents operate around the world, they imprint their material burdens on accumulating sediments."

The particle conspiracy involves many processes and reactions. Dissolved ions undergo electrostatic exchanges at charged sites on particle surfaces. Potentially stronger associations arise by specific binding with chemical groupings, especially hydroxyl (OH-) and carboxyl (COO-), a process strongly encouraged by the ubiquity of coatings of organic matter on particle surfaces. Incorporation into organic material, and into skeletal minerals produced by organisms, involves the major structural elements, such as carbon, calcium, and silicon. Also, many of the trace elements become caught up in the formation processes, either because they have essential biochemical roles, like the metals which occur in enzymes, or adventitiously, because of chemical similarities to the major elements or a tendency to form complexes with organic substances. As a result of these various processes the concentrations of many elements are regulated at levels greatly below those at which actual precipitation occurs. The various processes by which elements in solution at low concentrations become associated with particles already present in the system are collectively referred to as 'scavenging'.

11.9

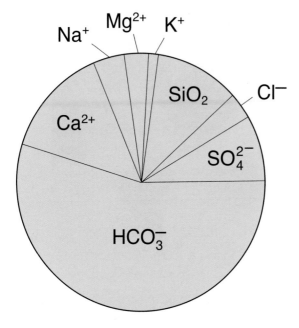

ocean with this drainage reside there for much longer times, on average, than do water molecules, so sea water becomes more concentrated in both elements (i.e., these accumulate in the ocean). Sodium, however, accumulates to a greater extent than does calcium. Elements of high geochemical reactivity, with MORTs less than that of the water supplied by rivers, are depleted in sea water relative to the concentration in which they are supplied in river water. The contrasting abundances of a wide range of elements in river and sea waters are shown in *Figure 11.11*.

The vertical and horizontal distributions of chemical constituents

The particle conspiracy not only affects the rates of turnover of the elements in the ocean, it also produces considerable variations in the concentration of some elements within the ocean, both vertically and horizontally. The ocean mixes internally on a time-scale of about 1000 years. This is short enough for the inputs of elements with long MORTs to become very thoroughly mixed within the ocean. Variations in concentration for these constituents do occur near their sources and where rainfall and evaporation change the salt content significantly. When normalised for salinity, however, the concentrations are uniform, within the limits

Elements differ widely in their tendency to undergo transfer from solution to solid phases; their ability to do so is often described as their 'particle reactivity' or 'geochemical reactivity'. This tendency is correlated with certain physicochemical parameters, such as ionic potential (the ratio of charge-to-radius of the element cation), which provide a rationalisation of geochemical behaviour in terms of fundamental chemical properties, such as periodicity. Differences in geochemical reactivity are apparent from a comparison of the average composition of the dissolved material in rivers entering the ocean (*Figure 11.9*) and that of sea water (*Figure 11.1*). Some differences (e.g., for chloride) are mainly because the constituent is dominantly supplied as a volatile rather than as the product of rock weathering. Features such as the much lower ratio of calcium-to-sodium in sea water than in river water, however, reflect the fact that constituents introduced by rivers accumulate to different extents in sea water; calcium is much more efficiently removed from the ocean than is sodium, because, as already seen, it enters into the formation of abundant sediments containing biogenous calcium carbonate.

An extremely useful, albeit crude, index of geochemical reactivity is the mean oceanic residence time (MORT) (*Figure 11. 10*). The MORTs for sodium and calcium are of the order of 10^8 years and 10^6 years, respectively. The MORT of water molecules in the ocean reservoir is dependent on their passage through the atmosphere and return through continental drainage and can be derived from the information in *Figure 11.4*; it is about 4 x 10^4 years. Sodium and calcium ions carried into the

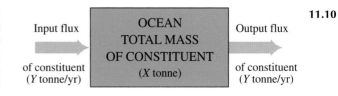

11.10

Figure 11.10 The mean oceanic residence time (MORT) of a constituent is derived by making the steady-state assumption that input and removal are balanced (see text) and is equal to X/Y years. Estimates of MORT are commonly based on the river input to the ocean. The model also assumes that a constituent becomes well mixed within the ocean reservoir prior to removal. This is not the case with the most particle-reactive metals (such as aluminium, iron, and manganese). For these elements, the calculated MORTs are of the order of 100–1000 years, which must be treated as nominal values.

Figure 11.11 Comparison of the average concentrations of various elements in sea water (c_{SW}) with those in the average river-water (c_{RW}) entering the ocean. The solid line represents equal concentrations of an element in sea water and in river water (the dashed lines show limits of one order of magnitude above and below the 1:1 relationship). Elements above the line accumulate in sea water to concentrations above those in the river water, in some cases to a very substantial degree (e.g., Na). Those considerably below the line are particle-reactive elements (e.g., Pb), which are markedly depleted in sea water relative to the concentration in the average river-water supply to the ocean (after Whitfield and Turner[24]; with permission from the authors and Macmillan Magazines Ltd).

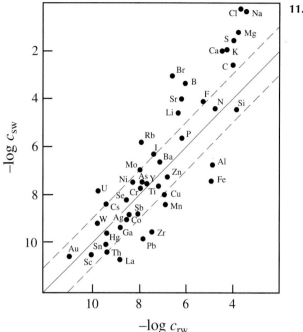

of analytical measurement, over almost the whole of the ocean. For example, the ratio of the concentration of sulphate to salinity in the Irish Sea does not differ detectably from that in the Pacific Ocean. Such constituents are termed 'accumulated' [*Figure 11.12(a)*]; like salinity, their concentrations are conserved in ocean mixing and they are, therefore, also referred to as conservative constituents. The major constituents tend to be conservative. The near-constancy of the relative proportions of the major constituents in sea water provides the basis for practical measurements of salinity for both the present-day approach, based on conductivity, and the main previous approach based on chlorinity (a defined quantity related to the concentrations of chloride and bromide).

Some minor constituents also show essentially conservative behaviour, but their low abundance in the supply to the ocean limits the oceanic concentrations that they can attain. These elements include lithium, caesium, and rubidium, which have chemical properties similar to sodium and potassium, and uranium, which has a low particle reactivity because it forms a strong complex with dissolved carbonate ions.

Elements with short MORTs are removed, however, too rapidly to become well-mixed. Their high geochemical reactivity means that, in the absence of local sources at the deep sea-bed (e.g., hydrother-

mal), they tend to decrease markedly in concentration away from the lateral and surface boundaries. These constituents are termed 'scavenged' [*Figure 11.12(b)*].

There is a very important further group of constituents with intermediate residence times that are long enough, on the face of it, for them to become quite uniformly distributed by the water-mixing processes in the ocean [*Figure 11.12(c)*]. Their particle reactivity is largely accounted for by associations with biogenous particles; they include the main micronutrients – phosphate, nitrate, and dissolved silicon. These constituents tend to be efficiently stripped from the surface waters, where most biogenous particles are produced. Decomposition of organic matter and dissolution of mineral skeletons deeper in the water column, or at the sedimentary interface, lead, however, to release of these elements from sinking particles.

11.12

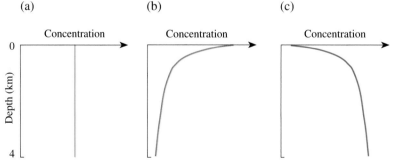

Figure 11.12 Schematic representation of idealised vertical profiles for elements representing the three general types of behaviour in the ocean: (a) accumulated (normalised to salinity); (b) scavenged; and (c) recycled. The factors which underlie these types of distribution are explained in the text and actual examples of types (b) and (c) are given in *Figures 11.14* and *11.15*, respectively.

175

Where the permanent pycnocline forms a stratified water column, the return of the dissolved material to the surface by water transport is much less effective than is the down-column transport when the element is linked to a particle. There is, therefore, an efficient 'biological pump' for these constituents from surface to deep water, so they increase in concentration with depth over much of the ocean, an increase which may be very pronounced [*Figure 11.12(c)*]; they are termed 'recycled elements'. A proportion of the sinking particles undergo deposition into the sediment, but the average atom of a recycled element travels with a particle into the deeper water and is released and returned to the surface layer many times before it is finally removed from the water column.

The organic matter produced by photosynthesizing organisms in the upper ocean is a complex mixture of compounds. Nevertheless, in the deeper waters, where recycling occurs, the increases in the concentrations of nitrate and phosphate and the decrease in the concentration of dissolved oxygen, which is used to oxidise the organic matter, occur in essentially constant proportions. This reflects the fact that, when averaged over time and space, the uptake of the main elements (carbon, nitrogen, and phosphorus) used to synthesize cellular material occurs in a common relationship, which reflects the mean composition of the mixture of compounds produced, such as carbohydrates, lipids, proteins, and nucleic acids. The atomic ratios in which carbon, nitrogen, and phosphorus are utilised are taken as 106:16:1; oxygen is utilised in the oxidation of organic matter in the ratio of 138 molecules per 106 atoms of carbon. These are known as Redfield ratios[18].

The types of distribution shown in *Figure 11.12* are idealised, but they also match quite closely the actual vertical profiles of constituents which show these behaviours to a marked extent (see *Figures 11.14* and *11.15*). All constituents show some particle reactivity and undergo recycling and scavenging, but for some the consequences, in terms of distribution, are less marked; in some cases they may be difficult or impossible to detect analytically. This may be because the down-column flux is low in absolute terms or because it is low relative to the dissolved concentration in the ocean reservoir. Only a slight vertical concentration gradient is detected for calcium in the ocean, despite the large flux of calcium carbonate from surface waters to the deep ocean, because of the high concentration of calcium in sea water. Few constituents of sea water are as abundant as calcium, however, and generally a major involvement in the particle conspiracy is reflected in clearly nonconservative behaviour.

The two main types of nonconservative behaviour are reflected in systematic horizontal variation in concentrations in the intermediate and deep waters of the major ocean basins[3]. A major part of the deep water in the ocean basins has a common origin in surface waters of the North Atlantic Ocean, particularly in the Norwegian Sea, where dense waters formed by surface cooling sink and flow south in the western Atlantic (*Figure 11.13*). The deep water, modified by additions from the Southern Ocean, flows northward in the Indian Ocean and in the Pacific Ocean. During this transit, the composition of the water is influenced by the particles which enter it from the surface. As the water becomes older, in the sense that more time has elapsed since it sank from the surface, more particles passing through on their way to the sediments scavenge the strongly particle-reactive constituents. The concentration of these constituents therefore decreases along the direction of the deep-water flow (*Figure 11.14*). In contrast, the recycled constituents increase in concentration (*Figure 11.15*), because the older waters have been subjected to more releases from biogenous particles sinking from the waters that overlie their route.

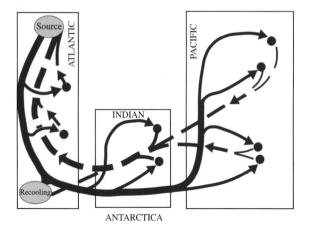

Figure 11.13 Schematic representation of the circulation of deep water in the major ocean basins (Broecker and Peng[3]; with permission from the authors). The solid lines show the flow of deep water, originating largely by sinking in the far North Atlantic Ocean. The dashed lines show the flow of surface water and the filled circles represent areas of localised upwelling.

Figure 11.14 Vertical profiles of some scavenged dissolved elements in the oceanic water column: (a) dissolved manganese and (b) dissolved aluminium in the northeastern Atlantic Ocean; (c) dissolved aluminium in the central North Pacific Ocean. Dissolved aluminium has a MORT of about 1000 years, and is therefore removed from the ocean on a timescale similar to the transit time of deep water from the North Atlantic Ocean to the North Pacific Ocean. There is thus a marked decrease in concentration between the two profiles. A strong sedimentary source shows up

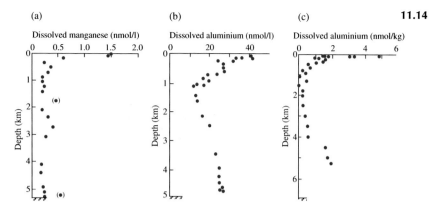

in the bottom waters in the North Pacific profile against the low deep-water concentrations. Surface input is dominated by dissolution of aeolian-transported mineral dust in profiles (a) and (c); this also adds significantly to the off-shelf transport of dissolved aluminium, which is important in the surface waters in profile (b). Dissolved manganese has a shorter MORT than aluminium and varies less in concentration in the deep ocean, except where *in situ* sources (e.g., hydrothermal) are significant. Compare these examples with the schematic representation in *Figure 11.12(b)*. Data are from (a) Statham *et al.*[19] (with permission from Macmillan Magazines Ltd), (b) Measures *et al.*[13] (with permission from the authors and Elsevier Science Ltd), and (c) Orians and Bruland[17] (with permission from the authors and Elsevier Science Ltd).

Figure 11.15 (a) Vertical profiles of some recycled dissolved constituents in three major ocean basins (squares, North Atlantic[4]; circles, southwestern Indian[16]; diamonds, central North Pacific[4]): the micronutrients are (i) silicon and (ii) phosphate, and the trace metals are (iii) cadmium and (iv) nickel. Within the overall feature of increasing concentration with increasing 'age' of the deep water, detailed differences in the profiles can be seen; in particular, nickel is less depleted in the surface waters than are the other constituents. A very close similarity in distribution is shown by phosphate and cadmium, reflecting the association of the metal, which has no known biological function, with organic matter sinking from the surface to deeper waters. The more gradual increase in concentration with depth for silicon, compared with phosphate, occurs because the skeletal material which carries silicon down the water column is recycled deeper in the ocean than are the soft organic tissues. Compare these examples with the schematic representation in *Figure 11.12(c)*. Figure from Morley *et al.*[16] (with permission from Elsevier Science Ltd). (b) The vertical distribution of concentrations of nitrate (μmol/kg) along a section through the western Atlantic, and Southern and Central Pacific

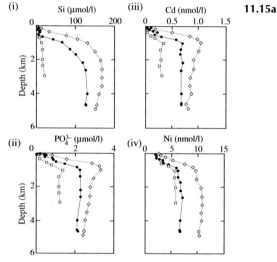

Oceans (after Sharp[20], based on GEOSECS data; with permission from the author and Academic Press). The section illustrates more fully the pronounced increase in concentration with depth and the increase along the direction of the major deep-water circulation.

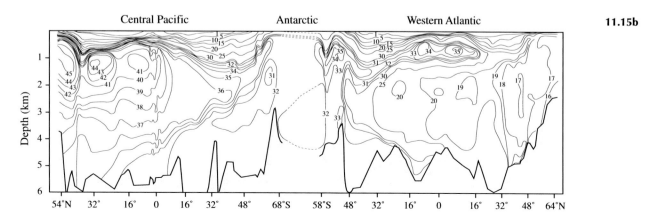

Marine Biogeochemical Processes and Global Environmental Conditions

It is clear from the foregoing that biological processes play a crucial role in the major sedimentary cycle and in the internal cycling of material within the ocean. Indeed, many geochemical processes are termed biogeochemical because of this role. Biological oceanographic processes, driven essentially by solar radiation which provides the energy for photosynthetic primary production of organic matter (see Chapter 6), have three consequences of the greatest importance for surface-Earth conditions.

One of these consequences is that many elements can occur in chemical forms which are thermodynamically unstable; their concentrations reflect the balance between their continuous biological production and their conversion into more stable chemical species. An example of this is the element nitrogen, which is supplied to the surface Earth as nitrogen (N_2) gas, a volatile which accounts for about 80% by volume of the atmosphere. Under the oxidising conditions which generally prevail at the Earth's surface, dinitrogen is thermodynamically unstable and, on that basis, should be oxidised to provide, ultimately, nitrate ions in the hydrosphere. Oxidation of nitrogen by ordinary chemical reactions occurs, however, at a negligible rate, because the bond between the two nitrogen atoms is exceptionally strong, representing an energy barrier to the process. This energy barrier is overcome during high-energy events in the atmosphere. Also, some organisms have the ability to fix nitrogen enzymatically, that is to convert the nitrogen molecule into ammonia, in which form it can be utilised biologically; this nitrogen eventually is converted into nitrate in the nitrogen cycle. These conversions are slow in relation to the magnitude of the atmospheric mass of dinitrogen, so the gas has a mean atmospheric residence time of about 10^6 years. Nevertheless, if biological processes did not reverse these reactions, the reservoir of atmospheric nitrogen would run down to a very low concentration over a period of some 10^7 years, with a correspondingly massive increase in the concentration of nitrate in the ocean. Yet, the major composition of the atmosphere has remained relatively constant over a significantly longer period[8]. The reversal through reduction of nitrate to nitrogen by organic matter, a process termed denitrification, occurs through bacterial respiration in certain aquatic and sedimentary environments, which are deficient in free oxygen. Marine environments of this kind, while they form only a limited part of the ocean system, account for a significant part of global denitrification.

Denitrification can act as a negative feedback mechanism in the nitrogen cycle, because increased aquatic concentrations of nitrate should lead to the production of more organic matter, the respiration of which by normal aerobic processes utilises oxygen. The maintenance by organisms of non-equilibrium conditions by feedback mechanisms, illustrated in a simplified form by this example, is a key element in the Gaia hypothesis[10,11] that the biosphere controls its environment in such a way that it is favourable for life processes (see also *Box 12.7*).

Further consequences of biological processes are that the distribution of constituents between the major reservoirs of material at the Earth's surface is very different from what it would be in a similar abiotic system, and that the distribution of many constituents within the oceanic reservoir is likewise different. These aspects are well-illustrated by reference to the marine carbonate system (Chapter 12). Most of the carbon dioxide released as a magmatic volatile during Earth's chemical evolution (see earlier) has been bound up as carbonates in sediments and sedimentary rocks, with a considerable fraction also locked away in sedimentary deposits as unrespired organic carbon compounds. If all the carbon dioxide was present in the gaseous form, the atmosphere would be fairly similar to that of Venus, where the surface pressure is about 75 atmospheres and carbon dioxide constitutes 95% of the atmosphere, compared with 0.03% of Earth's atmosphere. The radiation balance for Venus, in which the 'greenhouse' gases carbon dioxide and water vapour play a major role, leads to a surface temperature of about 450°C, compared with the mean value for Earth of 15°C.

The ocean is the major reservoir of carbon in the exchangeable carbon system, containing about 60 times as much carbon as the atmosphere. The way in which this reservoir exerts a major role in determining the steady-state partial pressure of carbon dioxide in the atmosphere, and the manner in which it has responded to increases in partial pressure as a result of human activities over the past 200 years, are among the most important questions in chemical oceanography today. The physicochemical and geochemical factors involved are complicated by the fact that inorganic carbon in the ocean is present in different chemical forms: as dissolved carbon dioxide, entering and leaving by gas exchange across the atmosphere–ocean interface, and as bicarbonate (HCO_3^-) and carbonate (CO_3^{2-}) ions, entering with river water as a result of the weathering reactions of carbon dioxide with continental rocks.

Dissolved carbon dioxide in the ocean reacts with water to form carbonic acid (H_2CO_3), which dissociates to produce bicarbonate and carbonate ions. It also reacts with carbonate ions derived from the river-water input to form bicarbonate ions. As a consequence, sea water can absorb more carbon dioxide than would be the case if the salts consisted entirely of, say, sodium chloride; with the present-day composition of the ocean, this factor increases the absorptive capacity of the ocean for carbon dioxide by a factor of about eight.

A further consequence is that the carbon dioxide dissolved in the mixed layer of the ocean equilibrates with atmospheric carbon dioxide rather slowly; virtually complete equilibration takes about 1.5 years. As a result, marked long-term disequilibria can exist regionally between the partial pressures in the atmosphere and in surface waters, particularly in areas of major upwelling and downwelling of water, although the exchange of gas into and out of the surface layer is in approximate balance for the whole ocean.

The biological pump exerts a powerful influence, drawing down carbon dioxide into the deeper ocean. This occurs through the uptake of carbon dioxide and its conversion into organic carbon compounds by photosynthesizing organisms in the surface waters, and the transfer of a fraction of the organic material into deeper waters, where the carbon dioxide is released by respiratory oxidation, mainly bacterial. The factors controlling the vertical distribution of the inorganic species in the ocean are particularly complex. In addition to the recycling processes, waters at different depths acquired different concentrations of carbon dioxide by exchange with the atmosphere when they were at the surface; also, the distribution of the total inorganic carbon among its principal species is dependent upon the *in situ* temperature and pressure, which affect the equilibrium constants for the system (Chapter 12). The importance of the biological pump can be seen, however, from calculations[3] which show that if deep water of average composition were brought to the surface and warmed to about 20°C, then in the absence of biological activity its partial pressure of carbon dioxide would be about 1000 parts per million (p.p.m.), compared with the present atmospheric partial pressure of about 350 p.p.m.

The distribution of inorganic carbon among its principal species in deep water also exerts a geochemical control upon the accumulation of calcium carbonate in sediments. In the surface ocean, the concentrations of dissolved calcium and carbonate ions correspond to a condition of supersaturation with respect to solid-phase calcium carbonate. The precipitation of calcium carbonate is, however, dominantly a consequence of the production of skeletal material by organisms, rather than a physicochemical phenomenon. As particles of calcium carbonate sink down the water column, they enter environments with higher pressure and lower temperature, both of which lead to an increased solubility of calcium carbonate. The concentration of calcium in sea water varies to only a small extent throughout the ocean, its behaviour being nearly conservative. In contrast, the concentration of dissolved carbonate ions shows a marked general decrease with depth and also decreases broadly with the 'age' of the deep water as it is transported through the Atlantic, Indian, and Pacific Oceans. These features reflect both the operation of the biological pump, transporting carbon dioxide to the deeper waters, and the high concentrations of carbon dioxide, derived from atmospheric exchange, in cold waters. The overall consequence of the increase in concentration of carbon dioxide is to convert some carbonate ions into bicarbonate ions. Thus, in some parts of the ocean, the concentration of carbonate ions in the deeper waters falls below that corresponding to saturation of the water with respect to solid-phase calcium carbonate.

There are two horizons in the water column which are significant in relation to the preservation of calcium carbonate particles in the bottom sediments. At the lysocline, the degree of undersaturation with respect to calcium carbonate is sufficient for dissolution of the particles to become significant (see *Figure 12.6*). If the total depth of water is less than the depth of the lysocline, particles accumulate without loss by dissolution. If it is greater than that of the lysocline, some dissolution will occur, and if the sea bed is at a depth much below the lysocline, calcium carbonate can undergo dissolution to such an extent that it ceases to be a significant sedimentary component. This latter horizon is termed the calcite compensation depth, calcite being the dominant mineral form of calcium carbonate produced by the surface-living organisms in the ocean. The difference in the depths of the lysocline and the calcite compensation depth is not constant; it depends primarily upon the gradient in concentration of carbonate ions in the water column overlying the sediment. On average, the depth at which the concentration of calcite in sediments falls to only a few per cent is about 700 m deeper than the lysocline.

The depths at which these horizons occur are typically some thousands of metres, but are not the same throughout the oceans, since they depend essentially upon the concentration of carbonate ions in the bottom waters (they are deeper in the Atlantic Ocean than in the Pacific Ocean). Such differences are reflected in the distribution of calcareous sediments below the world ocean (see *Figure 11.8*). Since a change in the concentration of carbon dioxide in sea water alters the concentration of carbonate ions, such a change also leads to greater dissolution or preservation of calcium carbonate in

11.16

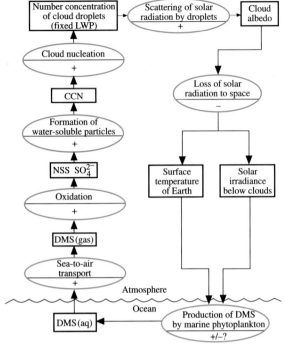

Figure 11.16 Postulated feedback mechanism linking global climate and production of dimethylsulphide (DMS) by phytoplankton in the surface layers of the ocean (from Charlson *et al.*[5]; with permission from the authors and Macmillan Magazines Ltd). The processes shown in the ovals affect the quantities shown in the rectangles; the sign in the oval indicates whether an increase in the preceding quantity leads to an increase (+) or decrease (−) in the next quantity. The mechanism can act as a regulator of climate if increased surface-Earth temperature and solar irradiance lead to an increased production of DMS by marine phytoplankton; this depends on a complex of factors, especially since production of DMS varies with the species composition of phytoplankton populations (NSS SO_4^{2-} = non-sea-salt sulphate; CCN = cloud-condensation nuclei; LWP = liquid water path).

sediments. This occurs, however, on longer time-scales than those applicable to the exchangeable carbon system.

It is apparent that the global cycle of carbon is crucially dependent upon the role of the ocean, in which the balance of the biogeochemical processes is finely tuned to the preservation of carbonate sediments over extensive areas of ocean floor and to the storage of carbon dioxide in the deep waters – to a much greater extent than would be the case for an abiotic ocean. Consequently, changes in the complex of interacting processes involved in ocean circulation and biological productivity may be expected to perturb the system significantly. Evidence from ice cores [see *Figure 12.1(a)*] shows that during the 160,000 years prior to 1800, when human activities began to increase the concentration of carbon dioxide in the atmosphere, the atmospheric pressure of carbon dioxide varied within the range of about 200–300 p.p.m. Minimum values occurred around the most recent glacial maximum, about 18,000 years ago. These are highly significant changes, but when it is remembered that there is so much more carbon in the ocean than in the atmosphere, the comparatively narrow limits of variation provide striking testimony to the efficiency of the ocean as a regulator of Earth's surface conditions.

Since man became a significant agent in the carbon cycle, the atmospheric partial pressure of carbon dioxide has increased to about 350 p.p.m. Of

the carbon dioxide added to the atmosphere by fossil fuel combustion, about 35% has been taken into the ocean (see Chapter 12). Understanding the role of the ocean in the carbon system is clearly a vital component, alongside knowledge of the role of the terrestrial biosphere, of our ability to model and predict global chemical and climatic change (see Chapter 3). Ultimately, however, the direction and extent of climatic change depend upon many interlinked factors, involving the cycles not only of carbon, but also of other major biogeochemical elements.

Two examples serve to illustrate this point. First, the volatile compound, dimethylsulphide, is released to the atmosphere as a result of its production in surface waters by some kinds of phytoplankton (see Chapter 6); it is the most abundant of a number of sulphur-containing gases which are released from the surface ocean as a result, directly or indirectly, of biological activity[1]. In the atmosphere, it is converted into sulphur dioxide, which forms a sulphuric acid aerosol that can act as a nucleus for cloud condensation. It has been postulated[5] that changes in the flux of dimethylsulphide as a result of changes in the temperature of the upper ocean could feed back to modify the temperature through a change in cloud cover (*Figure 11.16*); another example of the Gaia hypothesis of homoeostatic regulation of environmental conditions.

Second, another hypothesis links the aeolian transport of mineral dust to the recycling of carbon dioxide. At high latitudes, surface waters, unlike those of the stratified ocean, do not become strongly depleted in the principal micronutrients (nitrate, phosphate, dissolved silicon) needed for phytoplankton growth. Experimental work has shown that the addition of dissolved iron to such waters promotes an increased utilisation of these relatively abundant micronutrients and of carbon dioxide in photosynthesis (see Chapter 12). There is evidence from ice cores in Antarctica that during the most recent glacial period a much higher deposition of mineral dust (which is a potential source of dissolved iron for surface waters) occurred than in the interglacial periods. These findings have led to the hypothesis[12] that an increased efficiency of the biological pump in the Southern Ocean played an important role in the reduction of the atmospheric partial pressure of carbon dioxide during the glacial period.

Concluding Comments

In this chapter emphasis has been placed mainly on large-scale features in the distributions of chemical species and the role of the ocean in the context of the global environment. Knowledge in these areas has advanced rapidly during the past two decades, following the systematic measurements of constituents in the major ocean basins that was begun in the late 1960s under the Geochemical Ocean Sections (GEOSECS) programme. As explained in the Introduction, these advances have depended critically upon improved analytical methods. The ocean system is chemically driven by processes that occur on small scales, such as dissolved chemical species being scavenged at particle surfaces, or taken up and transformed by microalgae; the role of organic material in such processes is recognised, but little understood at a molecular level. Developments in ocean chemistry depend upon advances in these complementary approaches, and upon the new insights developing, for example, through the application of photochemistry and colloid science to marine environmental systems.

General References

In addition to the texts by Broecker and Peng[3] and Holland[8] listed below, several other books provide useful background:

Chester, R. (1990), *Marine Geochemistry*, Unwin Hyman, London, 698 pp.

Libes, S.M. (1992), *An Introduction to Marine Biogeochemistry*, Wiley, New York, 734 pp.

Volumes in the series *Chemical Oceanography* [Vols 1–4 (1975), Riley, J.P. and Skirrow G. (eds), and Vols 5–10 (1975–1989), Riley, J.P. and R. Chester (eds), Academic Press, London] provide accounts in depth of topics across the whole range of marine chemical interests.

References

1. Andreae, M.O. (1990), Ocean–atmosphere interactions in the global biogeochemical sulfur cycle, *Marine Chem.*, 30 1–29.
2. Baumgartner, A. and Reichel, E. (1976), *The World Water Balance*, Elsevier, Amsterdam, 176 pp.
3. Broecker, W.S. and Peng, T.-H. (1982), *Tracers in the Sea*, Eldigio Press, Palisades, 690 pp.
4. Bruland, K.W. (1983), Trace elements in sea-water, in *Chemical Oceanography*, Vol. 8, Riley, J.P. and Chester, R. (eds), Academic Press, London, pp 157–220.
5. Charlson, R.J., Lovelock, J.E., Andreae, M.O., and Warren, S.G. (1987), Oceanic phytoplankton, atmospheric sulphur, cloud albedo and climate, *Nature*, **326**, 655–661.
6. Davies, T.A. and Gorsline, D.S. (1976), The geochemistry of deep-sea sediments, in *Chemical Oceanography*, Volume 5, Riley, J.P. and Chester, R. (eds), Academic Press, London, pp 1–80.
7. Garrels, R.M. (1985), Sediment cycling during Earth history, in *Physical and Chemical Weathering in Geochemical Cycles*, Lerman, A. and Meybeck, M. (eds), Kluwer, Dordrecht, pp 341–355.
8. Holland, H.D. (1984), *The Chemical Evolution of the Atmosphere and Oceans*, Princeton University Press, Princeton, 582 pp.
9. Kempe, D.R.C. (1981), Deep ocean sediments, in *The Evolving Earth*, Cocks, L.R.M. (ed.), British Museum (Natural History) and Cambridge University Press, London and Cambridge, 264 pp.
10. Lovelock, J.E. (1979), *Gaia: A New Look at Life on Earth*, Oxford University Press, Oxford, 157 pp.
11. Lovelock, J.E. and Margulis, L. (1974), Atmospheric homeostasis by and for the biosphere: the Gaia hypothesis, *Tellus*, **26**, 2–10.
12. Martin, J.H. (1990), Glacial–interglacial CO_2 change: the iron hypothesis, *Paleoceanogr.*, **5**, 1–13.
13. Measures, C.I., Edmond, J.M., and Jickells, T.D. (1986), Aluminium in the northwest Atlantic, *Geochim. Cosmochim. Acta*, **50**, 1423–1429.
14. Meybeck, M. (1988), How to establish and use world budgets of riverine materials, in *Physical and Chemical Weathering in Geochemical Cycles*, Lerman, A. and Meybeck, M. (eds), Kluwer, Dordrecht, pp 247–272.
15. Milliman, J.D. and Meade, R.H. (1983), World-wide delivery of river sediment to the oceans, *J. Geol.*, **91**, 1–21.
16. Morley, N.H., Statham, P.J., and Burton, J.D. (1993), Dissolved trace metals in the southwestern Indian Ocean, *Deep-Sea Res. I*, **40**, 1043–1062.
17. Orians, K.J. and Bruland, K.W. (1986), The biogeochemistry of aluminium in the Pacific Ocean, *Earth Planet. Sci. Lett.*, **78**, 397–410.
18. Redfield, A.C., Ketchum, B.H., and Richards, F.A. (1963), The influence of organisms on the composition of sea water, in *The Sea*, Vol. 2, Hill, M.N. (ed.), Interscience, New York, pp 26–77.
19. Statham, P.J., Burton, J.D., and Hydes, D.J. (1985), Cd and Mn in the Alboran Sea and adjacent North Atlantic: geochemical implications for the Mediterranean, *Nature*, **313**, 565–567.
20. Sharp, J.H. (1983), The distributions of inorganic nitrogen and dissolved and particulate organic nitrogen in the sea, in *Nitrogen in the Marine Environment*, Carpenter, E.J. and Capone, D.G. (eds), Academic Press, New York, pp 1–35.
21. Thompson, G. (1983), Hydrothermal fluxes in the ocean, in *Chemical Oceanography*, Volume 8, Riley, J.P. and Chester, R. (eds), Academic Press, London, pp 271–337.
22. Turekian, K.K. (1977), The fate of metals in the oceans, *Geochim. Cosmochim. Acta*, **41**, 1139–1144.
23. von Stackelberg, U. (1987), General history and variability of manganese nodules of the equatorial North Pacific, in *Marine Minerals*, Teleki, P.G., Dobson, M.R., Moore, J.R., and von Stackelberg, U. (eds), Reidel, Dordrecht, pp 189–204.
24. Whitfield, M. and Turner, D.R. (1979), Water–rock partition coefficients and the composition of sea water and river water, *Nature*, **278**, 132–137.

CHAPTER 12

The Marine Carbonate System

M. Varney

Introduction

Global warming is more than just a topic of scientific interest. It has come to dominate the environmental agenda, to tax the minds of political leaders around the world, to affect the way energy companies do their business, and to concern the man in the street.

The main cause of global warming is the rise since 1900 in the content of the so-called greenhouse gases, the main culprit being carbon dioxide, CO_2. The current rise in atmospheric levels of CO_2 (*Figure 12.1*) is of great concern because CO_2 absorbs spectral components of solar radiation – it is a strong short- and long-wavelength, infra-red absorber (*Figure 12.2*). The present day concentrations of CO_2 are the highest in the recorded history of the Earth (compare *Figures 12.3* and *12.4*). The absorbed energy is partly dissipated as heat and warms the air, land, and sea. When CO_2 concentrations rise, the rate of heat loss to space is exceeded by the absorption of incident solar radiation by the atmosphere, which increases the amount of trapped heat at the Earth's surface – causing a rise in surface temperatures and in sea level (from the melting of glaciers and the thermal expansion of water). Rainfall patterns and storm activity are likely to change as a direct result. Our climate is becoming steadily hotter and we cannot predict what will happen in the future because we have no accurate comparison from the past. Predictions of changes to the Earth's climate are extremely limited due to our current lack of knowledge about the various interactions of CO_2 with land vegetation and the oceans.

The oceans play an important role in the climate system by regulating the amount of CO_2 in the atmosphere. This is one reason why it is important to understand thoroughly the workings of the marine carbonate system, which controls the movement of CO_2 between the ocean and the atmosphere. The marine carbonate system is also one of the key ingredients in the story of life in the sea, dissolved CO_2 being the primary source of carbon for marine plant life and the primary component (along with water) in the process of photosynthesis.

CO_2 is the third most abundant dissolved gas in

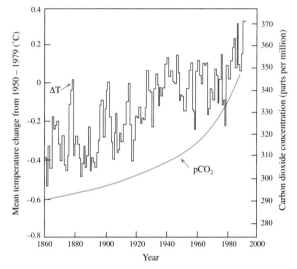

Figure 12.1 The relationship between temperature and CO_2 levels has been quite variable over the past 130 years. The true relationship is more complex than this illustration implies, and there has been considerable debate about whether changes in the atmospheric [CO_2] are the response to temperature fluctuations, or vice versa. Several hypotheses have been proposed, relating changes in [CO_2] to changes in biological productivity, in sea level, and in the circulation of surface and deep waters (including the relative importance of deep water-mass formation in polar latitudes, and of upwelling regions, where biological productivity is high). As these factors are all interrelated, the resulting models are complex and no definitive answer is yet available. It is reasonably certain that varying levels of CO_2 are not the main cause of temperature fluctuations – but once established they are likely to reinforce climatic changes, not initiate them. The concentrations of other atmospheric gases, such as methane, ozone, and freons, are also rising. We are therefore in the middle of a global experiment in which several geochemical cycles are being perturbed. Insights into the mechanisms of these cycles can be obtained from palaeoceanographic research (see also *Figure 12.3*). (Adapted from Schneider[9].)

Figure 12.2 The absorption of incoming solar (infra-red) radiation by the atmospheric gases H₂O, CO₂, and O₃. The amount of radiation emitted by the Earth's surface is shown by the dashed line. (Adapted from Coulson[2].)

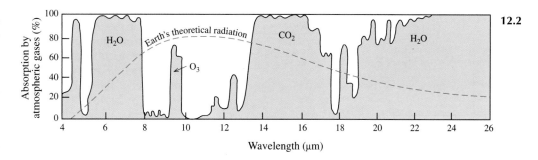

12.2

Figure 12.3 In polar regions, partial melting of surface snow and ice occurs during the summer months; refreezing in winter produces an ice layer on top of the old snow. Since snow is porous, small pockets of air are trapped in its pores when the surface layer of ice freezes. These 'bubbles' reflect the atmospheric composition at the time of entrapment. These annual markers, like tree rings, of ice and snow can be used to pinpoint the time of their deposition (see also *Box 8.3*). Observations from ice and sediment cores demonstrate that large-scale changes in atmospheric CO₂ levels occurred over geological time and are correlated with atmospheric temperature changes over the past 160,000 years. The correlation between atmospheric [CO₂] and temperature change (ΔT) is clear. Cold temperatures (during glacial periods) appear to be associated with low levels of atmospheric CO₂. The deglaciation events of about 140,000 years and 15,000 years ago are particularly obvious. These fluctuations are as expected – CO₂ is more soluble in cold than in warm water, and its atmospheric concentration should therefore be less during glacial (lower mean temperatures) than interglacial periods. (Adapted from Schneider[9].)

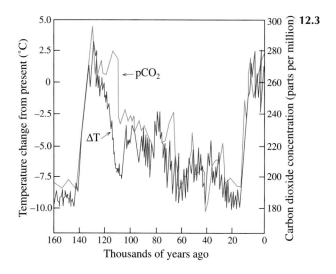

12.3

Figure 12.4 The annually-averaged atmospheric concentrations of CO₂ at Hawaii (dots) and the South Pole (crosses), showing a global increase of approximately 20 p.p.m. between 1958 and 1976. The curves are smoothed, piecewise fits to the data. (Adapted from Bacastow and Keeling[1].)

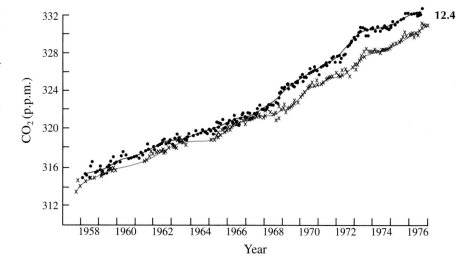

12.4

12.5

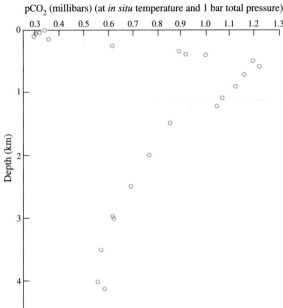

pCO_2 (millibars) (at *in situ* temperature and 1 bar total pressure)

Figure 12.5 The typical variation of CO_2 concentration with depth. The profile shows a very sharp sub-surface increase to a maximum at approximately 500 m, due to planktonic respiration processes which release the previously fixed CO_2 back to the water column. The profile was taken in the Eastern Pacific (28°20′N 121°41′W) during the National Science Foundation GEOSECS (Geochemical Ocean Sections) program in 1969. (Adapted from Takahashi *et al.*[10])

Box 12.1 The buffering capacity of sea water

The pH of sea water varies over a surprisingly narrow range, centred at pH 8±0.5. The most important point is that the dissociation of carbonic acid (H_2CO_3) forms a *buffering system*, which may be summarised by the general weak acid–conjugate base equilibrium, equation (12.2), which results in a solution pH given by equation (12.3), where K_a is the equilibrium constant for the dissociation reaction. Small additions of acids or bases alter the ratio of anion (HCO_3^-) to acid (H_2CO_3) only slightly, and have little effect on the pH of the solution. The buffering capacity is the extent to which the pH is changed by a given addition of acid or base. The higher the concentration of carbonic acid, the greater the buffering capacity.

$$H_2CO_3 \rightleftharpoons H^+ + HCO_3^- \qquad (12.2)$$

$$pH = pK_a + \log([HCO_3^-]/[H_2CO_3]) \qquad (12.3)$$

sea water, after nitrogen and oxygen (see Chapter 11). Most gases dissolve in sea water in proportion to their atmospheric partial pressures, but CO_2 is an exception. It is an extremely reactive gas and has an elevated aquatic concentration relative to its atmospheric partial pressure. CO_2 is intimately involved in biological processes, being consumed by plankton, which photosynthesise in the surface waters, so leading to the production of organic matter tissue (equation 12.1):

$$CO_2 + H_2O \rightarrow \text{'CH}_2O\text{'} + O_2 \qquad (12.1)$$

where 'CH$_2$O' is a general term for organic plant material. A consequence of primary production (see Chapter 6) is that the upper waters of the ocean are generally undersaturated in CO_2 over large areas. In contrast, upwelling of deep waters in the equatorial region and along the west coast of the American continent, for instance (see Chapter 3), brings water supersaturated with CO_2 to the surface. These waters are rich in CO_2 because sinking organic matter has decomposed in the deeper waters of the ocean. The amount of the gas that is dissolved in sea water is determined by an interplay of chemical, physical, and biological factors. In turn, CO_2 helps to maintain the acidity of sea water in the range of pH 8.0±0.5 (see *Box 12.1*).

The capacity of the ocean to absorb CO_2 from the atmosphere appears great (*Figures 12.5–12.7*), and ultimately it can dissolve orders of magnitude more than is already present. Estimating the true capacity is, however, difficult since nothing is at equilibrium and the system is highly dynamic. CO_2 forms carbonic acid with water, which then dissociates to form hydrogen carbonate (HCO_3^-) and carbonate (CO_3^{2-}), which are the main forms of dissolved carbon in sea water. A simplistic view of the dissolution of CO_2 into sea water is given in equation (12.4) [details are discussed later – for example, see equations (12.6)–(12.9)].

$$CO_2(g) + H_2O \rightleftharpoons H_2CO_3 \rightleftharpoons H^+ + HCO_3^- \rightleftharpoons 2H^+ + CO_3^{2-}$$
$$(12.4)$$

Reactions (12.1) and (12.4) illustrate the buffering ability of the marine carbonate system. An increase in atmospheric CO_2 increases the total amount of inorganic carbon within the sea. While this increases the buffering capacity, it also induces a slight increase in the ocean's acidity and thus acts to oppose further entry of the gas.

CO_2 concentrations increase with depth because CO_2 is used during photosynthesis and released again during respiration, and because the solubility of CO_2 increases with pressure (*Figure 12.5*). When

Figure 12.6 Total alkalinity, *A*, total dissolved inorganic carbon (*DIC*) concentration, and degree of calcite saturation as a function of depth in the equatorial Atlantic. Also indicated are the lysocline (see Chapter 11) at approximately 4600 m depth, where there is a perceptible amount of calcium carbonate dissolution, and the compensation depth at approximately 4900 m, below which all calcium carbonate should be dissolved. Since calcite dissolves at deeper depths than it should according to the calcite saturation index, this implies that plant and animal remains (which contain calcite) sink faster than they can be dissolved. (Adapted from Edmond and Gieskes[3].)

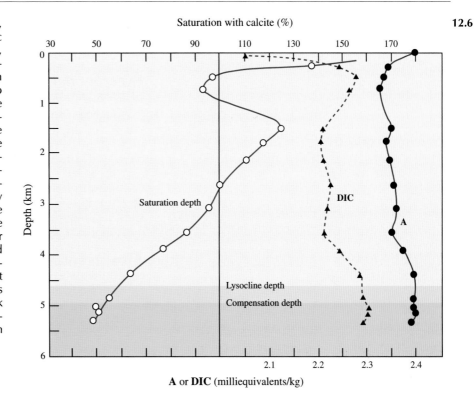

A or **DIC** (milliequivalents/kg)

Figure 12.7 The biogeo-chemical cycle of carbon. The numbers show the current estimates of the major reservoirs (in units of 10^{15} gC) and fluxes (in units of 10^{15} gC/yr). BP = transport of carbon to the deep sea by the 'biological pump'; PS = conversion of dissolved inorganic carbon (*DIC*) into particulate organic carbon (*POC*) by photosynthesis; *DOC* = dissolved organic carbon; RESP = conversion of organic carbon into *DIC* by respiration. (Adapted from Post *et al*.[8] and Moore and Bolin[7].)

12.7

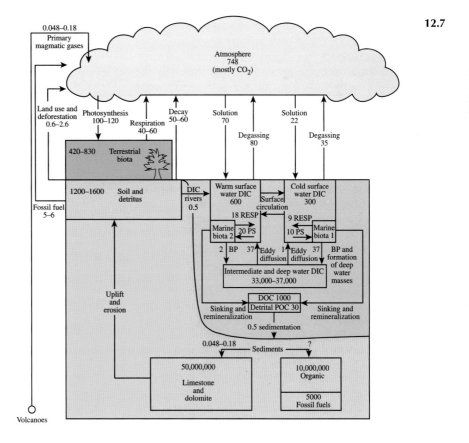

12.8

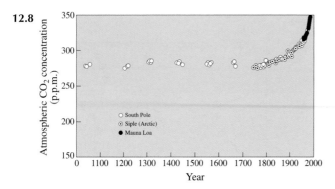

Figure 12.8 Extrapolated annually-averaged atmospheric levels of CO_2 at locations away from major terrestrial influences. Concentrations prior to 1960 are estimated from ice-core samples; post-1960, they are derived from data such as that in *Figure 12.3*. (Adapted from Post *et al.*[8])

photosynthesis occurs, CO_2 is stripped from the water – the pH then increases so the carbonate concentration ($[CO_3^{2-}]$) increases. CO_2 is generated by the oxidation of decaying organic matter in sub-surface waters [i.e., equation (12.1) moves to the left], this production being accompanied by a fall in pH, the loss of dissolved oxygen, and the liberation of nutrients by the material. The death and consequent sinking of phyto- and zooplankton detritus (see Chapter 7), the dissolution or precipitation of calcium carbonate (see *Figure 12.6*), and the formation of limestone are now, and have been in the past, inherently linked in the sea water carbonate system.

The global cycle of carbon is complex; it is difficult to explain adequately the processes, scales, and magnitudes involved. *Figure 12.7* shows some of the more important reservoirs and pathways, of which little is fully understood. Biogenic limestone is the single largest crustal reservoir of inorganic carbon. Sedimentary organic matter is the second largest, and is composed primarily of inorganic carbon fixed by marine photosynthesis. However, while these two reservoirs may be the ultimate sink for the atmospheric increase in CO_2, they have an effect only on geological time-scales. Geological uplift will eventually expose these ancient marine deposits to the atmosphere. Chemical weathering will oxidise the carbon back to gaseous CO_2, which will be taken up by plants, thereby closing the global biogeochemical cycle of carbon. The concentration of CO_2 in the atmosphere is about 0.03% (dependent on sampling height and locality). A long-term increase is evident, due to the release of (juvenile) CO_2 from the Earth, the domestic and industrial burning of fossil fuels (carbon previously fixed by marine phytoplankton and land plants),

various industrial processes (e.g., cement manufacture), forest fires, and changes in agricultural and land use patterns (*Figure 12.8*).

Experimental Measurement of the Carbonate Species

It would be useful to predict what the capacity of sea water might be to absorb some of the CO_2 increase in the atmosphere, and the likely consequences for terrestrial and marine life. It would also be useful to quantify all elements within the carbonate system using ship-board measurements at a minimum of experimental effort and cost. Curiously, despite the crucial importance of these measurements, no instruments have yet been developed to make them *in situ*. To interpret the marine carbonate system quantitatively we must first calculate the contributions of the individual components to the carbonate cycle.

$$CO_2(aq) + CO_3^{2-} \rightleftarrows 2HCO_3^- \qquad (12.5)$$
$$CO_2(g) \rightleftarrows CO_2(aq) \qquad (12.6)$$
$$H_2O + CO_2(aq) \rightleftarrows H_2CO_3 \qquad (12.7)$$
$$H_2CO_3 \rightleftarrows H^+ + HCO_3^- \qquad (12.8)$$
$$HCO_3^- \rightleftarrows H^+ + CO_3^{2-} \qquad (12.9)$$

Each reaction in the carbonate system can be characterised by an equilibrium constant (which varies as a function of salinity, pressure, and temperature). Carbonate, hydrogen carbonate, and dissolved CO_2 show considerable variations in concentration (spatially and temporally) which can be related to biological, chemical, and physical oceanographic processes. The solubility of the gas in its aqueous phase, $CO_2(aq)$, is enhanced by reaction with carbonate ions, equation (12.5). The position of equilibrium for this reaction (the equilibrium constant, K) is so far to the right of this equation that most of the CO_2 entering the ocean is rapidly converted into hydrogen carbonate, HCO_3^-. In fact, the above chemical reaction proceeds in the laboratory as fast as the solutions can be mixed. The major steps in the sea water carbonate equilibria are given in equations (12.6)–(12.9), where equation (12.6) is the air–sea exchange of atmospheric (g) to oceanic (aq) CO_2, equation (12.7) is hydration, equation (12.8) is the first ionisation, and equation (12.9) is the second ionisation.

$$K_0 = [H_2CO_3]/[CO_2(aq)] \qquad (12.10)$$
$$K_1 = [H^+][HCO_3^-]/[CO_2(aq)] \qquad (12.11)$$
$$K_2 = [H^+][CO_3^{2-}]/[HCO_3^-] \qquad (12.12)$$

Unfortunately, it is not possible to measure directly the concentrations of the two most important species, $[HCO_3^-]$ and $[CO_3^{2-}]$, without inadvertently shifting the above equilibria to one side or

Box 12.2 Definitions of alkalinity and dissolved inorganic carbon

The *alkalinity*, A_t, of sea water is the combined negative charge due to hydrogen carbonate (HCO_3^-) and carbonate ions (CO_3^{2-}), expressed in molal concentrations. Alkalinity is based on sea water being electrically neutral and is determined by titration (and therefore given the subscript t). In practice, the alkalinity is the *amount of acid (hydrogen ions) needed to convert all the anions back into their respective un-ionised acids*. The alkalinity equation expresses *electroneutrality* between the positively and negatively charged ions, equation (12.13), which can be rearranged to give equation (12.14), where [SA], the surplus alkalinity, is the concentration of all weak acids other than boric and carbonic acids. The second and third ionisation constants of boric acid are very low, so that boric acid is the major contributor to A_t from the borate system. In sea water, ($[OH^-] - [H^+]$) and the amount of SA are small, so equation (12.14) may be simplified, to give equation (12.15). SA cannot be ignored in anoxic regions, when sulphide, ammonia, and phosphate concentrations may be very high.

Typical values for the alkalinity of sea water are given *Table 12.1*. The *carbonate alkalinity, CA*, is the contribution to alkalinity from carbonate and hydrogen carbonate, equation (12.16), and is normally obtained by subtracting the borate contribution from A_t. The borate contribution, $[B(OH)_4^-]$, is obtained from the expressions for total boron, B_T, and the equilibrium constant for boric acid dissociation, K_B, equations (12.17)–(12.19). These equations are also a function of salinity, pressure, and temperature. The borate contribution is usually 3–5% of CA.

The *total dissolved inorganic carbon concentration, DIC*, is defined as the concentration of all of the dissolved forms, and is given by equation (12.20). Note that the terms ($[HCO_3^-]$ and $[CO_3^{2-}]$) refer to the total equilibrium concentrations and include contributions from ion pairing with Na^+, Ca^{2+}, Mg^{2+}, etc. (see *Box 12.3*). There are other weak acids and bases that are potential contributors toward the alkalinity of sea water, but under typical conditions their concentrations are extremely small; thus, their contributions are usually much less than 1% of A_t and can be ignored in the above calculations. (In the oceanographic literature, total alkalinity is sometimes referred to as *TA*, or simply *A*; carbonate alkalinity as A_C; total carbonate as TCO_2, *TIC*, or ΣCO_2.)

At the pH of most ocean waters, less than 1% of the inorganic carbon exists as ($[CO_2(aq)] + [H_2CO_3]$), and the H_2CO_3 concentration is only about 0.2% of that of $CO_2(aq)$.

It is important to remember the definition of alkalinity as the concentration of hydrogen carbonate and carbonate ions. It is not a measure of the pH or how alkaline sea water is. Once this is recognised, it is easy to appreciate that sea water alkalinity and acidity change in the same direction: where the DIC is high, so are alkalinity and acidity (low pH); conversely, where the DIC is low, so are alkalinity and acidity.

One of the few places where inorganic precipitation of calcium carbonate occurs is on the Bahamas Banks, where the sea is shallow and warm, and salinity is high (exceeding a salinity of 37). The warmer and more saline the water, the lower the solubility of gases, including CO_2. The concentration of CO_3^{2-} is also large, and often rises sufficiently for the water to be supersaturated with respect to $CaCO_3$, so that the inhibiting effect of the $MgCO_3$ ion-pair is overcome, and small crystals of calcium carbonate (in the form of aragonite) are precipitated. In these conditions, the term ($A_t - DIC$) is large. This helps to explain the apparent contradiction that when the DIC is high, calcium carbonate is more likely to dissolve, and vice versa.

$$A_t + [H^+] = [HCO_3^-] + 2[CO_3^{2-}] + [OH^-] + [B(OH)_4^-] + [SA] \quad (12.13)$$

$$A_t = [HCO_3^-] + 2[CO_3^{2-}] + ([OH^-] - [H^+]) + [B(OH)_4^-] + [SA] \quad (12.14)$$

$$A_t = [HCO_3^-] + 2[CO_3^{2-}] + [B(OH)_4^-] \quad (12.15)$$

$$CA = [HCO_3^-] + 2[CO_3^{2-}] \quad (12.16)$$

$$B_T = [H_3BO_3] + [B(OH)_4^-] \quad (12.17)$$

$$K_B = \{[H^+][B(OH)_4^-]\}/[H_3BO_3] \quad (12.18)$$

$$[B(OH)_4^-] = K_B B_T / (K_B + [H^+]) \quad (12.19)$$

$$DIC = [HCO_3^-] + [CO_3^{2-}] + [CO_2] \quad (12.20)$$

Table 12.1. Typical concentrations of carbonate parameters in the ocean.

Parameter	Typical oceanic concentration
pH	7.8–8.4
PCO_2	2.0–13.0 x 10^{-6} mol/kg
A_T	2.3–2.6 x 10^{-3} mol/kg

Box 12.3 Effect of ion-pair formation on the carbonate system

Oceanic surface waters are supersaturated almost everywhere with respect to calcium carbonate – otherwise many marine molluscs and crustaceans would not be able to form shells or hard exoskeletons. The possibility arises that the anthropogenic CO_2 being pumped into the atmosphere can be bound into the carbonate coral growths. This also raises the question of why spontaneous inorganic precipitation of calcium carbonate happens only infrequently. The reason lies in the inhibiting presence of Mg^{2+} ions. Much of the carbonate is in the form of $MgCO_3$ ion pairs. It requires the intervention of marine organisms to precipitate *calcite* or *aragonite*, which have the same chemical formula but different chemical structures. Aragonite is thermodynamically less stable than calcite, so aragonite dissolves more readily.

Ion-pair formation is usually neglected in the formulation of carbonate equilibria. It occurs principally between the cations Na^+, Ca^{2+}, and Mg^{2+} (the 'hard' cations with an inert gas electron structure), and anions such as F^-, HCO_3^-, CO_3^{2-}, and SO_4^{2-}. We do not consider the pairs $NaSO_4^-$, $MgSO_4$, and $CaSO_4$, although their formation is a considerable percentage of the total metal concentration (*Table 12.2*).

Borate forms ion pairs with $NaB(OH)_4$, $CaB(OH)_4^-$, and $MgB(OH)_4^-$; approximately 44% of the total borate concentration can be associated in this way. Although most of the carbonate and hydrogen carbonate are associated with Na^+, Ca^{2+}, and Mg^{2+}, they also associate with other ions that occur at lower concentrations in sea water (such as Pb^{2+}, Cu^{2+}, and Zn^{2+}).

The principal effect is to consume an amount of the 'free' carbonate and hydrogen carbonate in solution, i.e. equations (12.21)–(12.23). This implies there are a number of ion pairs, and each ion will associate differently. The existence of ion-pair formation is another reason why the apparent ionisation constants are dependent on ionic composition (not just ionic strength). Many estimates for the various formation constants are based either on thermodynamic theory (and extrapolation from a few measurements), or on actual potentiometric titration measurements using, e.g., ion-selective electrodes and pressure rigs.

$$[\text{Total species}] = [\text{free ions}] + [\text{ion pairs}] \quad (12.21)$$

$$[HCO_3^-]^T = [HCO_3^-]^f + [M.HCO_3] \quad (12.22)$$

$$[CO_3^{2-}]^T = [CO_3^{2-}]^f + [M.CO_3] \quad (12.23)$$

Table 12.2. Typical proportions of ion pairs that various chemical species form in sea water.

Chemical species	Percentage as 'free' ion
Free SO_4^{2-}	39
Na_2SO_4	37
$MgSO_4$	20
$CaSO_4$	4

the other. Instead, indirect measurements are made for which we either have existing methods or instruments (such as pH electrodes, or infra-red gas analysers). $[HCO_3^-]$ and $[CO_3^{2-}]$ can only be inferred through algebraic rearrangement of equations (12.6)–(12.9). Steps (12.7)–(12.9) have associated balance points (equilibrium constants) in their reactions, equations (12.10)–(12.12). The equilibrium constants, K_0, K_1, and K_2 are functions of salinity, temperature, and pressure, and increase with all three. Analytically, the equilibrium constants are the heart of an algebraic definition of the carbonate system.

The usual (carbonate-related) quantities measured in samples of sea water collected at sea include pH, total CO_2, alkalinity, or dissolved inorganic carbon (see *Box 12.2*), and the partial pressure of CO_2 in equilibrium with the solution. In the carbonate system, it is not possible to infer the component concentrations from one measurement alone, such as pH. Instead, *pairs* of measurements of different variables are made and the calculations manipulated to derive the appropriate concentrations. Extreme care must be taken to choose the correct methods and values, the experimental procedures used for the various definitions, their style of calibration, and other practicalities such as ionic composition, to avoid sources of error and possible confusion when handling the complex algebra associated with the carbonate system.

The theoretical development of the determination of carbonate species has been problematic, much of it because the measurement methods are sensitive to apparent changes in concentration (chemists refer to this as the *activity* of species). In general, the activities of dissolved ions change due to the differing composition of sea water around the globe and with depth. The carbonate equilibrium constants almost always need conversion for the salinity of the sample. In addition, all components in the carbonate system change their distribution (their position of equilibrium) with change in depth, due to the changes in the partial molar volumes (sea water is compressed by approximately 4% at 10 km depth). There is (unfortunately) little reliable partial molar volume data in the literature to enable theoretical prediction of the effects. Alkalinity and CO_2 can be regarded as conservative properties when the pressure changes. Alkalinity therefore varies directly with salinity, and is sometimes expressed as specific alkalinity when corrected to a constant salinity (often to a salinity of 35). The approach taken is a practical one: experimental methods have been developed that can be used on board ships, with no or few theoretical corrections necessary.

Can we Predict the Absorption of Atmospheric CO_2 by the Ocean?

Based on recorded changes in atmospheric compositions (*Figure 12.8*) and the output from fossil-fuel emissions (*Figure 12.9*), it appears that 50% of all anthropogenic emissions of CO_2 remain in the atmosphere. About 70% of the rest has been

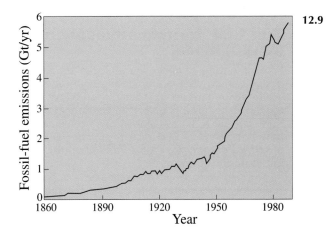

Figure 12.9 The estimated fossil-fuel emissions of carbon (in Gt/yr, 10^{15} gC/yr) from 1860 to recent. (Adapted from Marland *et al.*[6])

absorbed by the oceans, but the fate of the remaining 30% is not known. An increase in the terrestrial biomass (plants, trees, etc.) is thought to be the missing sink, but the estimates in the flux calculations are subject to large errors. *Figure 12.10* suggests that, overall, the ocean absorbs about 1.6 gigatonnes (Gt) of the 5.3 Gt of fossil fuel derived carbon emitted each year. As the annual atmospheric increase is 3.0 Gt, this leaves 0.7 Gt unaccounted for. This figure is about 4% of the net annual terrestrial primary production, or approximately 3% of the total CO_2 transferred from the atmosphere to the ocean.

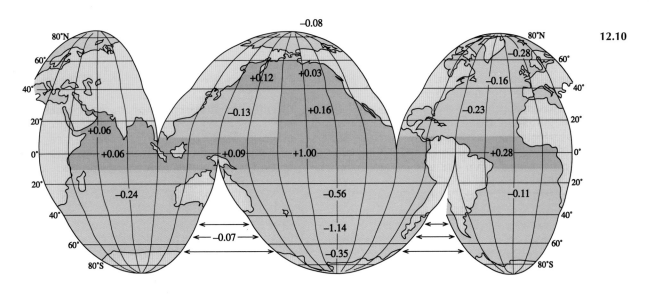

Figure 12.10 Averaged annual air–sea exchange of CO_2 across the sea surface (in units of 10^{12} gC/yr; orange shades indicate where CO_2 evades seawater, blue shade indicates net sinks). (Adapted from Takahashi[10].)

Box 12.4 Normal atmospheric equilibrium concentration (NAEC) model

When the rates of gaseous molecular transfer into and out of the sea surface are equal, the gases are at equilibrium; that is, the air and water concentrations are constant over time, but the gas molecules are freely and reversibly transported backward and forward across the air–sea interface. If one side of the interface has a higher or lower concentration of the gas, so that there is no longer an equilibrium, there will be a net transfer of gas molecules to counteract that difference. For example, the average partial pressure of CO_2 in surface water at 0°C is 15.2 Pa (150 µatm). Since this is less than its atmospheric partial pressure, 35.2 Pa (348 µatm), surface sea water at 0°C experiences an average influx of CO_2 from the atmosphere.

The degree to which any gas exerts a 'pressure' in water can be derived through the solubility relationships, based upon Henry's Law. The gaseous exchange can be represented by equilibrium (12.24), where $CO_2(g)$ and $CO_2(aq)$ represent the gaseous and aqueous concentrations of CO_2 molecules. The position of the equilibrium between the gaseous and aqueous phases is given by the equilibrium constant K, equation (12.25).

All gas concentrations are more usually expressed by their partial pressure, P, related to the gas law, $PV = nRT$, where V is volume, n is the number of gas molecules in that volume, R is the gas constant, and T is the temperature in degrees Kelvin. Therefore, the atmospheric concentration of CO_2, $[CO_2(g)]$, is given by equation (12.26), which is substituted into equation (12.25), to give the concentration of CO_2 in water, $[CO_2(aq)]$, equation (12.27).

K/RT is known as the Henry's law constant, K_H. As K_H is inversely related to temperature, gas solubility decreases with increasing temperature. Gas solubility also decreases with increasing salinity and hydrostatic pressure. K_H changes nonlinearly with salinity and temperature, and therefore a number of multiparametric empirical formulae have been developed to compute the concentration of the gas in sea water.

The equilibrium concentration is the gas concentration that would be attained if the water body were allowed to come to equilibrium at its in-situ temperature and salinity. This is termed the Normal Atmospheric Equilibrium Concentration (NAEC) model.

$$CO_2(g) \rightleftharpoons CO_2(aq) \tag{12.24}$$

$$K = [CO_2(aq)]/[CO_2(g)] \tag{12.25}$$

$$[CO_2(g)] = nV = P_{CO_2}/RT \tag{12.26}$$

$$[CO_2(aq)] = KP_{CO_2}/RT = K_H P_{CO_2} \tag{12.27}$$

The surface of the ocean undergoes gaseous exchange with the atmosphere. The direction of the exchange depends on the relative temperatures, and the difference in $[CO_2]$, between the water and the overlying air mass. If the sea surface waters have a lower partial pressure of CO_2 than that in the atmosphere, then the gas moves from the atmosphere into the sea. This gaseous invasion continues until either the partial pressures equalise (the position of normal atmospheric equilibrium concentration, NAEC – see *Box 12.4*) or the watermass sinks below the mixed layer before reaching equilibrium. The rates of invasion and escape (evasion) are influenced by meteorological conditions, surface waves and films, and contaminants in the sea surface. Diffusion of gases across the air–sea interface increases in stormy weather,

and the dissolved gases are carried to deeper levels mainly by turbulent mixing. The seasonal or main thermocline that separates the upper and lower water columns provides an impediment to vertical mixing. The upper water column can freely exchange gases with the atmosphere. The equilibration steps involve a number of processes, mostly physical mixing of water masses, and some others, such as diffusion across the air–sea interface and hydration of the gas, which are slow and possibly rate-determining.

Air–sea exchange tends to be a kinetically limited process compared to other processes that affect the partial pressure of CO_2 in the surface waters. The upper 100 m mixed layer of the ocean reaches equilibrium with the atmosphere in approximately one year. At high latitudes, warm water from the

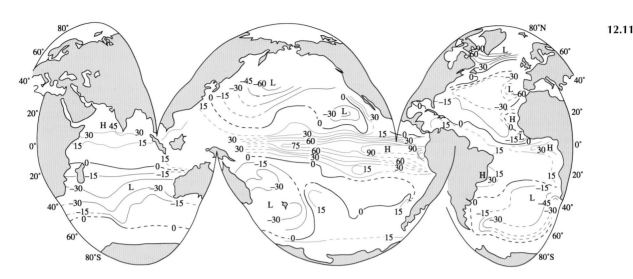

Figure 12.11 The averaged annual differences in the partial pressure of CO_2 between surface sea water and the overlying atmosphere (p.p.m.). Positive values indicate that sea water is supersaturated and negative values indicate that it is undersaturated with respect to CO_2. (Adapted from Keeling[4].)

equator carried north by the oceanic circulation becomes undersaturated in CO_2 upon cooling, and absorbs CO_2 from the atmosphere. The rate of invasion is enhanced through photosynthetic reduction of CO_2, especially in sub-polar waters, where nutrients are generally high and productivity is stimulated. It is therefore rare for surface waters to be at equilibrium with the atmosphere (*Figure 12.11*); equatorial waters tend to be supersaturated, and polar waters undersaturated. Indeed, it is even possible to stimulate photosynthetic activity to increase the 'draw-down' of CO_2 (see *Box 12.5*, and *Figure 12.12*).

Most of the CO_2 absorbed from the atmosphere is probably confined within the upper water column or mixed layer near the ocean surface. The limited mixing into deeper waters is the result of strong vertical stratification at high latitudes, where most of the anthropogenic CO_2 enters the ocean. The exceptions are the deep water formation areas (see Chapter 3) that receive wind-borne CO_2. This process is likely to have absorbed only a small amount of anthropogenic CO_2 so far. There are clear differences in the atmospheric levels and seasonal trends in CO_2 between the hemispheres. The 'draw down' of CO_2 by the ocean is limited by wind patterns and is therefore probably confined to some sea areas only, and may be accentuated during certain periods of the year (such as during the winter when the upper water column is cooled and mixes to greater depths).

The capacity of the ocean to absorb CO_2 is greatly augmented by the downward mixing and sinking of biogenic particulate organic matter

(*POM*) and calcium carbonate into deeper waters (see Chapter 7). As a result of this 'biological pump', which moves carbon from surface waters to deep waters, the bottom waters contain so much CO_2 that they are supersaturated on average by about 30% relative to the NAEC. As the dissolved inorganic carbon (*DIC*) content of the oceans continues to rise, water masses that are supersaturated with $CaCO_3$ (e.g., calcite, aragonite, etc.) become undersaturated [equation (12.5)] – then the $CaCO_3$ in the sediments begins to dissolve. A doubling of the present atmospheric CO_2(g) concentration would increase the *DIC* by about 5–6%, and double the [H^+] in surface waters (the effect would be smaller in deeper waters due to mixing). A possible consequence is that the production of calcite and aragonite by planktonic organisms in near-surface waters might be diminished as the water becomes less supersaturated – plankton communities may change.

In summary, although it is possible to measure the important components within the marine carbonate system, such measurements are analytically complicated and the theory is fraught with difficulties. Measurements are restricted to 'spot' samples that are made on board ship (or back in the laboratory). Ships and measurement systems are very expensive, so the cost of even a simple experiment can be prohibitively large. It follows that there are very few accurate records of geographical, spatial, or temporal changes in CO_2 in the ocean. Consequently, predictions are based on historical evidence and have a high level of uncertainty. Under these circumstances, any data that we do collect

Box 12.5 The removal of atmospheric CO$_2$ by iron enrichment of surface waters

Little effort has been made to alter the rate of fossil-fuel consumption or land use – despite the social, economic, and political impacts. This inactivity is partly due to the level of uncertainty regarding the current predictions of environmental change that will be induced by the increasing atmospheric concentrations of CO$_2$. It is also due to the perceived prohibitive costs of developing alternative energy sources and systems. As a result, attention has focused on removing the excess of CO$_2$ from the atmosphere, rather than cutting its production at source.

This has led to a proposal that marine phytoplankton growth might be stimulated by fertilising surface waters of the North East Pacific, the equatorial Pacific, and the Southern Ocean (an area of over 10% of the world's ocean). These waters contain abundant nitrate and phosphate, but support an unusually low biomass. The late John Martin believed that primary production in these nutrient-rich waters is limited by iron. Laboratory (microscale) experiments had already convincingly shown that nanomolar iron enrichments of high-nitrate, low-chlorophyll waters do, indeed, stimulate phytoplankton growth and biomass[12]. The 'bottle experiments' have also been successfully repeated in waters south of the Galapagos Islands (project 'IronEx' – see *Figure 12.12*), although the results differed from those expected. The observed changes in the partial pressure of CO$_2$ and in nitrate, fluorescence, and chlorophyll levels were considerably less than theory predicted, presumably due to various unquantified loss terms (the grazing by zooplankton fortunate enough to benefit from the increase in phytoplankton numbers, the export of organic carbon, etc.). However, mesoscale field experiments are notoriously difficult to undertake and interpret, and contain many unexpected and unquantifiable factors.

If the 'iron hypothesis' is correct, adding sufficient soluble iron to these waters should stimulate enough primary productivity to consume one-third to one-half of the anthropogenic CO$_2$ flux. Supporters of this hypothesis have described this as a rapid method for recreating sedimentary organic matter, to counterbalance the rate of fossil fuel destruction. However, in the extreme, it is also possible that sustained elevated levels of photosynthesis might also remove sufficient CO$_2$ from the atmosphere to induce periods of glaciation.

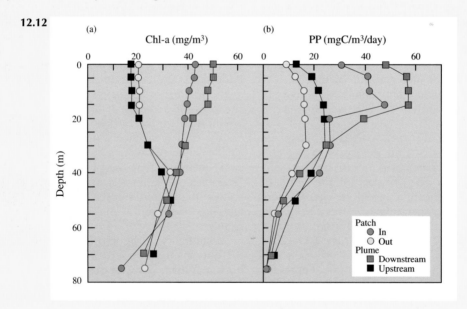

Figure 12.12 During project IronEx in November 1992, a patch of sea water was injected with iron, continuously tracked, and various chemical and biological parameters recorded. This was a particularly tricky operation involving aircraft overflights, radio-drogue buoys, and special chemical tracers (SF$_6$, for instance). The ship spent a total of 10 days sampling the patch of 'injected' iron. Here are shown comparisons of (a) chlorophyll concentration (Chl-a) and (b) primary production (PP), both in and out of the patch, and both upstream and downstream of the Galapagos Islands. There is a clear increase in both the chlorophyll concentrations (standing biomass) and the primary production (current activity of the plankton). (Data courtesy of the US JGOFS Steering Committee.)

Box 12.6 Dynamic or stable – a question of scale!

What is the time taken for a mass of water to re-establish carbonate equilibrium if the temperature were to suddenly decrease 1°C from 21 to 20°C? At this temperature, the concentration of dissolved CO_2 drops by 4%. The original CO_2 in solution associates with CO_3^- to form hydrogen carbonate ions in order to accommodate the shift in the equilibrium concentrations of all species. CO_2 has to invade surface waters because the water is out of equilibrium with the atmospheric CO_2 concentration. Using the basic equations for CO_2 equilibria, it is possible to calculate that the total inorganic carbon concentration would have to increase by 0.4%, or 8×10^{-6} mol/kg, assuming the original CO_2 concentration was 2×10^{-3} mol/kg. If we assume that the thermocline (the main barrier to mixing with deeper waters) is at 100 m, then the immediate effect is to cause a CO_2 deficiency of approximately 0.8×10^{-3} mol/kg total inorganic carbon. Because of the physics of gas exchange and other meteorological factors, the time taken to re-establish equilibrium with the atmosphere will be slightly longer than 1 year.

Quite a long time! In reality, as CO_2 enters the ocean the deficit between the air and the sea becomes less and the net flux across the interface falls exponentially, so it is difficult to calculate exactly how long it would all take. The oceans change temperature seasonally (this 'drop' takes place every six months). The mixing of waters between polar and equatorial regions happens over a longer time scale (e.g., years), so the equilibration of air and sea never quite reaches completion – hence the term dynamic!

have to be fully utilised, and we must make use of developments elsewhere to help in our research. Future research in this area must concentrate on the development of sensors that can be deployed (over monthly and annual periods of time) on buoys or submersible platforms. The present evidence suggests that the marine environment interacts strongly with the atmosphere in such a way that the climatic consequences of increasing levels of atmospheric CO_2 may be accurately modelled – provided that we

can find out more about the marine environment. This is a highly political and controversial area (see *Figure 12.13*, for example). Although the Earth may seem to be in some sort of homeostatic state of regulation (Gaia – *Box 12.7*) this should not lead us into a false sense of security. Short-term variations in atmospheric CO_2 may still create significant changes in meteorological conditions and global climate. This is an area that we have only just begun to research, and in which we have much to learn.

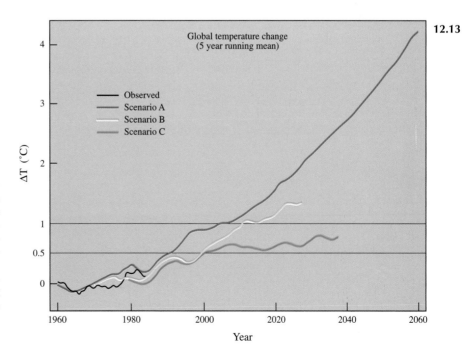

Figure 12.13 Predicted temperature increases, ΔT, for three different scenarios. In scenario A, the current rate of CO_2 emissions is continued. In scenario B, the emissions of CO_2 are held at their present values. Finally, in scenario C, the output emissions of CO_2 are cut such that atmospheric levels stabilise by the year 2000 AD. (Adapted from Libes[5].)

Box 12.7 The Gaia hypothesis

The biosphere appears to counteract naturally the artificial increase in atmospheric CO_2 by acting as a sink for it, and so buffering the greenhouse effect. The principal cause of the progressive fall in the ratio $[CO_2]/[O_2]$ is biological activity, removing CO_2 and releasing O_2 during photosynthesis. Relationships of this kind have led to the novel concept that the surface of our planet is actively maintained as a life-supporting environment by biological activity which acts as a feedback mechanism. This is James Lovelock's concept of the **Gaia** hypothesis, first proposed in the 1970s:

... without life's interference, CO_2 would accumulate in the air until dangerous levels might be reached.

It is the strong interaction of the geochemical cycles of the elements and the biosphere that *regulates* the environment, and Gaia's policy is always to turn existing conditions to its advantage. The biosphere actively maintains and controls the composition of the atmosphere so as to maintain an optimum environment for life and self-perpetuation. The geological fact that atmospheric conditions have remained practically the same over several tens or hundreds of thousands of years suggests that this dynamic system has been really quite stable!

General References

Broecker, W.S. and Peng, T.H. (1982), *Tracers in the Sea*, Eldigio Press, New York, 690 pp.

Butler, J.N. (1982), *Carbon Dioxide Equilibria and Their Applications*, Addison-Wesley, Reading, MA, 259 pp.

Lovelock, J. (1991), *Healing Gaia*, Harmony Books, New York, 1992 pp.

Skirrow, G. (1975), The dissolved gases – carbon dioxide, in *Chemical Oceanography*, Vol. 2, Riley, J.P. and Skirrow, G. (eds), Academic Press, London, Ch 9.

References

1. Bacastow, R.B. and Keeling, C.D. (1979), Models to predict atmospheric CO_2, in *Workshop on the Global Effects of CO_2 from Fossil* Fuels, Elliot, W.P. and Machta, L. (eds), Report CONF-770385, US Department of Energy, Washington, DC.

2. Coulson, K.L. (1975), *Solar and Terrestrial Radiation*, Academic Press, London.

3. Edmond, J.M. and Gieskes, J.M.T.M. (1970), On the calculation of the degree of saturation of sea water with respect to calcium carbonate under *in situ* conditions, *Geochim. Cosmochim. Acta*, **34**, 1261–1291.

4. Keeling, C.D. (1968), *J. Geophys. Res.*, **73**, 4547.

5. Libes, S. (1991), *Marine Biogeochemistry*, Academic Press, New York, 734 pp.

6. Marland, G.T., Boden, T.A., Griffin, R.C., Huan, S.F., Kancircuk, P., and Nelson, T.R. (1989), *Historical and Predicted Emmissions of Greenhouse Gases*, ORNL/CDIAC-25, Oak Ridge National Laboratory, Oak Ridge, TN.

7. Moore, B. and Bolin, B. (1987), The marine carbonate cycle, *Oceanus*, **29**, 11.

8. Post, W.M., Peng, T.H., Emmanuel, W.R., King, A.W., Dale, H., and DeAngelis, D.L. (1990), The biogeochemical cycle of carbon, *Amer. Sci.*, **78**, 314.

9. Schneider, S.H. (1989), The changing climate, *Sci. Amer.*, **261**, 38–47.

10. Takahashi, T. (1989), The effect of the marine carbonate system on climate, *Oceanus*, **32**, 29.

11. Takahashi, T., Weiss, R.F., Culberson, C.H., Edmond, J.M., Hammond, D.E., Wong, C.S., Li, Y.-H., and Bainbridge, A.E. (1970), Global effects of CO_2 from fossil fuels, *J. Geophys. Res.*, **75**, 7648–7666.

12. Wells, M. (1994), Pumping iron in the Pacific, *Nature*, **368**, 295.

CHAPTER 13:

A Walk on the Deep Side: Animals in the Deep Sea

P.A. Tyler, A.L. Rice, C.M. Young, and A. Gebruk

And now, how can I retrace the impression left upon me by that walk under the waters? Words are impotent to relate such wonders.

Professor Pierre Aronnax, in Jules Verne's *Twenty Thousand Leagues under the Sea*

Introduction

Water is a totally alien environment for us humans. Few of us do more than enter the sea and, as we swim, dip our heads briefly beneath the surface to take a momentary glimpse of a blurred and strange world. Even the best-equipped Self-Contained Underwater Breathing Apparatus (SCUBA) divers can penetrate no more than a few tens of metres, for to go deeper requires highly specialised equipment or very sophisticated and expensive submersibles. But imagine that, like Captain Nemo and the crew of Jules Verne's *Nautilus*, we could walk freely across the ocean floor and let us review what we might experience if we took the long journey across the Atlantic Ocean from Britain to the Bahamas.

If an extra-terrestrial visitor to the Earth wanted to take back a collection of animals from its most typical environment, he could do no better than sample the abyssal deep-sea floor across which our route takes us. For the world ocean covers 70% of the planet's surface and over 80% of it is more than 3000 m deep. Yet our own knowledge of what lives there is very recent. Although a few samples had been taken in relatively deep water in the first half of the nineteenth century, serious study of the deep-sea fauna began with the cruises of HMS *Porcupine* in 1868 and 1869[6], and received a further boost from the circumnavigation of HMS *Challenger* from 1872–1876 (see Chapter 1). The reports on the data collected from HMS *Challenger* filled 50 large volumes and included descriptions of hundreds of new species. But because of the coarse nets routinely used from HMS *Challenger*, and for the next 80 years or so, most of the animals collected from the sea floor were relatively large and lived on or just beneath the sediment surface, just the sort that we are likely to see. But we should be aware that this is just a fraction of the deep-sea fauna, for the vast majority is made up of tiny animals measuring millimetres at most and hidden from our eyes within the bottom muds beneath our feet. The richness of this tiny fauna became apparent only in the 1960s, when the Americans, Howard Sanders and Robert Hessler, started sampling the deep-sea floor with much finer nets and recovered an amazing variety of small invertebrates[5]. A consequence of studies using sampling apparatus to collect the smaller organisms is the recent suggestion[2] that the deep sea may have a species diversity equivalent to, if not greater than, that of tropical rain forests (see also Chapter 15). With this proviso in mind, let us plan our expedition. Although nets collect animals from the sea floor, the use of the deep-sea camera has allowed us to view this environment undisturbed. The photographs presented in this chapter come either from cameras mounted on trawls, from 'Bathysnap' (see Chapter 7), or from cameras mounted on submersibles, the last of which allows the operator to select the target.

Route Planning

Our route takes us from the intertidal (see Chapter 16), across the continental shelf to the west of the British Isles (*Figure 13.1*). As we cross the edge of the shelf at about 200 m depth, the slope of the sea floor steepens perceptibly and the already dim downwelling light diminishes rapidly. By the time we reach a depth of 300–400 m we are no longer able to detect any daylight from the surface and we enter a vast zone of perpetual darkness.

Our path takes us down the gentle northeastern slope of the Porcupine Seabight, named after HMS *Porcupine*, for this was one of the very first areas of deep sea to be sampled. We head first south and then west where, at a depth of about 3500 m, the Seabight opens out onto the Porcupine Abyssal Plain. A more direct route would have been down the eastern flank of the Seabight, but this area is riven with deep channels known as submarine canyons. The fauna here is different from that on our route, but the difficulty of the terrain makes sampling with surface-deployed gear hazardous so that this area, together with similar rough topography in the world's ocean, is best sampled from sub-

13.1

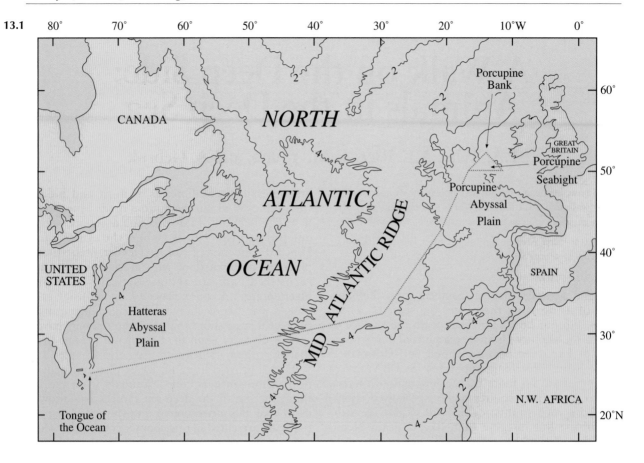

Figure 13.1 Chart of the route from the southwest of the British Isles to the Bahamas (compare with *Figure 1.9*). Contours are shown at 2 km and 4 km depth.

mersibles. An even more precipitous route, which might be taken by a party of more adventurous individuals, would be from the Porcupine Bank westward into rapidly deepening water. This descent to the floor of the abyssal plain would be almost like abseiling down a cliff face, for this particularly steep continental slope is typical of some thousands of kilometres off northwestern Europe and is one of the most spectacular geophysical features on the planet.

Whichever route is taken, the Porcupine Abyssal Plain (*Figure 13.1*) must be crossed westward over a vast and almost flat expanse, gradually deepening to almost 6000 m, but interrupted occasionally by small abyssal hills rising a few tens to a hundred or two metres above the general sea floor.

As we approach the middle of the ocean the sea bed starts to rise on the eastern flank of the mid-Atlantic Ridge to reach the crest at about 2000 m. Now we climb down into the narrow graben, or axial valley, and walk westward past lava flows and occasional hydrothermal vents (see Chapter 10). A second steep climb brings us out of the graben and onto the western flank of the mid-

ocean ridge which now descends steadily to the Hatteras Abyssal Plain with a general depth, like the Porcupine Abyssal Plain, of between 4000 and 5000 m. The 'landscape' here is also not very different, the monotony being broken only by the Bermuda Plateau, which just breaks through the surface at Bermuda itself.

The Hatteras Abyssal Plain extends all the way to the Bahamas and, as we reach the base of these islands, we enter a blind-ended channel, known as the Tongue of the Ocean, running between New Providence Island and Andros Island. We ascend a steep slope and, because of the clarity of the water, we begin to perceive daylight at about 500 m depth, considerably deeper than where we last saw daylight on the other side of the ocean. There is no real continental shelf in this region and our path takes us steeply upslope to the coral reefs fringing the Bahama Islands, where we finally reach dry land.

The View on the Route

The descent into the abyss

As we enter the sea our last view of dry land is the rocky shore common on the west side of the British

Figure 13.2 A rocky shore to the west of the British Isles. This is the last view of land as we enter the sea on our route to the west.

Isles (*Figure 13.2*). Our route takes us into steadily deepening water. The sea bed is composed of sands of varying grades that are commonly stirred up by the surge of storm waves breaking overhead or sculptured into ripples by the strong currents. The fauna contains a wide variety of species, many of them recognisable to the amateur naturalist[1]. At almost every step we are likely to see a fairly large bottom-dwelling animal, perhaps a scallop or one of its molluscan relatives, a starfish or brittlestar (sometimes as many as $300/m^2$), or a crab scuttling away to find shelter. We also disturb the occasional flat-fish, such as plaice, sole, or halibut, while around us we quite often see shoals of cod, whiting, and herrings. For, despite the ravages of overfishing, the continental shelf still supports a fish resource exploited by local fishermen (see Chapter 21), and much richer than we find in the deep ocean.

Beyond the shelf edge, the upper part of the continental slope on our main path descends gently into the Porcupine Seabight. With increasing depth the soft sediment becomes finer, reflecting the decreasing influence of disturbance from the surface and the reduced strength of the bottom currents. Large animals become less abundant, for the bottom is further away from the primary source of food, the thin surface layer which is lit well-enough to support the growth of the phytoplankton. Nevertheless, there is plenty of evidence in the form of tracks, trails, burrow openings, and mounds that large animals are here. And as we start to lose the animals with which we became familiar as we crossed the shelf, we begin to see new ones, the first of the true deep-sea fauna.

One of the most obvious is the red crab, *Geryon trispinosus* (*Figure 13.3*), a member of a family with many representatives occupying much the same depth zone in different parts of the ocean, some of them abundant enough to support commercial fisheries. *Geryon* is clearly responsible for many of the large burrows and mounds on this part of the slope. At the upper slope depths, around 500–1000 m deep, we also see strange plant-like colonies of the sea-pens *Kophobelemnon stelliferum* and *Pennatula*

Figure 13.3 The Atlantic red crab *Geryon trispinosus*, adopting an aggressive posture at 500 m depth in the Porcupine Seabight.

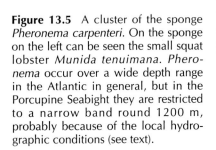

Figure 13.4 The sea-pens *Kophobelemnon stelliferum* (right) and *Pennatula aculeata* (left) anchored in soft sediment at 500 m depth in the Porcupine Seabight. The anemone-like polyps of the 30 cm high *K. stelliferum* are clearly visible, but on *P. aculeata* they occur as small polyps on each leaf.

aculeata (*Figure 13.4*), colonial octocorals with numerous anemone-like polyps. Many similar animals occur in shallow water, and these outliers of the deep-sea community represent a life-style, that of suspension feeding, which becomes less and less common with increasing depth, to be replaced by deposit feeding. Curiously, however, we are about to come across a remarkable abundance of animals by deep sea standards – and all suspension feeders. For as we approach the 1000 m isobath, we suddenly see patches of the hexactinellid sponge *Pheronema carpenteri* (*Figure 13.5*), spherical blobs the size of a large grapefruit, 'rooted' into the bottom by a mass of long glass spicules and with a large exhalant aperture in the top. Initially, we see the odd *Pheronema* here and there, but as we move on a few hundreds of metres, increasing our depth by only 10–20 m, the population density builds up

rapidly until we are standing amongst large patches, tens or hundreds of metres across and containing many sponges on each square metre, like a great prickly carpet. As we carry on down the slope this amazing population rapidly becomes more sparse until, by the time we reach a depth of 1300 m or so, it disappears completely, just as abruptly as it had appeared at the upper limit of the sponges' depth range. We do not know exactly why these sponges should be so abundant here, or why they should be so restricted in their depth distribution, for in other areas they occur over a much wider depth range. It may be a result of the local hydrography, a combination of internal waves and the bottom slope resulting in an enhancement of the near-bottom currents which, in turn, keep food material in suspension and therefore available to the sponges. The sponges also harbour a whole

Figure 13.5 A cluster of the sponge *Pheronema carpenteri*. On the sponge on the left can be seen the small squat lobster *Munida tenuimana*. *Pheronema* occur over a wide depth range in the Atlantic in general, but in the Porcupine Seabight they are restricted to a narrow band round 1200 m, probably because of the local hydrographic conditions (see text).

Figure 13.6 The sea cucumber *Paelopatides gigantea*, swimming over the sea bed by gentle undulations of its dorso-ventrally flattened body at about 2000 m depth in the Porcupine Seabight.

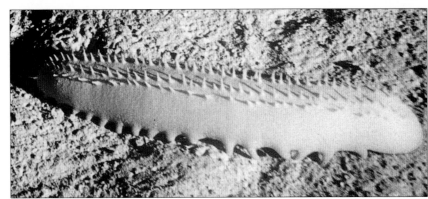

Figure 13.7 The sea cucumber *Benthogone rosea*, walking over the sediment surface at a depth of about 1500 m on the Goban Spur. The mouth is at the right, and the small tentacles can be seen collecting food particles from the sea bed.

community of other animals, ranging from a wide range of small organisms living in or among the sponge spicules to the small squat lobster *Munida tenuimana* (*Figure 13.5*) and the curious small spider crab *Dorhynchus thomsoni*.

As we move down toward the 2000 m level we come across occasional specimens of the holothurians (sea cucumbers) *Paelopatides gigantea* and *Benthogone rosea* (*Figures 13.6, 13.7*). These animals, typically 15–30 cm long, are deposit feeders, ingesting sediment as they move across the sea floor and extracting from it small particles of organic matter. They are representatives of one of the most typical deep-sea groups, for holothurians are found throughout the deep ocean and many different species are to be found.

As we pass from the Seabight and out onto the abyssal plain the general appearance of the sea bed depends upon the time of year. At most times it simply appears as a plain brownish-grey mud. But in the late spring or early summer, say during May or June, it looks markedly different. For then the sea bed is covered in a flocculent brownish green 'gunge', moved around by the near-bottom currents and therefore patchily distributed, being most concentrated in depressions and on the down-current sides of mounds (*Figures 13.8, 13.9*). This material is phytodetritus, a complex cocktail based mainly on the dead cells of normally surface-dwelling phytoplankton species, but also containing many other elements, particularly the carcasses or moulted skins of small planktonic animals and their faecal pellets (see Chapter 7). Phytodetritus sinks suddenly and rapidly to the sea floor over much of the north Atlantic, following the spring phytoplankton bloom, and thus introduces a seasonality to an otherwise monotonously stable environment. Like the discovery of the richness of the small infaunal animals of the deep-sea floor, this seasonal deposition of phytodetritus is a very recent discovery. Its existence was undreamt of prior to about 1980 and its discovery proved that the deep sea is a much less constant environment than had been thought for more than a century[4]. Its arrival, and subsequent very patchy distribution, may be one of the factors which results in the surprisingly high species diversity of the benthic fauna (see Chapter 15).

The alternative route onto the abyssal plain, down the precipitous slope to the west of the Porcupine Bank, has a quite different fauna. Here, in contrast to the soft sediments of the Seabight proper, the slope off the Bank is mainly exposed rock, black with a manganese coating (see Chapter 21), and with sediment accumulating only in the occasional areas where the slope is gentle. The fauna on the rock surfaces is quite unlike that in normal sediment-covered regions, for it is bathed by a northward-flowing current of 3–10 cm/s which is laden with food particles that support a rich suspension-feeding community. Typical fauna

13.8

13.9

Figure 13.8 Fresh phytodetritus collecting in depressions (about 20 cm in diameter), some of them feeding marks left by sea stars, at a depth of about 2000 m in the Porcupine Seabight in May 1982. Phytodetritus consists largely of dead and dying phytoplankton cells which picks up a variety of other particles as it sinks through the water column (Chapter 7). Much of it is recycled during its downward journey, but a significant proportion reaches the deep-sea bed as a seasonal pulse and forms an important food source for the bottom-living animals, in this case the sea-urchin *Echinus affinis*.

Figure 13.9 'Old' phytodetritus concentrated in biogenic depressions (about 50 cm diameter) on the Porcupine Abyssal Plain at a depth of 4850 m in September 1989. The patchiness produced by this phenomenon may at least partly explain the high biodiversity of the deep-sea floor communities.

include sponges such as *Euplectella* sp and the pompom-shaped sponge *Crateromorpha* (*Figure 13.10*). The cnidarians are richly represented here by gorgonians, the most dramatic being the bottle-brush *Thouarella* sp., *Iridiogorgia* sp., and the fan-shaped Paramuriceidae (*Figure 13.11*).

The last of these orientates normal to the prevailing current, allowing the maximum cross-section for particle filtration. Filter feeding is also found in the crinoids, the dominant echinoderm group in this habitat, of which the bright yellow *Anachalyspicrinus nefertiti* (*Figure 13.12*) and

Porphyrocrinus thallassae are the most spectacular examples. A similar life-style is adopted by the brisingid sea stars, such as *Brisingella multicostata*, which sit atop rocks in this zone to benefit from the accelerated flow over topographical highs (*Figure 13.13*).

The fauna of this part of the slope can be used to determine current direction (*Figure 13.14*). Non-random orientation of the crinoid *Porphyrocrinus thallassae* and the downstream winnowing of sediment behind a glacial erratic indicate the flow of Northeast Atlantic Deep Water along this slope[8].

13.10

Figure 13.10 The sponge *Crateromorpha* at about 2120 m depth to the west of the Porcupine Bank. The main filtering apparatus is the pompom at the distal end of the stalk, some 20 cm above the rock surface.

13.11

13.12

Figure 13.11 A branched gorgonian (ca. 1 m high) belonging to the family Paramureicidae at a depth of 2800 m on the western slope of the Porcupine Bank. The small anemone-like polyps found along each branch filter particles from the water column. These colonies, which may be a metre high, are orientated at right angles to the current so the broadest face of the colony faces the current, thus maximising particle capture.

Figure 13.12 The brilliant yellow crinoid *Anachalypsicrinus nefertiti* at 2480 m depth to the west of the Porcupine Bank. The star-shaped part of the animal filters particles from the water column, the stalk keeping this filtering apparatus well above the sea bed. The animal can be up to 25 cm high.

13.13

13.14

Figure 13.13 Three individuals of the multi-armed sea star *Brisingella multicostata* resting on a small rocky hillock at 2890 m depth to the west of the Porcupine Bank. The arms are extended into the water column to trap small animals and particles brought past the rock by the current.

Figure 13.14 Individuals of the bright red crinoid *Porphyrocrinus thalassae* attached to the side of a block of rock at 2310 m depth to the west of the Porcupine Bank. The crinoids are all orientated in the same way, showing that the current is from the right.

13.15

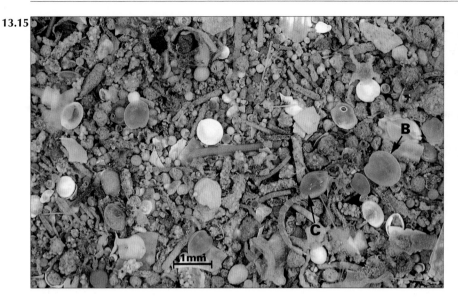

Figure 13.15 Small macrofauna from the abyssal plain. This is the residue left on the sieve after most of the fine sediment has been removed. The main macrofauna seen in this sample are bivalve mollusca (B) and the cumacean crustaceans (C). The small spherical structures are protozoans. (Courtesy of Dr J.D. Gage, Scottish Association for Marine Science, Oban, Scotland.)

The abyssal plain

At the foot of this steep rocky slope there is an abrupt transition to a much flatter bottom, typical of the edge of the abyssal plain, with soft sediment similar to that crossed on the main path at the mouth of the Seabight. The depth is about 4000 m here; although our path does take us deeper, we are no longer able to detect any significant slope. For we are now truly on the abyssal plain, with gradients typically of the order of 1:1000 or less. Apart from the occasional ghostly flashes of bioluminescence (see Chapter 14), it is, of course, totally dark. It is also very cold, with the temperature constantly hovering around 2°C. Finally, it is almost deathly still compared with the shallow coastal waters. For, although the currents still have a clear tidal component, they are relatively sluggish, rarely exceeding 10 cm/s. Here, we are just about as far from the source of food as it is possible to be in the ocean,

short of descending into one of the deep ocean trenches. All the incoming organic matter, including phytodetritus (see Chapter 7), has had to run the gauntlet of the mid-water animals in the 4–5 km of water above. Much of it has been recycled in those communities, with the inevitable loss of material at every step in the food web. So, above all, the abyss is an environment where food is at a premium and nothing is wasted.

An obvious indicator of this state of affairs is the relative paucity of the fauna. The hidden small animals living in the sediment (*Figure 13.15*) are about 100 times less abundant here than on the continental shelf. For the bigger ones, the difference is nearer to 1000! Curiously, despite the low animal abundances, we see more evidence of their activities here than in shallower regions. A combination of high current speeds, storm disturbance, and many animal 'feet' trampling over the sea bed rapidly

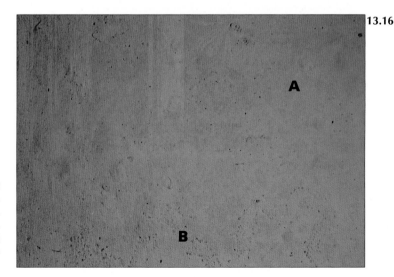

13.16

Figure 13.16 An area of about 20 m² of the Porcupine Abyssal Plain at a depth of 4850 m, which contains many small burrow openings, probably made by worms and crustaceans. The star-shaped marks (A) are the feeding traces of echiuran worms, while the meandering trail (B) was made by a sea cucumber.

13.17

13.18a

13.18b

Figure 13.17 A pair of the hermaphroditic sea cucumber *Paroriza pallens* at 1670 m depth on the Goban Spur; such pairs are probably formed to ensure successful fertilization between individuals of species with low population densities.

Figure 13.18 The sea cucumber *Oneirophanta mutabilis* at a depth of 4800 m on the Porcupine Abyssal Plain in September 1992. This is from a time-series (5 h intervals between frames) obtained with the deployed camera 'Bathysnap' and also shows a sea anemone and the remains of that season's phytodetritus.

destroys trails, faecal casts, and unoccupied burrows and mounds in coastal and upper slope waters. Here, in the tranquil abyss, these features last much longer. So there are many burrow openings in the mud, often with clear feeding lines radiating from them like spokes in a wheel (*Figure 13.16*). Many of these are made by echiuran worms, clearly much more abundant than their occurrence in trawl and sledge catches would indicate. Some of them are sizeable beasts, for the feeding traces made by their prbosces are 2 m or more across. We will also see many trails, a few centimetres across and meandering over the bottom for tens of metres (*Figure 13.16*). There are several quite distinct types, some being smooth grooves, some consisting of parallel rows of separate pits, and some much less regular, as if something has simply ploughed through the bottom mud more or less at random. Most of these trails are made by holothurians, for the sea cucumbers are without doubt the dominant invertebrate megafaunal group in the deep sea, crawling slowly across the sea floor and hoovering up the sediment to extract its small organic content. The basic cucumber shape familiar in shallow-water members of the group is represented here by species such as *Paroriza pallens* (*Figure 13.17*), but other deep-sea representatives modify the sediment in a bewildering variety of ways. These include the elegant white *Oneirophanta mutabilis*, with its short, peg-like 'legs' and long dorsal tentacles (*Figure 13.18*), and the large, purple *Psychropotes longicauda*, with its strange caudal 'sail' (*Figure 13.19*).

13.19

Figure 13.19 From the 'Bathysnap' sequence, at 4850 m on the Porcupine Abyssal Plain, this time showing the strange sailed sea cucumber *Psychropotes longicauda*. The function of the sail is unknown.

13.20a
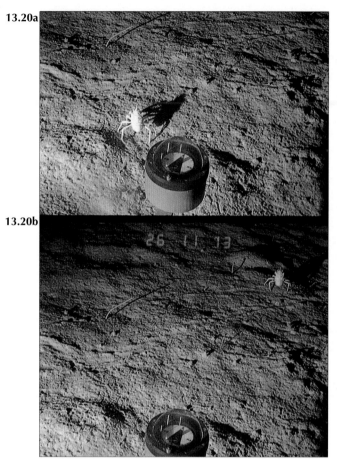
13.20b

13.21a

13.21b

Figure 13.20 A Bathysnap sequence of the galatheid crab *Munidopsis crassa*, moving away from the camera in a northerly direction at 4850 m depth on the Porcupine Abyssal Plain.

Figure 13.21 The penaeid shrimp *Plesiopenaeus armatus*, attracted to bait, in this case a mackerel wrapped in muslin, at a depth of about 4600 m on the Maderira Abyssal Plain.

Although the holothurians are by far the most common large organisms likely to be seen, we also occasionally come across other echinoderms, such as the starfish *Styracaster* or one of the stalked crinoids. We may also see the occasional large crustacean, though true crabs are never found as deep as this, few of them penetrating beyond a depth of about 2000 m. Here, in the abyss, their place is taken by species of the genus *Munidopsis* (*Figure 13.20*), which is well-represented in all the deep oceans of the world. They lack a common name, but are related to the squat lobsters of shallow waters. But the most spectacular crustacean found here is *Plesiopenaeus armatus*, an impressive brilliant red shrimp reaching a length of 30 cm or more (*Figure 13.21*), living generally on, or very close to, the sea floor, but also an accomplished swimmer. *Plesiopenaeus* must be good to eat, but the difficulties of fishing it at these depths would make its price prohibitive!

One thing we almost certainly would not see

during our abyssal walk is the carcass of a dead animal – and this is mostly because of the activities of another crustacean group, the amphipods, the group to which sand-hoppers and freshwater shrimp belong. They are represented throughout the oceans, both on the bottom and in mid-water, where they adopt a wide range of life-styles. But on the deep-ocean floor they are the scavengers *par excellence*. Food is at such a premium in the deep sea that any large lump, such as the carcass of a fish, squid, or whale, arriving on the bottom is attacked very rapidly. The first to arrive are the amphipods, mostly only a few millimetres long, but with some species reaching a respectable 10 cm or more. These voracious scavengers can reduce the carcass of a large fish to a well-picked skeleton in a matter of hours. Little wonder, then, that we are unlikely to see such a carcass. But if we did come across one before it had been completely destroyed it would be the centre of a frenzy of activity – or, at least, what passes for a frenzy in this curious environment. For many animals will want their share of the spoils. Most, including echinoderms, molluscs, and crawling crustaceans, move in relatively slowly to pick up any morsels left by the amphipods. The bonanza also attracts faster moving animals, on the Porcupine Abyssal Plain particularly the rat-tail or grenadier *Nematonurus armatus,* very widely distributed in the Atlantic at depths below about 2000 m (*Figure 13.22*). Curiously, if a fish carcass arrives on the deep-sea floor much further south in the eastern Atlantic, on the Madeira Abyssal Plain (see *Figure 9.9*), it is likely to attract large numbers of *Plesiopenaeus* (*Figure 13.21*) and rather few *Nematonurus,* partly because the relative abundance of the two species is reversed in the two localities.

If carcasses are short-lived and therefore rare in the deep sea, evidence of human activity is both abundant and long lasting. From the results of trawling in the abyssal north Atlantic, it is clear that the sea floor is littered with all manner of rubbish discarded from ships overhead. Such dumping is now banned by all the main maritime nations (see Chapter 22), but for hundreds of years it was the accepted way of disposing of waste at sea. As a result, bottles, cans, bits of pottery, and synthetic materials of all sorts, ranging from polythene sheets and bags, through netting and rope of man-made fibres, to cocktail sticks from the great days of passenger liners, are present. Although we might be lucky enough to come across a fascinating artefact from the days of sail, most of this was jettisoned during the past 30–40 years. The most common sea-floor artefact of all is almost entirely attributable to the century between about 1850 and 1950. During this period, when most ocean-going vessels were powered by coal-fired steam engines, the unburned residues from the fireboxes were routinely dumped over the side as clinker, ranging from small fragments no larger than the tip of your finger to great crusty lumps a metre or more across. Like the other ship-borne rubbish, clinker is naturally concentrated beneath the main shipping routes. But deep-sea biologists hardly ever take a trawl or dredge sample, even in the remotest parts of the ocean, that does not contain at least a few small fragments of clinker. So it seems as though, in little more than 100 years, mankind managed to affect, albeit only slightly, virtually every square kilometre of the largest and most remote environment on Earth – a sobering thought as we continue our way toward the very centre of the ocean.

The Mid-Ocean Ridge

As the rocks underlying the sea floor are younger nearer the mid-ocean ridge (see Chapter 8), the sedimentary veneer becomes thinner as we travel westward. Eventually, on the flanks of the Mid-Atlantic

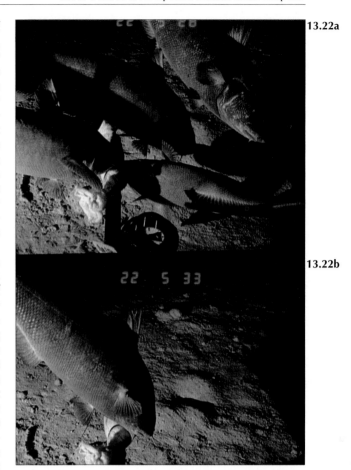

13.22a

13.22b

Figure 13.22 Rat-tail fish, *Nematonurus armatus,* attracted to bait at a depth of 4850 m on the Porcupine Abyssal Plain.

Ridge, rock again protrudes through the mud. In contrast to the steep rocky slope off the Porcupine Bank, these rocks have little or no fauna, for there is almost no suitable food in this region. The barrenness of the sea bed continues as we climb over the crest of rock into the median valley of the ridge system, contrasting with the almost unbelievable luxuriance of the rich populations in the few hundreds of square metres surrounding the occasional hydrothermal vents that we may pass. Unlike all the animals seen so far, these amazing vent communities are not dependent upon food material produced by photosynthesis in the overlying surface layers, but are supported instead by a quite different system based on the vents themselves. The superheated water gushing from the vents is laden with reduced inorganic compounds, particularly sulphides, that are used as an energy source by specialised anaerobic bacteria to build organic carbon molecules on which the rest of the vent community is dependent (see Chapter 10). These oases in the

13.23

Figure 13.23 A swarm of the shrimp *Rimicaris exoculata* at a hydrothermal vent at 3680 m on the Mid-Atlantic Ridge. Interspersed within the swarm of *Rimicaris* are the vent crab *Segonzacia* and another, less abundant, but large and yellow shrimp, *Chorocaris chacei*.

ly laden 'smokers' often deposit their chemical burden as they emerge, to form enormous natural 'chimneys' many metres high (see Chapter 10).

These sights would be amazing enough, but it is the biology that puts the final touch to the experience. Sometimes it is difficult to see the vent water because of the seething mass of life surrounding it. The shrimps here are similar to those seen earlier, but probably belong to a different species. If they seemed like individual bees before, now we have an enormous swarm. Hundreds of thousands of shrimps mill around the vent (*Figure 13.23*), so close-packed that it is almost impossible to follow the movements of any individual. What on Earth are they doing? Well, recent research by American, Russian, and French scientists has pieced together an amazing story of adaptation to this remarkable environment. First, in at least one of the shrimp species the eyes seem to have become modified to 'see' not light, but heat, for they appear to be sensitive to the infra-red emissions from the hot vent-water (see Chapter 14). So they can perhaps detect a vent from some distance away. This is important, because although the vent water emerges at very high temperatures, frequently in excess of 200°C, it cools rapidly in contact with the cold abyssal water so that, no more than a metre or two from a vent opening, it is virtually impossible to locate it from temperature alone. Next, the mouthparts of the shrimp are highly modified for a very un-shrimp-like diet of bacteria, which grow on the mouthparts themselves, on the surface of the carapace, and on the lining of the gill chamber; and the mouthparts can reach and scrape off the bacteria from all of these areas. Finally, the shrimps' behaviour keeps them in just the right conditions of temperature and water chemistry for their bacterial 'gardens' to grow most rapidly. That is what all the jostling is about.

As we wonder at this incredible sight, we should remember that the vent shrimps, and all the other animals which are associated specifically with vents, have a serious problem that most animals in 'normal' environments never have to face. For the vents are ephemeral, apparently lasting for no more than a few decades before the flow of life-supporting, hot, chemical-laden water suddenly slows down and stops. The adult animals around the vent site at this time are doomed. So all the time they must be making arrangements for the survival of their species by ensuring that their eggs or larvae

sea have been found in all the major oceans, surrounded by communities that are hundreds of times richer in terms of biomass than those on the neighbouring sea floor. Many vent species, ranging from bacteria to fish, are found nowhere else, but each individual vent community has a relatively low species diversity, being dominated by only one or two species which may occur in enormous numbers.

So what are we likely to see as we approach a mid-Atlantic vent? Well, long before the vent itself, the animals with which we have already become familiar become much more abundant. Then, as we get to within 20–30 m of the vent, we begin to see new and unfamiliar animals, including snails, decapods, and anemones. The decapods include orange–white shrimps, 5 cm or so long, that swim around like bees on a summer's day (*Figure 13.23*). Although the general abundance of life is dramatically greater than seen so far, it does not prepare us for life at the vents themselves. As our lights finally pick out the vents a few metres ahead, there is a sudden magnificent abundance of life. At the centre is the vent water, emerging from fissures in the rocks as columns rising far above our heads, either glass-clear and shimmering as the temperature contrast refracts the light from our torches, or belching forth like white or black smoke because of the heavy load of particles carried with it. These heavi-

are being cast to the watery winds so that they can colonise any new vents which suddenly start up in the same general area. This is one of the big puzzles of vent biology, and one which oceanographers are actively pursuing[7].

Return to the surface

Left with this, and other fascinating questions about vents in our minds, we climb out of the median valley, cross the western rim of the Mid-Atlantic Ridge and descend along the rocky western flanks. The climb across the Mid-Atlantic Ridge has been very long and rugged (approximately equivalent to a walk from Edinburgh to Rome, but with Alpine terrain the whole way) so the flat terrain and soft sediment we encounter on the desert-like Hatteras Abyssal Plain is a welcome change. The fauna here, like that of the Porcupine Abyssal Plain crossed earlier, is composed mainly of small animals buried in the mud, with only occasional large megafaunal animals crawling across the surface (see Heezen and Hollister[3]). The Hatteras Abyssal Plain is the largest spatial entity in the north Atlantic. Its monotony is broken only by the pinnacle that rises to become the Island of Bermuda and by some slight undulations that pass for abyssal hills. If we stroll to the far southwest corner of this giant plain, we encounter a deep canyon-like extension that cuts into the shelf supporting the Bahamas Archipelago. At the point of entry, between Abaco Island on the north and Eleuthera on the south, this canyon is known as Northeast Providence Channel, but it eventually

Figure 13.24 The giant isopod *Bathynomus giganteus* at 840 m depth off the Bahamas. In contrast to its terrestrial cousins, this species can grow up to about 15 cm long.

turns south and becomes a dead-end 'box' canyon, surrounded completely by islands and shallow banks, and known as the Tongue of the Ocean. Here we begin our very steep ascent toward the coral reefs and sea-grass meadows of the tropical western Atlantic.

As we hike up the slope, an isopod crawls by. Small isopods are so common and diverse in the deep sea that a single individual would not normally catch our eye. But this one is not like the gribble that consume our pier pilings or even like the pillbugs or woodlice that we find behind our couches and in our gardens. This lumbering giant, *Bathynomus giganteus* (*Figure 13.24*) is some 15 cm long. We do not know why a few isopods from the Antarctic and the deep sea display gigantism, but we do know that these giants have been around since before the dinosaurs. Indeed, during the Cretaceous age, when the White Cliffs of Dover were being deposited, giant isopods would probably have been encountered anywhere along our trans-Atlantic stroll; today, they are limited to a narrow depth range in the Gulf of Mexico and Caribbean region, indicating that the sea bottom has changed substantially over the intervening aeons.

At a depth of 900 m there is what appears to be a bright reddish-purple jellyfish, pulsating just above the bottom. As we approach, however, we find this beast to be very un-jellyfish-like in its anatomy. Whereas a jellyfish would have its mouth on the underside and surrounded by long stinging tentacles, this animal has its mouth on the top and surround by tentacles with the appearance of tiny florets of a cauliflower. The strange animal is a sea cucumber, *Enypniastes eximia* (*Figure 13.25*). Like its more sedentary relatives, it feeds by shovelling sediment into its mouth on the sea floor, but unlike other cucumbers, its body wall is not thick and

Figure 13.25 The swimming holothurian *Enypniastes eximia*. This brilliant red species swims above the sea bed, descending to the bottom to feed using the cauliflower-like tentacles.

13.26

Figure 13.26 The multi-armed sea star *Novodinea antillensis* in a typical feeding posture at 660 m depth off the Bahamas. Food, such as small crustaceans, is trapped by tiny pincers on the arms and moved to the mouth by the tube feet. The brilliant colour is typical of brisingid sea stars (see also *Figure 13.13*).

leathery. If we reach out and touch this animal, we find the body is soft and delicate. It can be damaged beyond recognition by only a moderate poke of the finger.

On a nearby rock outcrop there is a brilliant red starfish resembling *Brisingella*, which we saw on the slope west of Ireland. *Novodinea antillensis* (*Figure 13.26*), the Bahamian version, is perched on the highest available portion of the bottom, where it has access to the currents that pass by. Its posture is that of a filter-feeder, but close examination of its long arms reveals it to be a formidable predator of euphausiids and other swimming crustaceans. The entire body is covered with tiny pincers which

form a velcro-like surface capable of entrapping any small objects unfortunate enough to encounter it. When we brush the starfish arm with our own, it immediately grasps our hairs in hundreds of places. We have no trouble escaping, but a small crustacean whose legs are trapped is in a much worse position. Once captured, the prey is enclosed in a loop of the arm, grasped with the tube feet, and moved directly to the gaping mouth on the central underside of the disk.

Novodinea is not the only animal here that gives us a feeling of *déjà vu*. As we look at the stalked crinoids, the gorgonians, the hexactinellid sponges, and the other animals populating this rock outcrop,

13.27

Figure 13.27 The crowns of six individuals of the crinoid *Endoxocrinus* at 640 m depth off Egg Island in the Bahamas. Some of the arms are raised to be flicked rapidly downward, a mechanism which apparently they use to dislodge annoying crustaceans. Compare the number of arms seen on these individuals with the five seen on the species in *Figure 13.12*.

it is apparent that, given the same substratum and depth, faunas may be very similar in widely separated parts of the world ocean. The names of the players may be different on the two sides of the Atlantic, but the roles they play are very much the same.

Like flowers struggling toward the light in a shaded garden, stalked crinoids cluster along the ridges where currents are faster, thereby bringing more food to the waiting arms and pinnules (*Figure 13.27*). Each crinoid bends with the currents in exactly the same way as its neighbours; all have arms poised in a parabolic fan to entrap tiny particles in thick mucus. Occasionally, we see one of these arms abandon its normal posture to move up and down rapidly. It appears that the crinoid is waving to us, but a closer look reveals the fallacy of this interpretation. Small crustaceans occasionally dart toward the feeding groove on the oral side of the arm, possibly to steal some of the concentrated food that the crinoids have accumulated in their mucus. The crinoid responds to this irritation by flicking the offender away, much like a horse swishing flies with its tail.

On the soft sediment near this rock outcrop is one of the most common sea-urchins of this region, *Stylocidaris lineata* (*Figure 13.28*). If it is springtime, probably each individual will be paired with another of its kind. Sea-urchins do not mate, but instead cast their eggs and sperm into the sea, where fertilisation occurs. Normally, these deep-sea animals are so far apart that egg–sperm encounters are unlikely. They pair in the spring to increase the odds that sperm will find eggs to fertilise. A pure white snail living on this same slope (*Tugurium*

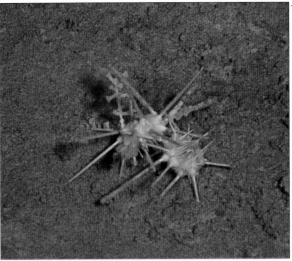

Figure 13.28 The sea-urchin *Stylocidaris lineata* on a rocky outcrop at 500 m depth off the Bahamas. During their reproductive season individuals of this species come together in pairs for spawning to aid their fertilisation success.

caribaeum) carries odd bits of broken shell and other refuse around the margin of its shell.

By 450 m depth, our dark-adapted eyes start to perceive a greyish glow of light from the surface. Two large sea-urchins catch our eye. The first, *Calocidaris micans* [*Figure 13.29(a)*], is the largest sea-urchin we have seen anywhere on the trip and the second, *Coelopleurus floridanus* [*Figure 13.29(b)*], is the most beautiful. The spines of *Calocidaris* are perfectly straight and have the consistency and lustre of ivory elephant tusks.

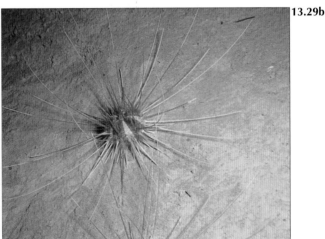

Figure 13.29 Protection in the deep sea. (a) The sea-urchin *Calocidaris micans* at 260 m on sediment off the Bahamas; the straight spines can be up to 20 cm long. (b) The striped sea-urchin *Coelopleurus floridanus*, at a depth of 450 m on sediment in the Tongue of the Ocean. The black, white, and red shell, and the elegant curve of the spines, make this a most striking animal.

Figure 13.30 The slit shell *Perotrochus midas*, on the underside of an overhang at 650 m depth off the Bahamas. The slit along the side of the shell gives the name, and the species is much prized by collectors.

Spines of *Coelopleurus*, by contrast, are tapering, red and white in colour, and flare upward from the colourful body in graceful arcs.

The terrain becomes ever steeper and, in most places, climbing equipment is required to scale the coral cliffs. As the light increases, sponges become larger, more colourful, and more diverse. There are also more and more soft corals, antipatharians, and gorgonians. Comatulid crinoids and brittle stars perch on large glass sponges to elevate their feeding structures in the flow. Basket stars use large gorgonians for the same purposes. Here, on the steepest portions of the cliff, there is a large snail with a shell that appears to be made of pure gold. This slit shell, *Perotrochus midas* (*Figure 13.30*), is prized by collectors, but finds refuge in this rugged environment where divers cannot reach and dredges become hopelessly entangled.

As we climb up the last piece of the cliff, plants appear and the light becomes strong enough to support symbiotic algae that provide much of the nourishment for hard corals. Colours explode as we leave the deep sea and enter the much better understood world of the tropical coral reef (*Figure 13.31*).

Our journey has taken us thousands of kilometres horizontally and through 5 km of depth. It has been equivalent to a march across North America, complete with the Rocky Mountains and Great Plains in the middle. During our stroll, we have focused on those animals that live out in the open and are large enough to see easily with the naked eye. We may have formed the impression that the deep sea contains relatively few species compared with a rain forest or a coral reef.

However, our impression is biased by a failure to examine the little creatures dwelling below the surface of the mud and the diverse creatures occupying the deep water column itself. If we were to count up all of the species, both seen and unseen, that we passed on our journey, our list would rival the species diversity of a tropical rain forest (see Chapter 15). The deep sea is at once the largest, most diverse, and least understood environment on Earth.

Figure 13.31 Typical coral reefs in the shallow sunlit waters of the Bahamas. A coral knoll is surrounded by gorgonians related to, but distinct from, those in *Figure 13.11*.

General References

Campbell, A.C. (1994), *Guide to the Fauna of Intertidal and Shallow Seas*, Hamlyn, London, 320 pp.

Gage, J.D. and Tyler, P.A. (1991), *Deep Sea Biology: a natural history of organisms living at the deep-sea floor*, Cambridge University Press, Cambridge, 504 pp.

Grassle, J.F. and Maciolek, N. (1990), Deep-sea species richness: regional and local diversity estimates from quantitative bottom samples, *Amer. Natural.*, **139**, 313–341.

Heezen, B.C. and Hollister, C.D. (1971), *The Face of the Deep*, Oxford University Press, New York, 659 pp.

Thomson, C.W. (1873), *The Depths of the Sea*, MacMillan, London, 527 pp.

References

1. Campbell, A.C. (1994), *Guide to the Fauna of Intertidal and Shallow Seas*, Hamlyn, London, 320 pp

2. Grassle, J.F. and Maciolek, N. (1990), Deep-sea species richness: regional and local diversity estimates from quantitative bottom samples, *Amer. Natural.*, **139**, 313–341.

3. Heezen, B.C. and Hollister, C.D. (1971), *The Face of the Deep*, Oxford University Press, New York, 659 pp.

4. Rice, A.L. (in press), Changing views on the biology of the deep sea floor: from sterility to unprecedented biodiversity, *Port Erin Marine Laboratory Centenary Volume*.

5. Sanders, H.L., Hessler, R.R., and Hampson, G.R. (1965), An introduction to the study of the deep-sea benthic faunal assemblages along the Gay Head-Bermuda transect, *Deep-Sea Res.*, **12**, 845–867.

6. Thomson, C.W. (1873), *The Depths of the Sea*, MacMillan, London, 527 pp.

7. Tunnicliffe, V. (1991), The biology of hydrothermal vents: ecology and evolution, *Oceanogr. Mar. Biol.: Ann. Rev.*, **29**, 319–407.

8. Tyler, P.A. and Zibrowius, H. (1992), Submersible observations of the invertebrate fauna on the continental slope southwest of Ireland (NE Atlantic Ocean), *Oceanol. Acta*, **15**, 211–226.

CHAPTER 14:

Light, Colour, and Vision in the Ocean

P.J. Herring

Introduction

The reflected surface colour of the ocean, whether seen from the deck of a ship or measured by an orbiting satellite, provides information about the contents, and processes, immediately below. Many of these near-surface processes, ranging from the absorption of red light by chlorophyll to the multi-hued visual signalling of reef fish, are dependent on colour. Interactions deep below the surface are also affected by colour; indeed, the colours of the animals caught in a research trawl give a good indication of the depth from which they have come. 'Colour' itself is a rather subjective concept; its perception is critically dependent upon both the light conditions of the environment and the visual systems of the observer. A more objective assessment requires knowledge of the spectral distributions of ambient, reflected, and absorbed light[11].

Like most research tasks, understanding the role of light and colour in the sea has not been straightforward. Ironically, many of the life activities of deep-sea animals have been assumed from the anatomy of dead specimens. This can make for some spectacular mistakes (such as earlier interpretations of light-emitting organs as eyes or ears), but in most cases the conclusions are probably reasonably close to the truth. However, it is salutary to remember that illustrations of the deep-sea fauna, portraying fish and other animals oriented horizontally, do so more by convention than by information. One of the great advances has been the recent use of remote vehicles, cameras, and manned submersibles to see what life in the deep sea *really* looks like (or, at least, looks like when floodlit!). A surprising number of fish, for example, routinely stand on their heads or their tails. Swimming brittle stars, luminous swimming sea cucumbers, metres-long siphonophores coiled in huge spirals, squids that shed their tails; these are but a few of the many unexpected behaviours that have been observed.

Much more about the interactions of light and animals can be learned only by careful experiments in the laboratory on perfect specimens. The laboratory may be on a ship, with all the limitations and discomforts attendant on working in a cramped, noisy, and disconcertingly mobile space, but with the advantage of immediate access to captured animals. It may be on shore, if the technical difficulty of maintaining the particularly delicate deep-sea species can be overcome. Such studies have shown that the eyes of deep-sea animals are irreparably damaged by brief exposure to daylight (or a submersible's floodlight) – so special cold and dark trawl buckets (or cod-ends) have had to be developed for experiments on the vision of these animals.

The brief account below draws on all these methods for its data, but previous experience indicates that our present interpretations will not do full justice to the extraordinary abilities of the deep-sea fauna.

Characteristics of Light in the Sea

Light in the sea derives from two sources, the Sun and the organisms. Sunlight (and reflected moonlight) illuminates at most only the upper kilometre of the ocean. Bioluminescence occurs at all depths. The intensity distribution of sunlight at the sea surface varies by only a factor of about two over the visible range of 400–700 nm (*Figure 14.1*). Moonlight and starlight intensities are, respectively, about 6 and 9 log units lower than sunlight, while the intensities of bioluminescent sources can approach that of moonlight[2].

Light below the surface differs from that above in both quality and quantity. A major change occurs at the air–sea interface; here, the combined consequences of reflection from the surface (at glancing angles of incidence), and refraction into it, reduce the angular distribution of light entering calm water to a narrow cone of solid angle 97°. A fish looking upward sees the world above the surface through this limited window (Snell's window). All it sees outside this window is back-scattered light from deeper water (by total internal reflection at the air–sea interface[11]). Any initial asymmetry in the radiance distribution within the window (due to the Sun's angle, for example) is rapidly degraded with depth, through the effects of scattering by both water molecules and suspended particles. Ice cover acts as a spectrally neutral diffuser and reflec-

Figure 14.1 Light at depth differs markedly from that at the surface. (a) The narrow spectral distribution of light at a depth of about 500 m in clear ocean water (*d*), compared with that of sun- and sky-light above the surface (*s*). (b) The symmetrical angular distribution of light in the ocean about the vertical axis. The length of the arrows indicates the relative radiance (intensity per unit solid angle) from point O in the respective directions. The radiance is maximum vertically downward and halves at 35° to the vertical. The vertically downward:upward ratio (*D:U*) in oceanic waters is about 200:1 (after Denton[2]).

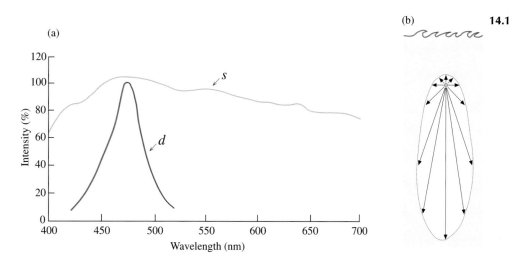

Colour

Shallow water

Life in the upper layers of the ocean is dominated by daylight. In the near surface layers the spectral consequences of absorption and scattering are relatively minor and the animals are exposed to a bright light containing all wavelengths. At the margins of the oceans this is associated with a structurally complex background of shore or reef within (or against which) individuals can hide. The bright colours of many reef fish are responses to this complex optical (and biological) environment. They allow their owners to display and signal to each other with the relative immunity of a safe haven of retreat. The body colours are effective only because the ambient light contains all wavelengths at sufficient intensity for different colours to be selectively reflected. They also presuppose that the observers have sufficient variety of visual pigments to allow them to discriminate the colours, or at least to recognise their contrasts.

Open ocean

In the open ocean the animals are faced with the same illumination, but with the radical difference that the background is uniform. There is no complexity of pattern or topography within which to hide. Bright colours could be used to send the same messages as in shallow waters, but, with nowhere to escape to, these beacons would become a dangerous liability. Disappearance is possible only by matching the background radiance. Animals living right at the air–sea interface, and vulnerable to predators from above, are frequently blue in colour, matching the

tor to produce a uniform, but dimmer, light field at the surface. Even in open water the light field soon becomes symmetrical about the vertical axis, with a downward radiance some 200 times the back-scattered upward radiance[2] (*Figure 14.1*). At increasing depths this angular distribution of radiance remains unchanged, but scattering and absorption reduce the light intensity in an exponential way. Both these processes are wavelength selective. Water absorption preferentially removes both long (red) and short (ultraviolet) wavelengths, rapidly resulting in near-monochromatic blue light, which even in the clearest oceanic waters is reduced by 90% for every 70 m of depth; this means that moonlight makes the same contribution to ambient light at 400 m as does sunlight at 800 m.

In coastal waters dissolved yellow material derived from plant decay (both marine and from river input) may absorb additional short wavelengths and result in a greener hue to both the downwelling and back-scattered light. Scattering by particles of very small size is inversely proportional to the fourth power of the wavelength of the light ($1/\lambda^4$); blue light at 470 nm is scattered five times more than red light at 700 nm. Larger particles scatter all wavelengths more evenly ($1/\lambda$). These characteristics determine the light conditions in the sea and the resulting adaptations of the organisms. The biological imperative for every marine animal is to survive and reproduce; this requires the ability to find food and (usually) a mate, and to avoid or deter predators. Sensory systems such as vision, and effector systems such as bioluminescence and colour, provide some of the means whereby this is achieved.

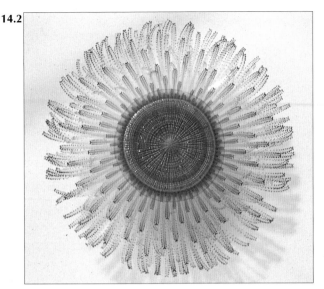

Figure 14.2 The 40 mm diameter coelenterate *Porpita* floats at the surface, where it is camouflaged by a blue carotenoprotein pigment.

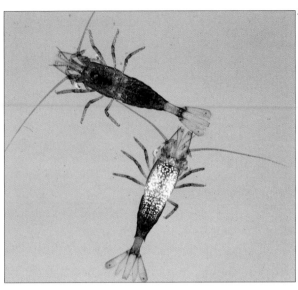

Figure 14.3 This 10 mm long surface living shrimp, *Hippolyte coerulescens*, is camouflaged by a blue carotenoprotein pigment and white reflective dorsal chromatophores. (Courtesy of the Southampton Oceanography Centre, England.)

upward scattered radiance of clear ocean waters. The colour is achieved in different ways by different species. Blue carotenoprotein or biliprotein pigments are commonly used (*Figures 14.2, 14.3*). Blue structural colours (selective diffuse or specular reflection) are other means of achieving the same result (*Figure 14.4*). The deep blue colour with which many upper ocean fish camouflage their dorsal surfaces has a similar structural basis.

Transparency and silvering

For smaller organisms in the upper waters, camouflage can be achieved by transparency. Many planktonic species, particularly the gelatinous forms, rely on this for their protection (*Figure 14.5*). For larger animals the complexity of the body tissues renders transparency impracticable. However, in the particular light environment of open water (with the brightest light from vertically above and a symmetrical radiance distribution about this axis), a mirror stood vertically in the water becomes invisible from any angle of side view. Many animals have taken advantage of this to mimic transparency by turning themselves into the equivalent of vertical mirrors. This is best

Figure 14.4 Blue colour can be achieved without pigment. In the isopod *Idothea metallica* (12 mm long), tiny reflective particles in the upper structure reflect blue light much better than red. The transmitted red is then absorbed by a dark pigment beneath the reflective layer. (Courtesy of the Southampton Oceanography Centre, England.)

14.5a
14.5b

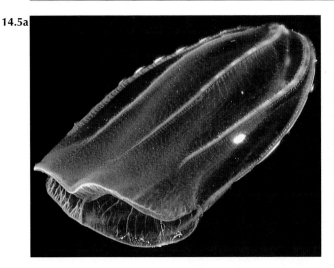
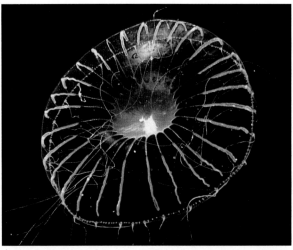

Figure 14.5 Many jellyfish are often almost invisible by virtue of their transparency. Typical examples are (a) the comb jelly *Beroe* (60 mm diameter) and (b) the medusa *Aequorea* (25 mm diameter). (Courtesy of Image-Quest 3-D.)

achieved by flattening the body, so that the flanks are vertical, and covering these with reflective material. Fish are consummate examples of this strategy, none more so than the hatchet fish (*Figure 14.6*). For their reflective material they use tiny crystals of guanine (an excretory product derived from nucleic acids), which are aligned parallel to the body surface. The crystals are arranged in multiple stacks, alternating crystal and cytoplasm (the watery matrix of the cell), whose spacing is such as to achieve constructive interference reflection. For a particular wavelength of light, λ, viewed at right angles to the stack of crystals, ideal interference reflection occurs[1] when the optical thickness of each layer is 0.25λ. With such a system, almost

100% reflectance is achieved with only 5–10 crystals. Without such spacing, five crystals would have a reflectance of only 20%.

One potential drawback of the system is that the best-reflected wavelengths shift toward the blue end of the spectrum as the angle of viewing becomes more oblique.

Effective silveriness requires reflection of all wavelengths, at all angles of view. To achieve this, the spacing of the stacks is adjusted so that either different colours are reflected from different stacks (which may be adjacent or superimposed on one another), or the spacing within the stacks varies in a regular way. Vertical flanks are not compatible with a muscular stream-lined body, so most fish

14.6

Figure 14.6 In the radially symmetrical light distribution in the ocean (*Figure 14.1*), a vertical mirror is invisible from the side. These hatchet fish (*Argyropelecus*, 25–50 mm long) have turned themselves into mirrors by flattening their sides and silvering them, using stacks of guanine platelets as interference reflectors.

14.7

Figure 14.7 Where some organs remain opaque they can still be camouflaged by silvering them separately, as is the case for the liver and eyes of the 40 mm long squid *Cranchia scabra.*

Even when much of the body is transparent, there may be particular tissues that remain opaque (e.g., eyes, red muscle, or digestive organs). These organs can still be silvered individually to achieve effective camouflage. Squid use the same strategy, but employ reflective platelets of protein rather than guanine crystals (*Figure 14.7*).

The concealment value of vertical reflective surfaces rapidly disappears as the surfaces are tilted. This property can be used to good effect to distract predators (e.g., the flashing of a twisting school of fleeing sardines or the eponymous silversides). Changes in body orientation can also be used to send optical signals to nearby members of the school.

The dramatic colour changes visible in some oceanic fish (e.g., the coruscating colours of a captured dolphin fish, *Coryphaena*) are brought about by very rapid changes in the spacing of the crystals in the reflecting cells. Each stack of crystals behaves like venetian blinds as contractile elements in the cells tilt the individual crystals, altering their distance apart and hence their reflected colour. Rapid colour changes, under similar control, take place in the reflective stripe of the freshwater neon tetra and in some damselfish.

Deep-water colours

If reflection is the saviour of animals in reasonably well-lit open water, it spells potential disaster for those in the dark of the ocean depths. Here, the reflection of a bioluminescent flash or glow could break the cover of an animal hitherto invisible

have a more elliptical cross-section. If the reflecting stacks in the skin or scales remain vertically oriented, despite the curved body surface, the effect of a vertical mirror is retained[1].

14.8

14.9

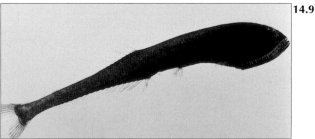

Figure 14.8 The uniform scarlet colour of the 70 mm long deep-living shrimp *Acanthephyra purpurea* is due to a carotenoid pigment that absorbs any incident blue light. We see it as scarlet in the white light of the camera flash; it is invisible in the deep sea, where blue bioluminescence is the norm and most predators only have blue-sensitive eyes. (Courtesy of the Southampton Oceanography Centre, England.)

Figure 14.9 The black melanin pigment of this typical deep-sea fish (*Gonostoma bathyphilum*, 110 mm long) plays the same camouflage role as does the scarlet pigment of the shrimp. (Courtesy of the Southampton Oceanography Centre, England.)

14.10

Figure 14.10 In midwater, where light from the surface is still important and day–night changes are substantial, the limited colouring of the 50 mm long shrimp *Sergestes* is mostly distributed in large dorsal chromatophores. This allows the animal to change its appearance according to the light environment.

against the black background. Thus, silvering is an anathema in deeper water, and is replaced by uniform matt colours of brown, purple, black, or scarlet (*Figures 14.8, 14.9*). This is a particularly clear example of the functional equivalence of subjectively very different colours. Dramatically different as these colours may appear to us in the sunlight on deck, they are all equally effective at preventing the reflection of any residual dim blue light filtering down from the surface, or of stray flashes of blue bioluminescent light. As the bioluminescence may come from any direction, the colouring is spread over the whole of the animal.

At intermediate depths, where daylight from the surface is dim but still significant, and day–night intensity changes are still important, a compromise is reached in which animals have some element of colour and some of silvering or transparency. This enables them to adjust their colouring quite markedly in response to changes in light intensity. Fish silvered in daylight have mobile dark pigment cells which disguise their silveriness at night. Shrimps have a 'half-red' appearance, in which the red pigment is present in large pigment cells, and are able to disperse or aggregate the colour as appropriate to the light conditions (*Figure 14.10*). In both cases, the pigment distribution is primarily dorsal, in response to the continued dominance of downwelling light.

As the bottom is reached, quite marked changes occur in the colours of the animals. Many animals are now grey, pale, or even white. There is no obvious rationale for this change, except that in any bioluminescent light the paler species present a lesser contrast when seen against the lighter sediment than do their heavily pigmented pelagic relatives only a few tens of metres above (*Figure 14.11*). Many of the animals on the bottom have large sensitive eyes, so vision clearly still plays an important role in this environment.

Many oceanic animals undertake substantial vertical migrations during their lifetimes. In general, the larvae and juveniles live at shallower depths than the adults, and thus experience gradual changes in the light conditions as they move deeper in the water column. Their appearance at any given stage of development reflects the depth at which they are living. Early shrimp larvae near the surface may be transparent, the juveniles at mid-depths half-red, and the adults at depth uniformly scarlet. Colour is clearly a key feature in the life of oceanic animals, but its appearance to the denizens of the deep is not always quite what it seems to the human eye in sunlight.

Figure 14.11 Many bottom and near-bottom animals are very pale, like this 220 mm long rat-tail *Nezumia*, at a depth of 1100 m off southwest Ireland. In the light of the flash these animals are not easily distinguishable from the pale sediment on the bottom; this may also be the case in whatever dim bioluminescence exists on the sea floor. Also here (just above and to the left of the fish) is a shrimp with a reflective eye (see also *Figure 14.18*) and (above and to the right) two large pot-like glass sponges. (Courtesy of the Southampton Oceanography Centre, England.)

14.11

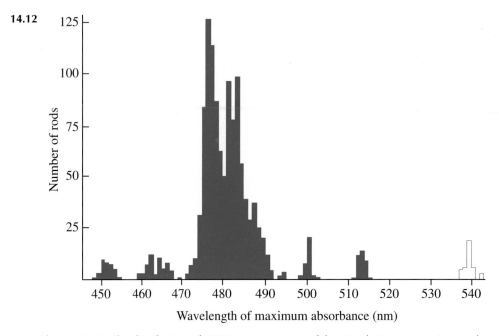

Figure 14.12 The distribution of 1370 measurements of the visual pigment maxima in the rods of 57 species of deep-sea fish shows a good match with the blue–green light in their environment (*Figure 14.1*). Most pigments are rhodopsins; the few longer wavelength porphyropsins are shown as open blocks (from Partridge *et al.*[13]).

Vision

The light environment of the ocean is matched by the visual adaptations of its inhabitants. Vision depends on the absorption of photon energy by the visual pigments and its transduction into a neural signal. The spectral sensitivity of the eye is determined by the absorption characteristics of the visual pigments in the retinal receptors. In vertebrates these receptors are single cells (rods and cones); in invertebrates, they are units (rhabdoms) formed from several cells.

Visual pigments

In general, the absorption maximum of the main visual pigment is a good match to the spectral characteristics of the environment. Thus, deep-sea fish usually have rod visual pigments with absorption maxima in the blue wavelengths around 480 nm (*Figure 14.12*), while shallow coastal species have maxima at longer wavelengths[11,13]. Near the surface the high intensity and broad spectral range of ambient light provide the opportunity for both colour vision and high acuity. The dominant visual task is to maximise the contrast present in the target area. When a *dark* object, or silhouette, is seen against the background of downwelling light, or horizontally against an infinite background of scattered light, the contrast is maximised by having a visual pigment which matches the spectral transmission of the water. When, on the other hand, a

bright reflective object is viewed horizontally, maximum contrast can be achieved by exploiting the spectral differences between the reflected and background light, using a visual pigment whose absorption maximum is offset from that of the background.

Visual pigments are formed by linking a protein, one of the opsins, to a vitamin A_1 or A_2 derivative (forming a rhodopsin or a porphyropsin, respectively). Porphyropsins absorb at longer wavelengths than do their rhodopsin partners. Although a few marine fish do have this pair of pigments, many of them lack the porphyropsin, but have more than one rhodopsin (i.e., vitamin A_1 with different opsins) and thus retain the potential for colour vision. Additional visual pigments may also be present, usually in different types of cone cell, and coloured filters or oil droplets may further differentiate the spectral sensitivity of individual receptors. In the most extreme cases (some mantis shrimps) there may be up to eight kinds of receptors, each with different spectral sensitivities. Recent work has shown that the shrimp *Systellaspis debilis* has a visual pigment which is sensitive to near-ultraviolet light, as well as one sensitive to blue–green light[4] (*Figure 14.13*).

Colour vision is also theoretically possible for fish with only one visual pigment, but with a retina containing multiple banks of rods. Each layer modifies the spectral nature of the light transmitted to

Figure 14.13 The shrimp *Systellaspis debilis* (60 mm long) has two visual pigments, one absorbing in the blue–green, the other in the near ultraviolet. Since these wavelengths have different transmission characteristics, their ratio could give the animal an indication of its depth. The dark spots on the thorax and abdomen are light-emitting organs.

the next layer, giving them, in effect, different spectral sensitivities.

Major changes in the light environment of fish occur during the lifetime of those, like the eel, which have a marine and a freshwater phase or, like the pollock, migrate into deeper water as an adult. These changes are compensated by visual pigment changes, either between rhodopsin and porphyropsin pairs or by opsin shifts between different rhodopsins. In either case the new suite of pigments is more appropriate to the visual tasks of the new environment.

The upward view

Light in the deep-water environment is dimmer, bluer, and highly directional. Animals in the upper few hundred metres are likely to sight prey or detrital particles within an angle of 35° from the vertical; here, the downward radiance does not drop below 50% of its maximum value (*Figure 14.1*). This limited, but brighter cone of view (70°), dominates the visual environment. Many animals at these depths have responded by evolving upwardly pointing eyes[9]. Every stage between fully lateral and fully upward eyes can be found in one or other species of mesopelagic fish, culminating in the extreme cases of *Opisthoproctus*, *Benthalbella* (*Figure 14.14*), and the hatchet fish *Argyropelecus*. Visual acuity (resolution) needs to be maximised in this direction and both amphipods and euphausiids have a gradation of forms whose eyes range from round, with a uniform acuity over the whole visual

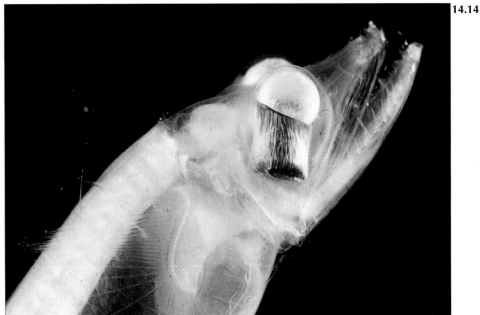

Figure 14.14 The tubular eyes of the 100 mm long fish *Benthalbella* provide a binocular overlap, allowing it to determine the range of prey as well as providing a large aperture for high sensitivity. (Courtesy of the Southampton Oceanography Centre, England.)

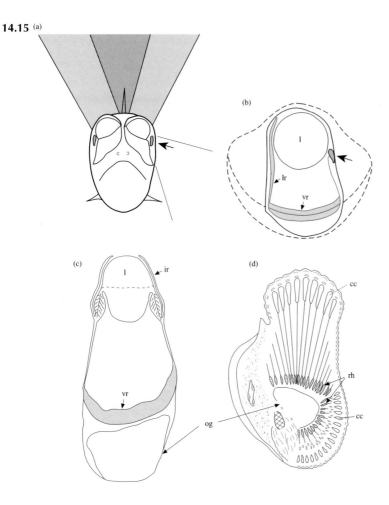

Figure 14.15 Some examples of tubular eyes are shown here. (a) The fish *Benthalbella* (seen from the front, lens diameter about 3 mm) showing the fields of view of the two eyes and their binocular overlap (dark hatched area). The lens pad [arrow, here and in (b)] extends the field by collecting light from the side and below. (b) A transverse section of the left eye superimposed on an outline of a 'normal' fish eye with the same size lens (l). A focused image is formed on the ventral retina (vr), and unfocused light from the lens pad is detected by the lateral retina (lr). (c) Sections of the eye (3 mm diameter) of the deep-sea octopod *Amphitretus* and (d) of the divided eye (2 mm diameter) of the euphausiid shrimp *Stylocheiron suhmii*, one portion of which looks upward, the other downward and to the side (cc, crystalline cones; ir, iris; og, optic ganglion; rh, rhabdoms). The fish eye is effectively the central part of a normal eye, and the octopod eye is optically similar. The euphausiid eye provides an upwardly directed ('tubular') region of high acuity (but narrow visual field); the lower region covers a wider visual field, but at lower acuity, in the much dimmer light from the side and below (after Land[9], from Marshall[12] and Locket[10]).

field (*Meganyctiphanes*), through double eyes with a narrow upper visual field of high acuity and a more extensive lower visual field of low acuity (*Phronima, Stylocheiron suhmii, Figure 14.15*), to those with effectively only an upward eye of high resolution and narrow field (*Cystisoma*).

Maximum sensitivity

At greater depths (and, of course, at night nearer the surface) the light levels become marginal for vision and the adaptations of the eyes of the animals are largely dedicated to maximising their sensitivity. This is determined by two factors: first, the illuminance of the retina and, second, the probability of photon capture by the receptors.

Image brightness

The illuminance (or image brightness) is a function of the aperture (pupil) and focal length of the eye. The aperture determines how much light enters. In the compound eyes of crustaceans, the aperture can be greatly increased by changing from an apposition type of eye to a superposition one[9]. Apposition eyes have the individual units (ommatidia, each containing one rhabdom) optically isolated from each other (usually by pigment). The eye has good resolution, but the effective aperture is that of a single facet. In superposition eyes, the units (ommatidia) are not optically isolated and the light from a large number of ommatidia can be focused on a single receptor. The aperture is now the facet area of all these ommatidia, giving a huge increase in sensitivity (*Figure 14.16*). Deep-sea shrimps of all kinds have such superposition eyes; often, their larvae, at shallower depths, have apposition eyes, which change into the superposition form as the larvae descend into deeper and darker water during their development.

The aperture of the fish eye is filled by the spherical lens (see *Figure 14.15*). Only in near-surface species is there an iris diaphragm to stop down the aperture in bright light. A larger aperture requires a larger lens, but the construction of the fish (and cephalopod) lens is such that the focal length is a constant ratio of the lens radius (2.5:1,

14.16 (a)

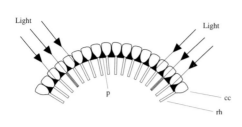

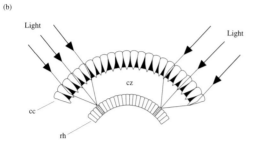

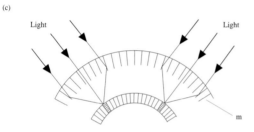

Figure 14.16 (a) A crustacean apposition eye, in which each unit (ommatidium) is optically isolated, with the receptor (rhabdom, rh) joined to the crystalline cone (cc). Only a very narrow beam of near-axial light reaches each receptor through its own cone; off-axis light is absorbed by the dark pigment (p). (b) A superposition eye, in which the receptors are separated from the cones by a clear zone (cz); the aperture is now formed by a large group of facets, greatly increasing the receptor illuminance. This kind of eye can be constructed using refractive cones, as in (b) (e.g., in euphausiid shrimps), or by reflective mirrors (m), as in (c) (e.g., decapod shrimps). (From Herring and Roe[8].)

Matthiessen's ratio). This means that any increase in aperture necessitates an equivalent increase in focal length, i.e., a larger eye. Large eyes are a feature of many deep-sea fish – but there is a practical limit to how large an eye can be, yet still fit on the head. A tubular eye overcomes much of this problem. A very large aperture is possible because only a tubular portion of the equivalently sized normal eye is retained. Fish such as *Dolichopteryx* have huge lenses, each almost half the width of the head; this is achieved by having a narrow tubular eye with a very limited field of view (*Figure 14.17*). Many fish with tubular eyes (including *Dolichopteryx*) have evolved secondary methods of increasing their visual field. Light guides, lens pads, and accessory reflectors capture some light from below the main visual field and convey it to an accessory retina. It is not focused, so no image is formed, but it does offer additional information to that available solely through the lens[10].

Photon capture

The probability of photon capture by the receptors is a function of the visual pigment density, its absorption maximum, the length of the light path, and the diameter of the receptors. Maximum sensitivity requires that the visual pigment match the ambient light very closely. Only a single visual pigment is normally present in the rods of deep-water fish. Cones are usually absent. The density of the visual pigment varies little between species, but the length of the light path through the receptors can be increased by lengthening the rods or having multiple banks of short rods[10]. A simple means of doubling the light path is to place a specular or diffuse reflector (or retinal tapetum) at the back of the eye.

14.17

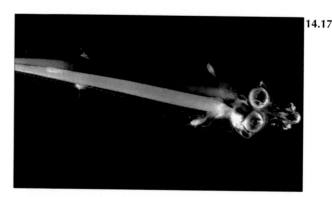

Figure 14.17 An extreme development of a tubular eye occurs in the 220 mm long fish *Dolichopteryx* (seen from above) in which the huge lenses filling the aperture provide the maximum possible retinal illuminance, while still just fitting on the head.

14.18

Many seals, fish, and crustaceans have such layers, which give the animal an intense eyeshine in bright light (*Figure 14.18*). In sharks and their relatives, black pigment covers the tapetum during the day and withdraws at night. The tapetum may be formed of layers of crystals, arranged for constructive interference of the ambient blue wavelengths, or of uniform granules or lipid spheres which give a more diffuse reflectance. Receptor diameter is effectively increased by pooling the output from a group of rods or rhabdoms into a single ganglion cell. This increases sensitivity, but at the expense of acuity, unless the focal length (i.e., eye size) is also increased to retain the same angular separation of the enlarged receptor units.

Eye reduction

In the depths of the ocean, where the only light is bioluminescence, some animals retain large sensitive eyes, but others lose much of the optical complexity present in shallower species and/or reduce the eye size to almost rudimentary proportions. In shrimp such as *Hymenodora*, all the focusing elements are lost and the eye retains only very complex rhabdoms and reflective material. In the fish *Ipnops*, the lens is also lost and the eye is little more than a flattened sheet of rod receptors.

The deep-water shrimp *Acanthephyra curtirostris* has tiny reflective eyes, compared with the larger pigmented ones of shallower relatives, such as *A. purpurea* and *S. debilis* (*Figure 14.19*; see also *Figures 14.8, 14.13*). The deepest-living euphausiid shrimps have equally tiny eyes, as do abyssal rattail fish. One remarkable shrimp (*Alvinocaris*) has completely lost its compound eyes, but has developed what appears to be a lens-less accessory visual system in the thorax. The visual pigment present

Figure 14.19 In very deep waters many animals reduce the size and organisation of their eyes. *Acanthephyra curtirostris* (65 mm long) has a much smaller eye than its shallower-living relatives. The eye no longer has any dark pigment and appears diffusely reflective (compare with *Figure 14.13*). (Courtesy of the Southampton Oceanography Centre, England.)

Figure 14.18 Many deep-sea animals have reflective layers in the eye behind the retina, increasing the chances of photon capture. In the shrimp *Plesiopenaeus* (130 mm long, attracted to a baited camera) this is visible as a bright eyeshine. (Courtesy of the Southampton Oceanography Centre, England.)

there may just be able to detect the black body radiation of the hot submarine vents where this animal lives[14] (see also Chapter 13).

Bioluminescence

The presence of functional eyes in many abyssal animals, well below the maximum possible penetration of daylight, testifies to the importance of biological light, or bioluminescence, in the ecology of the deep sea[7]. Fireflies and glow-worms are the well-known terrestrial examples of this phenomenon, but there are few others on land and only one (a New Zealand limpet) in fresh water. In contrast, the sea contains an immense variety of luminous organisms spread across 16 different major groups, or phyla. The organisms range from bacteria and dinoflagellates to squid and fish, and their distribution extends from the surface waters to the greatest depths. This rich variety of organism is matched by the astonishing range of their bioluminescent capabilities and functions. The conclusion to be drawn from the variety of luminous groups, and the different chemistries involved, is that the ability to emit light has evolved independently in many different organisms[5].

Mechanism

Bioluminescence is the harnessing of a chemiluminescent reaction by a living organism, with a sufficient quantum (photon) yield for the light to be visible. The overall reaction involves the oxidation of a substrate ('luciferin') catalysed by an enzyme ('luciferase'), so that some of the energy released by the reaction is emitted as light rather than heat. The energy can be transferred into a second, fluorescent, compound which then emits light at its own characteristic wavelength (always longer than that produced by the original reaction). There are

14.19

14.20a

14.20b

Figure 14.20 (a) The fish Malacosteus has two light organs near the eye, one (white) behind the eye and a larger one (visible only by virtue of its silvered upper edge) under the 5 mm diameter eye (courtesy of the S.O.C.). (b) When illuminated with ultraviolet light the contents of the dark brown light organ fluoresce an intense red. The white organ emits a blue biolumines-cence and the brown a deep-red bioluminescence. Their respective emission spectra are shown in (c). It is likely that the red light is produced by transferring the energy of the bioluminescence reaction to the red fluor within it (see text), and then further filtering the light through the deep brown surface of the light organ.

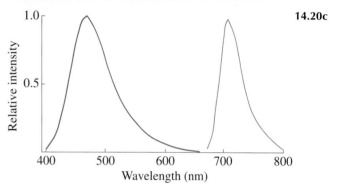

14.20c

many different kinds of luciferin, some restricted to particular groups of organisms, others spread across several different phyla. Relatively few have yet been identified. Most organisms make their own luciferin (although some may acquire it in their diet), but some fish, squid, and tunicates use symbiotic luminous bacteria as their source of light.

The brightest bioluminescence can only match the intensity of moonlight, so bioluminescence makes no significant contribution to the light of near-surface oceanic waters during the day. In turbid coastal waters, however, it may be the only light visible as shallow as 30 m, even during the day. At all other times and depths it is a major, and often the only, source of light in the sea.

Colours of bioluminescence

If the eyes of marine animals have adapted primarily to the characteristics of submarine daylight, with its blue transmission maximum in clear waters, it is likely that bioluminescence has followed this visual lead. The colours of marine bioluminescence are, indeed, very largely confined to blue–green wavelengths[6]. These, of course, are not only best perceived by other marine organisms, whether or not they are themselves bioluminescent, but also will be best transmitted through the water, both features

maximising their effective signalling range. Some animals, particularly hydrozoan jellyfish, add a green fluorescent protein to the luminous cells and emit green light. It is not clear whether this is a response to a more coastal environment, with its greener water transmission, or a means of increasing the quantum efficiency (i.e., the light yield) of the reaction.

It may be more important (for maximising the effective range) to have a bright green light than a dimmer blue one, if there is little difference in an observer's spectral sensitivity to the two wavelengths. There are a few other colours of bioluminescent light; the worm *Tomopteris* has a yellow luminescence and the fish *Malacosteus* (and a few others) has a deep red light, as well as a blue one. In the latter case, the red organ (under the eye) has a narrow bandwidth emission with a maximum at about 700 nm; this light is invisible to the dark-adapted human eye. A blue organ just behind it emits at 480 nm. The red light is probably produced by the same chemistry as the blue light, but with a combination of an additional red fluorescer in the light organ and a deep-red filter on its surface (*Figure 14.20*). The red organ cannot be very efficient; about 80% of the light it produces is absorbed by the filter in order to achieve the narrow bandwidth.

14.21

Figure 14.21 *Malacosteus* (175 mm long) has both blue- and red-sensitive visual pigments. The red sensitivity is further enhanced by a red reflector (tapetum) behind the retina, providing the red colour to the eye visible here in daylight.(Courtesy of Dr N.A. Locket, Adelaide University, Australia.)

14.22a

14.22

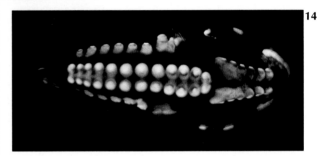

Figure 14.22 (a) The hatchet fish *Argyropelecus* (*Figure 14.6*), seen from below, showing the photophores arranged along its ventral margin, each containing a magenta-coloured filter. This results in the emitted bioluminescence being a clear blue, exactly matching the colour of light in the sea (courtesy of P.M. David, Southampton, England). (b) A luminescing specimen by its own light.

The long wavelength light must be very important to the fish, but it would not normally be able to see it if it had only a typical blue–green sensitive visual pigment. *Malacosteus* turns out to have a red-sensitive visual pigment as well. It also has a scarlet tapetum (see earlier) to maximise its sensitivity to these long wavelengths (*Figure 14.21*). It thus has a 'private' wavelength, which could be used either as a secure communication with others of the same species or to break the camouflage of red shrimps, whose colour only works in blue illumination and whose blue-sensitive eyes would not detect the red illumination. Red light of these wavelengths is rapidly absorbed by sea water, so it can only be effective over short visual ranges[3].

A few other marine animals produce light of more than one colour. The stalked sea-pen *Umbellula*, for example, has green luminescence on its stalk, but blue emission from the polyps at the top. Some squid produce green and blue light from different light organs, and can also change the emission spectrum from a single light organ. Another coelenterate, a sea anemone-like zoanthid, has colonies in which some individuals have green light and some yellow.

Luminous camouflage

The ecological value of most of the exceptions noted above is not yet clear, but the value of precisely controlling the colour has been well-established in the hatchet fish. This fish lives at depths where daylight is still important and, as already described, has mirror-like camouflage and tubular eyes. Despite being very laterally flattened, it cannot altogether avoid being seen in silhouette from below. Like many other fish at mid-depths, it eliminates this silhouette by having rows of light organs along its ventral surface [*Figure 14.22(a)*]. All its light organs point downward, except one, which points *into* the eye. By adjusting the luminescence shining into the eye to match the downwelling daylight, it simultaneously matches all its ventral lights to the surrounding light – and vanishes. However, the broad bandwidth light produced *within* the light organ is not quite the

Figure 14.23 Mature females of the pelagic octopod *Japetella* (100 mm long) develop a yellow-coloured bio-luminescent oral ring which degenerates after they spawn. Males have no luminous organ, so it is assumed that this provides a sexual signal.

14.23

same colour as light in the sea. The aperture of each light organ contains a purple filter pigment which corrects this spectral mismatch. After passage through the filter the luminescence has a narrow bandwidth blue emission with a maximum at about 475 nm, corresponding exactly to the spectrum of light in the sea [*Figure 14.22(b)*]. To complete the camouflage, the design of the reflectors in the light organs ensures that the angular distribution of the luminescence also matches that of submarine daylight[3], as illustrated in *Figure 14.1(b)*.

Other luminous defences

Camouflage is an example of a passive defensive function. The most common use of light in the sea is as an active defence, which can take the form of a single, short, bright flash (dinoflagellate), a volley of flashes (some fish), a wave of flashes moving over the body (sea-pens and medusae), or the discarding of sacrificial parts of the body which flash independently to distract a predator (scale worms and brittle stars). Another widely employed active defence is that of a squirted luminescence. Many shrimps, worms, a few squid and fish, and some medusae and ctenophores produce copious amounts of light in this form. Particularly among the ctenophores and medusae, it is not simply a cloud, but is composed of separately scintillating particles. Many of these gelatinous animals hang passively in the water fishing for prey; for any larger animal in the area, they collectively form a luminous minefield.

Sexual light signals

Small crustaceans, such as copepods and ostracods, have luminous glands whose secretions the animals kick away defensively as they swim. They can also be used for sexual displays. The pattern, timing, and trajectory of the luminous gobbets produced in the mating displays of males of the *Vargula* group of ostracods identify them to the waiting females, who swim up to join them. Syllid worms have analogous displays in many parts of the world (e.g., the Bermuda fireworm). Sexual differences in the size or position of the light organs of male and female lantern fish, stomiatoid fish, and some cephalopods also suggest that they have a sexual function (*Figure 14.23*). Female angler-fish have lures containing luminous bacterial symbionts, but the males do not. It is assumed that the lure attracts prey, but perhaps it also sends a specific signal to the males (*Figure 14.24*).

14.24

Figure 14.24 Female angler-fish, such as this 45 mm long *Chaenophryne*, maintain a culture of luminous bacteria in the very elaborate lure. Although regarded primarily as a means of attracting prey, it may also be a means of identifying the female to the non-luminous male.

14.25

Figure 14.25 The flashlight fish *Photoblepharon* is a shallow-water fish, some 90 mm long, which rises from dark crevices to feed on the reef at night. The light of the bacteria in the photophore beneath the eye is used to illuminate its plankton prey. The light can be turned off by pulling a shutter up over the light organ.

Luminous bacterial symbionts

Those animals utilising luminous bacteria have to constrain them in a particular gland or tissue, provide them with the right environment to thrive, dispose of dead ones, and ensure that the right bacteria are transferred to the next generation. Each host requires a particular species of bacterium, though many share the same species. Many hosts have light organs which are connected to the gut lumen and from which the symbionts can easily be cultured in the laboratory. As the natural (but largely accidental) bacterial flora of the gut contains some luminous species, acquisition of the right bacterium for the light organ is probably easy. The same bacterial species, and a few additional ones, are found 'free-living' on the surfaces of animals, on marine snow (see Chapter 7), and possibly free in the water. In the angler-fish and flashlight fish (*Figure 14.25*), there is no light organ connection to the gut and the bacteria are unculturable, so their origin is not known. Genetic data indicate that they are new species and very host-specific. Luminous bacteria glow continuously; the host has to develop a means of turning them off when necessary. Usually, this is achieved by pulling a shutter across the aperture of the organ, rotating the organ to face inward, or dispersing dark pigment over its surface, but control of the oxygen (blood) supply may sometimes be involved.

Photophore structure

The possible optical complexities of light organs fully match those of eyes[7]. The light-emitting cells sit within a pigment cup which limits the aperture, while a specular interference reflector of guanine or protein platelets, or a diffuse granular reflector, greatly increases the output efficiency. A lens focuses the light, a lamellar ring collimates it, and interference or pigmentary filters change its spectrum. Light guides spread the output over a larger area and light pipes even allow the light to be emitted some distance from the light organ. Many hundred light organs are present on some fish and squid (*Figure 14.26*); individual species may have three or

14.26

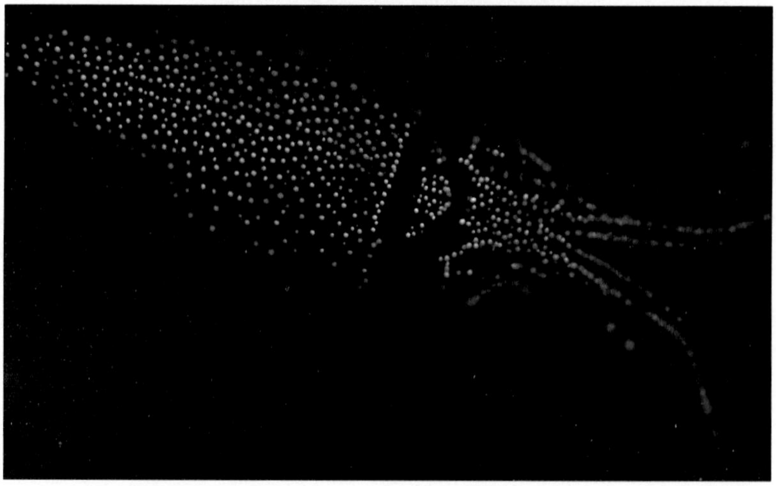

Figure 14.26 Some animals have hundreds of separate light organs. The 45 mm long Japanese firefly squid, *Watasenia scintillans* (seen here from below by its own light), has most of them arranged over its ventral surface as a camouflage, similar in principle to that of the hatchet fish. (Courtesy of Prof. Y. Kito, Osaka University, Japan.)

four structurally quite different types at different sites on or in the body. We do not yet know what kind of light many of these organs produce, nor what functions they serve. If light, colour, and vision in the ocean are regarded as an interlocking jig-saw puzzle, these are some of the pieces which have not yet been fitted in.

General References

Hastings, J.W. and Morin, J.G. (1991), Bioluminescence, in *Neural and Integrative Animal Physiology*, Prosser, C.L. (ed.), Wiley-Liss, New York, pp 131-170.

Herring, P.J. (1985), How to survive in the dark: bioluminescence in the deep sea, in *Physiological Adaptations of Marine Animals*, Laverack, M.S. (ed.), *Symp. Soc. Exp. Biol.*, **39**, 323-350.

Locket, N.A. (1977), Adaptations to the deep-sea environment, in *Handbook of Sensory Physiology*, Vol VII/5, Crescitelli, F. (ed.), Springer-Verlag, Berlin, pp 67-192.

Lythgoe, J. (1979), *The Ecology of Vision*, Clarendon Press, Oxford, 244 pp.

Marshall, N.B. (1979), *Developments in Deep-Sea Biology*, Blandford Press, Poole, 566 pp.

References

1. Denton, E.J. (1970), On the organization of reflecting surfaces in some marine animals, *Phil. Trans. Roy. Soc. B*, **258**, 285-313.
2. Denton, E.J. (1990), Light and vision at depths greater than 200 metres, in *Light and Life in the Sea*, Herring, P.J., Campbell, A.K., Whitfield, M., and Maddock, L. (eds), Cambridge University Press, Cambridge, pp 127-148.
3. Denton, E.J., Herring, P.J., Widder, E.A., Latz, M.I., and Case, J.F. (1990), The roles of filters in the photophores of oceanic animals and their relation to vision in the oceanic environment, *Proc. R. Soc., Lond. B*, **225**, 63-97.
4. Frank, T.M. and Widder, E.A. (1994), Evidence for behavioral sensitivity to near-UV light in the deep-sea crustacean *Systellaspis debilis*, *Marine Biol.*, **118**, 279-284.
5. Hastings, J.W. and Morin, J.G. (1991), Bioluminescence, in *Neural and Integrative Animal Physiology*, Prosser, C.L. (ed.), Wiley-Liss, New York, pp 131-170.
6. Herring, P.J. (1983), The spectral characteristics of luminous marine organisms, *Proc R. Soc., Lond., B*, **220**, 183-217.
7. Herring, P.J. (1985), How to survive in the dark: bioluminescence in the deep sea, in *Physiological Adaptations of Marine Animals*, Laverack, M.S. (ed.), *Symp. Soc. Exp. Biol.*, **39**, 323-350.
8. Herring, P.J. and Roe, H.S.J. (1990), The photoecology of pelagic oceanic decapods, *Symp. Zool. Soc. Lond.*, **59**, 263-290.
9. Land, M.F. (1990), Optics of the eyes of marine animals, in *Light and Life in the Sea*, Herring, P.J., Campbell, A.K., Whitfield, M., and Maddock, L. (eds), Cambridge University Press, Cambridge, pp 149-166.
10. Locket, N.A. (1977), Adaptations to the deep-sea environment, in *Handbook of Sensory Physiology*, Vol VII/5, Crescitelli, F. (ed.), Springer-Verlag, Berlin, pp 67-192.
11. Lythgoe, J. (1979), *The Ecology of Vision*, Clarendon Press, Oxford.
12. Marshall, N.B. (1979), *Developments in Deep-Sea Biology*, Blandford Press, Poole.
13. Partridge, J.C., Shand, J., Archer, S.N., Lythgoe, J.N., and van Groningen-Luyben, W.A.H.M. (1989), Interspecific variation in the visual pigments of deep-sea fish, *J. Comp. Physiol. A*, **164**, 513-529.
14. Van Dover, C.L., Szuts, E.Z., Chamberlain, S.C., and Cann, J.R. (1989), A novel eye in 'eyeless' shrimp from hydrothermal vents of the Mid-Atlantic Ridge, *Nature*, **337**, 458-460.

CHAPTER 15:

Ocean Diversity

M.V. Angel

Biodiversity is the term used to describe the rich variety of life found on Earth. It was the subject of the UNCED conference in Rio in 1992, which resulted in the signing of the International Convention on Biodiversity and the adoption of Agenda 21, which lays down the guidelines under which the Convention will operate. However, biodiversity is used to express this variety over a great range of levels of organisation. It can be applied to different types of ecosystems, or to express the number of species found locally or globally, or even to the amount of genetic variation within individual species. This can lead to confused thinking, unless the term's precise meaning is explicitly stated in each context. Our discussion concentrates on ecosystem diversity, and species richness and dominance within local and regional areas.

Diversity at these levels of organisation is the product of evolution. New species almost always evolve as a result of a subpopulation becoming isolated from its parent population for thousands of years, and being subjected to different selective pressures. Generally, the stock of species of the world has increased linearly over geological time

(based on the fossil record), although several mass extinctions have interrupted this trend (*Figure 15.1*). The causes of these mass extinctions are still a subject of vigorous debate[17]. However, since the beginning of the Mesozoic era the increase has been almost linear. If we are to understand the present patterns of biodiversity in our seas, we need to appreciate, first, how and why evolution has given the world such a rich compendium of species and, second, how ecological processes continue to maintain this richness.

The Origins of Disparity

The earliest traces of life that have been found so far are imprints of single-celled micro-organisms in sedimentary rocks laid down in the prototype ocean about three and a half billion years ago. The first to appear were prokaryotic cells or bacteria, which have no nucleus and are termed prokaryotes. Since they have the most ancient lineage of any living thing so far, these apparently simple cells show remarkable variations in their physiology, cell chemistry, and genetics. As we learn more about them, there are proving to be greater differences

15.1

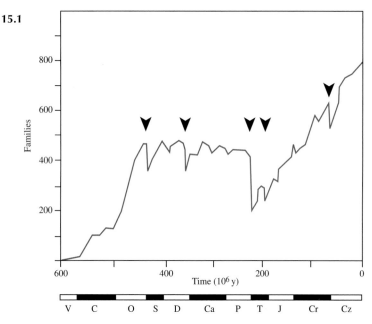

Figure 15.1 Diversity of marine families through the geological record since metazoan organisms first appeared. Note the steady increase until the end of the Ordovician (O in the lower scale), when the first of the mass extinctions occurred. Then the numbers of families remained roughly constant until the mass extinctions of the Triassic (T) and the beginnning of the fragmentation of the supercontinent Pangea (see *Figure 15.4*). Compare the steady increase in numbers of families since the Triassic with the patterns of continental drift shown in *Figure 15.4*. The one interruption was the mass extinction at the end of the Cretaceous (Cr), which more-or-less coincided with the split developing between Australasia and Antarctica, and resulted in the start of circumpolar circulation in the Southern Ocean (redrawn from Sepkowski[17]). (V = Vendian; C = Cambrian; O = Ordovician; S = Silurian; D = Devonian; Ca = Carboniferous; P = Permian; T = Triassic; J = Jurassic; Cr = Cretaceous; Cz = Cenozoic.)

Figure 15.2 Specimens of the large deep-living copepod species *Megacalanus princeps*. Copepods numerically dominate the vast majority of plankton samples, no matter at what depth they are collected. They outnumber all other animals of comparable size in any other ecosystem, including all insects, and yet there are only just over 1900 known species compared with perhaps a million insects. (© Heather Angel.)

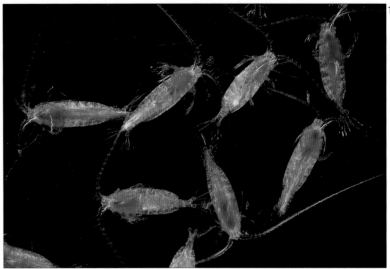

15.2

within the prokaryotes than within all other living organisms[9].

It took another two billion years before the first appearance of single-celled organisms *with* nuclei – the eukaryotes. Both the nucleus, which contains the genetic information, and other organelles within the cells – the mitochondria, in which the reactions occur that provide the cell with energy – are thought to have originated as a result of different prokaryotes forming symbiotic associations, which through time became a permanent and obligate relationship.

It was almost another billion years before the first multicelled organisms made an appearance, just before the beginning of the Cambrian era some 670 million years ago, and it was yet another 155 million years before multicellular forms began to invade the land during the late Silurian.

Considering its much greater geological age, it is hardly surprising that the fauna of the oceans is far more disparate than the terrestrial fauna, being made up of 28 phyla – the name given to the basic types of animals (*Table 15.1*), compared with just the 11 phyla represented in terrestrial faunas. However, is this greater disparity also repeated in a greater species richness in the oceans compared to the richness on land?

Terrestrial Versus Marine Species Richness

For pelagic faunas the answer appears to be no; terrestrial faunas appear to be far richer in species. For example, the most abundant and species-rich group of plankton are the copepods (*Figure 15.2*), of which there are just over 1900 species known from oceanic and brackish waters. Compare this with the most diverse inhabitants of terrestrial habitats – the insects (a group virtually absent from the oceans), in which just one order, the beetles,

includes several hundreds of thousands of known species, and whose true global richness is estimated to be in excess of a million. Much the same applies to the plants; there are an estimated 250,000 species of terrestrial green plants, but merely 3500–4500 in the oceans (*Figure 15.3*). Why are there such major differences in species richness?

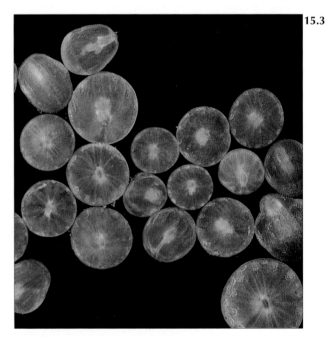

15.3

Figure 15.3 Giant phytoplankton cells (1–2 mm in diameter) of the monad *Halosphaera viridis*. Apart from the large algae that grow in shallow coastal waters, virtually all life in the ocean is dependent on the photosynthetic production of the 3500–4500 species of phytoplankton. Which should receive the greatest priority for conservation, one of the few phytoplankton species or one of the quarter of a million terrestrial green plants? (© Heather Angel.)

Table 15.1 Distribution of phyla in major habitats[12].

Phyla / Subphyla	Marine		Freshwater		Terrestrial		Symbiotic	
	Benthic	Pelagic	Benthic	Pelagic	Moist	Xeric	Ecto	Endo
Porifera	+++		+				+	
Placozoa	+							
Orthonectida								+
Dicyemida								+
Cnidaria	+++	++	+	+			+	
Ctenophora	+	+						
Platyhelminthes	+++	+	+++		++		+	++++
Gnathostomulida	++							
Nemertea	++	+	+		+		+	
Nematoda	+++	+	+++	+	+++	+	+++	+++
Nematomorpha								++
Acanthocephala								++
Rotifera	+	+	++	++	+		+	+
Gastrotricha	++		++					
Kinorhyncha	++							
Loricifera	+							
Tardigrada	+		++		+			
Priapula	+							
Mollusca	+++++	+	+++		+++	+	+	+
Kamptozoa	+		+				+	
Pogonophora	++							
Sipuncula	++				+			
Echiura	++							
Annelida	++++	+	++		+++		++	
Onychophora					+			
Arthropoda								
Crustacea	++++	+++	+++	++	++		++	++
Chelicerata	++	+	++	++	++++	+++	++	+
Uniramia	+	+	+++	++	+++++	+++	++	++
Chaetognatha	+	+						
Phoronida	+							
Brachiopoda	++							
Bryozoa	+++		+					
Echinodermata	+++	+						
Hemichordata	+							
Chordata								
Urochordata	+++	+						
Cephalochordata	+							
Vertebrata	+++	+++	++	+++	+++	+++	+	+

'Pluses' indicate approximate abundance of living described species:
+ = $1–10^2$
++ = $10^2–10^3$
+++ = $10^3–10^4$
++++ = $10^4–10^5$

15.4

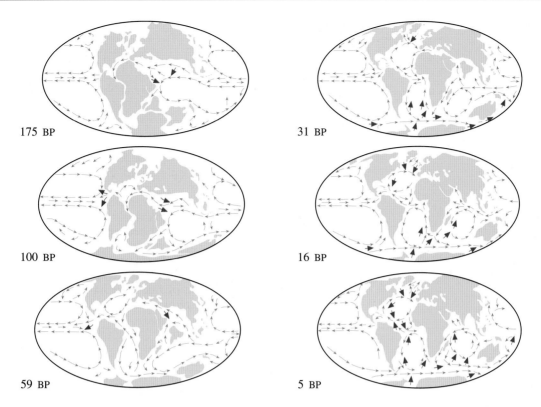

175 BP

31 BP

100 BP

16 BP

59 BP

5 BP

Figure 15.4 Reconstruction of the distribution of the continents and the main surface circulation (red arrows) of the oceans through the past 175 million years, following the fragmentation of the supercontinent Pangea (the blue arrows indicate where deep water forms). The distribution of old genera and families still retains the imprint of the ancient circulation patterns. Events like the interruption of circumequatorial and circumpolar circulation in the Southern Ocean have had a major impact on patterns of speciation and redistribution (modified from Parrish and Curtis[11]).

Factors Leading to Speciation

New species tend to evolve when populations are split and isolated. Isolation in both marine and terrestrial environments can arise from tectonic events, resulting from the drift of the continents altering the shape and morphology of ocean basins, and forming new islands and mountain ridges. Over the past 175 million years, plate tectonic movements have resulted in major redistributions of the land masses (see Chapter 20). Since the pattern of large-scale thermohaline circulation of ocean currents is determined by the interaction between the shape of the oceans and wind patterns (Chapter 2), ocean circulations during the geological past can be reconstructed with confidence. In the modern ocean, the distributions of major pelagic ecosystems match the patterns of the major ocean current gyres, and are likely to have done so in the past (*Figure 15.4*). Analyses of the present-day distributions of some of the more ancient families and genera of pelagic organisms reveal persistent echoes of these ancient circulations[18].

In addition to these major changes to ocean-basin morphology, other events have had major impacts on the distribution patterns of species. During the Miocene, for example, the Straits of Gibraltar repeatedly closed and opened. Each time the straits closed the sea dried out, leaving behind vast deposits of salt that today still underlie a drape of pelagic sediments beneath the Mediterranean Sea. When the Straits re-opened the sea catastrophically reflooded, lowering sea levels globally by tens of metres within relatively few years, maybe centuries. Such catastrophic lowering of sea level caused the deaths of many shallow-living species, and probably led to many extinctions. Another such event was the opening of the Isthmus of Panama about ten million years ago, which allowed the exchange of shallow-living faunas between the Pacific and Atlantic Oceans. As a result, many shallow-living tropical families, genera, and even species have circumequatorial distributions.

Throughout the Quaternary sea levels fluctuated by around 100 m as a result of the alternating glacial and interglacial stages, which appear to have been a response in the Earth's climate to

15.5

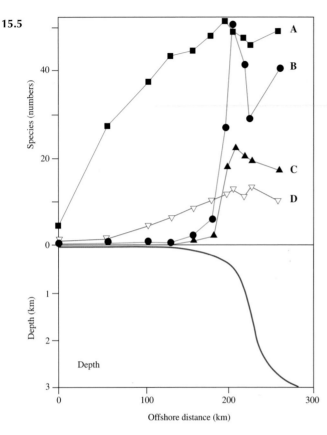

Figure 15.5 Changes in the numbers of pelagic species sampled across the shelf, shelf-break, and continental slope off Florida in the Gulf of Mexico: A, copepoda; B, mid-water fish; C, decapod crustaceans; and D, euphausiids. For these four pelagic taxa the numbers of species increase offshore and often peak at or close to the shelf-break (modified from Hopkins *et al.*[6]).

Fine-Scale Habitat Variability

Such large-scale events play an equally important role in speciation on land, but in addition there are much finer-scale processes at work. These result from terrestrial ecosystems being more finely structured, with much greater development of microclimates, and dispersal between the different patches being much more restricted. The physical structure and aspect of the landscape, the types of soils, and the availability of water and nutrients all contribute to the creation of fine-scale habitat mosaics in terrestrial ecosystems. These patches are dynamic; for example, forest clearings are created by trees dying or being razed to the ground by violent storms.

Animals and plants vary in their abilities to disperse, so populations of a given species may become isolated as the patches become too far apart for individuals to move between them, or a phenomenon such as a landslide or lava flow creates an impassable barrier. So, even within a mosaic of habitats, similar patches may be supporting slightly different assemblages of plant and animal species.

Since the dominant terrestrial plants are larger and longer-lived than their marine counterparts, many have co-evolved with invertebrates, especially with insects; for example, in Britain the oak tree alone has about 600 insect associates. Moreover, terrestrial plants by their physical presences, as well as their physiological activity, modify the habitats they inhabit; the environment inside a wood is very different from that immmediately outside.

There is some similar modification of habitats by the biota in the sea (for example, around coral reefs, in kelp forests, and over mussel beds), but the creation of such three-dimensional habitat structures by the organisms themselves is almost entirely limited to shallow coastal waters. The hermatypic corals, which create much of the structure of coral reefs, are limited to depths where their symbiotic algae can obtain enough light to photosynthesise.

In shallow seas, even where there is no biological structuring of the environment, the complex interactions between tides, waves, and the coastal morphology and geology create much finer-scaled mosaics of different habitats than occur in open-ocean habitats, particularly pelagic ones (see Chapter 16). A sampling transect across a continental-shelf sea and into the open ocean reveals rel-

eccentricities in its orbit around the Sun, the so-called Milankovitch cycles. Around the East Indies, in particular, fluctuating sea levels resulted in connections between the islands appearing and disappearing, so that the deep basins between the islands were isolated and then reconnected. Consequently, populations both on land and in the coastal seas were divided and then recombined.

This led to a series of biogeographical divisions between the islands' faunas and floras, such as the Wallace's, Weber's, and Lydekker's Lines, which are reflected, to some extent, in the marine faunas as well[4]. Similarly, in the Mediterranean there are glacial relict species which entered the sea during the most recent glaciation and are now not only totally isolated from their parent stock in the North Atlantic, but are also living in very different conditions.

During the past 22,000 years, since the height of the latest glaciation (note that recent calibrations of ^{14}C dating methods have resulted in this time interval being increased from 18,000 years), it is probable that at least some of these relicts have diverged sufficiently to become genetically incompatible. Thus, the isolation between populations in the ocean, that in time results in their diverging and evolving into new species, tends to be related to large-scale and persistent events.

atively few species at a given location over the shelf and a sharp increase in the number of species just beyond the shelf-break (*Figure 15.5*). However, if the sampling is extended to become regional in its coverage, the number of species caught inshore soon begins to exceed those caught in the open ocean; oceanic habitats are large-scale, whereas inshore and coastal habitats (which range from muddy estuaries to sandy shores and rocky beaches) are much finer in scale.

Another feature of coastal habitats is that they are geographically linear, so any break caused by a large estuary, a strait, or even a major piece of coastal engineering can act as a barrier to the dispersal of some taxa. Most (but not all) of the species which live between the tides or in association with the sea bed have dispersal phases as planktonic larvae, which may last from a few days to several months. Those with short planktonic phases, or none at all, may not be able to cross these barriers because of the prevailing currents. As a result, taxa which inhabit coastal waters are much more restricted in their distributions and are globally more diverse than similar taxa inhabiting the open ocean. This trend is illustrated below.

Recently, some human activities have begun to break down some of these barriers to distribution, with serious implications for the maintenance of marine biodiversity. For example, the building of the Suez Canal has led to increasingly large numbers of species exchanging between the Red Sea and the Mediterranean[13]. More recently, evidence has emerged that the transoceanic transportation of coastal species in the ballast waters of fast cargo vessels is disturbing local faunas[3].

Gradients and Ecotones in the Oceanic Water Column

In the water column of the deep ocean, there are important vertical gradients in many chemical and physical properties, including temperature, salinity, nutrient and oxygen concentrations, light, and hydrostatic pressure. The primary production on which all life depends is almost entirely restricted to the surface 100 m or so, where there is sufficient sunlight for photosynthesis (the chemosynthesis which occurs in the immediate vicinity of hydrothermal events is the main exception, see Chapter 10). Only about 1–3% of the organic matter produced by photosynthesis at the surface reaches the sea bed at depths of 4000 m. So, the general availability of food declines approximately

exponentially and depth profiles of biomass show that the pelagic biomass at a depth of 1000 m is a tenth that at the surface, and declines by another tenth at 4000 m (*Figure 15.6*). The only place that these vertical gradients combine to be steep enough to act as a barrier to the dispersion of a few species is across the seasonal thermocline. But even there some of the vertically migrating species are uninhibited in their movements, as they commute daily from the safety of day-time depths of 500 m or more, to spend the night feeding in the surface waters. During these migrations they may experience greater changes in temperature than occur throughout the year in the surface waters.

Horizontal gradients are almost always weaker, even at boundaries between different water masses, such as at the Antarctic Convergence. The one possible exception is in upwelling areas, where boluses of cold subthermocline water that have upwelled to the surface can create very sharp sea-surface temperature discontinuities. However, such fronts are far too ephemeral to have a long-term influence on diversity. Otherwise, the continual stirring of the water by currents and eddies blurs all such boundaries. As a result, there are almost no clearly defined ecosystems in pelagic environments; the assemblages of animals which merge almost seamlessly across the gradients are described as ecotones. Locally, patchiness and variability is high,

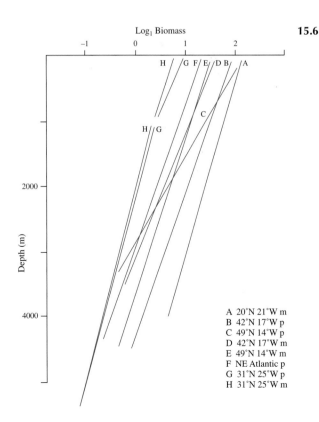

15.6

A 20°N 21°W m
B 42°N 17°W p
C 49°N 14°W p
D 42°N 17°W m
E 49°N 14°W m
F NE Atlantic p
G 31°N 25°W p
H 31°N 25°W m

Figure 15.6 Regression lines of biomass profiles of macroplankton (p) and micronekton (m) against depth at four locations in the northeast Atlantic (redrawn from Angel and Baker[2]).

15.7

Figure 15.7 An example of a specialist species in the oceanic plankton, the shell of a large (3 cm long) species of pteropod, *Clio recurva*, is covered with a colony of commensal hydroid *Campaniclava clionis*. (© Heather Angel.)

but this is not translated into large-scale variability.

Over the time-scales that are important to the creation of biodiversity, ocean waters mix rapidly; in the Atlantic and the Indian Oceans the waters exchange every 250 years and in the Pacific about every 500 years (see Chapter 3), times which are extremely short for evolutionary changes. So, perhaps a pertinent question is 'Are all pelagic taxa in the open ocean completely cosmopolitan?' It should be no surprise to find that they are, indeed, far more cosmopolitan than their benthic or coastal counterparts.

Specialists Versus Generalists

Another factor that may help to keep the numbers of pelagic species low is the lack of opportunity to specialise, to become adapted to a narrow range of environmental factors or a specific source of food (for example, many insect species are associated with a single food plant in terrestrial habitats). Palaeontologists have observed in the fossil record that specialist genera have much higher rates of turnover (extinction and speciation) than do generalist genera. In the pelagic realm, habitat variability is considerable over all spatial scales, being dominated by the changing conditions induced by oceanic turbulence and meso-scale eddies (see Chapter 4); this seems to have maintained a strong selective pressure for pelagic species to remain tolerant of these changes and hence remain as generalists.

Similarly, species inhabiting latitudes where the production cycle is more seasonal have less opportunity to specialise than do those inhabiting environments where the supply of food is more uniform and continuous. As always, there are some exceptions to every generalisation (*Figure 15.7* illustrates one). In addition, in the oceans food-web relationships appear to be very differently structured, tending to be based more on size of 'particle' than on quality. Thus, an individual may occupy very different niches throughout its life cycle, changing not only what it feeds on, but also the depth at which it lives. Terrestrial species tend to be much more conservative in the role they play in food-webs and in the space they occupy.

The contrast in species richness between oceanic and freshwater communities is indicative. Freshwater environments are more fragmented and, over geological time-scales, ephemeral. But freshwater microhabitats are distinct and favour specialisation; in the great lakes of Africa some of the fish are believed to have speciated within the past 500 years. Freshwater biomes contain a tiny fraction of the Earth's water and occupy a minute area of the land surface, and yet global species-richness of freshwater taxa is much greater than that of marine taxa. Moreover, given long-enough geological continuity, local species-richness can be exceptionally high. For example, in the catchment of the River Zaire there are over 690 species of fish, of which 84% are endemic[16], not so very many fewer than are known from the world's deep oceans. These freshwater fish are often highly specialised in their ecological requirements and have been reported to speciate within a few hundreds of years.

In oceanic fish, the main adaptations appear to be related to predator avoidance and finding enough food within specific depth zones; in an environment where hydrostatic pressure increases, the light environment changes systematically (Chapter 14), and availability of food decreases with increasing depth. In the tropics, water temperature may change more in 1 km vertically than in 2000–3000 km horizontally. So, as depth increases, there are systematic changes in both pelagic and benthic species composition, and in their physiological and morphological adaptations. If horizontal

15.8 (a)

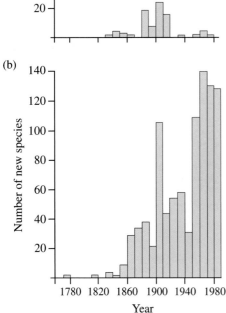

(b)

Figure 15.8 Histograms showing the number of new species of (a) euphausiids and (b) mysids described in each decade. These data suggest that the total of 86 known species of euphausiid is likely to be very close to the global total, whereas the 983 named mysid species (up until 1994) is probably well short of the global total. The difference between the two groups is that the euphausiids are all pelagic (and most are oceanic), whereas the mysids are predominantly found in shallow, inshore waters and are often associated with the sea bed (updated from Mauchline and Murano[10]).

Some Examples of Pelagic Species-Richness

Figure 15.8 illustrates the numbers of new euphausiids and mysid species described each decade. The euphausiid data clearly show that, since the period of the great exploratory expeditions around 1900, very few new species have come to light. Euphausiids are all pelagic crustaceans (*Figure 15.9*), the most famous of which is krill, *Euphausia superba* – the staple diet of many whales, seals, and penguins in the Southern Ocean (see also Chapter 21) – and they are predominantly oceanic. Mysids, which are also shrimp-like crustaceans (*Figure 15.10*), are mostly either inhabitants of shallow,

gradients in the ocean were as strong as those in the vertical, then diversity in the ocean might be expected to rival that of the land.

Where is Species Diversity Highest in Marine Communities?

This is not a straightforward question to answer, partly because very few surveys attempt to identify all the taxa present, even within the limited size range of organisms collected by a single sampler. Moreover, for certain size ranges known to be highly speciose, such as benthic macrofauna and meiofauna, the systematics of most of the component taxonomic groups are far too sketchy for inter-regional comparisons to be meaningful. Another problem which militates against making sensible inter-regional comparisons is the lack of standardisation of sampling methods and protocols of analysis. With the added problems created by the absence of any international databases incorporating taxonomic information, it is well-nigh impossible to compile all the relevant data; the true difficulty of making even rough attempts at answering this very basic question is all too apparent. For limited areas in which standardised sampling and analytical procedures have been used, and for those groups for which the numbers of species appears to be well-known and for which the systematics have remained quite stable, comparisons are possible.

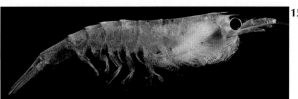

Figure 15.9 A typical euphausiid, *Meganyctiphanes norvegica,* from the North Atlantic. This species (35–40 mm in length) is known as the northern krill because it is a staple component in the diets of baleen whales in the North Atlantic. (© Heather Angel.)

Figure 15.10 A typical pelagic mysid, *Meterythrops picta* (15 mm in length), caught at a depth of 1000 m in the northeast Atlantic. (© Heather Angel.)

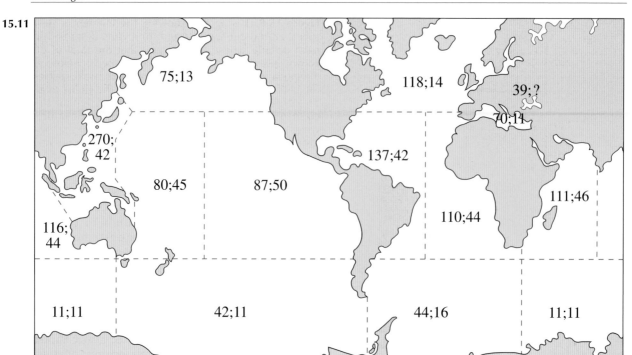

Figure 15.11 Zoogeographical regions used by Mauchline and Murano[10] in their review of mysids, together with the numbers of mysid:euphausiid species presently known from each. There is some suggestion of a latitudinal increase in euphausiid species toward the equator. The data for mysids are more ambivalent, but it is uncertain whether this is representative or an artefact created by the bias with which the group has been researched.

coastal, or inshore waters, or are associated with the sea bed. There are a few open-ocean pelagic mysid species, which, as might be expected, are far more cosmopolitan than their shallow-water relatives. The total of euphausiid species described up to 1994 is 86, and this is likely to be close to the actual total number in the world ocean. In contrast, the number of known mysids species (983) has continued to increase steadily as more coastal faunas around the world are studied, and the list must still be far from complete. This, of course, assumes that the presence of numerous cryptic species (those which cannot be separated using traditional morphological characters, but which are distinguished using molecular biological techniques) will not be revealed when the genetics of euphausiid species are investigated.

There is also an interesting contrast between the pattern of distribution of the mysids and that of the euphausiids. *Figure 15.11* shows one way of dividing up the world's seas into biogeographical regions, together with the numbers of mysid species known from each. There is a suggestion in the southern hemisphere of there being a poleward latitudinal decrease in the numbers of species present. However, this may result either from the relative undersampling of the Southern Ocean regions, or because there are fewer and less varied coastal areas around the Southern Ocean than at similar latitudes in the northern hemisphere. Similarly, the high numbers of species around Japan and the South China Sea may, indeed, be a true reflection a richer mysid fauna in this region relative to all the others – a richness that could be a result of either the Pleistocene oscillations of sea level or the rich variety of coastal environments in the island archipelagos of the region. However, it may also be an artefact resulting from the particular fascination the group has had for Japanese taxonomists. *Figure 15.12* shows the numbers of species of both mysids and euphausiids inhabiting one or more of the regions shown in *Figure 15.11*. It is the pelagic, open-ocean species that are trans- and inter-oceanic in their distributions and so occur in several of the zones, whereas most of the coastal, inshore, and littoral species are restricted to a single region. In the euphausiids, only 11 of the 86 species are neritic (10 of these 11 are restricted to a single region), whereas the vast majority of the 75 oceanic species occur in more than one region. Compare this with the 720 species (73%) of mysids which in habit coastal inshore waters and occur in just one region.

Figure 15.12 Histograms of the number of regions occupied by (a) euphausiids (*N* = 86) and (b) mysids (*N* = 983). The typically oceanic euphausiids tend to be much more widespread than the more coastal mysids.

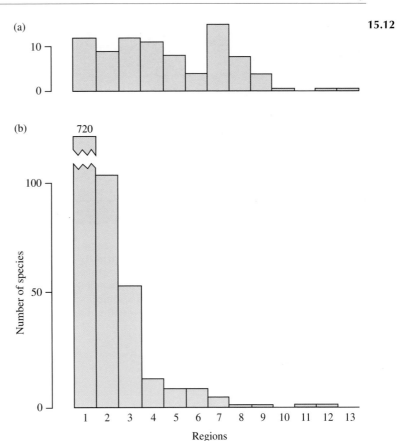

Seasonality

The latitudinal limits at 40° for the regions in *Figure 15.11* represent a significant change in the ecology of the oceans. This is the approximate poleward limit of oceanic waters in which the near-surface 100 m remain thermally stratified throughout the year. During winter at higher latitudes, the surface layers are cooled sufficiently for deep mixing to break-up the stratification and to renew supplies of nutrients in the surface waters. As a result, at temperate and subpolar latitudes not only is there a marked seasonal cycle in the primary production, but also the mean annual primary production is higher. At temperate latitudes, the production cycle is marked by a sharp increase in production in the spring, which occurs as the seasonal thermocline begins to be re-established. In most years there is often another smaller peak in production in the autumn, after the equinoctial storms have started to erode the thermocline but not destroyed it.

These increases in production may or may not be accompanied by a sharp increase in the standing crop of phytoplankton, i.e., a bloom (Chapter 6).

Such blooms can be readily detected in cloud-free conditions by colour scanners in satellites (Chapter 5). Using satellites to record the timing of the blooms has shown that toward subpolar and polar latitudes, the spring and autumn blooms converge to form a single summer peak. Following close after these peaks in production are peaks in sedimentary fluxes, which transmit the products of seasonal events in the surface waters to the benthic communities (Chapter 7). Generally, on the equatorial side of the 40° latitude line, production is not only far less variable seasonally, but is lower, except in upwelling regions where the physical conditions result in deep nutrient-rich water being mixed or upwelled toward the surface.

The annual amount of secondary production (i.e., how much animal growth occurs) is related to the amount of primary production, although the ecological efficiency with which material and energy is transferred from plants to herbivores and on to carnivores and the consumers of detritus varies between different ecosystems. Likewise, the standing crop of biomass also shows some relationship to primary productivity, but tends to be higher where the production cycle is markedly seasonal.

15.13

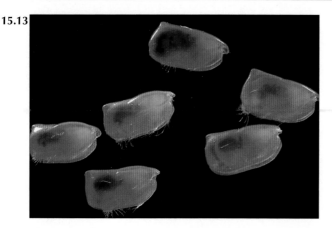

Figure 15.13 Specimens of the planktonic ostracod *Conchoecia valdiviae*, which inhabits depths of 600–1500 m in tropical and subtropical waters. Planktonic ostracods are often the second most abundant group of planktonic organisms in plankton samples, occurring from the surface to the greatest depths. (© Heather Angel.)

15.14

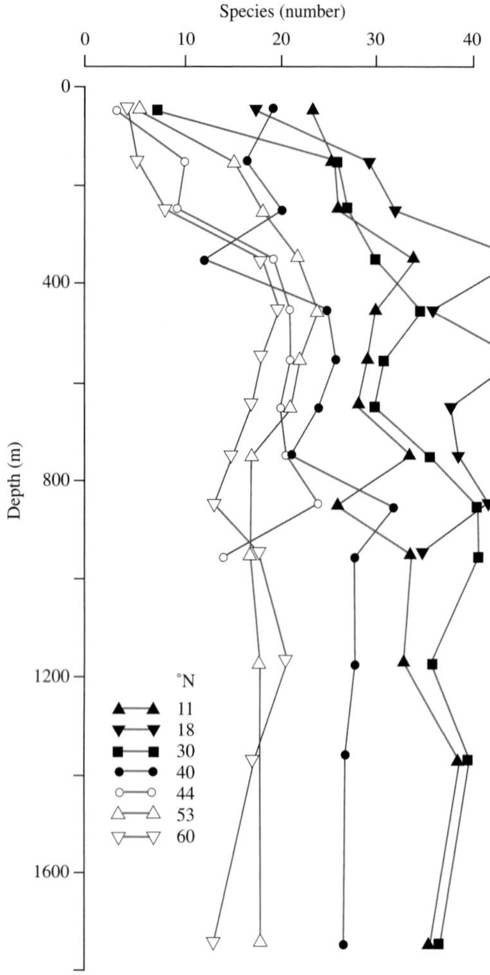

Species (number)

Depth (m)

°N
- ▲—▲ 11
- ▼—▼ 18
- ■—■ 30
- ●—● 40
- ○—○ 44
- △—△ 53
- ▽—▽ 60

These shifts in the annual levels and seasonality of production must play some role in the maintenance of diversity in the communities.

Latitudinal Trends in Species Richness

Species lists often indicate that tropical and subtropical faunas are richer in species than are temperate and boreal faunas. However, the crude distributional data for the euphausiids and the mysids presented in *Figure 15.11* give little indication of how these environmental changes may affect their diversity.

More detailed data are available for planktonic ostracods (*Figure 15.13*) which, while having the advantage of being sampled using the same technique and also analysed with the same protocols, are limited to the North Atlantic. Plots of the numbers of ostracod species caught in the top 2000 m within contiguous depth strata (100 m in the top 1000 m and coarser between 1000–2000 m) at stations between 10–60°N along 20°W (*Figure 15.14*) show that relatively few species are caught near the surface, and the numbers increase with depth to reach a maximum at around 1000 m. This is quite different to the trend for biomass (*Figure 15.6*). When these profiles of species number are superimposed a clear latitudinal trend emerges. At all the depths sampled, the ostracod communities are richer in species at latitudes <40°N, and are appreciably poorer at higher latitudes.

Analysis of the total numbers of species caught in three other pelagic groups in the same set of samples shows a similar latitudinal trend (*Figure 15.15*), although from these data it is impossible to determine whether the decrease across 40°N is smooth or stepped. This latitudinal trend in species richness runs counter to the trends in total annual productivity and integrated pelagic biomass along this transect. The maximum in numbers of species occurred near 18°N. At the time of the sampling, the boundary between the South and North Atlantic central waters was very close to this locality, and the communities included a mixture of both southern and northern faunal elements. Since this boundary moves latitudinally by 10° or so, it seems likely that this maximum in species richness will move with it.

Figure 15.14 Profiles of the numbers of planktonic ostracod species throughout the surface 2000 m along 20°W in the northeast Atlantic, showing how at latitudes <40°N there are more species at all depths than at higher latitudes. The change in average species richness appears to occur at the southern boundary of seasonal turnover in the near-surface waters, and in the regions where there is a spring peak in phytoplankton production[1].

Figure 15.15 Total numbers of species of four pelagic taxa caught at six stations along along 20°W in the northeast Atlantic; at each station 14 day and 14 night samples were collected systematically from the top 2000 m of water column[1].

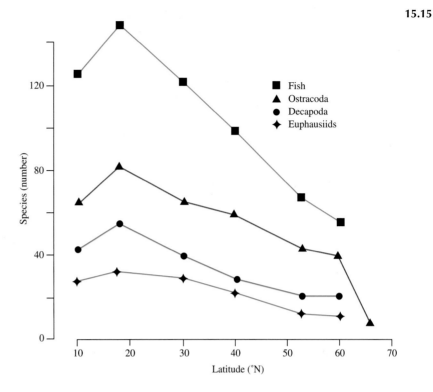

15.15

Species Richness in Benthic Communities

The depth profiles of species numbers for four taxonomic groups of benthic megafauna from the Porcupine Seabight region off southwest Eire (*Figure 15.16*) show much the same trends as for the pelagic ostracods, with peaks in species richness at around 1000 m in three of the groups. However, the data for decapod crustaceans appear to be somewhat anomalous because the most species-rich group, the brachyuran crabs, is almost entirely limited to shelf and upper-slope depths.

Earlier data from the eastern seaboard of the US for four groups of macrobenthos (i.e., animals passing through 4 mm but retained by 1 mm sieves) – polychaetes, gastropods, protobranchs, and cumaceans – implied (on the basis of a statistical method for comparing the numbers of species in samples of different sizes) that the maximum occurred in these groups rather deeper, at

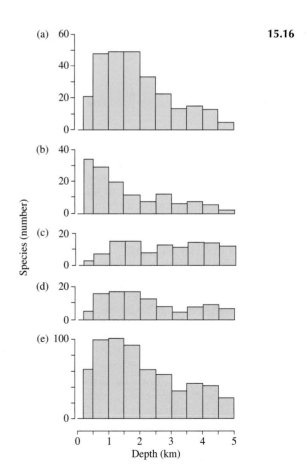

15.16

Figure 15.16 Depth profiles of the numbers of species of four abundant megabenthic taxa: (a) fish, (b) decapod crustaceans, (c) holothurians, and (d) asteroids (starfish); (e) is the sum of these four. These were sampled down-slope during a major sampling programme covering 5 years in the Porcupine Seabight to the southwest of Eire. All but the decapods have maxima in species richness at about 1–2 km depth (IOSDL database).

239

15.17

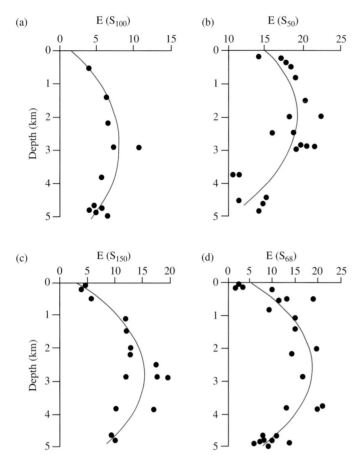

2000–3000 m (*Figure 15.17*). As for pelagic communities, both benthic biomass and the density of organisms decrease almost exponentially with depth (*Figure 15.18*), with a 10% reduction in biomass between 200 m and 2000 m; so, once again, the link between productivity and diversity is not straightforward, since the peak in species richness occurs where the biomass of the benthic community has fallen to nearly a tenth of that at the shelf-break.

Some recent evidence suggests that the species richness of benthic assemblages may match the rich disparity and may even be on a par with that of tropical rain forests. An intensive programme of sampling the benthic communities living on the sea bed at depths of 2000 m off the east coast of the US, involving the analysis of over 200 grab samples, revealed the presence of nearly 800 different species, of which over 50% were new to science[5]. When the sampling was extended both northward and southward along the slope, an extra species was added for each extra square kilometre that the sampling programme covered. If that rate is maintained throughout the global ocean, then the numbers of benthic species would be approximately equivalent to the area in square kilometres of the ocean floor – over 300 million. Moreover, this study did not include the meiobenthos, which contains the two most speciose groups in the oceans – the nematodes and foraminifers. Over 100 species of each of these groups were found in a cubic centimetre of mud scraped off the surface of the sea bed at a depth of just over 1000 m off the southwest of Ireland.

No one really knows how to extrapolate from the results of a few very localised samples to give a credible estimate of how many species occur on Earth, or even if this is possible. Many benthic species appear to have very extensive geographical distributions, so estimates that they exceed a million might seem excessive, but are they? It is worth noting that the huge estimates of the numbers of species inhabiting tropical rain forests are based on similar extrapolations from data obtained by misting just six trees with insecticides, and so are equally lacking in credibility! But perhaps, in this case, the extrapolations are not so wildly excessive given the very large numbers of plants, especially trees, that occur in rain forests and the strong ecological links between the plants and specialist insects.

Effects of the Seasonality on Benthic Diversity

Are changes in the production cycle reflected in changes in benthic diversity? We know that at temperate latitudes, where the production cycle is highly seasonally pulsed, the amounts of detrital material reaching the sea bed vary seasonally (Chapter 7). So, once again, we expect there to be substantial change in the benthic communities at latitudes of around 40°. Each year the deep-sea bed to the west of Britain, at depths of 4000 m, becomes carpeted with detrital material 6–8 weeks after the onset of the spring bloom. This provides a food bonanza for the bottom-living animals, some of whom appear to specialise in exploiting this detritus. No such bonanza is seen further south, either at 30°N on the Madeiran Abyssal Plain or at 20°N on the Cape Verde Rise. The animals which specifically feed on these detrital falls are missing at these latitudes, and there appears to be insufficient food available to support some of the larger animals.

So each region is inhabited by a different assortment of animals, with much fewer large species at the more southerly sites. For example, sea cucumbers – the holothurians – which are animals that feed on the enriched surficial sediment layer, become rare where the sedimentary flux is too low. So their species richness runs counter to the expected latitudinal trend seen, to some extent, in other groups (*Figure 15.19*). Fish abundances are also sharply reduced at the lower latitudes, but the specimens there belong to an unexpectedly rich assortment of species. The numbers of species taken in the limited sampling that has been done are lower than those taken at temperate latitudes. However, comparisons of the species counts in samples of a standard size indicate that sample diversity is as high as, if not higher than, those at temperate latitudes.

The difficulties of drawing generalisations are further illustrated by recent results comparing macrobenthic communities from undisturbed soft muddy sediments at depths of 30–80 m off Spitzsbergen (78°N), in the North Sea (55°N), and off Java (7°N), all sampled and analysed using exactly the same methods[7]. The results show that the diversity profiles at all three sites were indistinguishable, and gave no indication of a latitudinal trend (*Figure 15.20*). Is this observation merely a quirk of history? In other words, were the investigators unfortunate in choosing sites where, by chance, evolutionary history had resulted in the diversity being identical. This solution seems far-fetched, but is testable by repeating the comparison elsewhere. An alternative explanation might be that the processes that maintain the diversity at these three locations are not influenced by latitudinal

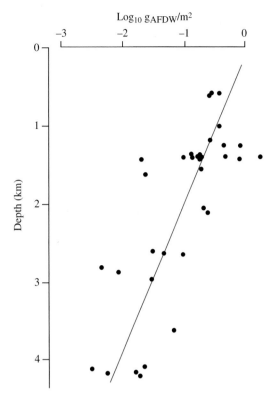

15.18

Figure 15.18 Profile of benthic biomass expressed as ash-free dry weight (AFDW) per square metre versus depth in the Porcupine Seabight region off southwest Eire[8]. The scatter is the result of the patchiness of the benthic communities, especially at depths of around 1250 m, where some samples contained several specimens of the sponge *Pheronema*.

forcing. For example, in such muddy environments, is there always an unlimited amount of organic matter around, irrespective of the production cycle?

Diversity and Productivity

There is an apparent paradox in the trends in pelagic diversity with depth and geographically, which run counter to the trends in productivity and biomass (standing crop) of animals, certainly in the water column and maybe on the sea floor. Populations in low-productivity regions seem to be characterised by being rich in species-richness without any one or two species being overwhelmingly dominant. In contrast, where productivity is high (and often more variable seasonally) far fewer species occur and one or two species tend to be numerically dominant. Does this explain why, when we fertilise the seas with our sewage and with the agricultural run-off of dissolved nitrates and phosphates in our rivers, the local productivity goes up, but the numbers of species goes down (a process called eutrophication)?

15.19

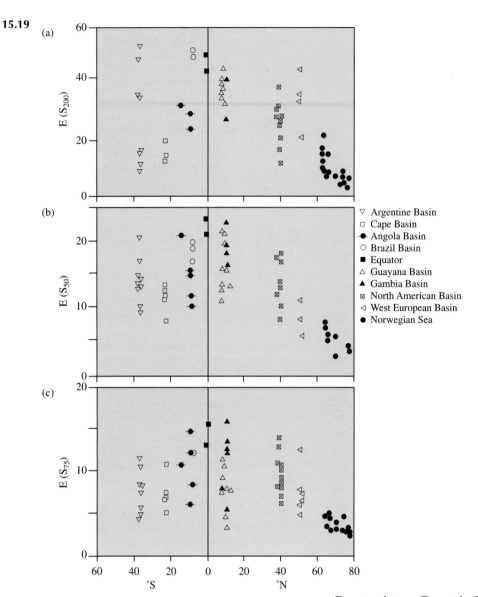

Figure 15.19 Numbers of three macrobenthic taxa expected to be caught in samples of *S* specimens from deep-ocean basins in the Atlantic Ocean: (a) isopoda, (b) gastropoda, and (c) bivalva (modified from Rex *et al.*[15]).

▽ Argentine Basin
□ Cape Basin
⦁ Angola Basin
○ Brazil Basin
■ Equator
△ Guayana Basin
▲ Gambia Basin
⊠ North American Basin
◁ West European Basin
● Norwegian Sea

Postscript – Genetic Diversity

Figure 22.3 shows the effects of several decades of dumping sewage sludge off Garroch Head in the Clyde on the biomass, number of species, and number of specimens of benthic animals across the dump site, with the central position showing a massive increase in specimens, a smaller increase in biomass, and a sharp reduction in numbers of species. Is such an impact the consequence of ecological disturbance regardless of its nature? It warns us that if we concentrate all our efforts on conserving those regions richest in species, then we will run the danger of allowing degradation to go unchecked in those regions where ecological processes are most important in keeping the Earth habitable.

One important type of diversity, genetic diversity, has received no mention in this chapter. Populations living in stable environments are expected to have low genetic diversity, in that they show relatively little variation in their DNA and hence in the structure of their proteins and enzymes. When observations were first made on oceanic species the results were expected to reveal very low levels of genetic heterogeneity, but the outcome has been quite the opposite. The genetic variability within pelagic oceanic species is proving to be among the highest. There are numerous hypotheses to explain this high variability, but little in the way of proof. For example, are the meso-scale eddies so prevalent in the ocean persistent enough to select different genetic strains, but not to

15.20

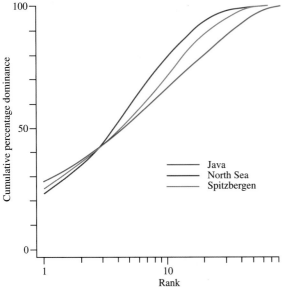

Cumulative percentage dominance

Java
North Sea
Spitzbergen

Rank

Figure 15.20 Plots of the relative abundances of species in soft-bottom communities (expressed as cumulative percentage dominance) at three sites, Java, the North Sea off Northumberland, and in a fjord in Spitzbergen. The similarity of these curves implies that the observed diversities, expressed as the relative abundances of species, in the bottom-living communities at these very different latitudes are virtually indistinguishable (modified from Kendall and Aschan[7]).

keep them isolated? Is this why pelagic species, despite having wide geographical ranges, do not show a greater tendency to split into races and subspecies? Or should the species concept be modified in some way to account for how oceanic populations diverge and recombine?

On land, within time-scales of millenia, geographical localities are fixed, but in the ocean 'place' is dynamic and so has much less significance, even for the many benthic species which have a dispersive phase to their life history. Genetic studies have hardly begun to scratch the surface of these problems. The removal of barriers to dispersion through human introduction, either purposeful for mariculture or by accident in the ballast waters of large bulk-carrying vessels, may well have a far greater impact at this level of organisation than we are prepared for, but will it matter? There is still so much to understand before we can evaluate whether maintenance of biodiversity in the oceans is a cause for real concern or a red herring.

General References

Angel, M.V. (1994), Biodiversity of the pelagic ocean, *Conserv. Biol.*, 7, 760–772.

Eldredge, N. (ed.) (1992), *Sytematics, Ecology, and the Biodiversity Crisis*, Columbia University Press, New York, 220 pp.

Norse, E.A. (1993), *Global Marine Diversity; A Strategy for Building Conservation into Decision Making*, Island Press, Washington DC, 383 pp.

References

1. Angel, M.V. (1994), Biodiversity of the pelagic ocean, *Conserv. Biol.*, 7, 760–772.
2. Angel, M.V. and Baker, A. de C. (1982), Vertical standing crop of plankton and micronekton at three stations in the North-east Atlantic, *Biolog. Oceanogr.*, 2, 1–30.
3. Carlton, J.T. (1989), Man's role in changing the face of the ocean: Biological invasions and implications for conservation of near-shore environments, *Conserv. Biol.*, 3, 265–273.
4. Flemminger, A. (1986), The Pleistocene equatorial barrier between the Indian and Pacific Oceans and a likely cause for Wallace's Line, *UNESCO Tech. Papers Marine Sci.*, 49, 84–97.
5. Grassle, J.F. and Maciolek, N.J. (1992), Deep-sea richness: regional and local diversity estimates from quantitative bottom samples, *Amer. Natural.*, 139, 313–341.
6. Hopkins, T.L., Milliken, D.M., Bell, L.M., McMichael, E.J., Hefferman, J.J., and Cano, R.V. (1981), The landward distribution of oceanic plankton and micronekton over the west Florida continental shelf as related to their vertical distribution, *J. Plank. Res.*, 3, 645–659.
7. Kendall, M.A. and Aschan, M. (1993), Latitudinal gradients in the structure of macrobenthic communities: A comparison of Arctic, temperate and tropical sites, *J. Exp. Marine Biol. Ecol.*, 172, 157–169.
8. Lampitt, R.S., Billett, R.S.M., and Rice, A.L. (1986), Biomass of the invertebrate megabenthos from 500 to 4100 m in the Northeast Atlantic Ocean, *Marine Biol.*, 93, 69–81.
9. Margulis, L. and Sagan, D. (1987), *Microcosmos: Four Billion Years of Evolution from our Microbial Ancestors*, Allen & Unwin, London, 301 pp.
10. Mauchline, J. and Murano, M. (1977), World list of the Mysidacea, Crustacea, *J. Tokyo Univ. Fish*, 64, 39–88.
11. Parrish, J. and Curtis, R.J. (1982), Atmospheric circulation, upwelling and organic-rich rocks in the Mesozoic and Cenozoic eras, *Paleogeogr., Paleoclim., Paleoecol.*, 40, 31–66.
12. Pearse, J. and Buchsbaum, A. (1987), *Living Invertebrates*, Blackwell Scientific Publications, Oxford, 848 pp.
13. Por, F.D. (1990), Lessepsian migration. An appraisal and new data, *Bull. l'Instit. Oceanogr., Monaco*, 7, 1–10.
14. Rex, M.A. (1983), Geographic patterns of species diversity in deep sea benthos, in *Deep-Sea Biology. The Sea*, Vol 8, Rowe, G.T. (ed.), Wiley Interscience Publications, New York and Chichester, pp 453–472.
15. Rex, M.A., Stuart, C.T., Hessler, R.R., Allen, J.A., Sanders, H.L., and Wilson, G.D.F. (1993), Global scale latitudinal patterns of species diversity in the deep-sea benthos, *Nature*, 365, 636–639.
16. Ribbink, A.J. (1994), Biodiversity and speciation of freshwater fishes with particular reference to African cichlids, in *Aquatic Ecology: Scale, Pattern and Process*, Giller, P.S., Hildrew, A.G., and Raffaelli, D.G. (eds), 34th Symposium of the British Ecological Society, Blackwell Scientific Publications, Oxford, 649 pp.
17. Sepkoski, J.J. (1992), Phylogenetic and ecologic patterns in the Phanerozoic history of marine biodiversity, in *Systematics, Ecology and the Biodiversity Crisis*, Eldredge, N. (ed), Columbia University Press, 220 pp.
18. White, B.N. (1994), Variance biogeography of the open-ocean Pacific, *Progr. Oceanogr.*, 34, 257–284.

CHAPTER 16:

Life in Estuaries, Salt Marshes, Lagoons, and Coastal Waters

A.P.M. Lockwood, M. Sheader, and J.A. Williams

Introduction

Any one location offshore generally presents relatively little change in its physical and chemical conditions over a short time-scale. The same cannot be said of near-shore waters, estuaries, and coastal lagoons, which provide some of the more varied and unstable environments on Earth. The organisms inhabiting such areas may experience wave action to differing degrees, tidal change, salinity, temperature, and oxygen variations, high sediment loads, and tidal currents of varying velocities on a daily or seasonal basis. Locations such as salt-marsh pools (*Figure 16.1*), estuarine creeks (*Figure 16.2*), and shallow sloping beaches (*Figure 16.3*), can experience quite wide ranges in one or more of their physical characteristics. Tidal progression (*Figure 16.4*) may further compound the difficulties for colonisers by imposing rapidity in the changes of factors such as salinity and temperature. Add to

these effects in the water column the diversity of types of substratum available for attachment or burrowing (*Figure 16.5*) and the impact of man (*Figures 16.1, 16.3–16.5*) and it is apparent that such waters can be inhabited only by species capable of responding to physical, mechanical, and physiological challenges.

Response to environmental fluctuations cannot, however, be restricted to mere passive tolerance; adaptive features must incorporate a whole gamut of positive measures, including biochemical, physiological, and morphological features.

Reproductive responses, too, must be involved; it is not much use having adult stages adapted to a particular environment if the motile young lack the facility to locate and colonise suitable habitats.

The conditions, then, are harsh, but few of the challenges presented by coastal waters have not been met. True, sites such as mobile shingle banks

Figure 16.1 Estuarine raised marsh on the River Test, Hampshire, England, showing tidal creeks and saline pools. The creeks experience substantial salinity variation during the tidal cycle, while the pools are exposed to sudden changes when the surrounding grass is submerged on high spring tides.

Figure 16.2 Closer view of the tidal creek shown in the middle distance in *Figure 16.1*. This is the site of the salinity measurements illustrated in *Figure 16.8*.

Figure 16.3 Estuarine beach areas are often impacted by anthropogenic influences, as well as experiencing physical changes.

16.4a

16.4b

Figure 16.4 Views of an estuarine site (a) at low water and (b) at high water on a neap tide. On high water spring tides, the raised marsh to the left would generally be submerged.

16.5

Figure 16.5 The variable habitats presented by a tidal creek. Rocks with fucoid algae provide refuge at low water for non-burrowing forms, such as the gammarid amphipods; the mud banks are burrowed by the annelid *Nereis diversicolor*, the amphipod *Cyathura carinata*, and by the non-feeding adults of the isopod *Paragnathia formica*. The prawn *Palaemonetes varians* and the mysid *Neomysis integer* may be found where deeper pools are left at low water.

16.6

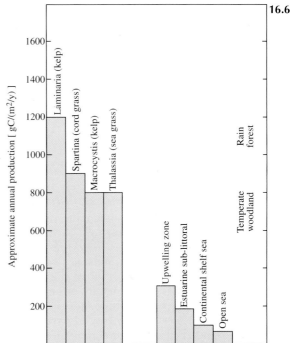

Figure 16.6 Estimates of primary production (gC/m²/yr) for various marine and terrestrial ecotypes. Note that values vary widely with both conditions and plant density. The figures indicate, first, that the productivity of sea-grasses and the fixed algae of the coastal edge can be as high as that of tropical rain forest and, second, that the unit area productivity of offshore waters, even in the nutrient-rich upwelling zones, tends to be considerably lower.

and salt-marsh pools have a restricted fauna, but no coastal environment is without any colonisers. So successful, indeed, are some of the inhabitants that shore and near-shore regions have the highest primary productivity per square metre of any marine area (*Figure 16.6*).

There is, of course, no single answer to the problems posed for organisms in the coastal and shelf seas, nor is there any single simple lifestyle. In this chapter, therefore, we select three diverse topics to illustrate the amazing versatility of estuarine and coastal forms. These include:

- The physiology of osmoregulation – the regulation of water and ion balance by organisms – to indicate the variety of solutions to the problems generated for living forms by variation in the chemistry (salinity) of the habitat.
- Animal–sediment interactions, to outline the importance of the geological and hydrodynamical regime for benthic animals.
- Patterns in community structure of zooplankton, to interrelate the changing populations with features of the physical environment and biotic factors.

16.7

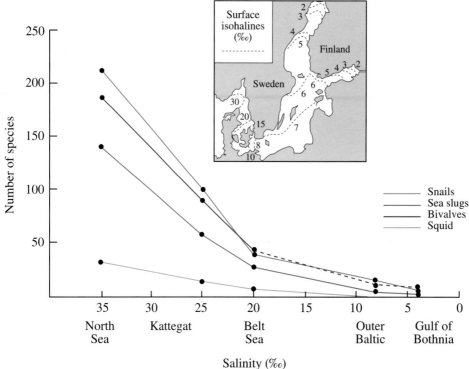

Figure 16.7 Waters with a salinity less than that of sea water tend to show depletion of faunal variety. In part this is for physiological reasons (see text), in part because of habitat restrictions. The figure illustrates such a decline in the number of species of four groups of mollusc in relation to the salinity in the Baltic (derived from numerical data by Jaeckel, see Remane and Schlieper[16]). The inset shows the salinity at different locations, measured in parts per thousand.

The Physiology of Osmoregulation

The body surface of most marine organisms is rather permeable to water and inorganic ions. Consequently, few species have the capacity to maintain the extracellular fluids at a concentration significantly different from that of the waters they inhabit. As there is also generally rather little tolerance to dilution by the cells of most marine forms, the consequence is that few species have the capacity to extend their range into the waters of dilute seas, estuaries, and coastal lagoons.

The point is well-illustrated with reference to the Phylum Mollusca. This group is represented by numerous species in the saline waters of the North Sea, but the dilute waters of the Baltic create problems resulting in the elimination of all but the hardy few (*Figure 16.7*). Even those forms which can penetrate into the middle Baltic, where salinity is only about 20% of that in the North Sea, are often but a shadow of their relatives from more saline waters. The common blue mussel *Mytilus edulis* in the waters south of Stockholm only grows to about half the size of those from the North Sea. Also, its shell, instead of being blue, robust, and opaque, is often fragile, partially translucent, and brownish. The Baltic Herring, the strömming, (beloved by Swedish epicures after it has matured packed in leaves below ground for some weeks) is similarly but a poor cousin, in terms of size, to the North Sea herring[16].

Osmotic and ionic levels also pose problems in terms of the penetration of animals into estuaries. Just how far various marine species can extend their range depends not only on the physiology and lifestyle of the organism, but also on the morphological and hydrological characteristics of the estuary. Where the estuarine shore is shallow sand or mud flats and the river flow is small by comparison with the tidal incursion, then the mid-to-upper shore generally only experiences undiluted water when immersed. Such areas may be colonised by animals with little tolerance of dilution, provided they can survive or avoid aerial exposure at low tide. By contrast, the channels carrying the river at low water may represent a greater challenge to colonisers by virtue of the rapid and profound changes in salinity that occur over the tidal cycle. This point is well-illustrated by data from a creek on the River Test in the UK, where salinity in the channel may vary from virtually fresh water to some 65% that of North Sea water and back again over a single tidal cycle (*Figure 16.8*).

The colonisers of such variable waters have but three choices:

- Balance passive osmotic uptake of water at low salinities with passive osmotic withdrawal of water at high tide; altogether a rather high-risk strategy that can only work over a restricted part of an estuary. Nevertheless, it appears that the little flatworm *Procerodes (Gunda) ulvae*, which typically lives around the neap tide high-

16.8

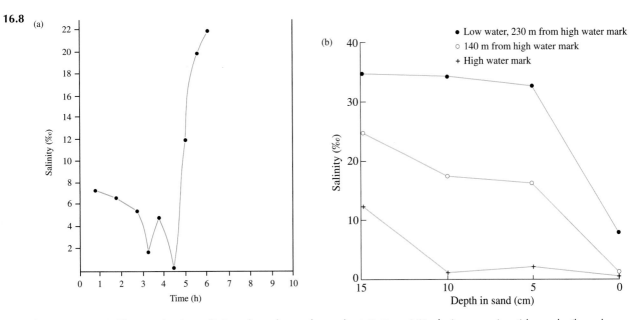

Figure 16.8 (a) Changes in the salinity of a salt-marsh creek at Totton, UK, during a spring-tide cycle (based on Ralph[14]). Note that the salinity varies from about fresh water to some two-thirds sea water within the cycle, imposing harsh physiological conditions on exposed fauna (*Figure 16.2* illustrates the actual site). (b) Salinity of the interstitial water of the sediment underlying a stream crossing the intertidal region of a beach (drawn from data in Reid[15]). The distances are measured from the high water mark. Note the decrease in salinity variation with depth in the sediment.

water mark in small streams, is largely such an osmotic yo-yo.

- Avoid the problem by burrowing. This option takes advantage of the much slower rates of change of salinity in the interstitial water than in the overlying medium, *Figure 16.8(b)*. The use of this strategy enables some burrowing annelids – segmented worms – to penetrate considerably further along estuaries than can epibenthic species exposed to the full vicissitudes of tidal salinity change.

- The third option is to regulate the body fluids independently, or at least partially independently, of the concentration of the medium. Virtually all the benthic and pelagic colonisers

of the region from mid-estuary to fresh water use some form of regulatory mechanism(s). The systems involved are considered shortly, but first mention must be made of the effects on colonisation of the configuration of estuaries.

Where the geomorphology is such that the estuary basin contains saline water at low tide, as is the case in the seaward reaches of most estuaries, then one of two situations arises. Either the water column is well mixed so that there is effectively no vertical stratification of salinity or, alternatively, especially where freshwater input is relatively large, a stratified water column occurs in which a more saline salt wedge penetrates upstream with fresher water above (*Figure 16.9*). The latter situation

16.9

Figure 16.9 Stylised representation of a stratified estuary, showing the wedge of more saline water penetrating upstream under outflowing dilute water.

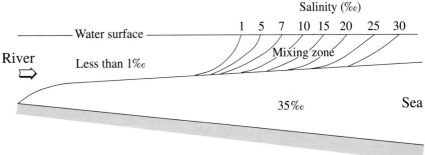

16.10

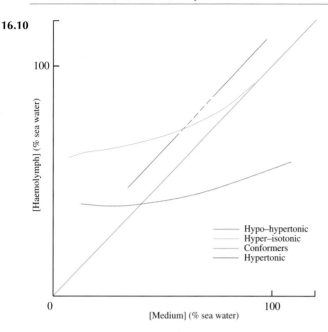

100

[Haemolymph] (% sea water)

Hypo–hypertonic
Hyper–isotonic
Conformers
Hypertonic

0   100

[Medium] (% sea water)

Figure 16.10 Comparison of the haemolymph–medium relationship for various types of regulation. Where the haemolymph concentration is above or below the isosmotic line, the animal actively regulates the concentration of its extracellular fluids.

tends to favour the upstream penetration of marine forms.

Osmoregulatory adaptations

All marine forms which colonise areas subject to dilution tend to have a relatively wider tolerance of salinity change than do their fully marine relatives (i.e., they are 'euryhaline' rather than 'stenohaline'). Tolerance is found both in species where the body fluid concentration tends to follow that of the surrounding medium as the latter changes (osmoconformers) and in those capable of at least some maintenance of body fluid concentration at levels differing from that of the medium (osmoregulators).

Osmoconformers

Soft-bodied forms, such as worms and molluscs, tend to have a high surface permeability and most are incapable of maintaining body fluids more concentrated than the medium. Those with any appreciable tolerance of dilution must therefore regulate at the cellular level. The problem is simple. The cells contain osmotic effectors, mostly inorganic ions and small organic molecules, and the cell walls are highly permeable to water. Consequently, on dilution of the extracellular fluid, water moves into and expands the cells, unless steps are taken to reduce the amount of internal osmotic effectors.

The need to retain the electrical excitability of cells such as nerves and muscles limits the degree to which inorganic ions can participate in effector reduction; a substantial part of any change is therefore generally accomplished through a decrease in the concentration of organic molecules. Much of the variation relates to the removal of non-essential amino acids, such as glycine, glutamic acid, alanine, and proline. This removal is effected either by conversion into organic acids, which can then be metabolised, or by allowing diffusion across the cell membrane.

The process is reversible, so when the medium becomes more saline inorganic ions play an important role in directing the activity of the enzymes involved in the conversion of organic acids into amino acids. Amino acids are also then taken up at an increased rate across the cell surface (see Gilles, 1979).

The majority of osmoconformers have a limited tolerance range.

Osmoregulators

In contrast to osmoconformers, osmoregulators maintain body fluids at a level differing from that of the medium over at least part of their tolerance range. *Figure 16.10* illustrates the three known types of relationship between concentration of the medium and concentration of body fluids. Hypotonic refers to something less concentrated than the reference medium, in contrast to hypertonic – more concentrated. Hyper-isotonic regulators, as the name implies, are conformers at high salinities, but hold the body fluids at a higher concentration than that of the medium at lower salinities. The shore crab *Carcinus maenas* is an example of such a regulator. Hypo-hypertonic regulators have body fluids which are maintained at a lower concentration than the medium at high salinities and at a higher concentration when the external concentration is low. They thereby keep the blood and cell concentrations within a relatively limited range, thus placing less of a requirement for osmotic control on the body cells themselves. Typical examples of such hypo-hypertonic regulators include the euryhaline bony fish and some species of prawn, isopods, and mysids. However, it is the brine shrimp *Artemia salina* which is the undisputed *prima donna* in this role, since it still maintains its internal concentration at a low level even when in saturated brine. The third, and final, class of regulation involves maintenance of hypertonicity to the medium over the entire tolerance range. So far this category is solely represented by the larva of the isopod *Paragnathia formica*. This larva is parasitic on euryhaline fish, such as young flounder, and potentially has an osmotic problem should the fish swim seaward. However, by being always significantly hypertonic to any medium to which it is acclimated, it gives itself at least some buffer before the risk of dehydration.

16.11

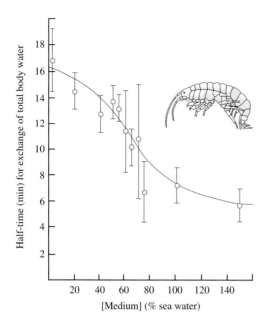

Figure 16.11 Variation in permeability to water with salinity in *Gammarus duebeni* (insert; after Lockwood *et al.*[10]). The graph shows the time for half the body water volume to be exchanged (using tritiated water as a marker), with the animals acclimated to different salinities. The half-time for exchange is taken as a reciprocal function of the permeability. The error bars represent ± one standard deviation.

Inevitably, hypertonic regulation carries with it the twin penalties of uptake of water by osmosis and loss of inorganic ions across the body surface and in the urine. Maintenance of the status quo of the body fluids therefore demands the expenditure of energy to counteract these effects. The larger the gradients and the greater the permeability the more substantial is the energy need, so it is advantageous to decrease the permeability. Unsurprisingly, there is a general tendency for coastal and estuarine forms to be less permeable than offshore marine species. In some cases the permeability to water can be varied at the level of the individual if the exter-

nal salinity changes. The amphipod crustacean *Gammarus duebeni* is a case in point. When in 2% sea water the permeability to water is less than half that when it is in sea water (*Figure 16.11*).

Even dynamic changes in permeability are not alone sufficient to maintain an osmotic gradient across the body surface. Additional processes are also required, which include:

- Conservation of ions in the body by the production of urine less concentrated than the body fluids.
- The active uptake of ions across the body surface[4].

Hypotonic urine

The representatives of many animal groups inhabiting fresh water have developed the capacity to excrete urine which is more dilute than the blood. By this means they can remove water brought into the body by osmosis while conserving valuable ions. By contrast, in brackish waters only fish and a few species of amphipod are known to have a significant capability in this respect (*Figure 16.12*).

Active transport of inorganic ions

The active regulation of internal ion levels is a feature of animal cells; derivations from this system have allowed transport of ion species, particularly the major ions sodium and chloride, across surface membranes[4]. In marine forms isosmotic with their medium such transport may be associated with the maintenance of some ions at levels of concentration differing from those in sea water. Additionally, ion transport sometimes appears to be associated with the need to bring water into the body for urine production or replacement of fluid lost by haemorrhage. Similarly, the expansion of some amphipods after moulting in sea water seems to be associated with increased ion uptake.

To adapt an ion-transport system, originally designed to transport water, to the new role of maintaining the body fluids hypertonic to the medium would seem to involve rather little further

Figure 16.12 Production of urine hypotonic to the haemolymph at low salinities in *Gammarus duebeni* (after Lockwood[9]). The solid dots represent individual urine samples taken directly after micturition from the urinary papilla. The hollow circles are blood samples. The osmotic pressure of the tiny samples (volumes as small as 10^{-3} mm³) are measured using the Ramsay cryoscope. Production of hypotonic urine, when the haemolymph is more concentrated than the medium, may be expected to assist in the conservation of inorganic ions in the body.

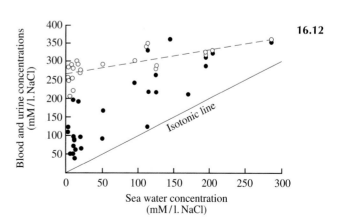

16.12

change; perhaps no more than a variation in the relative permeability of the body surface to ions and water. A small step in evolution, perhaps, but no less important for that. Had such a capability never evolved it is improbable that life as we know it in fresh water and on the land would ever have developed.

Animal–Sediment Interactions in Coastal and Shelf Environments

Although rocky substrata are not uncommon in littoral and near-shore areas, most of the sea floor in shelf and coastal environments is composed of sediments. Unlike the deep sea, where the substratum is predominantly fine mud, coastal areas are characterised by sediments that are extremely variable (clay, silt, sand, shingle, cobbles, and boulders or various admixtures of these), reflecting their proximity to sediment sources and the more dynamic sediment transport processes that occur in the shallow waters of the coastal and shelf regions.

Sediments, especially in shallow waters, are generally nutrient rich and teem with life. Benthic organisms (i.e., those living in or on the sea floor) are a diverse group, and species can be found in association with all types of sediment. While the greatest variety of species is usually found in sediments with a broad range of particle size, the greatest density of organisms and highest productivity tend to be associated with littoral, or near-shore, sediments in sheltered environments. Most estuaries fall into this category; the high animal productivity reflects the wide range of food resources available, both from within and imported from outside the system.

The organisms found in sediments range in size from bacteria to megafauna. Bacteria and smaller protozoa are usually associated with organic material and particle surfaces, and represent a food source for many of the larger organisms. Bacteria have great enzymatic versatility and are able to digest a wide range of organic materials which, though not directly available to these larger species, become available as microbial biomass.

Most of the animals living in sediment feed either directly on sediment (deposit feeders) or on particles collected from the water overlying the sediment surface (suspension feeders). On the continental shelf, deposit-feeding species reach their greatest diversity in fine, muddy sediments and their greatest abundance in sheltered inshore muds, such as those of estuaries, lagoons, salt marshes, and mangrove swamps. One reason for this relates to availability of food resource. Organic matter and bacteria within sediments are both often associated with particle surfaces and, since the surface area of fine particles is greater per unit volume than that of

coarse sediments, it follows that the available organic resource, organics plus bacteria, is likely to be greater in fine rather than coarse sediments.

Suspension feeders require a sediment which will provide stability and support, but are also dependent on particulate food resources in the near-bottom water. Both requirements are in turn dependent upon the local hydrodynamics and sedimentary transport processes.

The trophic structure (the relative importance of different feeding types) and species composition are thus affected to a substantial degree by the sediment structure and dynamics. However, benthic organisms are not merely passive responders to environmental influences, but are themselves able to actively exert effects on the sediments. Modification of the sediment can result from a number of biologically mediated processes. Of these, bioturbation, biodeposition, and biosecretion, as described below, are the most important. Each process acts to change the structure and properties of the sediment and this, in turn, may affect the range of potential colonising species.

Bioturbative processes lead to the movement of sediment as a direct or indirect result of the activity of organisms. Animals may select particles of a particular size to line their burrows or build tubes in which to live, and burrow walls may be compacted. Both processes may locally change sediment properties around the burrow. Animals burrowing, moving through the substratum, or simply growing, also displace sediment. However, the greatest rates of bioturbation are caused by large deposit feeders feeding on and passing sediment through their guts. Examples are found in most of the major benthic groups, notably the polychaete worms, the thalassid shrimps, the sea cucumbers (holothuroids), the heart urchins (spatangoids), and the bivalves. As a result of the activity of these deposit feeders, rates of sediment reworking may be as high as several kilograms per square metre per year, and the whole of the top few centimetres of sediment may be reworked several times over the same timescale.

Intensive reworking changes the sediment structure. Many of the finer particles are ingested and become bound together in faecal pellets, thereby effectively increasing the particle size (*Figure 16.13*). The resulting sediments have a higher water content and a lower bulk density, and so are more easily eroded by relatively weak currents[17]. An example of biologically mediated change in sediment structure is given in *Figure 16.14*, which shows seasonal changes both in sediment shear strength and in bioturbation rate by the polychaete worm *Arenicola marina*.

The development of graded bedding is another

Figure 16.13 A sample of surface sediment taken from a productive coastal lagoon in the UK. Almost all the surface sediment has been pelletised (individual faecal pellets are about 300 µm in length) by the feeding activities of the fauna.

16.13

frequent consequence of bioturbation[2]. This follows from deposit feeders, such as *Arenicola*, selecting finer particles to eat and expelling the faecal waste at the surface. The result is a surface layer of finer particles overlying coarser materials accumulating at the maximum feeding depth.

Fluid movement through sediments may also be enhanced as a result of the excretion and respiration of organisms. Indeed, the effect of such enhanced flow can be up to two orders of magnitude greater by volume than that of particle bioturbation. Together, the two processes alter the sediment chemistry and influence the depth of the redox layer (the boundary between the oxic and anoxic sediment).

Biodeposition results from the activity of suspension feeders that remove particles from the water column and usually expel faecal pellets onto the sediment surface, thereby increasing the rate of deposition of particles to the sediment. Many suspension feeders, and some deposit feeders, build tubes that extend above the sediment surface. For suspension feeders, raising the feeding structure above the surface in this way is advantageous in that both the water flow around the feeding appendages and the rate of encounter with food particles increase.

Tube structures have a marked effect on the stability of sediments, though the extent of the impact is dependent on their size and abundance[3]. Low tube densities may result in increased turbulence in the lee of each tube, with a consequent sediment scour and erosion at low or moderate current velocities. By contrast, high tube densities can result in a skimming flow over the top of the tubes, which reduces both flow and erosion at the sediment surface. Fine particles are able to settle in the low-flow regime between the tubes, often resulting in sediment accumulation and a fall in the mean

16.14

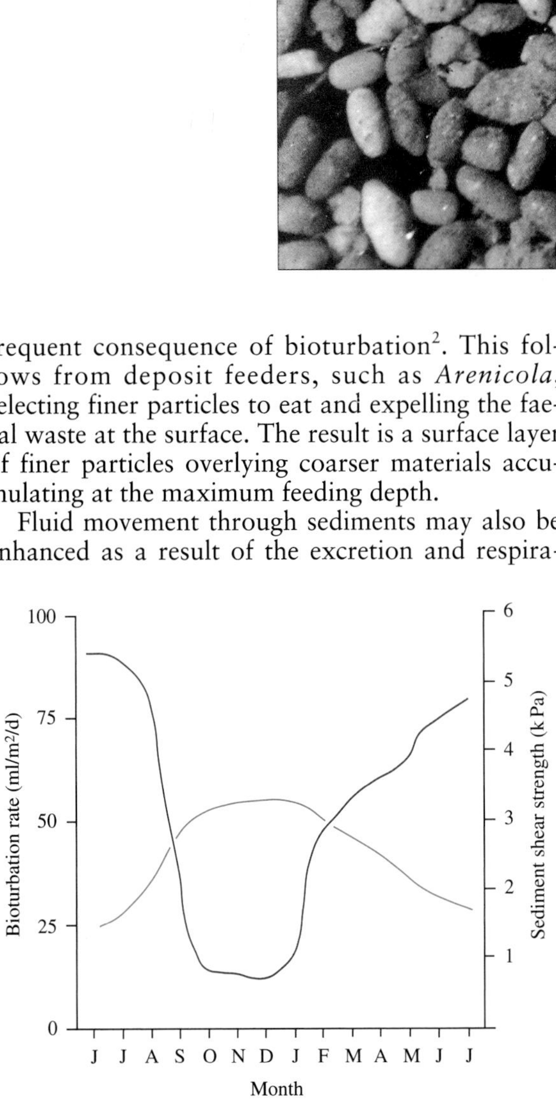

Figure 16.14 The seasonal pattern of bioturbation (blue line) by the polychaete worm *Arenicola marina* in a UK coastal lagoon. Changes in the shear strength (red line) of the top 5 cm of sediment mirror bioturbative activity.

Bioturbation rate (ml/m²/d) — *Sediment shear strength (kPa)* — *Month*

251

16.15

Figure 16.15 Tube beds of the polychaete worm *Pygospio elegans* in the Baie de Somme, France. The edge of the raised tube bed is being eroded to form a 'cliff' about 15 cm high. The densely packed tubes constructed by the worms are 0.5–1.0 mm in diameter.

sediment grain size. *Figure 16.15* shows such a high abundance of tubes, in this case the polychaete worm *Pygospio elegans*, on an estuarine sand flat at a density in excess of 100,000 tubes per square metre. At such high densities the whole bed becomes raised above that of the surrounding sand as a result of the accumulating fine sediment, together with the continual extension of the worm tubes. The end result is a change in the structure and properties of the sediment (*Figure 16.16*).

Biosecretion by benthic organisms is the release of organic materials, either as waste products or as functional secretions. These organic materials provide both a food resource for micro-organisms and a means for altering sediment chemistry. Of greatest importance with regard to change in sediment properties is the secretion of mucopolysaccharides, or mucus. Mucus can bind sediment particles together and thereby increase the stability of sediments. Many species produce mucus, but particularly important contributors are micro-organisms, especially bacteria and single-celled organisms. In littoral and shallow sublittoral areas, micro-algae, such as diatoms, are also contributors to the mucus-binding of sediment particles (*Figure 16.17*).

In polar, temperate, and certain tropical shelf and coastal environments, bioturbation, biodeposition, and biosecretion show clear seasonal patterns, which are reflected in annual cycles in the structure and properties of sediments (*Figure 16.14*). Also influenced by seasonal cycles are the planktonic communities, from which much of the organic matter providing the energy input for the benthic community ultimately derives.

Estuarine Plankton: Patterns of Community Structure

The planktonic community of the water column consists of an amazing spectrum of microscopic species, ranging from viruses, bacteria, and single-celled plants and animals to jellyfish, some crustaceans (*Figure 16.18*) and molluscs, and the early life stages of fish. Some are relatively non-motile, while others are capable of reasonably sustained movement. What connects the members of the plankton, however, is that their capacity to propel themselves tends to be small relative to the water currents and turbulence they experience. To some extent, therefore, their distribution is related to the mass movements of the water in which they live.

Together, the plankton members make up an

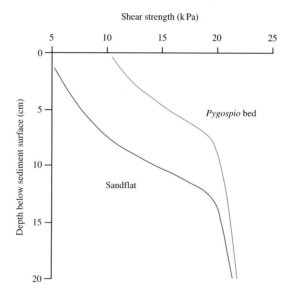

16.16

Figure 16.16 Changes in shear strength with sediment depth in *Pygospio* beds and in adjacent sediment containing low numbers of the polychaete worm. The dense aggregation of tubes (more than 100,000/m²) greatly alters the surface sediment characteristics and properties.

Figure 16.17 Diatom blooms on the surface of an estuarine mud flat. Such blooms are important in binding sediment particles by producing mucus.

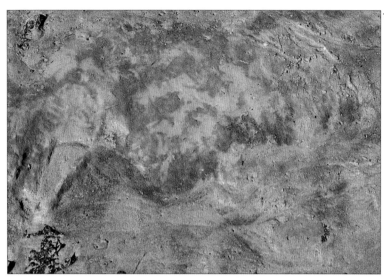

intricate food web which transfers the photosynthetic primary production of the phytoplankton in the upper water column through a number of levels within the zooplankton to the top pelagic predators[13]. Transfer of energy also occurs down the water column through the food web to sustain the benthic communities, in the ways already discussed.

Figure 16.18 Typical sample of crustacean zooplankton illustrating the relative size differences between (1) calanoid copepods, (2) decapod zoeae larvae, and (3) nauplii larvae – total body length of the copepod (1) is approximately 1.5 mm.

The plankton of estuaries and coastal water require similar environmental conditions for survival and reproductive success as do those in open water. In addition, however, they must have the capacity to respond to the much greater degree of environmental physicochemical variability of the near-shore waters, which imposes daily and seasonal patterns that in turn influence their population dynamics[6]. Adaptational responses to the environmental changes are likely to involve complex energy-demanding processes, but off-setting this disadvantage, as far as the phytoplankton are concerned, is the high availability of nutrients in estuaries, which leads to high productivity (*Figure 16.6*).

Similarly, the many environmental and biological factors described earlier all combine to restrict the variety of zooplankton species that occupy these waters by comparison with the varieties of the open sea. By contrast, the higher levels of primary production make for high levels of abundance of those zooplankton species which are able to respond to the physicochemical variability of estuaries. High overall levels must not be taken to mean that there is a uniformly high population throughout estuaries. Patchiness is a common feature, with rapid differences in population densities occurring over short distances. Great care has to be taken when sampling with standard nets to ensure the correct interpretation of the distribution of the various elements in the plankton.

In temperate latitudes there are dramatic seasonal changes in both productivity – the rate of increase of the weight of biological material per unit area per day (mg/m^2/d) – and standing crop – the weight of biological material per unit area (mg/m^2). The classic pattern for coastal waters involves a low phyto- and zoo-plankton abundance and production during the winter. In spring, there is a very rapid production of phytoplankton, the

16.19

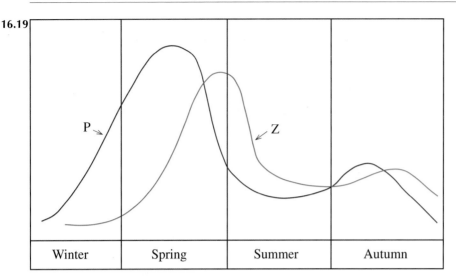

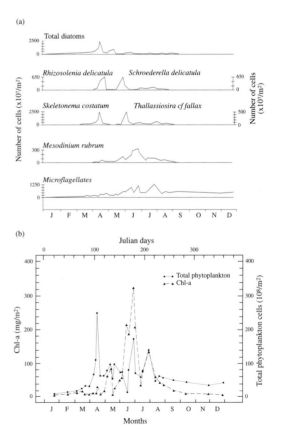

Figure 16.19 Idealised relationship between phytoplankton (P) and meso-zooplankton (Z) seasonal abundances in temperate coastal waters. The temporal pattern does not represent the relative abundances of phyto- and zoo-plankton.

spring bloom, which is followed by a summer decline in the standing crop of phytoplankton as the population of zooplankton grazers rises (*Figure 16.19*; see also Chapters 6 and 15). These grazing zooplankton decline in turn during the summer as a result of food limitation, known as bottom-up control, and/or the development of predator populations, known as top-down control. Often, both phytoplankton and zooplankton exhibit a smaller peak of abundance in coastal waters in the autumn. As previously indicated, the overall situation in estuaries tends to be more complex, as can be illustrated with reference to Southampton Water in England.

Here, riverine and other inputs ensure that nutrient supply is less limiting than in offshore waters. In consequence, the spring phytoplankton bloom is prolonged and is represented by a succession of different species; typically, there is an initial succession of diatom species, which are numerically dominant in March–April. Following on is an increase in the dinoflagellate and microflagellate community, which continues at high densities throughout the summer until September–October [*Figure 16.20(a)*]. Overall, therefore, despite fluctuations in abundance associated with the species succession, there is an essentially prolonged bloom of primary producers through from spring to autumn. This extended bloom is well-illustrated by the levels of photosynthetic pigment chlorophyll-a in the water column over the annual cycle [*Figure 16.20(b)*].

Also, often observed coincident with peak chlorophyll levels in early summer are blooms of the photosynthetic ciliate *Mesodinium rubrum*.

The high levels of primary production throughout the spring–summer period mean that the herbivore-dominated zooplankton community is probably not significantly food-limited during this time. Nevertheless, these zooplankton populations do show both spatial and seasonal variations.

Two elements contribute to the zooplankton

16.20

Figure 16.20 (a) Phytoplankton species succession in the middle part of Southampton Water, England, indicating the dominant Dinoflagellate species and total diatom, microflagellate, and *Mesodinium rubrum* temporal signals. The cell abundances are integrated throughout the upper 6 m of the water column. (b) Seasonal pattern of chlorophyll-a and total phytoplankton cell density.

Figure 16.21 Seasonal patterns of meso-zooplankton in the mid-estuary area of Southampton Water, England, for 1990–1991. The community is presented in terms of (a) abundance and (b) the percentage composition of the most significant elements of the meso-zooplankton.

(a)

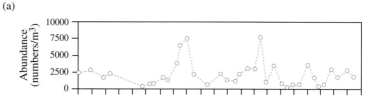

(b)

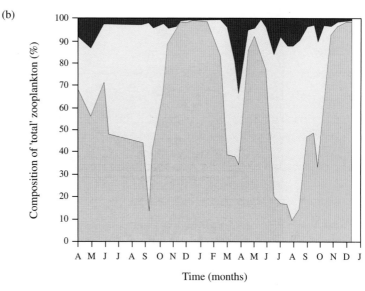

grazers, the 'holoplankton', which spend their whole life cycle in the plankton, and the 'meroplankton', which are the larval stages of benthic forms. Overall, in Southampton Water the holoplankton, and in particular the copepod crustaceans, dominate for much of the year.

Following low winter densities of zooplankton there is a rapid rise in abundance throughout the estuary in spring coinciding with increasing temperatures and abundance of phytoplankton. The zooplankton population then declines in June through to August, often by as much as 75–90%. A second and smaller peak of abundance sometimes occurs during autumn, between September and late October. The calanoid copepods, which typically constitute 80–95% of the meso-zooplankton, mirror the pattern of the overall population change[18] (*Figure 16.21*). The nauplii (copepod and barnacle crustacean) larvae population are the second most dominant component of the zooplankton of Southampton Water. Their numbers are relatively high throughout the spring and summer, but do not exhibit the summer decline shown by the adult copepods. The nauplii, therefore, can make up 45–80% of total zooplankton abundance during the summer (*Figure 16.21*). The remainder of the meso-zooplankton is made up of meroplanktonic larvae of benthic polychaetes, molluscs, and crustaceans, and of pelagic gelatinous species, all of which appear and disappear sporadically during the spring through to autumn.

While this basic temporal pattern is presented throughout the estuary, there is also a spatial component to the overall community structure, with the lower part of the estuary exhibiting a more heterogeneous species composition than the upper. Changes in abundance, as well as in species composition, are also apparent[11]. Thus, while copepods dominate the upper estuary, their numbers decline toward the mouth where they represent only 50% of the total zooplankton population.

Considering, therefore, that the sustained phytoplankton bloom should provide sufficient food for the meso-zooplankton, it is probable that factors such as food limitation do not play a major part in regulating the meso-zooplankton community. What can be deduced about other factors which might lead to the summer decline of the calanoid copepods, in particular within the estuary?

Two factors which could contribute to this summer decline are, first, the *Mesodinium* red tide and, second, predation by gelatinous species.

Figure 16.22 LANDSAT image of *Mesodinium* red-tide bloom in the upper reaches of Southampton Water, England; the sub-surface bloom is represented by a red–purple signal. Sites 1 and 2 refer to typical upper and mid-estuary sampling stations. The estuary runs from NW–SE, with the Solent toward the bottom right.

Coincident with the decline in copepod numbers is the appearance in the estuary, in high numbers, of the photosynthetic ciliate *Mesodinium rubrum*, which forms a non-toxic red tide bloom, particularly in the upper estuary[7] (*Figure 16.22*). During late May to early August, *Mesodinium* is patchily distributed throughout the estuary. In bloom conditions, particularly in June and July, cell densities can rise to more than 5000 cells/ml, and in these circumstances zooplankton density and distribution may be influenced by the effect of *Mesodinium* on the water column. *Mesodinium* undergoes diurnal vertical migrations. During the day it is typically located in a layer just below the surface and its photosynthetic activity can supersaturate the upper water column with oxygen. Downward dispersal of the bloom at night, coupled with respiration, can, in contrast, result in the deeper waters becoming sub-oxic [*Figure 16.23(a)*]. There is substantial anecdotal evidence that these red tides result in reduction of fish numbers. Critical analysis of the situation suggests, however, that *Mesodinium* does not significantly exclude either zooplankton or juvenile fish from the vicinity of blooms[5]. There is, indeed, no obvious inverse relationship between zooplankton numbers and either *Mesodinium* cell density or the reduced oxygen profile in the water column, as can be seen by comparing *Figures 16.23(a)* and *16.23(b)*. Any cause-and-effect relationship between *Mesodinium* blooms and summer decline in zooplankton can therefore be questioned. Red tides, caused by other species, have, however, produced more severe effects elsewhere (see Chapter 22).

Also coincident with the period of reduced meso-zooplankton abundance during the summer is the appearance of populations of gelatinous predators, jellyfish, and ctenophores, which have distinct peaks of abundance between March and late September[11]. These organisms are acknowledged as important predators in the marine plankton ecosystem[1,8]. All the gelatinous predators found in Southampton Water are generalist feeders, in that they catch and consume a wide range of both species and prey size.

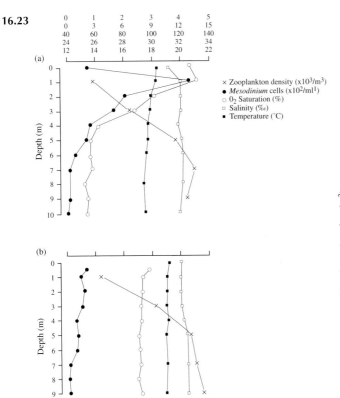

× Zooplankton density (x10³/m³)
● *Mesodinium* cells (x10²/ml¹)
○ O₂ Saturation (%)
□ Salinity (‰)
■ Temperature (°C)

Figure 16.23 Vertical profile of the water column during (a) *Mesodinium* bloom conditions and (b) non-bloom conditions. The nature of the distribution pattern and abundance of meso-zooplankton is similar in both situations.

Figure 16.24 Typical seasonal abundance pattern of gelatinous meso-zooplankton predators in mid and upper Southampton Water, England. Species are not drawn to scale.

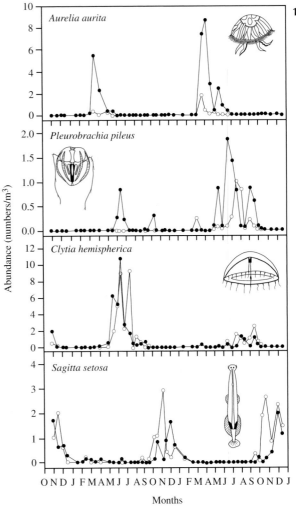

● Upper Southampton Water
○ Mid Southampton Water

The jellyfish *Aurelia aurita* is invariably the first to appear in the plankton, being seen first as the young ephyra stage in early March and growing on until the end of June[12] (*Figure 16.24*). This species has a large body size when mature and typically accounts for a large proportion of the gelatinous biomass of the upper estuary (85–90%). Following *Aurelia*, the ctenophore *Pleurobrachia* and the jellyfish *Clytia* appear in the water column between late April and mid-September. The chaetognath *Sagitta* is also present sporadically throughout much of the year, but is really an autumn–winter predator in the estuary.

Throughout the estuary the greatest period of gelatinous production is between April and July with a peak in May. At this time it has been calculated that the impact of these species is such that they remove some 5–25% of the copepod zooplankton on a normal basis and up to as much as 70% when there are predator swarms[12]. The evidence therefore suggests that the gelatinous predators play a significant top-down role in structuring the decline in meso-zooplankton abundance in the summer.

Conclusion

At the start of this chapter it was claimed that estuaries, coastal zone, and shelf seas present some of the most demanding habitats on Earth. Touching, as we have, on only one aspect of the physiological problems, one aspect of benthic existence, and the population sequencing of one estuary does but poor justice to the theme. Every neritic (coastal and shelf-sea) region has its unique conditions and spatial and temporal biotic responses. The elucidation of the processes that enable the biota to adapt and respond to such a range of conditions continues to provide the excitement, challenge, and stimulus for further study of coastal waters in all their aspects, both now and in the foreseeable future.

General References

Gilles, R. (ed.) (1979), *Mechanisms of Osmoregulation in Animals: Maintenance of Cell Volume*, John Wiley and Sons, Chichester, 667 pp.

Gray, J.S. (1974), Animal–sediment relationships, *Ann. Rev. Oceanogr. Marine Biol.*, **12**, 223–261.

Gupta, B.L., Moreton, R.B., Oschman, J.L., and Wall, B.J. (eds) (1977), *Transport of Ions and Water in Animals*, Academic Press, London, 817 pp.

Lopez, G., Taghon, G., and Levinton, J. (eds) (1989), *Ecology of Marine Deposit Feeders*, Lecture Notes on Coastal and Estuarine Studies, **31**, Springer-Verlag, Berlin, 322 pp.

Parsons, T., Takahashi, M., and Hargrave, B. (1984), *Biological Oceanographic Processes*, 3rd edn, Pergamon Press, Oxford, 330 pp.

Pequeux, A., Giles, R., and Bolis, L. (1984), *Osmoregulation in Estuarine and Marine Animals*, Springer-Verlag, Berlin, 221 pp.

Raymont, J.E.G. (1980), *Plankton and Productivity in the Oceans*, Vol. 1, *Phytoplankton*, Pergamon Press, Oxford, 489 pp.

Raymont, J.E.G. (1983), *Plankton and Productivity in the Oceans*, Vol. 2, *Zooplankton*, Pergamon Press, Oxford, 824 pp.

Remane, A. and Schlieper, C. (1971), *The Biology of Brackish Water*, John Wiley and Sons, New York, 372 pp.

Sommer, U. (ed.) (1989), *Plankton Ecology – Succession in Plankton Communities*, Springer-Verlag, Berlin, 369 pp.

Williams, R. (1984), An overview of secondary production in pelagic ecosystems, in *Flows of Energy and Materials in Marine Ecosystems*, Fasham, M.J. (ed.), NATO Conference Series IV, Vol. 13, Plenum Press, New York, pp 361–405.

Yingst, J.Y. and Rhoads, D.C. (1980), The role of bioturbation in the enhancement of microbial turnover rates in marine sediments, in *Marine Benthic Dynamics*, Tenore, K.R. and Coull, B.C. (eds), Univ. S. Carolina Press, Columbia, pp 407–422.

References

1. Alldredge, A.L. (1989), The quantitative significance of gelatinous zooplankton as pelagic consumers, in *Flows of Energy and Materials in Marine Ecosystems*, Fasham, M.J. (ed.), NATO Conference Series IV, Vol. 13, Plenum Press, New York, pp 407–433.

2. Cadee, G.C. (1976), Sediment reworking by *Arenicola marina* on tidal flats in the Dutch Wadden Sea, *Neth. J. Res.*, **10**, 440–460.

3. Ekman, J.E., Nowell, A.R.M., and Jumars, P.A. (1981), Sediment destabilisation by animal tubes, *J. Marine Res.*, **39**, 361–374.

4. Gupta, B.L., Moreton, R.B., Oschman, J.L., and Wall, B.J. (eds) (1977), *Transport of Ions and Water in Animals*, Academic Press, London, 817 pp.

5. Hayes, G., Purdie, D.A., and Williams, J.A. (1989), The distribution of Ichthyoplankton in Southampton Water in response to low oxygen levels produced by a *Mesodinium rubrum* bloom, *J. Fish. Biol.*, **34**, 811–813.

6. Jerling, H.L. and Wooldridge, T.H. (1991), Population dynamics and estimates of production for the Calanoid Copepod *Pseudodiaptomus hessei* in a warm temperate estuary, *Estuar. Coast. Shelf Sci.*, **33**, 121–135.

7. Kifle, D. and Purdie, D.A. (1993), The seasonal abundance of the phototrophic ciliate *Mesodinium rubrum* in Southampton Water, England, *J. Plank. Res.*, **15**, 823–833.

8. Larson, R.J. (1986), Seasonal changes in the standing stocks, growth rates and production rates of gelatinous predators in Saanich Inlet, British Columbia, *Marine Ecol. Prog. Ser.*, **33**, 89–98.

9. Lockwood, A.P.M. (1961), The urine of *Gammarus duebeni* and *G. pulex*, *J. Exp. Biol.*, **38**, 647–658.

10. Lockwood, A.P.M., Inman, C.B.E., and Courtenay, T.H. (1973), The influence of environmental salinity on the water fluxes of the amphipod crustacean *Gammarus duebeni*, *J. Exp. Biol.*, **58**, 137–148.

11. Lucas, C.H. and Williams, J.A. (1992), A preliminary examination of the seasonal succession of gelatinous predators within the zooplankton community of Southampton Water, *Porcupine Newsletter*, **5**, 77–83.

12. Lucas, C.H. and Williams, J.A. (1994), Population dynamics of the Scyphomedusae *Aurelia aurita* in Southampton Water, *J. Plank. Res.*, **16**, 879–895.

13. Olsson, P., Graneli, E., Carlsson, P., and Abreu, P. (1992), Structuring of a post-spring phytoplankton community by manipulation of trophic interactions, *J. Exp. Marine Biol. Ecol.*, **158**, 249–266.

CHAPTER 17:

Artificial Reefs

K.J. Collins and A.C. Jensen

Introduction

An artificial reef is a structure placed in the sea, either to protect or enhance the existing habitat or to create a new type of habitat for marine animals and plants. This may lead to the enhancement of local fisheries. The reef design, material used, and the site chosen are all influenced by the required function of the reef.

Natural reefs

Many people's image of the marine environment is that of a beach extending underwater, the sea bed being formed from large expanses of sand. Reality is different; sea floors range from fine muds through sand and gravel to boulders and rock. While on land, rocks and cliffs are among the more barren areas, underwater they are covered in a profusion of marine organisms. Fixed seaweeds need light to grow and hence are restricted to shallow water. More widely distributed are sedentary or encrusting animals, such as sponges, hydroids, anemones, and bryozoans. By living on elevated hard surfaces, exposed to currents, these animals increase their opportunities to catch plankton while minimising the smothering effect of settling sedi-ments.

Most animals with sessile adult stages produce pelagic eggs and larvae that live in the water column in order to distribute their offspring to areas suitable for colonisation. Consequently, any new hard surfaces are quickly occupied. Mid-water fish are often attracted to rocks or wrecks that give some shelter from tidal currents and predators, and increase feeding potential. Where such hard substratum contains crevices, these can also provide shelters for crabs and lobsters. Rocks and reefs therefore offer valuable settlement sites for seaweeds and invertebrates, and shelter for mobile bottom-dwelling forms and fish, so providing additional feeding opportunities for many animals. Any extension of such a habitat, say by the deployment of an artificial reef, increases the quantity of animals able to live within a given area of sea bed.

Artificial reefs

The use of artificial reefs as fishing sites has a long history, presumably arising from chance observa-

tions of fish being attracted to objects placed in the water. In Sardinia, tuna have been caught for hundreds of years in complex floating net traps weighted with stones. At the end of each season the stones are cut loose and fall to the sea bed. Fishermen noticed how many fish species were attracted to these accumulating piles of weights. Similarly, accidental shipwrecks have long been popular fishing sites. Such structures (*Figures 17.1–17.3*) provide niches for marine life, attracting many fish species, which take advantage of the ship's structure to shelter from currents and predators and to find food.

17.1

Figure 17.1 Fish are attracted to shipwrecks, which provide shelter from currents and predators. Here, shoals of sardines swarm on the wreck of a freighter, near Sardinia, Italy.

17.2

Figure 17.2 Wrecks also provide settlement substrate for epibiota, as seen by a diver looking at growth on the railings of a wrecked freighter, near Sardinia, Italy.

Artificial reefs are incidentally created by engineering works, such as harbour breakwaters (*Figures 17.4–17.6*) and supports for bridges (which are often built on muddy or sandy sea beds). Such structures provide habitats for species that could not live on the open sea bed, and attract mobile benthic species, such as cryptic fish, lobsters, and crabs. Oil production platforms also attract numerous fish species and provide a good settlement surface for animals and plants. In the Gulf of Mexico, the states of Louisiana and Texas are involved in an active artificial-reef creation programme utilising obsolete platforms. These are either relocated to shallow water and sunk, or toppled *in situ*. The money saved by the oil companies in disposal costs is ploughed back into the management of these structures for recreational fishing – over 4000 structures are available to anglers. In the UK there is interest in using North Sea oil production platforms as artificial reefs, now that they are beginning to reach the end of their useful working life (25–30 years).

17.3

Figure 17.3 The excavation of the historic wreck of the Mary Rose (sunk 1545) provided a range of new habitats in the muds and silts of the Solent (southern England). Large shoals of pouting congregated around the excavated timbers.

Figure 17.4 Coastal defence structures, such as these concrete tetrapods off Funchal, Madeira, provide new surfaces for colonisation by algae and encrusting animals.

Figure 17.5 This breakwater in Monterey, California, provides a convenient spot for Californian sea-lions to bask in the sun and groom. (Courtesy of Jane Jensen.)

Figure 17.6 Below water the breakwater boulders provide good anchorage for a giant kelp forest (Monterey, California). The boulders are also settled by a profusion of sessile animals, such as these colourful solitary corals.

17.7

Figure 17.7 Extensive government support for artificial reef construction in Japan has led to the development of complex artificial reef structures. Favoured materials are concrete, steel, and plastics. This unit is made of concrete containing coal fly ash. (Courtesy of Tatsuo Suzuki, Hazama Corporation, Japan.)

Fisheries

Artificial reefs around the world

Probably the majority of artificial reefs have been built by artisan fishermen in tropical countries. The purpose of such a reef is to increase catches in local fishing grounds using simple, readily available materials, such as rocks, trees, bamboo, and scrap tyres. For example, 1600 pyramid bamboo modules have been made and deployed in clusters of 50 by local fishermen in the central Visayan Islands of the Philippines. The catches of a wide range of fish from these reefs exceeded construction costs within the first year, providing an annual harvest of 8 kg/m^2.

The Japanese are the world leaders in artificial reef technology for commercial fishery enhancement and have been creating artificial reefs since at least the eighteenth century. Currently, Japan is in the third phase of artificial reef development, that of creating entire fishing grounds where there had been none before. This programme commenced in 1974 with the goal of diverting Japanese fishing effort from distant water fishing to mariculture and resource management in Japanese waters. Government investment has been substantial; for example, in 1988 US$150 million was allocated to subsidise the construction of 2.2 x 10^6 m^3 of fishing reefs. Over 17 x 10^6 m^3 of reefs have been deployed, covering almost 10% of the sea bed shallower than 200 m. Materials used are generally 'prime materials', i.e., concrete, steel, and glass-reinforced plastic, although some stabilised ash is also now used. The engineering and design aspects of Japanese artificial reefs are well refined (*Figure 17.7*). Quality standards regarding building materials, design, location, and construction have been produced, which must be met if structures are to qualify for government certification and therefore a subsidy toward deployment costs. However, the biological appraisal of artificial reef performance is not so well advanced. Some have concluded that there are insufficient biological and economic data to enable judgement of the cost-effectiveness of many of the operations.

In the US, the artificial reef programmes of many Maritime States are run for the benefit of recreational sports fishing, SCUBA (Self-Contained Underwater Breathing Apparatus) diving, commercial fishing, waste disposal, and environmental mitigation. An example of the latter is seen in the construction of the Pendleton artificial reef by the Southern California Edison company. This 120 hectare reef was built offshore to compensate for the possible loss of a giant kelp *(Macrocystis)* habitat due to power station construction.

American experience of reef construction dates back over 100 years, in which time a variety of (mostly waste) materials have been used, including concrete, rock, construction rubble, scrap tyres, cars, railway carriages, and ships. The US has a national artificial-reef plan, but no government funding commitment. Funding has come from the Federal Aid in Sport Fish Restoration Program, which may provide up to 75% of reef construction costs, with individual States providing the rest (in 1987, more than US$140 million was provided by the Federal Aid Program). Besides the government plan, individual states have local artificial reef programmes and Interstate agreements promote a rational approach to the deployment of artificial reefs. Florida, with over 200 sites, has the greatest number of artificial reefs in the US.

Elsewhere, artificial reefs have been developed according to local requirements, using materials judged to be suitable. Only Japan and the US have a national development plan. Countries such as Malaysia and the Philippines use waste tyres to build many of their artificial reefs. This is also the case in Australia, where frequently reefs have been built from 'materials of opportunity', such as tyres and redundant ships. These reefs are used primarily as a focus for recreational angling, with some SCUBA diving. In Taiwan many fishing vessels (made obsolete by government policy to reduce the size of the fishing fleet) were sunk to provide new habitats.

In Europe artificial reefs were pioneered along the Mediterranean coast in the late 1960s. At present, most reefs are still associated with scientific research. Italy, France, and Spain have been the most active reef-building countries since 1970. Currently, Spain is placing more artificial reefs into its coastal waters than any other European country (about 100 reef complexes are in place or planned). Reef building has, until recently, been carried out nationally, with little cross-border cooperation. This is changing; in 1991, Italian artificial reef scientists formed an Italian reef group to encourage liaison between research groups, and an association of Mediterranean artificial reef scientists now exists. Current initiatives include a network of European artificial-reef research scientists to establish a coordinated direction for reef research within the European Union.

Artificial reefs for fisheries enhancement

Many fish species are attracted to objects in the water, which provide an orientation point and, possibly, some shelter from sunlight or currents. This behaviour has been exploited by fishermen using fish-attracting devices (FADs). Fish attracted to the FAD are more easily caught because the fishing effort can be concentrated in one place. For example, artisan fishermen may use simple bamboo rafts to support bunches of coconut palm leaves in the water below. More modern and complex FADs are often reminiscent of kites that are 'flown' in mid-water. Such devices make existing fish stocks easier to exploit; in the Philippines, increases in tuna catches from under 10,000 tonnes in 1971 to 266,000 tonnes in 1986 are attributed to the use of FADs. While increasing catches, FADs make no contribution to biological productivity and so must be used with caution if overexploitation is to be avoided.

Fisheries enhancement, especially of fin-fish fisheries, is probably the area in which artificial reefs meet their greatest challenge. The attractiveness of a reef to many, although not all, species of fish, both bottom dwelling and free swimming, is in little doubt. However, the contribution to the overall biological productivity of free-swimming fish species is difficult to assess. The use of artificial reefs to attract and concentrate a population of fish, so facilitating exploitation, is a powerful tool in fisheries management; one that can be used to increase exploitation or, by placing a reef in a protected area, provide sanctuary.

Artificial reefs essentially provide habitat. If the fish attracted to the reef gain some positive advantage from the reef structure, then this can be set against the negative impact that fishing has on fish numbers. The greater the benefit from the provision of a reef (e.g., if a fish uses the reef as a feeding or spawning site), the more likely it is that the reef will provide some net gain to the fish population. If this translates into an increase in the numbers of fish, and therefore an increase in the total number available for exploitation, the fishery population can be said to be enhanced. Due to the mobility of many commercially valuable fish species this is very difficult to quantify. With territorial species, which spend part or all their life cycle on a reef, the value of the structure is more easily defined.

Examples of non-commercial fish species that apparently benefit from association with an artificial reef can be seen on one in England (in Poole Bay, central south coast). Large shoals of pouting (*Trisopterus luscus*), a species which spawns offshore, congregate around the reef during the day (*Figure 17.8*). They disperse at night to feed on

17.8

Figure 17.8 Shoals of pouting congregate around units of the Poole Bay artificial reef, UK, during the day, gaining shelter from tidal currents.

Figure 17.9 In early summer, male corkwing wrasse build nests of seaweed between the blocks forming the Poole Bay artificial reef. They defend a territory around the nest and are not easily intimidated by observing scientists.

Figure 17.10 The corkwing wrasse constructs a complex nest from seaweeds, seen here in its mouth, taken from algae growing on the reef and also drifting past in the current.

nocturnally active small crustaceans which inhabit the surrounding sea bed. The main advantage that they gain from the reef is shelter from tidal currents, reducing the amount of energy needed for swimming, and so allowing increased growth. Another fish species, the corkwing wrasse *Crenilabrus melops*, is territorial, staying within, or close to, the reef structure. In May–June each year since 1990 males have built seaweed nests between the reef blocks (*Figures 17.9* and *17.10*). Females lay eggs within these nests, which are maintained and guarded by the male until the eggs hatch some 3–4 weeks later. Wrasse have been observed feeding on the reef epifauna (animals living on hard surfaces), such as barnacles. For this species the reef provides food, shelter, and a site for reproduction.

The potential fishery value of artificial reefs is more easily demonstrated with less mobile, but valuable commercial species, such as molluscs and crustacea. Artificial reefs in the Adriatic Sea provide settlement sites for large numbers of mussels, an important commercial species throughout Europe. Because mussels produce such large num-

bers of larvae, the artificial reefs are not being colonised at the expense of other rocky areas, but in addition to them. In the Adriatic, this fisheries enhancement case appears to be well-proven; mussel harvests boost the commercial catches from artificial reefs, giving a three-fold return on deployment costs over 7 years. Net proceeds for fishermen operating within the artificial reefs have been shown to be 2.5 times that of activity outside the reefs. Elsewhere in Europe, attention has focused on lobsters as a commercial species under increasing fishing pressure that may benefit from habitat provision in the form of artificial reefs.

Case study: artificial reefs and lobsters

Man-made shelters ('pesqueros' in Cuba, 'casitas cubanas' in Mexico), have been used to provide small, temporary refuges for spiny lobsters (*Panulirus argus*). These function as a focus for fishing effort, rather like a FAD for fin fish. Canada, Israel, and the UK, interested in other species of lobsters, have focused attention on artificial reefs as a specific lobster habitat. Canada built the first artificial reef specifically for lobster research in 1965. Over the following 8 years the lobster population of the artificial reef was monitored by diving scientists. The reef was initially colonised by large specimens of the American clawed lobster (*Homarus americanus*), which were thought to have outgrown their burrows, so being forced to roam to seek new shelter. By 1973 the size-frequency distribution of the artificial reef population was similar to that on natural reefs in the area. It was concluded that the standing crop on the reef might be increased by a different pattern of rocks. However, a cheaper source of reef material or a multiple-use reef was required before an artificial reef could be considered an economically viable proposition, so the research was halted.

In Israel, efforts focused on the non-clawed slipper lobster, *Scyllarides latus*, an important commercial species found off the Mediterranean coast. Research showed that slipper lobsters preferred horizontal shelters with two narrow entrances on the lower portion of the reef. Using shelters to hide from predators is believed to be an important defense mechanism for these animals, so the presence of the artificial reef provided a new and suitable habitat for colonisation. Slipper lobsters migrate into deeper water as the inshore water temperature rises in summer, but tagged individuals were seen to return to a coastal tyre reef during spring (*Figures 17.11* and *17.12*) over the project period of 3 years. In the long-term, populations of these heavily exploited animals could be protected against fishing effort by building appropriately designed artificial reefs in protected areas, such as underwater parks and reserves.

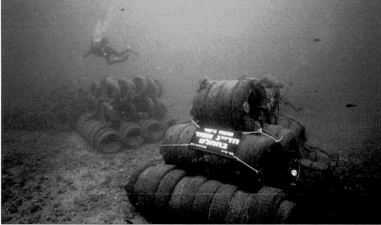

17.11

Figure 17.11 Scrap tyres provide a resilient material for constructing artificial reefs. However the tyres must be securely bound together and weighted, as in this example off the Israeli Mediterranean coast at Haifa. (Courtesy of Ehud Spanier, Centre for Martime Studies, University of Haifa, Israel.)

17.12

Figure 17.12 The spaces within tyres provide ideal shelter for fish such as these squirrel fish. This Red Sea species has migrated through the Suez canal into the Mediterranean Sea. (Courtesy of Ehud Spanier, Centre for Martime Studies, University of Haifa, Israel.)

Work continued since 1989 on the experimental reef in Poole Bay, England, found that lobsters (*Homarus gammarus*; *Figure 17.13*) appeared on the reef within 3 weeks of its deployment. Tagging studies were initiated in 1990, and data to June 1994 show that lobsters have found the artificial reef a suitable long-term habitat, the longest period of residence standing at 4 years. This can be compared to a maximum age in the range of 10–20 years, sexual maturity being reached at 4–5 years of age. Conventional tagging of lobsters below the fishery minimum landing size of 85 mm carapace length (ca 250 mm total length) in a nearby fishery revealed that these lobsters do not undertake any seasonal migration, and the range of most movements is less than 4 km in magnitude. The use of a novel electromagnetic telemetry system has started to reveal complex local movement behaviour. These data reveal that lobsters are mostly active at night and frequently change their daytime shelter. The internal galleries and tunnels of the conical reef units (1 m high, 4 m diameter) made from randomly stacked, cement-stabilised, pulverised fuel ash blocks (40x40x20 cm) were often occupied by more than one lobster. An animal was also monitored leaving the reef site for up to 3 weeks and

Figure 17.13 The randomly stacked blocks of the Poole Bay artificial reef provide a wide variety of crevices and tunnels. Here a lobster emerges from the entrance to a gallery within the reef unit where it spends most of the day.

then returning. Only the larval stage of the lobsters life cycle is planktonic; the juvenile and adult lobster live either in burrows in the sediment or in shelters in rocky sea bed for the rest of their lives. Diver observations and evidence from pot-caught lobsters suggests that the reef can support all aspects of the benthic life cycle; berried (egg carrying) females utilise the shelters and release their larvae from them, some reproducing more than once on the reef. Lobster larvae have been taken from the waters above the artificial reef and a wide size-range of juvenile and adult animals has been captured and/or observed by diving scientists.

Artificial reefs have been shown to support effectively three species of commercially important lobster. Research in the UK has shown good survival of hatchery-reared juvenile lobsters released into the wild and subsequently recruited into the fishery. It seems feasible that an artificial reef could be 'seeded' with hatchery-reared juveniles and that they would live to become part of the fishery. At present, the maximum densities of lobster population that can be achieved are not established, but data for *H. americanus* suggests that the Canadian quarry rock reef supported one lobster per 6 m² while the Poole Bay reef is thought to hold one *H. gammarus* per 2 m². Since neither structure was designed to maximise lobster habitat there is a potential for improvement. The artificial reef densities can be favourably compared to results (in the order of one lobster per 30 m²) of diver surveys of natural reefs. Animal density is strongly correlated to the number of suitable shelters available. Lobster territorial behaviour also influences the usage of habitat. It seems more than possible that in the future a reef could be designed to provide lobster shelters in various sizes to minimise 'off-reef' movement, caused by the need to seek a new shelter after increasing in size following moulting. Already predictions have been made of the number and size of shelters in a reef made up of spherical boulders (a starting point calculation for more realistic material shapes), which can be linked to results describing the habitat requirements of lobsters.

Aquaculture

Artificial reefs can serve several roles in fisheries management, from the enhancement of wild fisheries to more intensive aquaculture systems.

Settlement of mussels on reef structures in the Adriatic is described above. Stock density can be intensified by adapting the well-established 'suspended rope' system of mussel cultivation, practised throughout the world, stringing seeded ropes of mussels between artificial reef units. Suspended rope is also used for kelp culture in Japan.

In Japan, specially designed artificial reefs are used for the culture of kelp, urchins, and abalone.

Combined systems for all three have been constructed where the kelp supplies food for the urchins and abalone. There is interest in aquaculture systems to develop abalone as a cultured species in Europe; artificial reefs may play a part in this.

Many salmon farms in northwest Europe have started to utilise the 'cleaning' capabilities of species of wrasse, such as the corkwing and goldsinny, to remove sea lice from the bodies of the captive salmon as an 'environmentally friendly' alternative to treating the fish with chemicals. Currently, most of the wrasse are caught in the wild, so localised populations are depleted. As wrasse are territorial rock-fish and spawn and nest in rocky habitats, there is a potential to artificially enhance wrasse habitat near salmon cages. This would provide a self-sustaining source of 'cleaning' fish near the site of salmon farms. The potential of artificial reefs and their epifaunal filter-feeding community to act as 'biofilters' to remove waste (uneaten food and faecal material) close to aquaculture facilities in the Baltic Sea is being investigated. Mussels, acting as filters, could play an important role in lessening the impact of aquaculture on the environment.

Biodiversity Management

The contribution that artificial reefs can make in biodiversity (see Chapter 15) management is that of habitat manipulation. This has the potential to increase the number of species in an area and provide purpose-designed habitats for target species.

Provision of new habitat

Artificial reefs are usually constructed to provide elevated hard substrata where formerly there was none. Artificial reefs mimic natural reefs, but can be built to provide greater surface area, elevation, current shadow/disturbance, or niches/crevices to favour target species.

An early European artificial reef study was that by the Association Monégasque pour la Protection de la Nature off Monaco in the 1970s. Artificial reefs (2 m³), made from hollow blocks or tiles cemented together, were laid on a muddy sea bed within a marine reserve. These attracted a good settlement of epifauna and provided a habitat and refuge for spiny lobster. Following on from this work, specifically designed cave habitats for rare red coral have been successfully developed by the Institut Océanographique at Monaco.

The placing of artificial reefs on a mud–sand sea bed smothers and kills infauna (animals living in the sediment) directly under the reef (see also Chapter 22, references to the Garrock Head waste-dumping ground). Some 100 infaunal species were replaced by more than 250 epifaunal species within two years in the Poole Bay artificial reef. The infaunal populations in undisturbed sediment around

Figure 17.14 After 4 years underwater, the Poole Bay artificial reef density and variety of colonising animals and plants closely resemble the biological communities seen on local natural reefs.

Figure 17.15 Colonisation of the Poole Bay artificial reef was rapid. Within 1 year the surfaces were entirely covered by a variety of hydroids, bryozoans, and sponges.

the reef were unaffected by reef deployment. The lowest estimates of epifaunal biomass per unit area were equivalent to that of the previous infaunal biomass. However, the greater surface area available on the reef (2.5–3.0 times ground area lost) gave a higher biomass estimate than from the sea bed that the reef covered. This new habitat led to a rapid increase in species numbers and diversity. After 4 years the epibiota (*Figures 17.14* and *17.15*) provided a food source for molluscs, lobsters, crabs, and fish; these, in turn, provided food for cuttlefish and predatory fish. The reef is providing a valuable site for reproduction, such as nesting

Figure 17.16 Artisan fishing communities in Kerala, Southern India, have used locally available materials, such as bamboo, to construct artificial reefs which are deployed on their village fishing grounds immediately offshore. (Courtesy of Steve Creech, Hampshire, England.)

Figure 17.17 Smaller concrete slabs, cast in the sand on the beach, have been assembled into artificial reef units and are being paddled out to sea balanced on canoes made from three tree trunks lashed together (CWARP, Coal Waste Artificial Reef Program). (Courtesy of Steve Creech, Hampshire, England.)

by corkwing wrasse or the laying of egg masses by whelks (*Buccinum undatum*).

Restoration of damaged habitat

Modern techniques have increased the environmental impacts of fishing; overexploitation of stocks by modern, powerful vessels and heavier fishing gears (such as 'rockhopper' trawls; it is estimated that every square metre of the North Sea is trawled 3–5 times annually), have led to the destruction of some sea-bed habitats. As an example, in southwest India development aid was provided to equip fishermen with trawlers to increase fish harvests. The coastline of Kerala province is one of the most densely populated areas of India. Artisan fishing communities depend on catches immediately off the beach on which they live, up to 3 km offshore, the range of their log canoes. The trawlers, based in the northern part of the province, decimated the artisan fisheries further south, which, as catches declined, were no longer economic to operate. In an attempt to restore damaged fishing grounds the coastal communities have used a variety of local materials (stones, cast concrete and bamboo, *Figures 17.16* and *17.17*), with some success, to restore bottom-habitat diversity and fish catches.

In the Maldives, the lack of construction materials has led to the use of coral (as blocks and aggregate) for building on the low-lying islands. However, the loss of the coral reefs that provide a living coastal defence barrier ultimately threatens the entire archipelago. In an experiment, concrete structures normally used for coastal defence in Europe have been laid, and living corals transplanted onto them in an attempt to re-establish coral reefs. Corals have grown and fish have been attracted back to what had become a coral rubble desert.

It seems ironic that concrete should be used to restore the reef destroyed in the quest for building materials.

Around the UK coast there are several offshore sites where waste could be dumped under licence (see also Chapter 22). Off Blythe (northeast coast of England), coal fly-ash from coal-burning power stations has been dumped for many years. This has resulted in the smothering of rocky outcrops with a sterile blanket of ash, covering ideal lobster and crab habitat, and fish-feeding grounds. While dumping is legislated to cease in 1995, it will be many years before this area will return to full productivity. One restoration method would be to continue dumping coal ash as stabilised ash blocks (tested in the Poole Bay reef) to restore habitat diversity and accelerate the recolonisation process.

Protection of existing habitat

In the Mediterranean Sea, most artificial reefs have been placed as nature conservation and/or habitat protection structures. At least 150 artificial reefs have been deployed for habitat protection by Israel, Italy, France, and Spain. Reef complexes range in size from a few hundreds of square metres to several square kilometres. Their prime role is to prevent the destruction of sea-grass meadows by trawling (*Figure 17.18*) – sea-grass is a valuable habitat for many commercial species of fish.

One example of such a protective reef or barrier is found off the new port of Loano (northwest Italy). Here, *Posidonia* sea-grass has been protected by an artificial barrier some 3 km^2 in area, deployed in 1986. Most of the anti-trawling barrier consists of 350 1.2 m^3 cubes placed on the sea bed. In the centre of the artificial barrier a number of concrete block 'pyramids' (*Figures 17.19–17.21*),

268

Figure 17.18 Illegal trawling of sensitive marine areas in Spain is being counteracted by these anti-trawling units. The steel railway lines protruding from the solid concrete blocks snag and destroy nets. (Courtesy of Technologia Ambiental, SA, Madrid, Spain.)

Figure 17.19 Another anti-trawling reef, off Loano, northwest Italy, uses pyramids of 2 m concrete cubes with holes to encourage cryptic animals.

Figure 17.20 The Loano artificial reef was deployed to protect an area of sea-grasses, which are important nursery grounds for juvenile fish.

Figure 17.21 The Loano artificial reef units provide a variety of habitat types. This massive bryozoan colony being inspected by the diver is typical of animals found on shaded overhangs on natural rock outcrops.

Figure 17.22 In Taiwan, high structural-strength coal-ash concrete has been developed to construct complex artificial reef units. These modules are about to be deployed from a barge. (Courtesy of Kwang-Tsao Shao, Institute of Zoology, Academica Sinica, Taipei, Taiwan, ROC.)

Figure 17.23 The open structure of the Taiwanese reef modules attracts shoals of fish. (Courtesy of Kwang-Tsao Shao, Institute of Zoology, Academica Sinica, Taipei, Taiwan, ROC.)

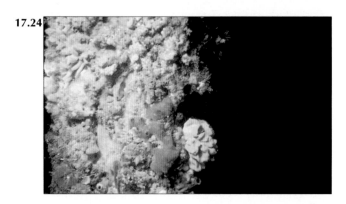

Figure 17.24 The surfaces of the coal-ash concrete modules are colonised with encrusting animals. (Courtesy of Kwang-Tsao Shao, Institute of Zoology, Academica Sinica, Taipei, Taiwan, ROC.)

made from 2 m³ blocks, provide additional height to the structure and are a focus of scientific research. The whole barrier construction is a marine reserve created by the Comune di Loano (local government), the regulations of which are supported by active policing of the reserve zone. Research scientists have documented the colonisation of the barrier structures by a variety of epifauna and epiflora and the arrival of shoals of fish attracted to the barrier. The reintroduction of spiny lobster and grouper into the area is planned.

Waste Materials

The use of waste materials in the construction of artificial reefs has economic advantages over the use of rock, concrete, and steel, but carries a potential environmental risk; there should be stringent environmental compatibility assessments before any material, including wastes, is allowed to be used for reef construction.

'Stabilisation' is one technique of ensuring that a finely divided material does not release potentially harmful compounds into the environment. The majority of stabilisation studies have involved cement stabilisation of coal ash, including pulverised fuel ash, and flue gas desulphurisation sludge containing gypsum. Initial research was pioneered in New York in the late 1970s, leading to the Coal Waste Artificial Reef Program. Besides addressing the engineering problems posed by the effect of sea water on solidified ash in sea water, this group laid the foundations for full environmental impact studies. These investigate the block chemistry, monitor colonisation of the material, and look for evidence of bio-accumulation in the reef-associated biota. Similar studies have followed in the UK, Italy, and most recently in Hong Kong. Coal ash combined with calcium carbonate residues has been studied in the Bohai Sea, China.

The largest scale experiments have taken place in Japan and Taiwan (*Figures 17.22–17.24*), where high structural-strength, stabilised coal-ash was developed. This enabled the construction of complex structures, as opposed to the simple blocks used in the other experimental ash reefs. There are ambitious plans in both countries to scale up the experiments to the full-scale utilisation of coal-fired power station outputs. The most exciting is the sea mount concept from Japan, which envisages the construction of structures in about 1000 m of water to divert nutrient-rich oceanic currents by placing a barrier in their path and so causing an upwelling of productive sea water (*Figure 17.25*).

Experiments with oil ash (from oil-fired power stations) stabilisation were undertaken in Florida (*Figure 17.26*), along similar lines to the coal ash studies. In New York, work has now been extended to the use of stabilised municipal waste incinera-

Figure 17.25 Artist's impression of a planned 'Super Ridge', to be deployed in deep waters off Japan to create an upwelling of nutrient-rich water, which will provide the basis for aquaculture. This will be constructed from the cement-stabilised coal-ash output of a coastal power station. (Courtesy of Tatsuo Suzuki, Hazama Corporation, Japan.)

17.25

tor ash. This latter material is also the subject of research in Florida, Bermuda, and the UK.

Ash from power stations is used in terrestrial building applications (road bases, as filler in cement, and in blocks), but a considerable quantity (about 50% in the UK) is still dumped (a typical 2000 MW coal-fired power station produces about 800,000 tonnes of ash per year). The research into construction of stabilised-ash artificial reefs shows that it would be environmentally acceptable to use this waste material in the sea, contributing positively to the resolution of the land-dumping problem.

While research is continuing into the use of ash waste, relatively little attention is being given to the effects of compounds leached from other waste materials, such as tyres, that have been used in artificial reef construction for several decades. Reef construction is an obvious solution to the disposal of large numbers of this extremely durable product, but the enthusiasm for solving a serious waste disposal problem must be tempered with the caution of environmental impact analysis.

The principal international legislation covering the deposition of waste and other matter in the ocean is the London Convention, 1992 (formerly the London Dumping Convention – see also Chapter 22). Placement of material for the construction of artificial reefs is not covered by the Convention. However, aware of the range of materials that have been used for such purposes, the London Convention Scientific Group has recommended that the guidance prepared for the interpretation of the Annexes to the Convention in relation to dumping at sea contains all the considerations needed to assess the placement of an artificial reef or structure. The newly revised regional

17.26

Figure 17.26 In this successor to the American CWARP programme, oil ash and coal ash were combined with cement to produce artificial reef blocks which were deployed off the eastern Florida coast. The surfaces were initially colonised by barnacles and the units attracted shoals of fish. (Courtesy of Fred Vose, Department of Fisheries and Aquatic Sciences, University of Florida, US.)

Oslo–Paris Convention, covering the northeast Atlantic area, has included placement of matter, such as for the construction of artificial reefs, within its purview and is establishing a set of technical guidelines for the practice.

The Future

Artificial reefs have a long history and are used world-wide to promote fisheries, but the basic understanding of their role in marine ecosystems is limited. There still is a fundamental argument as to whether artificial reefs increase the total productivity or simply attract stock from surrounding areas. It is likely that the answer is species specific; in Japan, no increase in flat-fish catches was caused by the deployment of reefs, just a redistribution of the fish population. On the other hand, octopus landings and catch rates were increased by 1.8 kg/m^3 of reef per year after 50,000 m^3 of artificial reefs were deployed. This was considered a net increase in total stock, as there was no evidence of redistribution. As octopus are often habitat-limited, the creation of new habitat with suitable dens was thought to be the major influence behind the stock increase.

As is apparent from our discussion, much of the work on artificial reefs has been largely descriptive, with some quantification of fish numbers to compare with natural reefs. On a world-wide basis, greater fish densities are reported from artificial rather than natural reefs. The species assemblages are usually similar for a given location. There is a need to improve on the quality of ecological data, measure the productivity of reefs, study the interrelationship of different types of organisms, and thus, eventually, determine the energetics of an artificial reef system.

We have described how artificial reefs can serve a number of roles in fisheries management, protecting or providing suitable habitat for particular stages in the life cycle of fishery species. There is still a need to understand the behaviour of commercially exploited species to find out how artificial reefs can best be constructed to supply their requirements. The utilisation of reefs in fisheries management will rely not only on scientific data, but also on the perceived social benefit of maintaining artisan fisheries, the cost-effectiveness of construction against value of catch, and national legislation (e.g., that allowing an organisation to own and manage a reef and sea bed).

Artificial reefs can be seen as a positive way of managing and/or contributing to the marine environment. We are likely to see an increased demand for reef construction to provide artificial reefs for recreational angling and SCUBA diving, commercial fishery enhancement and/or protection, aquaculture structures, and breakwaters for coastal protection.

General References

D'Itri, F.M. (ed.) (1986), *Artificial Reefs – Marine and Freshwater Applications*, Lewis Publications, Chelsea, Michigan, USA, 588 pp.

Duedall, I.W., Kester, D.R., and Park, P.K. (eds) (1985), *Wastes in the Oceans, Vol. 4: Energy Wastes in the Ocean*, John Wiley and Sons, New York, 818 pp.

Seaman, Jr., W. and Sprague, L.M. (1991), *Artificial Habitats for Marine and Freshwater Fisheries*, Academic Press, San Diego, 285 pp.

Stone, R.B. (1985), *National Artificial Reef Plan*, US Department of Commerce, National Oceanic and Atmospheric Administration, National Marine Fisheries Service, NOAA Tech. Memorandum NMFS OF.6, 39 pp + Appendices 49 pp.

Those wishing to learn more of the development and present state of knowledge about artificial reefs might also look at the proceedings of the following conferences:

Third International Artificial Reef Conference, 3–5 November 1983, Newport Beach, California, *Bull. Marine Sci.*, 37(1), 1–402, 1985.

Fourth International Conference on Artificial Habitats for Fisheries, 2–6 November 1987, Miami, Florida, *Bull. Marine Sci.*, 44(2), 527–1082, 1989.

Fifth International Conference on Artificial Habitats for Fisheries, 3–7 November, 1991, Long Beach, California, *Bull. Marine Sci.*, 55(2+3), 265–1359, 1994.

General Fisheries Council for the Mediterranean (1990), *Report of the First Session of the Working Group on Artificial Reefs and Mariculture*, 27–30 November 1989, Ancona, Italy, FAO Fisheries Report, No.428, FAO, Rome, 162 pp.

Indo-Pacific Fishery Commission (1991), *Papers Presented at the Symposium on Artificial Reefs and Fish Aggregating Devices as Tools for the Management of Marine Fishery Resources*, 14–17 May 1990, Colombo, Sri Lanka, RAPA Report 1991/11, 435 pp.

Waste material artificial reefs session, Second International Ocean Pollution Symposium, 4–8 October, 1993, Beijing, China, *Chem. Ecol.*, 10(1), 1–189 (1995).

CHAPTER 18:

Scientific Diving

J.J. Mallinson, A.C. Jensen, N.C. Flemming, and K.J. Collins

Introduction

A team of competent divers can offer field-work assistance to virtually all branches of oceanographic science, whether for pure or applied research, scientific archaeology, or fisheries management. Underwater, they measure and observe a wide variety of physical, geological, and biological processes, such as the microhabitat that exists on the underside of the polar ice (*Figure 18.1*), submarine freshwater springs billowing out from the sea bed, ice-age stalactites in drowned caves, and the most secret and intimate behaviour of marine animals. These processes, objects, and events may be difficult, more expensive, or impossible to observe otherwise. Scientific divers have worked in many parts of the world's oceans, including Rockall in the open Atlantic, the coral reefs of Belize, and even in the highest lakes in the world on the slopes of the Himalayas.

Scientific divers may themselves be scientists who have been trained to dive, or divers who have received training in scientific techniques. Divers are able to position accurately, deploy, and set-up, maintain *in situ*, and later recover delicate instrumentation on the sea bed; such instruments can remain in place for extended periods, monitoring natural processes in the least invasive manner. Some tasks may require specific skills, such as marine-life identification, recognising and mapping small-scale underwater topographic features, measuring artefacts in relation to a fixed datum point, or the underwater construction of apparatus.

The ideal of the scientific diver is to conduct observational and experimental work under the sea with the same accuracy as would be achieved in the laboratory or on land, making decisions and judgements in real time in response to events that occur and opportunities that arise.

The development and widespread availability of SCUBA (Self Contained Underwater Breathing Apparatus) since the late 1950s has expanded the scope of scientific activity underwater. Marine scientists use SCUBA as a means of reaching their research sites. The freedom to work untethered is of enormous value; the diver *in situ* is able to think and adapt to changing conditions, conducting work that is more complex and detailed than anything that can be achieved by the present genera-

18.1

Figure 18.1 At the British Antarctic Survey Signy Research Station, samples are collected from the unique environment beneath the ice. In order to dive under the winter sea-ice in Antarctica, a hole is cut using a chain saw and the diver descends on a lifeline to ensure a safe return to the access hole. The diving equipment is adjusted to safeguard against freezing up in the cold water. (Courtesy of G. Wilkinson, British Antarctic Survey, Cambridge, England.)

18.2

tion of underwater Remotely Operated Vehicles (ROVs) (*Figure 18.2*).

The International Scientific Committee of the World Federation of Underwater Activities (CMAS) encourages the highest standards of diving safety and promotes innovative research techniques for all extremes of diving conditions around the world. These standards are published by UNESCO[1]. Most scientific divers initially train as recreational divers with one of the international diver-training organisations, such as the British Sub-Aqua Club (BSAC), the Professional Association of Diving Instructors (PADI), or the National Association of Underwater Instructors (NAUI). All three are used world-wide, with different countries favouring one or another. Once quali-

fied, the diver can become involved in a variety of scientific activities, from primary observational surveys to very detailed tasks designed and carried out by research scientists. The level of training required and the need for supplementary training varies from country to country, but all scientific diving follows well-established codes of practice and safety procedures to ensure that the task in hand is undertaken in safety. An estimated 150,000 scientific dives take place world-wide each year.

Observation and Recording

Quantitative observation and recording requires accurate position fixing (see Chapter 20) and divers to make estimates of the occurrence and frequency of different phenomena. The divers use quadrats,

Figure 18.3 The small-scale spatial distribution of oysters is determined by counting individuals in each of the four 0.5 m² sections of a 1 m quadrat. The quadrat is moved across the sea bed until an area of 10 x 10 m has been covered, giving a statistically acceptable matrix of 400 units.

18.3

Figure 18.4 A diver's sketch of the sea bed off southern California shows the variation in biota from the beach of La Jolla Shores to the edge of the Scipps Canyon. There is little life in the surf zone of the big Pacific breakers, flat-fish and echinoderms extend across the sand flats, and macro-algae with associated cryptic fish and crustacea colonise the rocky outcrops at the start of the drop-off into the deep water of the canyon.

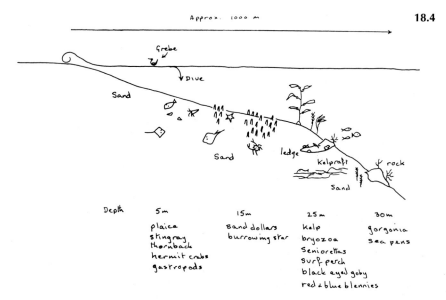

usually of 0.5 x 0.5 m, or 1.0 x 1.0 m, to concentrate their study and estimates into small manageable areas (*Figure 18.3*). Once the animal or feature has been identified and counted in several quadrats, it may be assumed that this density of occurrence can be extrapolated over a larger area of similar type. This method, in practice, requires the divers to measure many quadrats in a systematic pattern, or deliberately random pattern, in order to check that they have obtained a truly representative measure of the organism's or feature's variability.

On a slightly larger scale, communities on different substrata underwater can be recorded and mapped as they exist in nature (*Figure 18.4*) – a far cry from the mangled heaps of biota that are tipped out onto the deck of a survey vessel from a bottom trawling net. Divers can quantify their observations by counting individuals or applying a semi-quantitative scale to a range of organisms.

Divers can map extensive areas of sea bed by recording biological and topographical features as they swim along a tape measure laid across the sea bed. Such mapping has been used in coastal-zone management to ensure that uses, such as commercial fishing, tourism, and resource extraction (oil, aggregate, etc.), do not conflict with each other or cause excessive damage. This is of particular importance in ecologically sensitive areas; destruction by any activity cannot be assessed unless a baseline has been established (*Figure 18.5*).

The movement of the water itself is also a subject for direct examination *in situ*. Intense thermoclines develop at moderate depths in many parts of the world where the water temperature decreases rapidly through a vertical interval of a few tens of

Figure 18.5 Frontier Tanzania, a joint programme between the Society for Environmental Expeditions and the University of Dar es Salaam, has combined scientific expeditions involving young people with extensive mapping of the marine environment around Mafia Island (off the east coast of Tanzania). Expedition members are trained in survey techniques, as well as in fish and other marine-life identification, to produce data that directly contributes to the formulation of management plans for the proposed Mafia Island Marine Park. This park will be a multi-user, multi-zone park with areas designated for limited resource use, tourist activity, fishing, scientific research, and conservation. These last two types of area, to be left undisturbed, are often ecologically important sites for fish and coral reproduction and/or recruitment. (Courtesy of M.A. Baldwin, Frontier Tanzania.)

centimetres, or even a few centimetres. Sometimes, in the Mediterranean, a diver can descend through a thermocline that is so sharp that it is like dipping one's foot or finger into a cold bath. These rapid

18.6

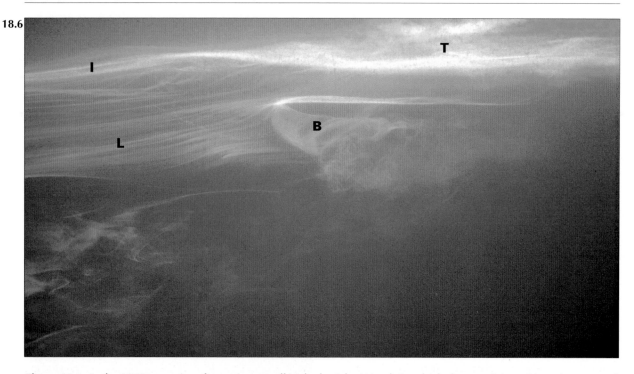

Figure 18.6 In the 1960s, a series of experiments off Malta by John Woods in which dye was injected into the seasonal thermocline revealed a number of unexpected features of small-scale flow. Persistent streakiness provided proof that the flow is mainly laminar, whereas previously it had been assumed to be continuously turbulent. Undulations revealed high-frequency internal waves trapped on ocean fine-structure sheets of enhanced stability frequency. The orbital motion of water displaced by these waves created shear across the sheet, occasionally provoking Kelvin–Helholtz instability, which produces billows that are typically 0.2 m high. The billows in turn generate a patch of turbulence which persists for about 10 minutes. Here all four phenomena are shown: laminar flow streaks, L; an internal wave, I; a billow, B; and turbulent mixing, T. (Courtesy of J.D. Woods, Imperial College, London University, England.)

transitions are occasionally visible in nature because films of dust or dead organic particles settle on the layer where the density changes (cold water is denser than warm water). Divers have studied the internal waves which travel along these surfaces by releasing fluorescene dye which spreads along the surface in undulating sheets (*Figure 18.6*).

These experiments showed that the flow in the seasonal thermocline is overwhelmingly laminar rather than turbulent, which led to a paradigm shift in oceanography with profound implications for

naval operations, climatology, and plankton ecology of the upper ocean.

Scientific Marine Archaeology

Many scientific techniques are used in archaeology. Underwater archaeology takes observation and mapping to its extreme, as valuable information about the context and position of the artefacts and

18.7

Figure 18.7 In 1967, a remote archaeological site at Asopos in the south of Greece was surveyed using snorkelling equipment and an underwater camera. Standing stone walls were photographed at a depth 3.5 m underwater, in a style and pattern which could not be identified at the time. It was 12 years before the authorities granted a permit for the site to be surveyed in detail by a large team of divers; most of the underwater city was surveyed, but by 1979 the strange walls in deep water had completely disappeared, presumably destroyed by winter storm waves. The photographs had been published in 1969 so the walls' existence is indisputable.

features uncovered can be lost if not accurately recorded quickly. Changing sea levels have caused the inundation of many coastal cities and settlements through the ages. Once underwater, structures eventually deteriorate, some through sudden catastrophe, while others gradually erode away (*Figure 18.7*).

Well-preserved shipwrecks have revealed a mass of historical information, unknown from land excavations. Working long hours, often in difficult conditions of poor visibility, darkness, and cold, diagrams are produced that can be used to reconstruct events or items that cannot be brought to the surface, such as impressions in the sediment of long-since corroded iron weapons or structural features of a fragile ship (*Figure 18.8*). Study of the preservation of submerged timbers provides information (often all that is available) about the long-term presence and effects of marine organisms. The depth to which artefacts of known dates are buried provides information about sedimentation rates.

Sampling

Observations, particularly when meticulously recorded, are an excellent starting point for environmental investigation. Most research programmes, however, require data beyond the recording capability of an observing diver, so physical samples are needed for laboratory analysis. Collection by diving is an expensive option to choose in terms of time and money and has distinct limitations, especially when many samples are needed from a large area. Remotely operated grabs deployed from ships sample soft mud and sand, but as the sea bed becomes harder these are less effective. The diver is a 'thinking sampling device', and will bring the essential skills of discriminating observation, judgement, selection, and decision-making to the work underwater. In other words, the diver can choose a sampling method to suit the job and adjust techniques to achieve the best results. If the sea bed is mostly mud with occasional rocks, the diver can, without positional bias, sample the mud with a core tube and the rock with a hammer, recording the relative proportions of each (*Figure 18.9*). A corer dropped from the surface would sample (or not), whatever it landed on.

In some situations a diver is the only possible option. A specific material or species may have to be searched for, recognised, and sampled with care. Delicate sessile specimens can be detached from their substrate with minimum damage, or samples for chemical analysis collected by hand to avoid contamination. In addition, a description of the surrounding sea bed can also be made to accompany the sample to the laboratory, so that the non-diving scientist is given a preserved specimen, a photograph of it in life, and a description of the

18.8

Figure 18.8 The *Mary Rose*, a Tudor warship sunk on the South coast of England in 1547, was buried in mud with one side virtually missing. From measurements and drawings made by divers excavating inside the hull and burrowing underneath it, a cradle was constructed to exactly fit and support the fragile structure as it was lifted from the sea bed. At no time, before it was lifted out of the mud, could the entire ship be seen or measured.

18.9

Figure 18.9 A diver collects core samples (for infaunal analysis) of a known volume from a stony sea bed, which are not easily obtained remotely. A metal tube is hammered into the sea bed to a known depth. The tube is then dug out and capped at each end for recovery. If necessary, a compass can be used to determine the tube's magnetic orientation and it can be kept upright as it is brought to the surface. Several replicates are usually collected to ensure a representative sample.

18.10

Figure 18.10 A diver, with appropriate training, can identify marine fauna and flora on the sea bed. Detail of a specimen's undisturbed behaviour and the habitat in which it lives can be observed and recorded. It can be collected and brought to the surface with a photograph of that actual specimen in its natural environment.

habitat and community of which it was a part (*Figure 18.10*).

Sampling by divers need not be restricted to the shallow sea bed. 'Blue-water diving' in oceanic waters allows planktonic species (normally the larger ones, like jellyfish) to be taken from the water column without the damage caused by capture in nets. These animals can then be maintained in aquaria to reveal information about their life cycles and physiology that would have been impossible to obtain from stressed and damaged net-caught specimens.

Divers are very skilled at penetrating caves and wrecks where access for ROVs and sampling devices is difficult or impossible. By exploring deep

within submarine cave systems, divers have collected biological and geological samples yielding information new to science (*Figure 18.11*).

Photography and Video

The viewpoint of the diver can be brought to a wider audience by the use of photography and video. Early Hass and Cousteau films have inspired many a budding scientific diver and the realism of more recent filming techniques is probably responsible for today's popularity of sport diving and marine conservation. The ability to retain images has a number of advantages. Photographs and films allow a better insight of the subject to those who do not dive. They can also provide proof of exis-

Figure 18.11 The vast and complex underwater cave systems of the Caribbean 'Blue holes' were created by freshwater flows through the limestone bedrock of islands during periods in the past when sea levels were much lower. Samples of stalactites and stalagmites from the caverns were collected and dated by the uranium–thorium method. These records provide evidence of periods when the caves were in a subaerial environment and when they were under the sea, so that the changes of sea level during the late Ice Age can be determined accurately. Only a diver can collect water and rock samples from 20 m underwater, 200 m into the Lucaya cave system of Grand Bahama. (Courtesy of R. Palmer, Technical Diving Ltd, England.)

18.11

tence to the cynical and a lasting record of a disappearing or transient feature. Actual images of live animals and plants now complement line drawings and description in identification guides.

In the hands of a skilful diver, photography can be a research technique in itself, as well as one to support other recording and sampling methods. Photogrammetry, used extensively in archaeology, comprises a series of photographs, taken from a fixed grid, which are pieced together to give a detailed, accurate photograph of a large area. Similarly, stereophotography, using two cameras mounted above the site, provides three-dimensional images from which measurements can be taken. Time-lapse, a video as well as photographic technique, is used to determine changes over time in diversity and growth. Divers can install equipment above a marked station to record the rate of increase in the size of a sponge or encrusting bryozoan over weeks or months, or to record the changes in diversity of a community from season to season. Animal behaviour can be monitored with minimal disturbance to the subject. While animals may respond to the presence of a diver, a piece of apparatus is quickly accepted as part of the sea bed (*Figure 18.12*).

Divers do not generally work at depths greater than 50 m for safety reasons. Their time underwater is also limited by a number of factors, such as human physiology, equipment, location, logistics, and, ultimately, financial cost. Scientific divers can work directly with their subject *in situ* or bring their work to the surface for the study to be continued more conventionally. Advances in the equipment available for diving continue to expand the scope of the diver and therefore the progress of underwater science.

18.12

Figure 18.12 A diver installs a time-lapse video recorder above a lobster pot to investigate the efficiency of this fishing method. The camera runs intermittently for a few seconds at a time over 24 hours. At night a lamp is automatically switched on during recording. The lobster's approach to the pot is captured and its behaviour inside the trap monitored.

General References

Flemming, N.C. and Max, M.D. (eds) (1990), *Scientific Diving: A General Code of Practice*, CMAS, UNESCO, Paris, 254 pp.

Woods, J.D. and Lythgoe, J.N. (eds) (1971), *Underwater Science – An Introduction to Experiments by Divers*, Oxford University Press, Oxford, 330 pp.

National Oceanic and Atmospheric Administration (1991), *NOAA Diving Manual: Diving for Science and Technology*, 3rd edn, US Department of Commerce, Washington DC.

Dean, M., Ferrari, B., Oxley, I., Redknap, M., and Watson, K. (eds) (1992), *Archaeology Underwater: The NAS Guide to Principles and Practice*, Nautical Archaeology Society, London, 336 pp.

Reference

1. Flemming, N.C. and Max, M.D. (eds) (1990), *Scientific Diving: A General Code of Practice*, CMAS, UNESCO, Paris, 254 pp.

CHAPTER 19:

Marine Instrumentation

G. Griffiths and S.A. Thorpe

Introduction

An ability to measure is central to any science. Data provide quantified measures of the composition and contents of the oceans and their supporting sediments and rocks, and are needed to test or develop hypotheses; in recent times, the data have been needed to provide starting points ('initial data') to run models to forecast natural changes and the consequences of anthropogenic inputs to the ocean or the atmosphere.

Marine research requires investment in ships and, for the science to develop, technological developments, the construction of new or more accurate instruments, and novel measurements to provide further insights into ocean processes. Ships have always been essential for marine studies (see Chapter 1), and will be for many decades into the future, but they are expensive so other methods of observing the ocean are evolving. The most striking of such methods to appear in the twentieth century is satellite remote sensing.

The information about the upper parts of the ocean obtained in recent years by Earth-orbiting satellites is immense (see Chapter 5). Measurements are made over vast areas of the ocean surface, often in regions seldom visited by research vessels, and are repeated at regular intervals so that variability over periods of months and years can be studied. Satellites have greatly improved the accuracy of position fixing (Chapter 20), but satellite sensors are only able to obtain measurements at or near the ocean surface. The movement and properties of the waters deeper than a metre or so, in particular those responsible for carrying heat around the globe, are often undetected or poorly resolved by satellite sensors, so we are forced to rely on other, more 'conventional', methods of sampling and measurement to probe the remote ocean depths, usually (but not always, as we explain later) involving the use of ships.

The deployment and recovery of instruments from ships at sea carries a reliance on seamanship and navigation, and on the efficient operation of winches and cranes. There are risks to those involved, particular in rough weather when safe working on deck is difficult (if not impossible), and delicate instruments or sensors may be damaged in handling and as they are lowered over the side, perhaps swinging before entering the water. With experience, risks are reduced, but are inevitably high when weather changes suddenly, as the research vessel is unexpectedly struck by a large wave, or when data are required in just those conditions in which hazards are greatest.

Once in the water, hazards to instruments are still present. Their design must be appropriate to the conditions in which they operate, perhaps to pressures 500 times that at the sea surface (e.g., on an abyssal plain at a depth of 5 km), in a salty and corrosive environment. Collapse of buoyancy almost inevitably results in the loss of both recorded data and instruments. Near the surface, instruments are subjected to large and variable forces applied by breaking waves, and may be subjected to 'marine fouling' by organisms settling on and attaching to sensors (see inset in *Figure 19.12*). There are problems, too, of recovering samples from depth without damage due to reduced pressure or contamination, and of preserving samples at depth, possibly for extended periods of time, before recovery to the surface for study. Damage and loss of moorings by trawling is a severe risk in areas of intensive fishing. The 'dramatic' events which occur and result in damage are (sadly for the reader's entertainment) not those recorded by photographs, since the focus of attention and activity is then on the safety of the instruments, scientists, and crew. However, some of the consequences, the damage to instruments, are recorded [see *Figure 19.30(b)*].

Given the large expense of marine operations and the importance of the data, the reliability, accuracy, and precision – demanding careful calibration – are paramount. The resolution of the data and the sampling intervals must be matched to the sensors and fitted to the purpose for which measurement is intended. There is little point in sampling with high frequency if the sensors themselves do not react to rapid change, or if high-frequency data are known to be of no use. Data storage and subsequent analysis are expensive and internally recording instruments are limited by log-

Figure 19.1 The research FLoating Instrument Platform (FLIP) moored in the deep sea while participating in an experiment to study surface waves. FLIP is a 120 m manned spar buoy operated by the Marine Physical Laboratory of the Scripps Institution of Oceanography in La Jolla, California. It is towed to an experiment site in a horizontal position. Ballast tanks are then flooded, inducing the 'flip' into the vertical position as shown. With 87 m of the cylindrical hull below the surface, FLIP is extremely stable. In a seaway, typical vertical motions are less than 2% of the wave height. A crew of five operate the platform, hosting up to 10 scientists for 'cruises' of as long as 35 days. Doppler sonar (see text and *Figure 19.28*),

mounted beneath the water line, is used to profile the sea-surface and upper-ocean velocity fields. Conducting wires to measure waves, current meters, conductivity, temperature, and depth instruments (CTDs) to measure the thermal, salinity, and density of the ocean, and other instruments involved in the study of air–sea interaction (see Chapter 2) are suspended from light-weight booms deployed when FLIP is vertical. (Courtesy of Dr R. Pinkel, Scripps Institution of Oceanography, La Jolla, California, USA.)

ger capacity and available battery power. In many cases, 'real-time' data display and analysis are an advantage, if not a requisite, so that more intensive measurements or a different sampling strategy can be implemented when particular events or processes occur; study in the ocean, unlike in the laboratory, does not have control over change, and preparation is needed to respond to opportunities to make measurement as they occur. Herein lies skill, excitement, and reward to the observational scientist in the natural environment.

This also illustrates a fundamental difference in the science from that conducted entirely in the laboratory; experiments in the sea often cannot, except in statistical ways, be repeated, and a customary axiom of science, that the conditions in experiments should be described in such detail that they may be precisely repeated by others, cannot apply in the natural environment where replication in identical conditions is rarely possible.

Precise measurement in the ocean, for example of turbulent motions, may require 'stable platforms' which do not move or vibrate. To solve such difficult measurement problems is sometimes impossible by the available means, and new platforms or techniques may be the only answers (e.g., *Figure 19.1*). The challenge to instrument designers is enormous! We describe in this chapter some of the instruments that they have cleverly devised. While the range of instrument types is illustrated herein, the total numbers and variety are great, so

we have had to be very selective in making a choice. The selection is not comprehensive and there is no implication that the 'best' instruments are included, although some are.

International protocols for data quality and collection in the ocean may lag behind atmospheric ones (from which meteorological predictions are made) by as much as 30 years; this is the case even for basic physical properties. Detailed and continuous measurements of biology and chemistry in the water column are rarely possible. The ability to forecast the oceans, to predict the important part which they play in climate change (Chapter 3) or how they will respond to man-made wastes or pollution (Chapter 22), is correspondingly less precise; this is magnified because the most energetic motions which must be resolved and measured are of a much smaller scale than those in the atmosphere (Chapter 4), so far more measurements are needed to detect their presence and quantify their effect.

Thus, an immense challenge is posed to marine scientists – how are they to make measurements sufficient to verify and initialise predictive models of the ocean? In this chapter we provide a statement of some of the recent progress in developing and using instruments in the ocean, and then discuss the future prospects of improving sampling techniques and attaining the goal of a Global Ocean Observing System (GOOS) to provide a firm basis for predictions in the twenty-first century.

Figure 19.2 A deep-ocean hydrographic instrument package consisting of a CDT, a 24 x 10 litre-bottle rosette multisampler, fluorometer, and transmissiometer. (Courtesy of S. Hall, SOC.)

tion, its latitude, longitude, heading, and speed. High-precision satellite navigation – the Global Positioning System (GPS) – is used routinely on oceanographic research ships, giving position accuracy to about 40 m world-wide (see Chapter 20).

Acoustic instruments on ships provide a remote-sensing capability. Echo-sounders provide accurate depth measurements, both beneath and on both sides of the ship's track (Chapter 20). Acoustic Doppler sounders give measurements of the currents in the water below the vessel (see later).

Most research cruises make use of automated surface sampling – where sea water from an intake on the ship's hull is pumped to the laboratories for analysis. Temperature, salinity, fluorescence (indicating chlorophyll in phytoplankton), optical transmission (indicating particulate matter), and the nutrients, including nitrate, phosphate, and silicate, may be measured in this way.

Volatile chemical substances must be measured immediately, such as the chlorofluorocarbons (CFCs) used as tracers of water pathways through the ocean (Chapter 11). Some substances can change their concentration within the samples as a result of biological activity; such samples must be deep frozen on board ship for later analysis.

'In Situ' *Instruments*

We begin with instruments which take samples or measurements of the sea water or sediment surrounding them.

Instruments connected directly to a vessel

Instruments on board research vessels

The earliest scientific measurements at sea were made from vessels while underway. Medieval navigators estimated ocean currents by observing the drift of their vessels, comparing the observed position with the estimated (or dead-reckoned) position, and ascribing the difference to the currents. By the 1820s the major surface currents of the world's oceans had been discovered and charted by hydrographers (see Chapter 1).

Today, instruments fitted to the ship still provide fundamental measurements of the ocean. Foremost among these is the measurement of the ship's posi-

Lowered instruments

Knowledge of the global-scale circulation of the deep ocean has been obtained from measurements

Figure 19.3 Precision thermometers still have a role – today's instruments use platinum resistance sensors and LCD displays rather than the mercury-in-glass type, the mainstay of oceanic temperature measurements for over a century. (Courtesy of M. Conquer, SOC.)

Figure 19.4 Used for obtaining large diameter cores for geochemical and geotechnical analysis (see, for example, Chapter 8), a 2 m long Kastenlot box-section corer is deployed from the midships winch of RRS *Discovery*. (Courtesy of SOC.)

19.4

of temperature and salinity using equipment lowered from stationary ships. From the early nineteenth century to the late 1950s, oceanographers were dependent on bottles closed at depth to collect water samples for salinity determination on board ship, and on mercury thermometers for accurate temperature measurements. Exercising great care in handling and calibration and with a practised eye, the best of these deep-sea thermometers could be read to an accuracy of some 0.003 K. The arrival of the transistor and integrated circuit revolutionised temperature measurements. It became practical to use electrical-resistance thermometers, to process the tiny electrical signals within a lowered instrument, and to send the results back to a display on the ship.

The early Salinity, Temperature, and Depth (STD) instruments, developed in about 1960, were not as accurate as the mercury thermometers, but they provided continuous, rather than discrete, measurements of both temperature and salinity from surface to sea floor; they contributed to the discovery that the density of the ocean rarely increases smoothly with depth, but is irregular with microstructures on the scale of 1 mm to 10 m (see *Figure 18.6*).

The STD instruments also measured salinity – previously, salinity could only be determined by collecting a sample of water in a reversing water bottle, and analysing the sample using chemical or electrical methods back on board the ship.

Developed from STD instruments in the early 1970s, the Conductivity Temperature Depth instrument (CTD) provides more accurate measurements, and is still today the work-horse of physical oceanographers (*Figure 19.2*). Its temperature accuracy is now better than that of mercury thermometers, though modern mercury equivalents are still used as a check (*Figure 19.3*). But the conductivity measurements that give salinity still need to be calibrated against water samples. Although this is a chore, water samples cannot yet be dispensed with as they are still the *only* source of data on many chemical compounds of interest.

Bottom-sampling instruments

Samples of sediments from the ocean floor are obtained by a broad class of devices known as grabs and corers. Some corers are general-purpose tools, others are designed for specific subsampling and analysis needs. The most basic tool is the gravi-

ty corer. In its simplest form the design is usually a large weighted head with a tubular or box-section barrel, which is driven into the sediment. A catcher mechanism closes as the corer is pulled out of the sediment and this retains the core during retrieval. Gravity cores are prone to compression as well as disturbance, and the depth to which a core can be taken is restricted to a few metres due to friction.

Box corers (e.g., *Figure 19.4*) are of a more sophisticated design, and of a larger diameter to reduce friction. The larger diameter also allows subsamples to be taken, and because these corers are lowered onto the sea bed, rather than dropped from a height (as are gravity corers), the sediment–water interface is preserved, which is a great advantage.

Piston corers were designed to take long cores – up to 50 m. The action of a piston within the barrel reduces the internal friction, effectively sucking the sediment into the core barrel. Piston corers, like gravity corers, miss the sediment–water interface, which – being light unconsolidated material – is 'blown away' by the pressure wave ahead of the core barrel.

For biological and geochemical analysis this

19.5

Figure 19.5 The multicorer takes eight short cores for biological and geochemical purposes (see Chapter 11). The stand rests on the sea floor and the corer is lowered into the sediments. (Courtesy of SOC.)

interface needed to be sampled, a need that led to the design of multicorers (*Figure 19.5*), with a grid of short tubes mechanically pushed into the sediment, retaining the sediment–water interface and causing minimal disturbance.

Towed sensors and samplers

The traditional oceanographic measurement station, where the ship is stopped and wires are lowered to make measurements, is costly in ship time. To save time, engineers and scientists have devised instruments and vehicles that may be towed behind a ship to give continuous coverage while underway. Following tows of instruments at constant depth came the need for vehicles which could sample the water column more completely.

Canada led the way with the Batfish™ in the early 1970s, a vehicle that could be towed at 5 m/s while undulating between the surface and 400 m. With a payload of a CTD and other instruments, it

19.6a

19.6b

Figure 19.6 Towed vehicles do away with the need for the vessel to stop and lower instruments. The SeaSoar undulator travels from the surface to 500 m depth in a horizontal distance of 2 km at a ship speed of 9 knots (about 4.5 m/s). It carries CTDs and other physical, chemical, and biological instruments to measure the structure of the upper ocean (see, for example, *Figure 4.10*). (a) Deployment through the stern 'A' frame of RRS *Charles Darwin* is straightforward in calm seas, but can be difficult in a gale (a spare vehicle is in the right foreground). (b) The vehicle being deployed – SeaSoar's ability to carry several instruments makes it especially valuable for research spanning more than one discipline. Increasingly, the vehicle is used to make measurements for biological oceanographers, including ocean colour, phytoplankton pigments, and zooplankton counts using optical and acoustic sensors.

19.7

Figure 19.7 Introduced in the early 1970s, the UOR – known commercially as the Aquashuttle – can be towed from ships of opportunity at speeds of up to 10 m/s. The vehicle can either be self-contained, following pre-set undulations and storing data internally, or it can be controlled from the ship through a conducting cable to give real-time data. It can be launched by one person and can record its data internally. It is ideally suited for deployment by non-experts on merchant ships, so extending the geographical coverage of data collection beyond the tracks of research vessels in a most cost-effective way. (Courtesy of Plymouth Marine Laboratory, Plymouth, UK.)

19.8

19.9

Figure 19.8 The Rectangular Midwater Trawl (RMT) 1+8 acoustically controlled mid-water net system consists of a pair of rectangular trawl nets within a single frame. The outer net has an opening area of 8 m² with a mesh of 4.5 mm to catch micronekton, the inner net an opening area of 1 m² with a finer mesh (320 μm) is for plankton. Acoustic commands from the ship control a mechanical release gear to open and close the nets. Data on temperature, depth, speed and distance travelled, and net position are telemetered acoustically to the ship. (Courtesy of SOC.)

Figure 19.9 The epibenthic sledge is designed to skim over the surface of the sea bed and collect the organisms living on the sea floor, immediately above it, and in the upper few centimetres of the sediment. An acoustic pinger indicates when the sledge has reached bottom, when the opening–closing mechanism has operated, and when the sledge leaves the bottom. Cameras and electronic flash may be mounted on the sledge to provide images of the nature of the bottom and to give some indication of the efficiency of the net in capturing animals. (Courtesy of SOC.)

gave new insights into the structure of the upper ocean at 1 m vertical and 1 km horizontal resolution. This idea was developed into the UK SeaSoar vehicle, which can undulate from the surface to 500 m in a distance of less than 2 km. Several expeditions have used SeaSoar for tows of over 10,000 km, with the greatest danger being during recovery or deployment (*Figure 19.6*).

Supporting SeaSoar at sea requires a team of electronic and mechanical engineers, as well as data-handling specialists, to enable the wealth of information to be processed aboard ship in real-time.

For many experiments this is too heavy a burden (too large a team to support financially or to accommodate onboard ship); for these, simpler vehicles are available to carry out similar measurements. The Undulating Oceanographic Recorder (UOR; *Figure 19.7*) is one such device for covering the upper 200 m.

Research laboratories have developed many other types of specialist towed instruments for sampling, many of which are unique, serving the needs of individual researchers or small teams – examples are shown in *Figures 19.8–19.11*.

19.10a

19.10

19.11

Figure 19.10 A thermistor chain (a) laid out under cover before deployment, and (b) being deployed. This chain of thermistors to measure the thermal structure of the upper ocean is 400 m long and consists of 100 sensing 'pods', each providing a measure of temperature, and 28 measures of pressure so that the depth of the measurements are well-established. A 2 tonne sinker weight keeps the chain near-vertical under tow at a speed of 4 knots (about 2 m/s). The chain uses digital data communication and contains only six wires, two for power supply, two for data, and two for control. Sampling the 100 pods takes only 0.9 s. This particular chain has been used on 14 ocean measurement surveys, and more than 30 weeks of continuous data have been collected, comprising some 38,000 km of track. Much of the work has been directed toward the detection and tracking of fronts and eddies in the area between Iceland and the Faeroes, where waters of the North Atlantic meet those of the Norwegian Sea. (Courtesy of Dr J. Scott, Defence Research Agency, Dorset, UK.)

Figure 19.11 The Lightfish is an instrument to measure multispectral reflectance at high spatial resolution. There are six sensors pointing up and down, the six white ports in the upper black bar, within which are mounted the irradiance sensors sampling at 410, 440, 490, 520, 550, and 670 nm. It is towed just below the water surface from a ship, collecting data from ocean transects which can be compared with observations from airborne or satellite sensors; this provides 'ground-truth' calibrations of these sensors for ocean colour estimates, which can be related to the near-surface distribution of phytoplankton (see Chapter 14). Data are logged onto a ship-board computer, together with other relevant data such as temperature, fluorescence, and transmittance. (Courtesy of Dr A. Weeks, SOC.)

Figure 19.12 (a) Deployment of an Aanderaa™ current meter mooring from the foredeck of RRS *Discovery* (courtesy of SOC). (Inset) A recovery beacon that enables the ARGOS satellite system to track and locate the mooring if it surfaces prematurely (courtesy of M. Conquer, SOC). (b) A release mechanism with biofouling after recovery from a 12-month deployment. (Courtesy of M. Conquer, SOC.)

19.12a

Instruments fixed in position

Current meter moorings

Obtaining information on the behaviour of the ocean over time-scales of more than a few days requires scientists to resort to self-recording instruments moored to the sea floor, or left drifting. Ingenious non-electronic self-recording instruments were designed before the arrival of the transistor, but the rapid advances of recent years owe much to the microprocessor and to high-energy density batteries.

The highly successful Aanderaa™ current meter was designed in the early 1960s. Using mechanical encoding of the rotation of a Savonius rotor and the direction of a large vane, it owed a great deal to the weather vane and anemometer.

Nevertheless, the instrument became the standard for measuring deep-ocean currents, and in a solid-state form is still popular today. Capable of being deployed for periods of a year and more, its simplicity, reliability, and relative cheapness make it an almost ideal oceanographic tool.

In depths of up to some tens of metres, typical of waters close to shore in the shelf seas, equipment may be moored to the sea bed and to a surface buoy, enabling simple recovery.

Such a simple mooring is not practical in the deep ocean, where the buoyancy is usually placed beneath the surface, well away from the influence of surface waves and shipping or fishing (see *Figure 19.12*).

The problems are then, first, to find the mooring and, second, to retrieve it. A single instrument, the acoustic release, gives the answer. When the continually listening unit hears a coded sound-pulse from a ship it sends out a reply signal that indicates the range from the ship to the mooring.

As the ship homes in, another signal activates the release mechanism to separate the mooring from its anchor, so the mooring returns to the sea surface.

When rising the release provides a beacon signal, often augmented on the surface by radio transmitters, flashing lights, or radar reflectors.

As moorings contain increasingly more valuable

19.12b

equipment, and have a small probability of early failure, satellite position-indicating transmitters may be fitted to the buoyancy. These alert the laboratory to moorings that have surfaced prematurely, giving position and drift information that aids their recovery.

Figure 19.13 A surface meteorology buoy, with anemometers (for mean wind speed and direction), sea and air temperature sensors, a 3 m path acoustic current meter, a buoy motion package (to give wave height and directional spectrum), and radio and satellite data telemetry (see Chapter 2 for a discussion of the way the atmosphere affects the ocean). (Courtesy of A. Hall, SOC.)

Figure 19.14 A Wavecrest buoy, developed by the Netherlands company Datawell, being deployed. The buoy follows the sea surface and sensors measure the components of acceleration, which, integrated twice, give the wave-height variations. The buoy is moored to the sea bed using a compliant tether – usually a length of thick rubber line – to avoid mooring forces affecting the record. (Courtesy of C. Griffiths, Dunstaffnage Marine Laboratory, Oban, Scotland.)

Moorings measuring meteorology and surface waves

Surface meteorological buoys (*Figure 19.13*) combine accurate sensors, replicated for reliability, with data telemetry via satellite and terrestrial radio links (see also *Figure 3.24*). Additional sensors that measure the motion of the buoy give estimates of wave period, height, and direction. An example of an instrument designed purely for wave measurement is shown in *Figure 19.14*.

Bottom-mounted instruments

Benthic landers are instrument packages that provide observations near the sea floor. They may be deployed from the ship's warp, when the experi-

ment duration is short, or may be autonomous to give measurements over long periods.

Geophysical experiments make extensive use of sea-bottom seismographs that record the signal associated with natural earthquakes or from induced sound sources, such as explosives and air guns (*Figure 19.15*; also see Chapter 8).

Scientists studying the geochemistry of the sediment–water boundary require water samples from within the sediment (the pore waters). Bottom landers have been built that contain hydraulically driven syringes to penetrate the sediments, take samples at a range of depths, and retract in readiness for recovery. Other sensors, such as those to measure pH and oxygen concentration within the sediments,

Figure 19.15 The Digital Ocean Bottom Seismometer (DOBS) is a self-recording listening station with an in-water hydrophone and geophones in contact with the sea floor. Sound generated from ship-towed air guns or from dropped explosive charges reaches the DOBS directly through the water and also through the ocean floor, through which the sound speed is much higher. From the characteristics of the different propagation paths, the nature of the sea floor can be inferred. (Courtesy of SOC.)

Figure 19.16 This sampler contains hydraulically driven syringes that penetrate the sea floor to take sediment and pore-water samples at a range of depths. Other sensors, such as pH and oxygen probes, may be fitted to the lander. (Courtesy of SOC.)

Figure 19.17 Work on research vessels proceeds through the night – here deploying the Sediment Transport and Boundary Layer Equipment (STABLE) benthic instrument platform with current and high-frequency acoustic sediment transport sensors. (Courtesy of J Humphrey, Proudman Oceanographic Laboratory, Bidston, UK.)

can be included on these landers (*Figure 19.16*).

Movement of sediment along the sea floor helps to shape much of the coastline, so several instrument platforms have been designed to study sediment concentration and transport close to the sea bed. Rapid turbulent motions, critical in causing fine sediment to become suspended, need to be measured along with the concentration or mass of the suspended matter. Measurements are made several times a second. Instruments based on the principle of electromagnetic induction are suitable and can be made sufficiently robust. A voltage sensed by electrodes on the instrument is generated by the movement of the conducting sea water through a magnetic field produced by an internal solenoid. These current meters can be combined with acoustic probes to measure the sound scattered from the suspended particles in bottom landers (*Figure 19.17*).

Increasing emphasis is being given to obtaining data over many years from the deep ocean to monitor natural and man-made change. Some locations lend themselves to regular visits by research ships, others are either too remote or visited too infrequently. New bottom-mounted instruments have been designed to solve the problem of obtaining data over several years from remote locations. The MYRTLE (Multi Year Tide and Sea Level Equipment) package releases data podules from the sea bed that, on the surface, return their data via satellite to the laboratory (*Figure 19.18*).

Figure 19.18 Sea-level changes of centimetres can be observed from the deep-ocean floor by the MYRTLE package. Designed for operation over 5 years, MYRTLE releases data 'podules' to the surface, where, as they drift, they telemeter data to the laboratory via ARGOS satellites. The package shown here is being deployed on the continental slope off the Antarctic Peninsula, the southern boundary of Drake Passage. With another MYRTLE package on the northern slope off South America, the slope of the sea level between the two sites provides a measure of the transport of water through the Drake Passage by the Antarctic Circumpolar Current, and of its variability (see Chapter 4). (Courtesy of Proudman Oceanographic Laboratory, Bidston, UK.)

19.19a

19.19b

Figure 19.19 Here are two examples of free-fall instruments. (a) EPSONDE[4], a tethered 'semi-free-fall' device as shown in the top right-hand sketch. Data from the slowly falling probe is transmitted up the cable to recorders on the deployment vessel. The cable is sufficiently light and thin that it does not affect the smooth descent of the instrument through the water. The central photograph shows the instrument being deployed at sea. The lower right-hand corner shows the shear probes used for measuring subcentimetre velocity fluctuations in the turbulent ocean, together with thermistors and a thin-film thermometer to measure temperature (courtesy of Dr N. Oakey, Bedford Institute of Oceanography, Dartmouth, Nova Scotia, Canada.) (b) A High Resolution Profiler (HRP) during recovery. This instrument, developed by a team at Woods Hole Oceanographic Institution, Woods Hole, Massachusetts, is internally recording and is free to fall through the water with full ocean-depth capability[7]. Once the maximum depth has been reached, ballast weights are released by a solenoid mechanism on command from an on-board instrument computer and the HRP returns to the surface. The spikes or whiskers at the top are to increase the drag and so reduce the instrument fall speed. The instrument cradle to reduce damage in deployment and recovery is visible to the left. The instrument carries sensors similar to those of EPSONDE (courtesy of Dr J. Toole, Woods Hole Oceanographic Institution, Woods Hole, Massachusetts, USA).

Free-fall instruments and ocean microstructure

The discovery of ocean microstructure in the 1960s drew attention to the turbulent mixing which must accompany it and offered the challenge to design instruments to measure velocity down to the scale of millimetres, at which energy is lost by viscous dissipation. Osborn[5] and Siddon's solution was to use a pair of piezoceramic bimorph beams – allowing flow-induced pressure fluctuations to be registered as voltages – embedded together in a 6 mm diameter cylindrical probe which, when carried steadily through the water pointing ahead of the supporting body, responds rapidly to lift variations and therefore to the two components of velocity normal to the direction of motion of the probe. These small components can be determined once the orientation of the probe and its mean speed through the water are known. Smooth, vibration-free, slow steady motion of the instrument through the water at about 1 m/s has been achieved by decoupling the instrument from the research vessel and allowing it to free-fall before recovery by line [*Figure 19.19(a)*], or by the instrument's releasing ballast which leaves it buoyant [*Figure 19.19(b)*]. [Alternatively, microstructure probes can be mounted on a very stable platform, such as a submarine[6] or a mooring (*Figure 19.20*).] Accompanying measurements of pressure (from which the mean speed – the fall speed – is found), temperature, salinity, and tube orientation observations are usually made from the free-fall instruments when velocity microstructure is measured. Measurements have been made right to the sea surface by constructing instruments which, having released ballast, rise upward with sensors extended on mountings above the tube containing the power supply and the recording and control electronics.

Figure 19.20 Operating free-fall devices to measure microstructure is very expensive because of the cost of maintaining ships at sea, typically $US15,000 per day. Cost reduction and the need for very long-term measurements prompted the development of a moored instrument for the study of ocean mixing. The Tethered Autonomous Microstructure Instrument (TAMI), shown here during deployment, carries sensors (protected by the ring at the right of the instrument), which detect currents as small as 0.1 mm/s and temperature changes of 10^{-5} K over a sampling volume of diameter about 5 mm. On-board computers process the measurements and record data on disk drives for 6 months. The instrument is anchored in water up to 5500 m deep. A 1.5 m diameter syntactic foam float (about 670 kg buoyancy) holds the mooring taught while the instrument (about 90 kg buoyant) floats some 7 m above on a line attached to the shackle on the spar shown (just below the fin on the instrument tube). The tube swivels into the prevailing current. An acoustic command to a release located about 20 m above the anchor on the sea bed returns all but the anchor to the surface for recovery. The instrument was developed by Dr R. Lueck at the University of Victoria, Canada. (Courtesy of Dr R. Lueck, University of Victoria, BC, Canada.)

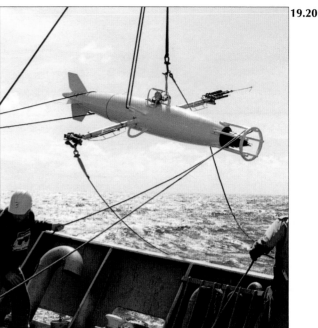

19.20

Measurements using tracked devices

Surface drogue floats and drifters

Drifters or floats released at sea that carry requests to the finder to report the time and location of recovery have for centuries provided information about the mean current drift at the sea surface. A major question was always whether drifters were driven by the wind or by the currents.

Radio and satellite navigation and telemetry has led to a resurgence in using surface drifters. The ARGOS system, developed by France and the US, uses low-power satellite transmitters on buoys to give positions with to an accuracy of <1 km, and up to 32 bytes of data per transmission. Shown in *Figure 19.21*, drifters are designed with a minimal area above water, and are fitted with drogues at depths of tens to hundreds of metres to tag and follow the ocean currents and eddies (Chapter 4). These buoys are used in large numbers. From 1978–1981 some 300 were released in the Southern Ocean. The buoys measure sea-surface temperature and barometric pressure, as well as currents. The World Ocean Circulation Experiment (WOCE) called for some 3800 buoys to be deployed globally from 1990–1996.

Deep drifting buoys

New insights into the strength and variability of currents in the deep ocean came from the invention (in the early 1950s, by Dr John Swallow) of the neutrally buoyant float. The first experiments in 1955 proved the idea – that an aluminium tube, negatively buoyant at the surface, would gain buoyancy as it sank, because its compressibility is

19.21

Figure 19.21 An ARGOS satellite-tracked buoy and its drogue on deck. The drogue consists of concentric tubes of nets 18 m long, the outer tube of 0.5 m diameter. These drogues have been used down to 800 m. The buoy contains sensors for temperature and for detecting the presence of the drogue; weathering many storms over operating lives of years eventually leads to the loss of drogues from many buoys. If this is not detected, false conclusions may be drawn about the currents or ocean eddies (see Chapter 4). (Courtesy of M. Conquer, SOC.)

less than that of sea water. Through careful balancing, the float could be made neutrally buoyant at any pre-set depth. Fitting an acoustic transmitter to the float enabled it to be tracked from a ship using triangulation methods (Chapter 20). Examples of the tracks of such floats are given in *Figure 4.6*.

19.22

Figure 19.22 A WOCE-style RAFOS float being held before deployment on R/V *Knorr*. This version of the float was recently developed by the University of Rhode Island and the Woods Hole Oceanographic Institution, and is manufactured by Seascan Inc. RAFOS floats are ballasted to be neutrally buoyant at a pre-selected depth, where they drift for as long as several years, showing the water motions (see Chapters 3 and 4). The floats receive acoustic signals from moored sound sources, record the times of arrival so that their position can be determined, and (at the end of their mission) surface and transmit the data, including temperature and pressure, back to the laboratory via the ARGOS satellite system. (Courtesy of Dr T. Kleindinst, Woods Hole Oceanographic Institution, Woods Hole, Massachusetts, USA.)

For a period of two decades the basic Swallow float was refined, which led to the development of the SOFAR float in the early 1970s. The SOFAR channel is an acoustic wave guide in the ocean, arising from the combined effects of temperature and pressure on the speed of sound (see later). These floats communicate over ranges of 1500 km and more, with autonomous listening stations, thus eliminating the need for ships to attend the floats. A derivative of the SOFAR float – termed RAFOS (see *Figure 19.22*) – reverses the roles of the float and listening station; the listening is done on the float, and the sound sources are moored and recoverable. This reduces costs, since sound transmitters are far more expensive than receivers.

The next logical step was to make floats independent of a tracking network. The Autonomous Lagrangian Circulation Explorer (ALACE) was developed by the Scripps Institution of Oceanography, along with the Webb Research Corporation, in the late 1980s (see *Figure 19.23*). These modern floats are used in large numbers; the WOCE experiment has commitments for over 1000 to be used from 1990–1996.

19.23

Figure 19.23 An ALACE float, showing the satellite telemetry and location antenna. Instead of obtaining position fixes from moored sound sources, the ALACE float periodically increases its buoyancy and rises to the surface every 20 days. There, by using the ARGOS platform location and data transmission satellite system, the ALACE float can be located and data sent to shore before the float returns to its pre-set depth. The major challenge in designing the ALACE float was to devise a reliable, high-efficiency buoyancy control system that could function at ambient pressures of up to 200 bar. The buoyancy change is performed by changing the effective volume of the float. An electrically driven hydraulic pump moves oil from a reservoir internal to the float to an external bladder. An example of the float tracks is shown in *Figure 3.13*. (Courtesy of Dr R. Davis, Scripps Institution of Oceanography, La Jolla, California, USA.)

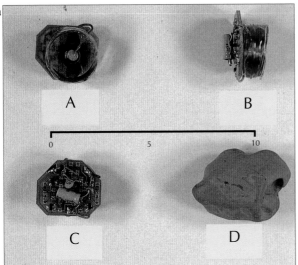

Figure 19.24 The 'electronic pebble' is a tracer developed to monitor the mobility of shingle on beaches. (a) The system comprises synthetic pebbles consisting of a battery-operated circuit (A, B, and C) encapsulated in resin (D), which emits a coded train of magnetic pulses (scale bar, 0–10 cm). These are placed on the beach at low tide and can then be detected and decoded by a portable receiver (b) or a towed receiver (c) when in close proximity. Their movement indicates the motion of similar pebbles. (Courtesy of Dr M. Workman, SOC.)

Movement of shingle

An example of a specialist development is the 'electronic pebble', used to track the movement of shingle on beaches (*Figure 19.24*).

Remote Sensing Instruments

Excluding airborne and satellite remote sensing (Chapter 5), there are two principal methods of remote sensing employed in the study of the ocean, subsurface acoustics, of which the most familiar is the conventional echo-sounder, and land-based radar. The use of optics is limited to ranges of only a few metres generally, because of the high attenuation of light by sea water[3] compared with the tens of metres to thousands of kilometres distances (depending on frequency[8]) over which sound can propagate.

19.25

19.26

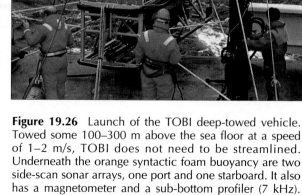

Figure 19.25 GLORIA, the Geological Long Range Inclined Asdic, is housed in a launch/recovery gantry on the stern of RRS *Discovery*. For more than 20 years GLORIA has been used for long-range side-scan sonar surveys, during which time it has surveyed about 6% of the ocean floor. The instrument can survey 20,000 km² a day, and over a 7-year period covered the entire 7 million km² of the US Exclusive Economic Zone. GLORIA images show large features on the ocean floor, such as volcanoes and sea-mounts, channels, sediment slides, and geological hazards. Examples of GLORIA sonographs are given in Chapters 9 and 10. (Courtesy of SOC.)

Figure 19.26 Launch of the TOBI deep-towed vehicle. Towed some 100–300 m above the sea floor at a speed of 1–2 m/s, TOBI does not need to be streamlined. Underneath the orange syntactic foam buoyancy are two side-scan sonar arrays, one port and one starboard. It also has a magnetometer and a sub-bottom profiler (7 kHz). The vehicle can also carry CTDs, transmissiometers, and chemical sensors, which are useful for locating hydrothermal vent sites. Examples of TOBI sonographs are given in Chapter 10. (Courtesy of SOC.)

Acoustic remote sensing

Sonar can be used in two principle ways, 'active' or 'passive'. In the 'active' mode, sound is produced by an instrument, usually in short pulses, and the reflections are recorded by the same instrument or by another. The range of the reflecting 'targets' (for example, the depth of water below the ship carrying the sonar) can be determined from the time between emission of sound and its return, and the known speed of sound in sea water, a time of approximately 1500 ms. 'Passive' sonars receive sound over a broad band of frequencies and can be used, for example, to quantify the occurrence or detect the presence of physical processes that produce sound, such as rainfall (producing sound at about 12 kHz) or the cracking and break-up of winter ice.

Survey of the sea bed

This is the subject of Chapter 20. *Figures 19.25* and *19.26* provide examples of modern deep-sea survey instruments.

Bottom-mounted single instruments using active or passive sonar at frequencies of 1 MHz or more (with a consequent sonar range of 30 m or less), are also used to detect and quantify sediment in suspension, often alongside electromagnetic current meters (*Figure 19.17*) to measure the vertical distribution of current and the stress of the turbulent flow on the sea bed.

Acoustic Doppler Current Profilers (ADCPs)

The Doppler principle – the change in pitch of a wave due to relative motion between the source and receiver – was first used successfully in the early 1960s by Kocsy and others at Miami to measure currents. However, it was not until the early 1980s that the technique became widely adopted for both ship-mounted and *in situ* instruments. Its main advantage over conventional current meters is that it is a remote sensing method. Sound from a transmitter is directed along a narrow beam away from the instrument and is scattered back toward the instrument, often from particles or zooplankton being carried by the current. A receiver compares the pitch of the received and transmitted frequencies, and from their difference, or Doppler shift, calculates the current at different distances from the instrument – hence the 'profile'. Modern ADCPs can measure currents accurate to 1 cm/s in up to 128 depth slices, each typically 4–8 m thick; they can be used from a ship (*Figure 19.27*) to provide a detailed picture of the currents of the upper ocean.

Moored and bottom-mounted ADCPs can provide high-resolution current measurements in distance and time. They contribute to studies of internal waves in the open ocean and have been used extensively through Arctic ice floes, enabling measurements that could not otherwise be obtained. The ADCP also has another use – it may be used to

Figure 19.27 A four-transducer cluster for the acoustic Doppler current profiler on the hull of RRS *Discovery* used in the survey of ocean currents and eddies (see Chapter 4). Acoustic transducers on ships' hulls need careful positioning. Too near the bow and bubbles from the bow wave interfere with sound propagation from the transducer; too near the stern and the transducers pick up noise from the propellers. (Courtesy of R. Bonner, SOC.)

infer the distribution of zooplankton in the open ocean from the strength of the acoustic back-scatter signal.

Acoustic study of the upper ocean

Upward-pointing sonar and hydrophone arrays have been used to investigate the dynamics of the upper ocean, especially turbulence and breaking waves; examples are given in *Figures 19.28–19.29*.

Long-range acoustics

The speed of sound in sea water depends primarily on temperature (increasing by about 4 m/s per K) and pressure; except in Arctic waters, the minimum sound speed occurs in mid-water, typically at some 800 m depth, where sound is trapped and chan-nelled (the SOFAR channel, see above). Low frequency (<100 Hz) sound propagates around the global ocean over distances exceeding 15,000 km, unless impeded by land masses. Changes in the arrival time of coded pulses of sound, which can be measured very accurately, provide information about the changing structure of the water in the propagation path. Over ranges of 1000 km, acoustic 'tomography' can provide information about the structure, variation, and propagation of meso-scale eddies; while over 10,000 km ranges, experiments repeated over a few years should provide information about the variation of temperature averaged over the acoustic path, and hence about changes in ocean climate. This is an objective of the Acoustic Thermometry of Ocean Climate

Figure 19.28 An array of six long-range Doppler sonars mounted on the hull of the research platform FLIP. With FLIP in the vertical orientation (as shown in *Figure 19.1*), the four two-panel sonars in the foreground point downward and outward, defining the edges of a four-sided pyramidal measurement array. The larger six-panel sonars are directed horizontally, just below the sea surface. Each beam transmits pulses of 20–30 ms duration at a frequency of 67–80 kHz and a peak power of 2 kW. The sound scatters from zooplankton drifting with the water. From the Doppler shift of the echo, the radial component of water velocity, and the characteristics of internal waves and upper-ocean structure can be determined to ranges as great as 1.2 km. This system was developed in 1979 and has been used extensively since. (Courtesy of Dr R. Pinkel, Scripps Institution of Oceanography, La Jolla, California, USA.)

19.29

Figure 19.29 A hydrophone array, used to 'listen to' and track the motion of breaking waves, which play an important part in the transfer of momentum and gases between the air and the sea (see Chapter 2), being recovered using a Zodiac inflatable boat. Recovery of equipment at sea is often a difficult and potentially dangerous operation, requiring careful attention to stringent safety precautions. For further details of the equipment and its use, see Farmer and Ding[1]. (Courtesy of Dr D. Farmer, Institute of Ocean Sciences, Sidney, British Columbia, Canada.)

19.30a

19.30b

Figure 19.30 Acoustic transducers used in a successful pilot experiment to propagate coded pulses of sound at 57 Hz from close to Heard Island in the Southern Indian Ocean in late January and early February 1991 to receiving stations around the world's ocean, some over 15,000 km away. Here, the acoustic sources are shown (a) ready to be lowered off the support ship (fairings were fitted to smooth the flow around the cylindrical transducers) and (b) on recovery after several days of severe weather. Fairings were torn away, exposing the transducers to the full force of the Southern Ocean swells. The experiment was designed to test the feasibility of developing an 'acoustic thermometer', using acoustics to measure the changes in mean ocean temperature (sound speed increases with temperature). An ocean warming of 4 mK per year would result in an increase in sound speed of about 0.02 m/s per year which, at a range of 15,000 km reached by sound in about 10,000 s, would lead to a reduction in travel time of about 0.13 s per year, which is easily measurable (see text and *Figure 3.16*). Plans have been made to establish a global network of transducers and listening stations to monitor climate change in the ocean. (Courtesy of Dr A. Forbes, Scripps Institution of Oceanography, La Jolla, California, USA.)

(ATOC) program, of which the Heard Island Experiment (*Figure 19.30*) was a forerunner[2].

Radar

Land-based radar provides another powerful tool to study the surface of the ocean. One example of its use is illustrated in *Figure 19.31*.

Future instruments and methods

The development of sensors

The greatest demands for automated methods of ocean monitoring and survey are now from the disciplines of biology and chemistry. As knowledge of the ocean advances, the limitations of nets (biolo-

Figure 19.31 The Ocean Surface Current Radar (OSCR), deployed at Crammag Head Lighthouse on the Galway Peninsular, W. Scotland. In the foreground are three elements of the 85 m long, 16-element receiver array. Each element stands 2.25 m high. Behind the receiver array are four elements of the transmitter array, which are 5.5 m high. The lighthouse contains the radar hardware and computer. The OSCR radar transmits electromagnetic waves in the HF radio band at 27.0 MHz, corresponding to a wavelength of 11 m, which are scattered from the waves on the sea surface. Scattering comes mostly from waves with half the radar wavelength, 'Bragg scattering', in a manner similar to that of X-rays from a crystal lattice. The frequency (Doppler) shift of the returning radar waves is measured by the receiver, and the speed of sea waves can thence be measured. This is a combination of their movement through the water (which is well-known from wave theory) and the speed of the radial component of the surface current, which can be estimated by subtraction. The use of a multi-element array produces a narrow beam, which can be digitally steered through 90° of azimuth. The OSCR system collects data from 1 km² 'cells', with a maximum of 700 data cells and a maximum range of 40 km. Two radars are used with orthogonal coverage of the required survey area to resolve the two components of surface vector current. (Courtesy of Mr R.D. Palmer, SOC.)

19.31

gy) and laboratory analyses (chemistry) become more obvious and, indeed, will limit scientific progress. Improvements can only come about through new chemical and biological sensors.

One approach is to create new instruments by miniaturising equipment found in the laboratory, reducing its power consumption, and removing any need for human intervention. These are testing requirements, but may be achievable for:

- **Flow cytometry**, a technique developed for medicine, in which the characteristics of single cells are analysed at rates of thousands per second using laser optical methods. Developments in solid-state lasers and fibre-optics now make it feasible to consider marine *in situ* flow cytometers. The biomass of single-celled organisms in the ocean is immense and poorly known, so this technique is potentially very powerful.
- **Broadband acoustics and miniature low-light cameras**, used in medicine for imaging, may provide a means of studying zooplankton species distribution and behaviour, building on the present use of acoustics for estimating abundance.
- **Flow injection chemical analysis**, a standard laboratory technique, is being adapted for *in situ* use. Micromachining, with pumps and valves based on electrostrictive materials – materials that change their physical dimensions in response to an applied electrical signal – could provide a route to miniature analysers that require minimal amounts of reagents and maintenance.

The other approach is to devise novel sensors or adapt existing sensors, for example:

- **Ultraviolet and infra-red absorption** – some ions exhibit an absorbance at well-defined wavelengths in the ultraviolet or infra-red regions. Nitrate concentration in the ocean is particularly important, and may be determined through ultraviolet absorption. New ultraviolet lamps, narrow ultraviolet filters, and *in situ* signal processing are now making this possible.
- **Biosensors**, already used in medicine, integrate a chemical substance or enzyme with a semiconductor to give a transducer that converts directly the concentration of a substance into an electrical signal (e.g., sensors for glucose are now commonplace). As a technique, this has potential to add to our knowledge of the microbiological and organic chemistry of the ocean.

Novel vehicles and platforms

The use of submersibles that carry scientists and instruments to investigate the ocean depths is becoming routine (*Figure 19.32*). The advent of cheap disposable fibre-optic cable enables miniature, remotely operated vehicles (ROVs) to be designed for applications such as the acoustic and optical imaging of zooplankton. The fibre-optic link provides real-time control of the vehicle and video data from sensors (*Figure 19.33*).

The task of making routine measurements in the ocean to meet the demand of present computer models is growing beyond the capability of the world's fleet of research vessels, even with ROVs.

19.32a

19.32b

19.32c

19.32d

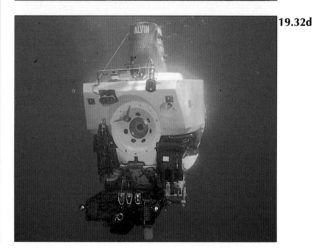

Figure 19.32 Some of the submersibles used for investigation of the deep ocean. (a) The French submersible, *Cyana*, capable of operating to depths of 4000 m. The hull is titanium, and the submersible carries a pilot, navigator, and one scientist. It is usually equipped with underwater lights, an arm, claw, and video. (Courtesy of Professor P. Tyler, SOC.) (b) The Russian *Mir I* submersible, capable of operating to 6000 m carrying two pilots and a scientist. The hull is titanium. The submersible is here carrying a (red) sediment trap, and still and video cameras; it has been used to investigate hydrothermal systems in the North Atlantic (see Chapter 13). (Courtesy of Professor P. Tyler, SOC.) (c) The US *Johnson Sealink* submersible in preparation for deployment. It is capable of operating to 900 m with a pilot and scientist in the front compartment and with a technician and second scientist in an aft compartment. The sphere is acrylic. In addition to lights and cameras, the submersible has a manipulation claw. (Courtesy of Professor P. Tyler, SOC.) (d) The US Navy-owned Deep Submergence Vehicle (DSV) *Alvin* operated by the Woods Hole Oceanographic Institution. A typical 8 hr dive takes two scientists and pilot to 4500 m with 4 hours on the bottom for observation, photography, and experiments. Three video and two 800 frame cameras are usually carried, together with two hydraulic arms and instruments such as corers, temperature probes, water samplers, and a biological sample pump. (Courtesy of Woods Hole Oceanographic Institution.)

Figure 19.33 The Remotely Operated Vehicle (ROV) Jason, named after the Greek hero who searched for the Golden Fleece. It is owned and operated by the Woods Hole Oceanographic Institution and has been widely used in studies of the deep ocean floor. An early version was operated from DSV *Alvin* [*Figure 19.32(d)*] during the expedition, which discovered the wreck of the 'Titanic' in 1986. Launched from the support vessel to which it is connected by a 10 km long fibre-optic cable, Jason carries cameras and sonar from which signals are transmitted back through the cable to the pilot, who steers the vehicle from the support ship, and to the scientist directing operations and the research team. Images and navigation data are recorded for subsequent analysis. (Courtesy of the Woods Hole Oceanographic Institution.)

19.33

Computer power is rising exponentially, continually increasing the gap between modelling and observations. Autonomous vehicles and systems provide a low-cost way to scale up our observations. The ARCS vehicle from ISE Research Ltd, Canada, the Odyssey vehicle from MIT/Woods Hole Oceanographic Institution, and the GEC-Marconi AUV have successfully demonstrated autonomy for scientific data-gathering over limited depths and for short periods. The UK Autosub project aims to provide scientists with a deep-diving vehicle able to collect routine physical, chemical, biological, and geophysical data over transects of several thousands of kilometres (*Figure 3.15*).

General References

Griffiths, G. (1992), Observing the ocean – recent advances in instruments and techniques for physical oceanography, *Sci. Progr.*, 76, 167–190.

Jerlov, N.G. (1968), *Optical Oceanography*, Elsevier, Amsterdam.

Leaman, K.D. (1990), Physical oceanographic measurement techniques at sea, in *The Sea: Ocean Engineering Science*, Vol. 9B, Mehaute, B. and Hanes, D.M. (eds), J. Wiley & Sons, New York, pp 1163–1192.

Urick, R.J. (1975), *Principles of Underwater Sound*, 2nd edn, McGraw-Hill, New York.

References

1. Farmer, D.M. and Ding, L. (1992), Coherent acoustical radiation from breaking waves, *J. Acoust. Soc. Amer.*, **92**, 397–402.
2. Heard Island Principles (1991), The Heard Island experiment, *Oceananus*, **34**, 6–8.
3. Jerlov, N.G. (1968), *Optical Oceanography*, Elsevier, Amsterdam.
4. Oakey, N.S. (1988), EPSONDE: an instrument to measure turbulence in the deep ocean, *IEEE J. Oceanogr. Eng.*, **13**, 124–128.
5. Osborn, T.R. (1974), Vertical profiling of velocity microstructure, *J. Phys. Oceanogr.*, **4**, 109–115.
6. Osborn, T., Farmer, D.M., Vagel, S., Thorpe, S.A., and Cure, M. (1992), Measurements of bubble plumes and turbulence from a submarine, *Atmosphere–Ocean*, **30**, 419–440.
7. Schmitt, R.W., Toole, J.M., Koehler, R.L., Mellinger, E.C., and Doherty, K.W. (1988), The development of a fine- and micro-structure profiler, *J. Atmos. Oceanogr. Tech.*, **5**, 484–500.
8. Urick, R.J. (1975), *Principles of Underwater Sound*, 2nd edn, McGraw-Hill, New York.

CHAPTER 20:

The Sea Floor – Exploring a Hidden World

P. Riddy and D.G. Masson

Introduction

Almost since the human race began to interact with the sea, a knowledge of the sea-bed structure and composition has become interwoven with the lives of those who work in the marine environment. At first an experiential and intuitive knowledge probably allowed boatman and sailors to find sources of food and, avoiding hazardous rocks and reefs, to return safely to shore. Later, as foraging trips went further afield and perhaps exploration began, some sort of record or simple chart would have been made of where shoals and deeps occurred. Eventually, weighted lines were used to measure depth; when such measurements were recorded along with the location at which they were made, bathymetric surveying was born. Position-fixing and depth measurement have thus always been intimately related, although the criers of 'by the mark' when 'heaving the lead' were usually more concerned about how much water was beneath their keel (safety) than exactly where they were (their position, see *Figure 1.3*). The days of navigation or position-fixing by sextant [(*Figures 20.1(a)–20.1(c)*] and depth measurement by weighted line are largely over, although these techniques illustrate well the basic tools of bathymetric surveying (indeed, they may still be used where resources to purchase advanced technology are limited).

Today, we are also interested in the wide range of material of which the sea floor is made, its potential value (see Chapter 21), and how it is shaped and organised by the overlying water. So the depth, the variations in sediment type, and the outcropping rocks are together useful in the characterisation of the sea floor. Other factors, such as the relationship between the species which live on and in the sea bed (see Chapter 13) and the sediments/rocks which are present (see Chapter 8),

20.1a

20.1b

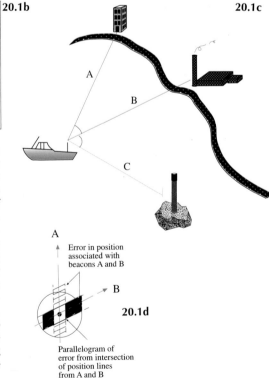
20.1c

20.1d

Figure 20.1 Sextants (a) are traditional navigation instruments which have been in use (b) since 1731. The diagram (c) illustrates the use of sextants for position-fixing in good visibility up to a few kilometres from the landmarks. Two sextants are used to measure simultaneously the angles between three landmarks or objects of already-known position – further calculation allows two intersecting position circles to be drawn[6]. This technique can give accuracies of a few metres, but using sextants from the deck of a boat (an unstable platform) requires practice! (d) Lines A, B, and C are known as position lines. Each position line has an associated error which is dependent on the position-fixing system being used. This diagram shows the effect of errors associated with two of the position lines. Instead of the position being defined at a single point, the vessel's position lies somewhere within the parallelogram. No matter what position-fixing system is used, an error will always be associated with each position line. For high-accuracy positioning at least three position lines are normally used to try and minimise the effect of errors.

Figure 20.2 GLORIA 6.5 kHz side-scan sonar image of a submarine volcano in the eastern Pacific off South America. Strong sonar targets are white, and acoustic shadows are black. The volcano mouth is approximately 2 km in diameter.

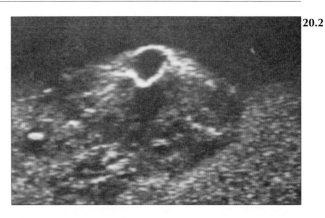

contribute to the more complete picture. The subsurface structure of the sea floor influences the visible surface structure. The chemical composition of the sediments can influence the range and diversity of life present and the nature of the sea floor. These factors, in turn, influence or are influenced by the overlying currents. In this chapter, we focus on the basics of the methodology and technology used to investigate the geological and physical characteristics of the sea floor in shallow and deep water environments, and illustrate some of the features found (*Figure 20.2*).

Marine Geophysical Surveying

The goal of the marine geophysical survey is to provide an integrated picture of the surface, the subsurface structure, and the composition of the sea bed. Most such surveys involve remote sensing of the sea bed by tools deployed at or near the sea surface. Surveys follow an exploration or search

track over the area of interest, with the appropriate sensors being deployed to continually record all the data (*Figure 20.3*).

The structures of the sea bed and ocean basins revealed by geophysical survey techniques are described and illustrated in Chapters 8 and 9. Several bottom-survey tools use sound waves to obtain information from the sea floor, while others measure the magnetic field, the gravity field, or the electrical field. Any system which remotely senses the sea floor requires a compromise between accuracy and/or resolution, and range – this becomes

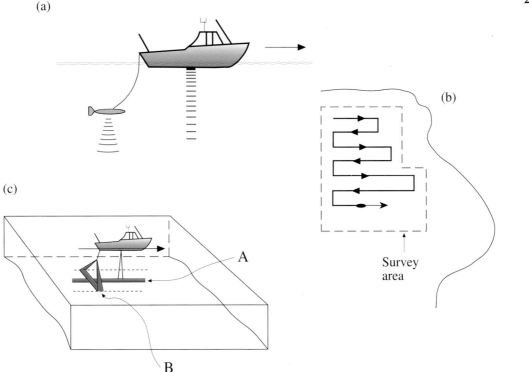

(a)

(b)

(c)

A

B

Survey area

Figure 20.3 A survey design strategy involves (a) the selected equipment being deployed from the vessel, and (b) the vessel following a survey track designed to optimise coverage of the area. (c) A – profilers collect a narrow 'line' of data from directly beneath the vessel; B – sonar and swath systems collect a broad band of data across the vessel's track.

Table 20.1 Electromagnetic spectrum and position-fixing systems. Only typical parameters for each system are given. In particular, short-range systems can be given improved range capability by elevating or enlarging the antennae and increasing the output power of the transmitter, although power transmitted is limited by legislation. With very high-accuracy systems, the movement of the vessel and the position of the measuring devices with respect to the 'antenna' become important factors in the measurement of position. (Based on Ingham[6].)

Frequency	Wavelength	Type of electromagnetic position-fixing system[a] with examples	Typical working accuracy
Radio waves			
10 kHz	30 km	**Very long range (world-wide):**	
		Omega (10–15 kHz)	2–4 km
30 kHz	10 km	**Long range (200 km):**	
		Loran C; Lambda (100 kHz)	0.2–2 km
		Decca (112–196 kHz)	40–400 m
		Pulse 8 (100 kHz)	16–30 m
300 kHz	1 km	**Medium range (80–150 km):**	
3 MHz	100 m	Hyperfix (1.6–3.4 MHz)	8–15 m
30 MHz	10 m		
300 MHz	1 m	TRANSIT satellite (150 and 400 MHz)	200 m
		GPS satellite (1572.42 MHz)	15 m
		GPS satellite with SA on	100 m
		GPS in differential mode	1–10 m
		Syledis (406–488 MHz)	8–15 m
Microwaves			
3000 MHz	100 mm	**Short range (0–80 km):**	
		Radar (3000–10,000 MHz)	>1 m
		Trisponder (9000 MHz)	1–5 m
30 GHz	10 mm		
		Microfix (5.48 GHz)	1–5 m
300 GHz	1 mm		
Radiated heat			
3000 GHz	0.100 mm		
30,000 GHz	0.010 mm		
Infra-red			
300,000 GHz	0.001 mm	Infra-red distance measurers	a few cm
Visible light			
3,000,000 GHz	0.0001 mm	Lasers	a few cm

[a] Short range systems measure range(s) and sometimes bearing; others are predominantly hyperbolic in operation.

clearer later when the systems are described. Descriptions of devices to obtain samples of the sea bed (e.g., cores) are omitted as they are described in Chapter 19.

Position-Fixing – The Position of the Vessel

We begin with navigation or position-fixing princi-ples and systems, as other data are of little use if we do not know where they come from!

Position lines

Position lines can be defined in a variety of ways, the accuracy of each position-fixing system being calculated from the error associated with, or 'thick-ness', of the position line [*Figure 20.1(d)*]. All methods involve the measurement of angles or

Figure 20.4 Types of position-fixing systems and their use (see also *Table 20.1*). Longer range EM hyperbolic systems (A) can provide accuracies of 10–500 m at ranges of 50–2000 km. All-weather satellite systems (B) provide 24-hour positioning to around 100 m (differential users can be accurate to within 1 m). Sextants used with landmarks (C) in good visibility provide high accuracy close to shore, and are used in mid-ocean for astronavigation. Range–range and hyperbolic systems (D) provide 1–10 m accuracy at ranges up to 80 km. Optical and EM-bearing systems (E) give accuracies of 0.01–10 m at ranges of 2–20 km. (Based on Ingham[6].)

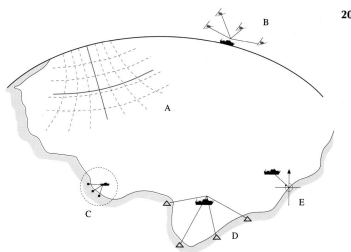

ranges, or a combination of the two. Further information on the wide range of methods and instruments used for position-fixing is given by Ingham[6] and Forssell[4]. Today's requirements for high-accuracy, all-weather position-fixing systems have led to the development of sophisticated electronic systems based around the transmission of electromagnetic (EM) waves at frequencies which range from radio waves to visible light (*Table 20.1*). Such systems (*Figure 20.4*) offer a variety of accuracies and ranges, but until the arrival of satellite systems high accuracy was only possible over short ranges.

EM waves are briefly discussed in *Box 20.1*. Of the various systems, the Navstar Global Positioning System (GPS, see below), is fast becoming the most used world-wide.

Global position-fixing system (GPS)

This is a satellite-based system developed for the US government to provide a high-accuracy, all-weather position-fixing capability for 24 hours per day anywhere on the globe.

The system has a number of components, illustrated in *Figure 20.5*. Users can be given different levels of access to the system, the highest level of accuracy normally being reserved for the military. The receiver-position calculations are based on the satellites' orbital parameters as transmitted by the satellites.

A modification of the these signals, known as Selective Availability (SA), has degraded the potential accuracy of the system from 15 m to >100 m, although full accuracy is still available to authorised users by decoding.

20.5

Space segment

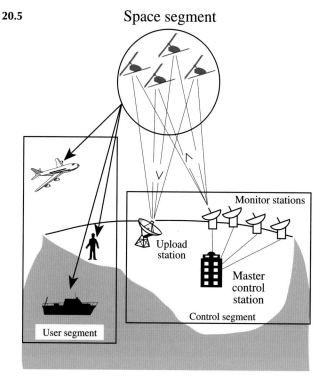

Figure 20.5 GPS components and operation. Satellites moving in known orbits, about 20,200 km above the Earth's surface, transmit information on their position continuously. This is received by users and allows them to calculate their position – the satellites become the landmarks of the user. The satellites are monitored and updated through the control segment network, which can lead, for example, to correcting deviations in a satellite's orbit. Various errors can be coded into the satellite transmissions to give an error in the calculation of position for users not equipped with the appropriate decoding information; this typically results in positional errors of about 100 m (see text).

Box 20.1 What is an electromagnetic wave?

An electromagnetic (EM) wave is a combination of oscillating electric and magnetic fields which transport EM energy. EM waves are able to propagate in a vacuum at a speed of approximately 299,776 km/s, and cover a wide frequency band which includes visible light (*Table 20.1*). EM waves are reflected or absorbed by objects and refracted by changes in the refractive index of the medium through which they travel. Waves undergo a velocity change when refracted as they travel through different parts of the Earth's atmosphere. This results in a potential source of error in measuring the ranges from satellites, which becomes important in the calculation of accurate positions.

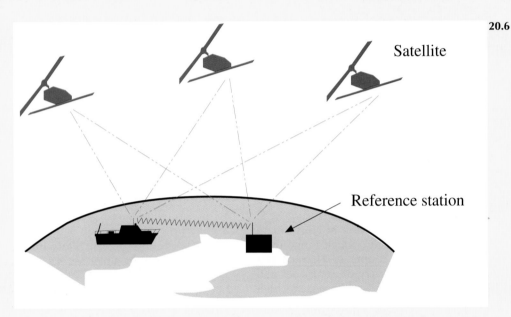

20.6

Satellite

Reference station

Figure 20.6 Differential GPS (DGPS) involves corrections based on the known position of a reference station being calculated and transmitted to the vessel. The vessel uses the correction information to recalculate and correct its known position, which has been calculated from directly received satellite signals. This provides positional accuracy of less than a metre.

Differential GPS (DGPS)

This development of GPS uses a reference station with knowledge of its own position to calculate the errors in the satellite signals caused by SA and by atmospheric conditions (*Box 20.1, Figure 20.6*). The errors, or an error correction, are then transmitted in real-time from the reference station to the vessel, where they are applied to give much greater accuracy than is provided by GPS. The accuracy of DGPS partly depends on how close the reference station is to the survey. Where the reference station is within 30 km of the survey vessel, 'instantaneous' accuracies of less than 1 m are consistently obtained. When used in a special geodetic mode, centimetre accuracies can be obtained. Some companies have set up networks of reference stations to routinely provide corrections that allow a positional accuracy of around 5 m over an area extending hundreds of kilometres from the base stations.

The position of the sensor

Some survey sensors are towed at the end of a cable, which can be several kilometres in length (*Figure 20.7*). If the position of the sensor is not well-known, it is impossible to produce an accurate seabed map of the property being measured. In practice, acoustic devices on board or placed on the sea bed (e.g., an array of accurately positioned transponders or navigation beacons) may be used to measure the relative position of a sensor package.

The Sea Floor

What are we looking for? It may be the size, orientation, shape, and composition of natural features (or natural obstacles), or the location of manufactured objects, such as sunken vessels and mines. Surveys of shallow-water environments tend to be more detailed and to find wide local variations of features, sediment type, and their distribution. In

20.7

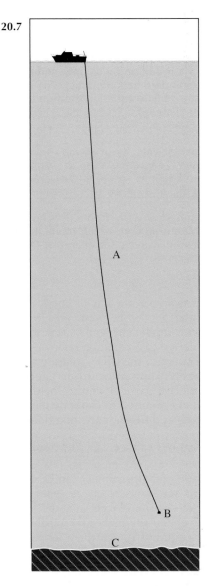

Figure 20.7 Surveying at depth in the ocean involves a tow cable (A) and a sensing package (B), which is held near to the sea bed (C). Scaled to a vessel length of 225 m, the water depth here is 3150 m, but sensors can be deployed at 5000 m or more.

20.8

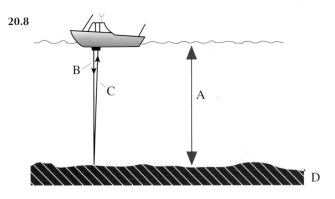

the deep-water environments, which cover a substantial area of the Earth's surface, there is generally more interest in the larger scale changes in sediments and features (see Chapters 8 and 9). In either environment, the minimum expected size of the feature of interest usually determines the type of system used, the survey line spacing, and, consequently, the type of position-fixing system. Sedimentary features vary in size from sand ripples of a few centimetres in wavelength and height, to the massive slumps and turbidite features described in Chapter 9. Small features are often found superimposed on larger features, which may be important in understanding the general evolution of an area. Changes in sediment type can be detected by changes in the related reflective properties of the sea floor, as can be seen, for example, in *Figure 9.13*. Deep underwater canyons are easy to identify, but difficult to survey accurately. The ocean liner *Titanic* remained hidden in 5 km of water for over 40 years before Bob Ballard[1,2] used sophisticated sonar systems and submersibles with cameras to find and explore the wreck.

Principles of Surveying Equipment

The echo-sounder

The use of instruments which remotely sense the sea floor is well-illustrated by the echo-sounder, which is used for measuring water depth (*Figure 20.8*). A short pulse of sound at a given frequency is transmitted into the water column by an acoustic transducer, either on the vessel's hull or towed at a known depth beneath the surface. The transducer can both transmit and receive sound pulses. The pulse reflected back from the bed is collected; its travel time to reach and return from the sea bed and its amplitude are measured. The amplitude of the returning pulse relative to that transmitted depends on the depth of water (since water attenuates sound energy) and the reflection characteristic (or reflectivity) of the sea bed. The size, shape, reflectivity, and orientation of the sea-bed features to the arriving sound pulse also affect the returning pulse amplitude.

Some systems use the characteristics of reflected pulses to obtain other information, such as the type of sediment which makes up the sea bed. The relationship between a sound *pulse* and a sound *wave*

Figure 20.8 In echo-sounding, the time a pulse of sound takes to travel to the sea bed and back is measured; since the speed of sound in water is known to be about 1500 m/s, it can be converted into depth. The amplitude (strength) of the returning, reflected signal depends on the depth and absorbence of the water and the type of sea bed – e.g., rock will give a stronger reflection than sand. (A) Depth of water, (B) path of outgoing pulse, (C) path of reflected pulse, and (D) sea bed.

Box 20.2 What is a sound wave?

Energy in the form of a sound wave can be transmitted through water, since it is a compressible media. The speed (*v*, m/s) is proportional to the water density as modified by salinity, temperature, and depth, as described by equation (20.1), where *S* is salinity (standard salinity units), *T* is temperature (°C), and *D* is depth (m)[7]. The equation indirectly describes the density variation through the water column, density having the most significant effect on the velocity of sound.

$$v = 1492.9 + 3(T - 10) - 6*10^{-3}(T - 10)^2 - 4*10^{-2}(T - 18)^2 + 1.2(S - 35) - 10^{-2}(T - 18)(S - 35) + D/61 \qquad (20.1)$$

Many instruments make the inaccurate assumption of a constant velocity of sound in sea water, although the error introduced is often not important in the qualitative interpretation of results. When high accuracy is required, measurements of the temperature and salinity in the survey area can be combined to estimate the variation of sound velocity with depth, and hence to correct depth and range measurements.

Although sound waves travel outward from their source as spherical wave fronts, it is convenient to represent the passage of sound waves through the water column as rays which travel in straight lines. Sound waves are absorbed, refracted, and diffracted like other waves; for example, they obey Snell's Law when being transmitted through layers of sub-bottom rocks.

is illustrated in *Figure 20.9*, and the propagation of sound waves is described in *Box 20.2*.

Water depth is measured as a profile beneath the vessel. Since the transducer sends out a sound beam that is roughly cone shaped, the depth is measured from returns (frequently the first return) from within the cone; this can lead to inaccurate results. Multibeam sounders (discussed later) have been developed to provide a swath of depth measurements each side of the vessel, and to improve the resolution of depth and sea-bed features. Accurate measurement is important in areas where the bathymetric data are used to produce charts for shipping or, for example, in planning a construction project. Changes in sea-floor level or in the shape of the topography may indicate convergence or divergence of sediment flux or of varying current speeds.

All major bodies of water experience tidal changes in water level, caused mainly by the gravitational attraction of the Sun and Moon. This can vary from a few centimetres in enclosed seas, like the Mediterranean, or in the deep oceans, to over

10 m in, for example, the Bay of Fundy. In order to compare water depths from one area with another and to provide tidal charts for mariners, depths are generally referred to a common reference level (e.g., Ordnance Datum in the UK). Accurate tidal corrections, probably using a local tide gauge together with a numerical model of the local tides (see *Figure 1.6*), are required to calibrate measurements of depth made at different parts of a tidal cycle.

What size features can we see? The resolution of features

The resolution of an instrument can be defined as is its ability to see two objects as separate (*Figure 20.9*). The closer the objects are, the harder they are to resolve, and the higher the resolution of the instrument required to see them as separate. To improve the exact identification of reflected sound waves, marine instruments usually transmit a stream of short pulses of sound, each of a few cycles grouped together. Each instrument's transmissions are characterised by a pulse rate and fre-

20.9 (a)

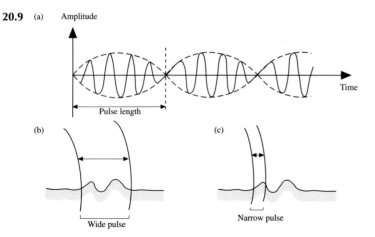

Figure 20.9 (a) Pulses consist of several wave cycles grouped together. (b) The wider (actually longer) pulse reaches both features at the same time, so they appear as one reflection. (c) Both features are seen separately by the narrower (actually shorter) pulse. The sea-bed resolution of a system depends on the length of the sound pulses transmitted from the sensor, and is generally taken as half the pulse length.

20.10

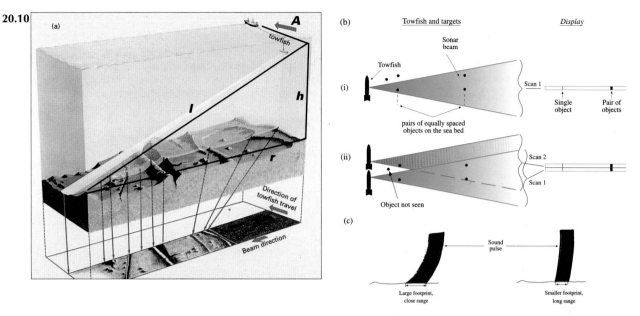

Figure 20.10 The principles and resolution of side-scan techniques are shown. (a) The system records successive scans of the sea bed and builds up a picture line by line. In this case, the conversion of the time into distance using the speed of sound produces a slanted range of features from the transducer; this has to be converted into the true range using the equation $r^2 = l^2 - h^2$, where r is the true range, l is the slant range, and h is the height of the towfish. This conversion assumes the sea bed is flat and the velocity of sound is constant. Any velocity correction would have to account for the more difficult-to-define spatial velocity variation. The along track distance (A) is determined from vessel navigation. (b) Here, the side-scan towfish position while emitting consecutive pulses is illustrated. Resolution along-track is complex, being basically a function of the beam pattern, proximity of objects to the towfish, and the display device. (i) Two objects may appear as one of slightly higher intensity-of-reflection than a single closer object. The further objects appear as one more intense object on the display. (ii) The two closer objects are still not resolved with successive pulses, as they appear as one continuous object on the display device. With a larger delay between pulses (long-range settings), the towfish can move sufficiently far forward to miss small objects at a close range. (c) For resolution across track, the outgoing pulse forms a larger footprint at short ranges than at long ranges.

quency to which its receiving electronics is 'tuned', thereby improving the quality of reception and reducing the effect of other sound sources. The resolution is determined by the frequency, wavelength, beam shape, and pulse length of the transmissions which characterise the instrument.

Instrument resolution is only part of the story – detail is also lost if we take a picture which represents 75 m of sea bed and compress it down to a 30 cm print. The overall resolution of the system, of both transducers and recording devices, is important [e.g., see *Figures 20.10(a)–20.10(c)*].

What size objects are visible? The effect of background

Have you ever tried to find a particular pebble on a pebbly beach? If the feature or object has the same reflective properties as its neighbours and is at the limit of the system resolution, then it will be impossible to spot. Similarly, sea-bed features must have a different reflective characteristic to their neighbours to stand out. For example, wood saturated with water is difficult to distinguish from soft sediments by using underwater acoustic measurements.

Instrumentation

Deep and shallow water

Owing to the difference in the scale of investigations in shallow and deep water (see above), systems tend to be designed for one or the other. A high-resolution, deep-water system generally uses similar internal electronics to its shallow-water counterpart, but is deployed in a different way, and, to work at depth, must be designed to withstand substantial pressure. Lower frequency sound used by longer range systems suffers less attenuation in the water column, and is more suitable for deep-water environments. *Table 20.2* lists a number of types of system and gives examples of the operating ranges which can be expected.

Side-scan sonar

The standard echo-sounder (see p. 305) only makes measurements directly beneath the vessel, which is an inefficient use of ship time. Side-scan sonar is used to take the equivalent of a mosaic of aerial photographs of the sea floor (see Chapter 5), but using sound [*Figure 20.10(a)*]. The general principles of resolution are outlined above and in *Figures*

Table 20.2 Surveying instrumentation. The data given can only serve as a guide, since the parameters depend on the configuration of each system, and the type of sea bed. The compromise between system resolution and depth of penetration is clear. Chirp systems are a relatively recent development in profiling technology, and employ a swept-frequency pulse and sophisticated processing of the return signal to achieve high resolution, while giving good penetration of the sea bed.

System	Frequency (kHz)	System Resolution (cm)	Penetration (m)
Echo-sounder	200	5	0–0.5
Echo-sounder	35	20	0–2
Pinger	3.5	20	20–75
Boomer	1–6	40	50–100
Chirp systems	**0.5–12.5**	**10**	**40**
Sparker	0.1–1	200	100–200
Air gun	0.05–0.1	200–5000m	100–tens of km

20.11a

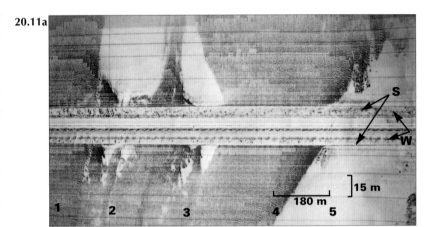

20.11b

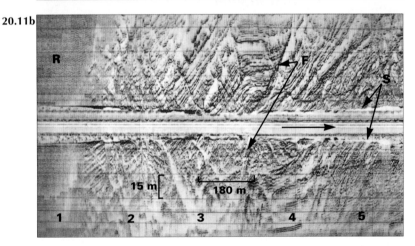

Figure 20.11 Two annotated sections of side-scan traces recorded in shallow water. The grey scale is calibrated for the strength of reflections, from black (maximum) to white (minimum), and is typical of this type of system. Both traces were recorded during the same survey with consistent system settings. (a) This is from an area of soft sediment. The numbers 1–5 are fixed reference points marked on the trace at one minute intervals. Note the distance scales marked on the trace and the consequent scale distortion of the features – along track distances are proportionally larger than across track distances. The bed has distinct zones which are rippled (darker areas) and unrippled, respectively, interspersed with other dark areas, ambiguous because of the lack of ripples. The sea bed in this area is predominantly sand with varying quantities of shell fragments. The sand ripples are of small amplitude, typically less than 0.2 m, and the apparently narrow boundary between rippled and unrippled areas extends over several metres. (S, sea-bed reflection; W, water-column noise.) (b) The trace shows a rocky area which has undergone considerable folding and faulting, and is directly adjacent to an area of rippled sand (R). Note the fault indicated by the strong reflection (F) running diagonally across the centre of the trace. The layers of rock are being viewed from 'end on', dipping at a steep angle down into the page.

Figure 20.12 (a) The GLORIA side-scan vehicle during deployment from its dedicated launch cradle. The vehicle is 7.75 m in length and weighs approximately 2 tonnes in air, but is neutrally buoyant in sea water. It is towed at a speed of between 8–10 knots. (b) GLORIA 6.5 kHz side-scan image showing a linear chain of submarine volcanoes in the Pacific Ocean, west of San Francisco (scale bar is 20 km) – strong sonar targets are white, acoustic shadows are black (note this is opposite to those in *Figure 20.11*).

20.12a

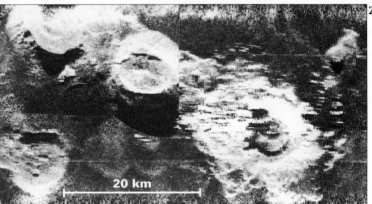

20.12b

20 km

20.10(b) and *20.10(c)*. In the interpretation of side-scan pictures, a number of other factors influenced by the range of operation have to be kept in mind.

The images built up, as illustrated in *Figure 20.10(a)*, are named sonagraphs and represent the **intensity of reflection** plotted in the two space coordinates, range and distance, along the vessel's track. The intensity depends on the reflective nature of the sea floor as well as on the orientation of the features [*Figure 20.10(a)*].

Features may be compressed or expanded in the along-track direction unless a correction is applied to allow for the speed of the vessel (*Figure 20.11*). Other distortions of the image can be caused by electronic noise and interference, and the movement of the towfish in the water[5].

Sonagraphs provide much information about the nature of the sea bed, (e.g., texture, composition, and the orientation of features). Their interpretation requires experience and some knowledge of an area in order to decide what features are likely to be real. Natural features, for example, tend to follow particular patterns or shapes and have a reflectivity which is often different to human artefacts.

GLORIA (Geological Long Range Inclined Asdic)

GLORIA, or Geological Long Range Inclined Asdic (*Figures 20.12* and *20.2*), is a unique side-scan sonar system, developed at the UK Institute of Oceanographic Sciences, for rapid imaging of large areas of sea floor. It operates at a frequency of 6.5 kHz and has a depth-dependent swath width with a practical maximum of 45 km. Towable at speeds of up to 5 m/s, up to 20,000 km² of sea floor can be covered in a single day (an area the size of Wales or Massachusetts). Over a 25-year period it has evolved from a large and difficult-to-deploy single-sided sonar vehicle into a much more compact vehicle which produces both side-scan imagery and swath bathymetry (see later) on both sides of the vehicle track. The neutrally buoyant vehicle has its own handling system and is configured with buoyancy at the top and transducers below to give good roll stability. It is towed from the nose at a depth of 40–80 m, depending on the ship's speed and its tow-cable length (usually about 400 m). The resolution of GLORIA, as with other side-scans, varies with range, being largely dependant on the horizontal beam width (about 2° for

20.13

(a)

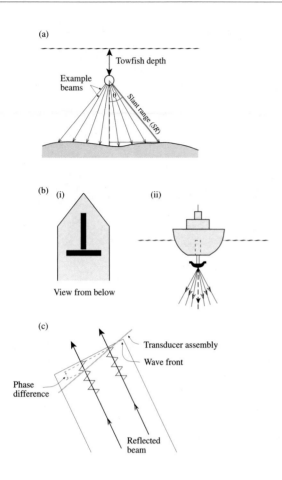

(b)

(c)

Figure 20.13 The more common types of deeper water *multibeam* systems (e.g., Simrad EM12, Seabeam, Krupp Atlas Hydrosweep) use a cross-shaped array of transducers to form a series of beams fanning out perpendicular to the ship's track. In these systems, the beam pattern is controlled by transmitting from the along-ship arm of the cross and receiving on the across-ship arm. Multibeam systems are generally more accurate by a factor of 5–10 than the interferometer systems, but give lower swath widths. (a) The travel paths of a number of example beams – continuous depth measurements are available over the full width of the swath. Simplistically, depths are calculated by converting the beam travel time into a slant range (*SR*) and using the equation $D = SR\cos\theta + T$, where D is depth, T is the towfish depth, and θ, the beam angle to the vertical, is known or measured. (b) Examples of multibeam transducer array configurations: (i) line array (discussed in the text); (ii) cylindrical array, which can be directly hull-mounted, or placed on a retractable unit to avoid turbulence associated with the hull. (c) In interferometer measurements, a reflected wave-front strikes the transducer assembly at an angle. By measuring the change in phase as the wave moves across the transducer(s), its angle of incidence can be measured, allowing the beam angle, θ, to be calculated. The angle of incidence of the wave front is related to the distance of its reflecting origin from the ship's track.

GLORIA). At mid-range, it can resolve features about the size of a football pitch in a water depth of 5 km, i.e., an area around 120 x 60 m.

Swath-sounding systems

Echo-sounders provide depth along a profile; side-scans provide both depth beneath the vessel from a straight down–up return and a picture of an area based on the variation in the intensity of reflections. It would be more efficient if one system could provide depth and intensity measurements at several ranges (not only below the vessel). This is achievable by swath sounding.

A development of the echo-sounder/side-scan technology, swath sounders measure depth along a swath which extends each side of the vessel's track [*Figure 20.13(a)*]; depths are calculated for each

point along the swath. The shallow-water versions of these systems can provide full coverage and an overall accuracy of better than 10 cm in depth and 5 cm in range over the swath. The configuration of a system determines the depth range in which it is used, with maximum swath widths (i.e., maximum ranges) of shallow-water systems being typically 5–8 times the depth of water beneath the transducers. Most multibeam systems do not provide side-scan data, although these are available with some systems.

There are two main types of swath sounders, multibeam and interferometer, both of which provide line-by-line measurement of depth along the swath. Transducers can be hull- or towfish-mounted and are operated in a number of configurations,

Figure 20.14 Three-dimensional visualisation of a section of the mid-Atlantic Ridge based on data from a Hydrosweep multibeam swath system. The image shows a series of deep basins (dark blues and purples) within the central mid-ocean ridge rift valley, which crosses the image from bottom left to top right. The higher rift valley shoulders are seen toward the edges of the image (yellows and greens).

20.14

20.15

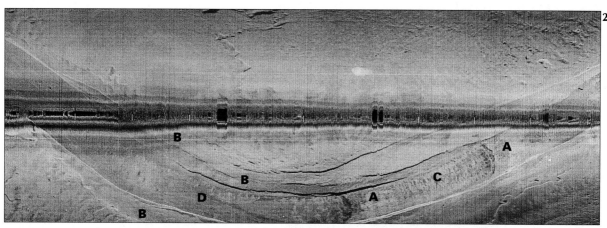

Figure 20.15 TOBI 30 kHz side-scan sonar image of a deep-sea channel on Monterey Fan, off the western US – strong sonar targets are white, acoustic shadows are black. The channel, here at a depth of about 4100 m, is between 1–2 km in width and up to 50 m deep. Note the terraced walls (B) and 'waterfalls' (A) within the channel (D), showing strong similarities with similar features associated with subaerial rivers (C is a sediment wave field).

of which two examples are illustrated in *Figure 20.13(b)*. A number of overlapping beams are electronically formed from the same set of transducers to produce the *multibeam* scan necessary for swath bathymetry (e.g., one system produces 52 beams from four transducers). The technology is changing rapidly[3], and more recent developments have produced compact, relatively portable systems which still function in depths of <1 m.

Essentially a development of side-scan systems, *interferometer* systems (e.g., GLORIA, SeaMARC II) use a set of transducers to measure the interference pattern of reflected sound [*Figure 20.13(c)*]. This sound 'interferometer' allows the accurate measurement of the travel time of sound waves reflected from small, adjacent areas of sea floor. Travel time is again converted into distance and, knowing the beam characteristics and direction, the depth can be calculated. The intensity of reflection can also be used to provide more information about sediment type or sea-floor orientation.

A three-dimensional relief map, based on swath bathymetry, is shown in *Figure 20.14*. Swath bathymetry provides an 'ordnance survey' or topographic-type map, in contrast to the 'aerial photograph' provided by side-scan sonar. Near geological-quality interpretation can be achieved when the results of both are combined using computational tools, such as a Geographical Information System (GIS).

The Towed Ocean Bottom Instrument (TOBI)

TOBI is one of a family of deep-towed instrument platforms, which includes, among others, the original Scripps deep-tow, the American SeaMARC 1, and the French SAR system. By operating within a few hundred metres of the sea floor, such vehicles can provide resolution far higher than that of surface-towed vehicles such as GLORIA. The penalty is that this mode of operation can require the use of many kilometres of tow cable, resulting in very slow tow speeds. TOBI (*Figure 19.26*), has a swath width of 6 km, can cover 400–600 km² per day, and can operate at depths of 6000 m, allowing its use over all but a few percent of the sea floor. *Figure 20.15* shows a TOBI sonagraph.

Able to carry a wide range of instrumentation, TOBI currently carries a side-scan sonar operating at 30 kHz with a beam width of about 0.8°, giving a resolution on the scale of a few metres. Other instruments carried include a sub-bottom profiler (7 kHz), and a transmissiometer to allow the measurement of temperature and depth.

Sub-bottom profiling

Sub-bottom profilers are used to explore the even more invisible world beneath the sea floor. Still using reflected sound waves, the frequency of operation is chosen such that the sound can penetrate the sea floor and be reflected from interfaces between different types of rock or sediment [see *Table 20.2* and *Figure 20.16(a)*]. As in the conventional echo-sounder, the system records a profile of data along the vessel's track; a three-dimensional picture of the subsurface structure requires the collection of many closely spaced profiles. The speed at which sound waves travel varies in rocks of different type and density (see *Table 20.3*). This wide range of velocity increases the potential for errors in the conversion of travel time into depth of a reflecting layer beneath the sea bed (depth = travel time times velocity). The variation of velocity with-

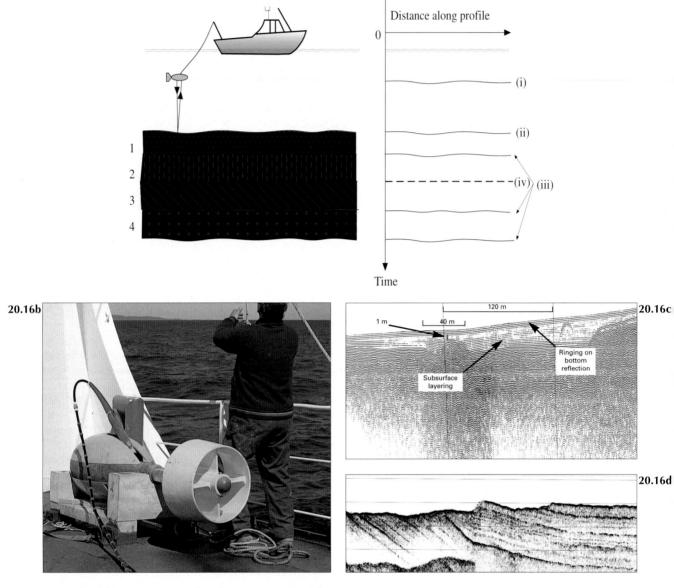

Figure 20.16 (a) Layers 1–4 represent rock layers of different types and/or density. In sub-bottom profiling a stream of pulses travels through the water column and penetrates the sea floor. Reflections can occur at the interfaces between rocks of different density, providing the conditions for reflection defined by Snell's law are met. The return time of reflected pulses is measured, and their travel time converted into depth to give a point-by-point profile of the position of the reflecting boundary; reflections occur at (i) the sea surface, (ii) the sea bed, and (iii) the sub-bed. When the density of adjacent layers is very close, no reflection occurs, as illustrated by the absent reflector at level (iv), between layers 2 and 3. (b) Institute of Oceanographic Sciences' echo-sounder and 3.5 kHz towfish. The tow cable is streamlined by the attached 'fairing'. Such towfish typically hold a set of four transducers and can weigh in excess of 100 kg. (c) A pinger profile from Lac de Cazaux et Sanguinet, Aquitaine, southwest France, taken with a 3.5 kHz ORE system. The profile shows a filled river channel cut through a horizontally layered sand stratigraphy and buried beneath fine, flocculated lake sediments. (Data acquired by Southampton Oceanography Centre.) (d) Shallow seismic image of Bouldner Cliff, West Solent, UK, taken using a swept frequency (2–8 kHz) Chirp system (System Design, Geochirp©, Geoacoustics Ltd). The sequence represents inclined reflectors of lithified Oligocene clays, marls, and limestones. The 780 m long section shows a penetration of 25 m and a final resolution of approximately 75 cm for individual beds (<10 cm resolution is possible with this system). (Data acquired by Southampton Oceanography Centre.)

in the water column is significantly less. These systems provide a profile of reflections from beneath the sea floor [*Figures 20.16(c)* and *20.16(d)*], from which the geology of an area can be explored and its evolution ascertained (see the examples in Chapter 9).

As explained in *Table 20.2*, in surveys requiring detailed information from close to the sea floor (e.g., investigation of the thickness of gravel deposits, the pre-site construction survey for a coastal or offshore installation, and the relationship between sea floor and subterranean topography), high-resolution, shallow-penetration profiling system are required. By contrast, low resolution and low frequency systems are used where deep penetration is needed, as in oil exploration and to characterise the Earth's structure.

Satellite and airborne systems, and underwater photography

Images recorded from a variety of sensors deployed in satellites (Chapter 5) or aircraft offer information based primarily on reflected energy in the EM spectrum. Such images provide a wide variety of information related to the structure, composition, and contents of the near-surface water. Since the penetration depth of EM waves into sea water is very limited, little direct information about the sea floor can be obtained from them. However, in shallow-water environments, sea-bed reflections have been observed; indeed, in fairly shallow coastal areas the circulation patterns seen in the images often reflect sea-bed features (e.g., see *Box 5.3*). Underwater photography or video offers an exciting opportunity to provide a visual image of areas

Table 20.3 Speed of sound in different rock types. Compressional wave velocities are given. The wide range reflects variations in seismic velocity due to, for example, the degree of water saturation in sediments. In the interpretation of seismic sections, selection of the appropriate velocity from the wide range possible for a particular rock type makes the conversion of pulse return times into depth complicated. In practice, a variety of techniques are available to measure a depth profile of the velocity variation.

Rock type	Range of seismic velocities (km/s)
Sediments	1.0–3.5
Sandstones	2.0–6.0
Limestones	2.0–6.0
Salt	4.5–5.0
Granite	5.5–6.0

'photographed' with sound (see, for example, Chapters 10 and 13). There are various types of equipment, ranging from hand-held devices (cameras, probes) used by divers (Chapter 18) to remotely driven tethered systems used on submersibles. Optical systems provide images with high resolution on a scale not possible with acoustic instruments, but with the disadvantage that the operational range is restricted to a few metres.

General References

Belderson, R.H., *et al.* (1991), *Sonagraphs of the Sea Floor, a Picture Atlas*, Elsevier, Amsterdam, 185 pp.

Forssell, B. (1991), *Radionavigation Systems*, Prentice Hall, New York, 392 pp.

Fish, J.P. and Carr, A.H. (1990), *Sound Underwater Images*, Lower Cape Publishing, Orleans, MA, 189 pp.

Ingham, A. (ed.) (1975), *Sea Surveying*, Vols 1 and 2, Wiley, London, 306 and 233 pp.

Kayton, M. (ed.) (1983), *Navigation, Land, Air, Sea and Space*, IEEE Press, New York, 465 pp.

Urick, R.J. (1983), *Principles of Underwater Sound*, 3rd edn, McGraw-Hill, New York, 423 pp.

References

1. Ballard, R.D. (1985), How we found the Titanic, *National Geogr.*, **168**(6), 696–717.
2. Ballard, R.D. (1987), Epilogue to the Titanic, *National Geogr.*, **172**(4), 454–465.
3. de Moustier, C. (1988), State of the art swath bathymetry survey systems, *Internat. Hydrogr. Rev., Monaco*, **65**(2), 25–54.
4. Forssell, B. (1991), *Radionavigation Systems*, Prentice Hall, New York, 392 pp.
5. Fish, J.P. and Carr, A.H. (1990), *Sound Underwater Images*, Lower Cape Publishing, Orleans, MA, 189 pp.
6. Ingham, A. (ed.) (1975), *Sea Surveying*, Vols 1 and 2, Wiley, London, 306 and 233 pp.

CHAPTER 21:

Ocean Resources

C.P. Summerhayes

Introduction: The Ocean as a Resource

Humans have used the ocean as a resource since their appearance on Earth. Early humans fed from its shores, gathering seaweed and shellfish, and hunting seals. As time went by they learned to travel short distances over it and, eventually, to fish from boats. Later, people learned to exploit the ocean for trade, creating their ports from fine natural harbours. They found other uses for it, too, such as driving mills with tidal power. Maritime nations used the ocean to transport armies, and as a battleground.

With the dawn of the industrial age, man's engineering ingenuity led to the steady seaward march of onshore industries. Mines extended out beneath the sea, in the UK tapping coal off northeast England and tin off Cornwall. Large structures, like bridges and piers, were built offshore. Dredgers began mining sand, gravel, and minerals from beneath the waves. Telephone companies laid cables across the ocean bed, linking far-flung continents by instantaneous communication. And in the twentieth century, oil and gas production began offshore.

Nowadays we regard the oceans as resources of food; of renewable energy (from tides, waves, and the ocean's thermal structure); of fossil fuel (oil and gas); of minerals, from diamonds to mundane sand and gravel; and of chemicals, fertilisers, and medicines. In this chapter we discuss these resources. With the continued growth of the human population, it is inevitable that we will have to use more of them, and use them more efficiently, as time goes by. We also use the sea in other ways. It is a major tourist resource; we like to swim in it, sail on it, and dive in it. We use it as a highway for trade, and as a space for putting things in, like submarine telephone cables, or various kinds of wastes (see Chapter 22). It can be a source of riches, through salvage and the discovery of sunken treasure. And, in the future, we may expect to see some of our grandchildren living in purpose-built cities in it.

What is now realised is that to exploit the resources of the ocean we have to learn to manage them better, so that they are kept truly sustainable. Some things, like wave and tidal power, will always be sustainable, no matter what we do to our environment. Living renewable resources, such as fish, however, will not; nor will mineral resources. Governments have made a start – first, by enacting legislation to define ownership of the resources; second, by enacting legislation to regulate fishing practices so as to conserve stocks[22]. Under the United Nations' *Law of the Sea*[1], countries have a right to claim Exclusive Economic Zones extending

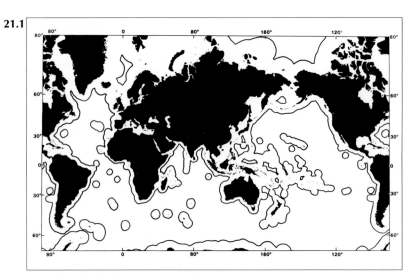

Figure 21.1 The locations and extents of Exclusive Economic Zones extending 370 km (200 miles) offshore, within which nations may claim right to all renewable and non-renewable resources. (© Institute of Oceanographic Sciences Deacon Laboratory, Wormley, England.)

Figure 21.2 A school of mackerel, photographed off Cornwall with a towed underwater camera from the research vessel *Clione*. (Courtesy of John Ramster, MAFF, UK.)

Figure 21.3 Bringing the catch in. (Photo, Tom Stewart; © Zefa Pictures.)

370 km (200 nautical miles) offshore[3,9,16] (*Figure 21.1*). Further seaward, the *Law of the Sea* dictates how the resources of the global commons are to be managed and shared. Ownership is the first step along the road to sustainable development, with all that this implies for conservation and a managed environment[5].

Growing use of the ocean as a resource will lead to the growth of ocean forecasting as a marine industry – users will require more knowledge. The offshore environment will have to be surveyed in detail if it is to be used properly[24,38], and the ocean-ic environment will have to be monitored as the basis for detecting and forecasting change (see Chapter 3).

Food from the Sea

Fish have always been part of the human diet (*Figures 21.2–21.4*). These days fish are caught not only to eat, but also to turn (the less edible ones, anyway – about a third of the catch) into fish-meal for pigs and poultry, or into fertiliser (*Tables 21.1*

Figure 21.4 Scottish fishing boats in Pittenweem Harbour, Fife, Scotland, on the north coast of the Firth of Forth.

Table 21.1 Disposition of world fish production[20] – percentage of world total catch in live weight.

	1981 (%)	1991 (%)	Change (%)
Fresh	21.7	22.6	+4.2
Freezing	23.8	25.0	+5.0
Curing	12.8	10.9	−14.8
Canning	14.5	12.9	−11.0
Feed	27.2	28.6	+5.1

Table 21.2 World value of fishery exports[20].

	1981 (%)	1991 (%)	Change (%)
Fresh/chilled fish	10.9	13.5	+23.8
Frozen fish	18.6	18.7	+0.5
Fish fillets (fresh/frozen)	9.8	10.9	+11.2
Cured fish	9.1	5.6	−38.5
Crustaceans (fresh/frozen)	20.6	22.3	+8.3
Molluscs (fresh/frozen)	6.6	8.3	+25.8
Canned fish	13.7	11.0	−19.7
Canned crustaceans/molluscs	4.1	5.5	+34.2
Fish-meal	6.3	3.9	−38.1
Fish oil	0.3	0.3	0

Table 21.3 World fish-catch statistics[19] (million tonnes).

	1985	1986	1987	1988	1989	1990	1991	Growth (%) (1985–1991)
World total	86.3	92.8	94.4	99.0	100.2	97.4	97.0	+12.4
Marine waters	75.7	81.0	81.7	85.6	86.4	82.8	81.8	+8.1
Inland waters	10.6	11.8	12.7	13.4	13.8	14.6	15.2	+43.4
Inland waters, main subtotals								
Freshwater fish	8.6	9.6	10.4	11.0	11.4	12.1	12.6	+46.5
Diadromous fish*	1.39	1.44	1.59	1.61	1.68	1.69	1.71	+23.0
Marine fish	0.08	0.08	0.08	0.10	0.09	0.11	0.17	+112.5
Crustaceans	0.26	0.29	0.35	0.37	0.41	0.42	0.44	+69.2
Molluscs	0.26	0.27	0.28	0.27	0.27	0.29	0.30	+15.4
Selected species, subtotals								
Cod, hake, haddock	12.46	13.57	13.79	13.63	12.90	11.83	10.47	-16.0
Krill	0.19	0.46	0.37	0.37	0.40	0.37	0.23	+21.1
Squid	1.79	1.75	2.32	2.29	2.72	2.33	2.56	+43.0
Mussels	0.96	1.00	1.13	1.29	1.31	1.32	1.33	+38.5
Salmon	1.17	1.09	1.10	1.17	1.45	1.45	1.64	+40.2
Seaweeds	3.88	3.86	3.55	4.13	4.38	4.32	4.91	+26.5

* Migratory between fresh and salt water.

and *21.2*). Some of the less edible ones are processed by extracting surimi gel, a fish protein with a texture like that of shellfish, which forms the basis for production of synthetic crab and lobster meat, so adding considerable value to the original catch[26].

Like the human population, the world fish-catch initially grew exponentially. Now that the catch approaches 100 million tonnes a year, the rate of growth has slowed and may be close to a plateau. *Table 21.3* shows this plateau in the catch from marine waters, *Table 21.4* shows the main species caught, and *Table 21.5* shows the main fishing nations. Many biologists feel that the catch has reached a limit beyond which further fishing will damage the ocean's ecosystem, perhaps irreparably[4,17,42]. Some people suggest that further growth in the production of seafood should come from other organisms, like squid or krill, the catching of which could provide another 100 million tonnes[17,42]. *Table 21.3* shows the growth in squid and krill catches compared with the decline in traditional species (cod, hake, and haddock). The problem is that nobody knows enough about the ocean and its ecosystems to be able to predict the level of catch that can be maintained, no matter what creature is caught.

Table 21.4 World fish-catch by principal species, 1991[19].

Top ten species	Million tonnes	Cumulative percent
Alaskan pollack	4.9	4.9
South American pilchard	4.2	9.0
Anchoveta	4.0	13.1
Chilean jack mackerel	3.9	17.0
Japanese pilchard	3.7	20.7
Skipjack tuna	1.6	22.3
Silver carp	1.4	23.7
Atlantic herring	1.4	25.1
European pilchard	1.4	26.5
Atlantic cod	1.3	27.8

Table 21.5 World fish-catch by principal producers, 1991[19].

Top ten fishing nations	Million tonnes	Cumulative percent
China	13.1	13.1
Japan	9.3	22.4
Former USSR	9.3	31.6
Peru	6.9	38.5
Chile	6.0	44.5
USA	5.5	50.0
India	4.0	54.0
Indonesia	3.2	57.2
Thailand	3.1	60.3
South Korea	2.5	62.8

Figure 21.5 Effects of climate control on changes in pelagic fish catches for Ecuador, Peru, and Chile, 1965–1989. Note the decline off Peru and corresponding increase off Chile between 1970 and the warm El Niño event of 1982–1983, following which the trends reversed. The El Niño event of 1982–1983 appears to represent the peak of a much broader warming period that diminished greatly the fish catch off Peru. Catches ranged from over 14 million tonnes (10% of total world fish-catch) in the late 1960s to 7 million tonnes in recent years. Species composition shifted from 90% anchoveta at the catch peak through less than 10% during the mid-1980s to 50% in the mid-1990s. Although this analysis takes no account of changes in the number of fishing vessels or method of fishing, it makes the point that climatic change can be very important. (Based on Sharp and McLain[37], *Figure 3*; redrawn at the Institute of Oceanographic Sciences Deacon Laboratory, Wormley, England.)

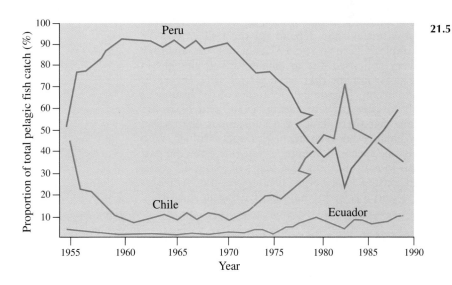

317

21.6

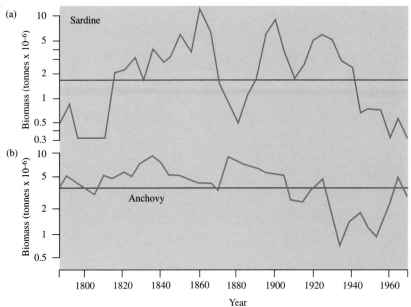

Figure 21.6 Changes in biomass (tonnes x 10⁻⁶) of (a) sardine and (b) northern anchovy off California, showing the climate-induced sardine–anchovy flip-flop typical of environments dominated by upwelling currents. Sardine periods are believed to indicate warmer climates. (Based on Cushing[11], *Figure 30*; redrawn at the Institute of Oceanographic Sciences Deacon Laboratory, Wormley, England.)

With the expected population growth, there is no doubt that if fish are to remain a part of our diet we will have to learn to manage the oceans sustainably. The only sensible way to increase the catch without damaging the stocks is to farm fish rather than hunt them. To do this on a large scale would require learning a great deal more than we currently know about fish behaviour and the response of fish to changes in the natural environment (see also Chapter 17).

One of the problems of the fishing industry is that natural changes in the environment can cause major changes in fish stocks, devastating the local communities that depend on the fish catch. An example is Peru in the 1970s, where the anchovy fishery collapsed due to climate-driven changes in the environment that were related to a severe El Niño event[37] (*Figure 21.5*). Studies of the history of fish catches off California, Peru, and Namibia show that there have been substantial changes in the balance between anchovies and sardines, with only one species dominant at any one time; the patterns look cyclical on time-scales of about 40 years, and existed long before fishing began[11,37] (*Figures 21.6* and *21.7*). In the North Atlantic, historical

studies show major changes in the herring population in the North Sea (*Figure 21.8*), and in the cod population on the Grand Banks and off Greenland (*Figure 21.9*). Historical analyses show that these populations grew during the world wars, when fishing levels were reduced, and shrank afterward; clearly, overfishing causes the populations to decline (see *Table 21.3*, cod, hake, and haddock). However, the local patterns of their distribution suggest that subtle alterations in the environment driven by changes in climate can be equally

21.7

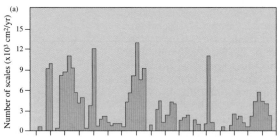

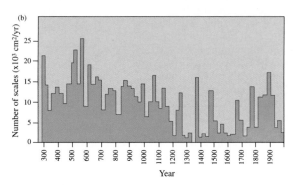

Figure 21.7 Fluctuations in the (a) sardine and (b) anchovy populations continue back through time off California, based on studies of the number of fish scales in laminated sediments from the anoxic Santa Barbara Basin. These clearly indicating a control on population by climate rather than by fishing. (Based on Sharp and McLain[37], *Figure 4*; redrawn at the Institute of Oceanographic Sciences Deacon Laboratory, Wormley, England.)

318

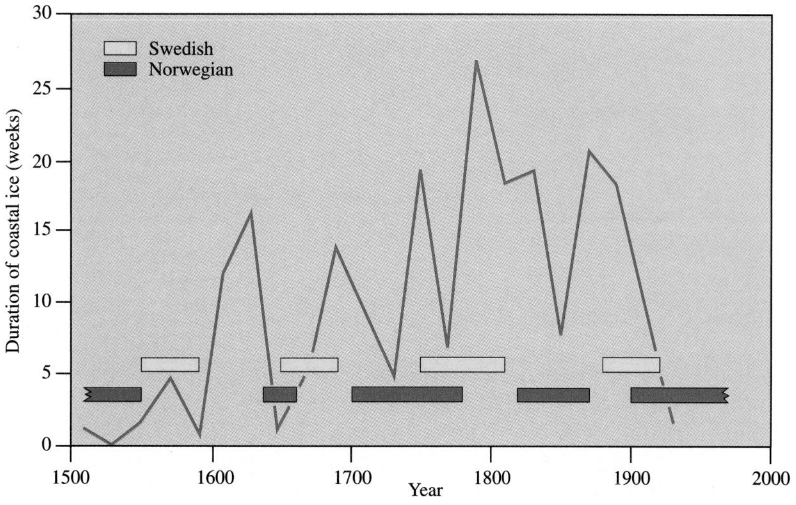

Figure 21.8 Climatic control of herring in the northern North Atlantic, based on herring catch. During cool periods, coastal ice stays longer on the north coast of Iceland and the herring move south from Norway to Sweden. In warm periods, when there is less ice off the north coast of Iceland, the herring move north from Sweden to Norway. (Based on Cushing[11], *Figure 29*; redrawn at the Institute of Oceanographic Sciences Deacon Laboratory, Wormley, England.)

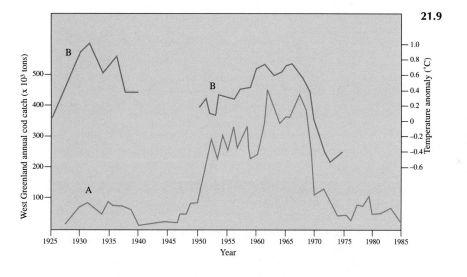

21.9

Figure 21.9 The rise and fall of the West Greenland cod fishery: (A) the catch of cod; (B) surface temperature anomalies for the Greenland coast south of Fredrikshab. Cod appeared on the offshore banks between 1912–1923, initiating a gradual build-up of stock, which then crashed in the late 1960s. This overall pattern is attributed to climatic factors, particularly the local warming of 0.2–1.0°C, which itself is a reflection of the regional warming of the North Atlantic at the time. (Based on Dickson and Brander[13], *Figures 2* and *3*; redrawn at the Institute of Oceanographic Sciences Deacon Laboratory, Wormley, England.)

Figure 21.10 A fin whale at the Whaling Station at the head of Hvalfjorder (whale fjord) on the west coast of Iceland in 1978; the station operated until the current pause in commercial whaling, taking about 250 fin whales per year. (Courtesy of Tony Martin, Sea Mammal Research Unit, England.)

21.10

Table 21.6 Whale catches (thousands)[19].

	1985	1986	1987	1988	1989	1990	1991
Blue and fin whales	7.9	6.5	6.3	0.68	0.61	0.65	0.66
Sperm and pilot whales	83.2	153.8	135.6	132.7	168.4	105.8	50.0

important in determining stock levels from place to place[11].

Clearly, we need to understand and to be able numerically to forecast the natural variability in the system in order to exploit the living world more effectively. By itself, however, such an understanding will be inadequate; we also need to prevent overfishing. The difficulties in policing fishing operations make effective control unlikely. This is another argument in favour of adopting the methods of the farmer, rather than those of the hunter, to ensure the true potential of the ocean for feeding the world's growing population is achieved.

When technology was primitive, fishing was indiscriminate; early man took what he could get. Advances in technology brought greater control, and an ability to target particular species systematically, no matter where they might be found.

Specialist fisheries arose, like the cod fishery of the Grand Banks, or the herring fishery of the North Sea. Whaling (*Figure 21.10*) provides a good example of the evolution of a hunting-based fishery[4]. It can be done by primitive tribes in coastal waters, and in Europe it began to develop into an industry by the twelfth century in northern Spain. Local overfishing drove whalers further afield, to Newfoundland and Spitzbergen, where whaling had become big business by the mid-seventeenth century. The business boomed in the nineteenth century, when technology made whaling easier and turned it into a global industry. It reached its peak in the 1930s and 1940s, but the combination of rapid technological advance, making the job easier, and the limits of the whale population lead to decline by the 1960s. Large-scale commercial whaling is now a thing of the past, although some

21.11a

21.11b

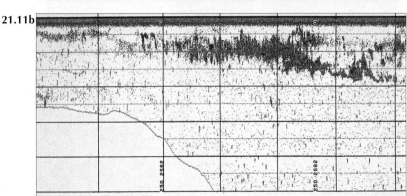

Figure 21.11 (a) Antarctic krill, *Euphausia superba* (size approx. 2 cm), plays a key role in carbon cycling in the Southern Ocean. (b) The echo-sounding trace, with horizontal scale lines spaced 25 m apart, shows a swarm of krill as a large widespread acoustic signal (in blue and red) centred at about 50 m (the red horizontal line) over the edge of the Antarctic continental shelf (shown at the lower left). Whales feed on these dense layers of krill and other zooplankton. (© British Antarctic Survey, England.)

21.12a  21.12b

Figure 21.12 (a) Development by the Japanese Marine Science and Technology Centre (JAMSTEC) of an artificial sea floor (20 × 20 m), which can be raised to the surface (right) and lowered below the surface (left) to and from its working depth of 4 m; it forms a platform for growing abalones, and its tanks contain black rock-fish. (b) On the prototype in Ryohri Bay, Japan, the abalone are fed with kelp and measured at the surface once a week. (© Mineo Okamoto, JAMSTEC, Japan.)

nations, notably Norway and Japan, continue to catch large numbers of whales[33] – *Table 21.6* shows the decline in catch in recent years. The creation of a whaling reserve around Antarctica in 1994 should help to renew and maintain whale populations.

In the case of herring, the drift net was replaced by the purse-seine net in the 1960s in Europe, which led to overfishing and bans on fishing in the North Sea at various times. Perhaps inevitably, the response of the local industry to a decline in one stock is to go for another. Declines in the herring catch by European fisheries were offset by a tremendous rise in industrial fishing for sand-eels and the like for fish-meal and fertiliser. The consequent decline in sand-eels is thought to be the cause of larger than usual numbers of deaths of sea-birds, which feed on them[32]. New fisheries are already developing to harvest new species of fish, like the Grenadier, from deep water, or different creatures, like squid and krill[42] (*Figure 21.11* and *Table 21.3*).

The lessons from the whale, cod, and herring fisheries show that, without protection, living resources cannot survive the technological shift from primitive to industrial fisheries. Bigger and better ships, with more sophisticated, expensive, and durable equipment, are depleting a supply that once seemed boundless. With the help of satellites, sonars, computers, and refrigeration, combined with global-range ships, humans will sweep the seas more or less clean of fish if they continue unchecked.

Mariculture

One answer is mariculture, farming the sea (*Figure 21.12*). Oysters have been cultivated in Asia for over 2000 years. The Chinese developed freshwater aquaculture over 1000 years ago, and wrote the first fish-farming textbook in 475BC; it described the farming of carp (fresh water), milkfish (brackish), and mullet (marine)[3]. Mariculture has been common for centuries in Southeast Asia, where fish farmers seed ponds with the eggs of fish and crustaceans, transferring the larvae and juvenile forms to larger ponds and feeding them on algae and plankton[4]. World mariculture production is currently around 10 million tonnes, excluding freshwater environments – fin fish account for about 45%, shellfish for about 25%, and seaweeds for the rest. The Chinese are the largest producers. Japan derives about 15% of its total ocean produce from mariculture, including such exotic items as sea-urchin roes and seaweed (which is grown on nets[35]); note the growth of seaweed shown in *Table 21.3*, which also shows the growth in fish taken in inland waters (much of it from mariculture). The UN Food and Agriculture Organisation (FAO) estimates that by 2000AD aquaculture (including freshwater fisheries) will account for 20–25% by weight of the world fisheries production, and about 50% of the value[3].

21.13a 21.13b

Figure 21.13 (a) Long-line mussel farm in Loch Etive, Scotland. The black objects are plastic floats from which hang the ropes on which mussel spats settle (© Jim McLachlan, McLachlan Shellfish and Fish Farming Equipment). (b) At harvest time, after around 3 years, there may be 250 kg of mussels per rope (© Nicki Holmyard, Association of Scottish Shellfish Growers).

Around Europe, oysters and mussels have been cultivated in coastal areas for centuries (*Figure 21.13*; note the growth in mussels shown in *Table 21.3*). Since the late 1960s, salmon farms have grown apace in the fjords of Scotland and Norway (*Figure 21.14*); the salmon industry in Scotland now equals the size of the beef and lamb industries there[3]. The market is expanding and the outlook is good – note the growth in salmon shown in *Table 21.3*. Interest is growing in farming other species, especially halibut and turbot, and in developing salmon and other farms in the open sea rather than in lochs[3]. The advantage of the open sea is that the farms could be larger, and are not at such great risk from local pollution.

One variety of mariculture that does not use farms is ocean ranching, in which juvenile fish are released into the sea to be caught at a later stage. Salmon ranching is a commercial success in Japan and Alaska[28]. It has been shown that plaice ranch-ing could work on the Dogger Bank in the central North Sea, where plaice hatchlings transplanted from the Dutch coast grew more rapidly and suc-cessfully[42]. Unfortunately, this operation has not been taken up commercially.

Legislation

Fishing can be controlled by legislation, although not without difficulty. It is a global business (*Table 21.5*), so the ability of the large fishing countries, like Russia, South Korea, and Japan, to fish just about anywhere means that the stocks of fish off countries with less fishing capacity could be deplet-ed without any recompense. To counter this prob-lem, most fishing countries agreed in 1977 to a 370 km (200 mile) limit, to keep to themselves the right to fish their own waters and to prevent over-fishing[35]. It was a declaration of extended geo-graphic limits around Iceland that kept British fish-ing vessels out of Icelandic waters and led to the so-

Figure 21.14 Salmon farm in Loch Creran on the west coast of Scotland. Young salmon, bred in fresh water, are transferred to the octagonal, 4.5 m (15 foot) diameter sea cages at 18 months and fed on high-protein fish food for about 2 years. Netbags (4–6 m deep) are suspended from the floating cages; tides help to flush effluents away. The farmer here is putting wild wrasse (a 'cleaner' fish) into the cages to keep the salmon free of lice. Total production from this one farm is around 40 tonnes/year. (© Jim Buchanan, Association of Scottish Shellfish Growers.)

21.14

Figure 21.15 A desalination plant – removing salt from sea water provides drinking water for people in arid countries. (© Zefa Pictures.)

21.15

called cod wars between the two countries in the mid 1970s[3,35]. The European deep-water fishery was severely affected by the extension of the Canadian fishery exclusive zone to 200 miles over the Grand Banks. Factory ships from other nations can circumvent the problem by 'setting up shop' seaward of the limit and paying local vessels for their catches from within the zone[35].

In some countries Governments have set quotas on the size of fish or amount of catch in order to control fishing and preserve stocks. Unfortunately, controls work poorly as they are virtually impossible to police effectively. In protest against quotas, French fishermen burned down the Town Hall of Rennes, in Brittany, in 1993. Tempers will continue to run high as the hunting culture clashes with the implementation of the principles of sustainable development.

We are at the beginning of what Borgese[4] has called the Blue Revolution, in which sustainable techniques, such as farming, will take over from hunting as the method of harvesting food from the sea. Farming in the coastal zone will grow in volume and expand in area to become common everywhere, with more and more species being farmed. Farms will move offshore to take advantage of space, on the one hand, and local supplies of nutrients (pumped from cold subsurface waters), on the other hand.

Chemicals and Medicines from the Sea

The sea is a vast storehouse of dissolved minerals. Unfortunately, most of the dissolved constituents are disseminated in such tiny amounts that extraction will never be profitable. Only a few are abundant enough to be extracted on a commercial basis, the most common of which is sodium chloride, or common table salt. What many people do not

realise, unless they live in arid coastal areas, is that one of the principal extracts of salt water is fresh water. Distilling plants to produce fresh water from salt water have been common on ocean-going ships for well over a century. Desalination plants have become increasingly common on land, especially in arid coastal regions like the Red Sea and the Persian Gulf (*Figure 21.15*). At present, they make commercially unattractive the once-popular idea of obtaining fresh water from icebergs towed north from Antarctica[9].

Chemicals

About 3.5% of the weight of salt water is dissolved solids; sodium chloride accounts for 71% of this (see Chapter 11). It has been a prime ingredient for cooking and a principal article of trade for well over 5000 years. Roman soldiers were part paid in salt (*salarium argentium*, from which the word salary derives). Humans probably first came across natural salt in dried-up lakes and coastal ponds. It was not a great leap of imagination to create artificial coastal ponds, so bringing the process of evaporation under control, and the basic approach has not changed to the present day (*Figure 21.16*). However, raw sea-salt derived in this way can be impure and rather bitter, containing iron, calcium, and magnesium compounds as well as sodium chloride (to obtain pure salt, several stages of crystallisation must be used). Evaporation accounts for around one-third of the world supply of salt, most of the world's sea-salt being produced by India, Mexico, France, Spain, and Italy[16].

The only two elements commercially extracted from the sea on a large scale are magnesium and bromine[16]. After oxygen, hydrogen, chlorine, and sodium, magnesium is the next most common element in sea water. In recent years, some 18% of the world magnesium production of 1.8 million tonnes has come from sea water, mostly produced

21.16

Figure 21.16 Harvesting salt from salt-pans in Portugal. (© Centro de Coridade Nossa Senhora.)

in the US. Magnesium is an exceptionally light metal and has a variety of uses in structures of various kinds. In the extraction process, sea water is mixed with dolomite rock (a calcium–magnesium carbonate), which precipitates the magnesium from sea water as a hydroxide. This is then converted into magnesium chloride, and the magnesium and chlorine are separated electrolytically.

Bromine is the ninth most common element in sea water (see Chapter 11). It is used in antiknock compounds in petrol, for instance, and is produced as a by-product of both salt making and the production of magnesium from sea water.

Uranium, the twenty-ninth most common element in sea water, is 20,000 times less abundant than bromine in the sea. Even so, many countries carry out research into extracting uranium from sea water as a means of ensuring stable supplies of energy in the future. Japan, in particular, has a sizeable interest, as befits a country without oil reserves. However, the current supplies of uranium on land make sea-water extraction far from commercial.

Medicines

Medicines have long been obtained from land plants, and nowadays several modern medicines are also produced from marine plants and animals[9]. The great diversity of marine species makes it highly likely that many could have medicinal properties, so the study of these organisms for medicinal purposes is expanding. The discovery of other new compounds is hindered, however, by the limited accessibility of the marine plant and animal world beneath the sea.

A wide variety of pharmaceuticals are marine-derived. For instance, marine biologists in New Zealand have identified a compound in an extract from the native green-lipped mussel that appears to relieve symptoms of arthritis. Capsules containing crude extracts from the mussel have been sold as food supplements under the name Seatone for 20 years in 65 countries. The active component is a glycoprotein which blocks the action of neutrophils, white blood cells that trigger the immune system into action at sites of infection or tissue injury. Heart stimulants may be obtained from certain sea anemones. The antiviral compound idoxuridine and the antitumour compound arabinosylcytidine have been developed from compounds found in a marine sponge from the West Indies. A variety of red algae (seaweed), *Digenia simplex,* is used to expel intestinal worms. A variety of brown algae yields a choline derivative which can lower blood pressure[7,9].

Marine toxins, produced commonly as part of an animal's chemical defence system, show great promise, not only as pharmacological compounds, but also as models for the development of new synthetic chemicals[7,9]. For instance, lophotoxin is a lethal substance recently discovered in several species of seawhips of the genus *Lophogorgia*. The novel chemical structure of lophotoxin and its ability to induce neuromuscular blocking of a certain type indicate that it may represent a new class of neuromuscular blocking agents with unique pharmacological properties. Several highly poisonous fish, in particular the puffer fish, porcupine fish, and sunfish, contain tetrodotoxin, a very potent neurotoxin which has some use as a local anaesthetic and muscle relaxant in terminal cancer patients.

The blood of the horseshoe crab (*Limulus polyphemus*) is perhaps the best-known source of a marine-derived compound with pharmacological properties[30] (*Figure 21.17*). In 1955 researchers discovered the causative agent for clotting *Limulus* blood. The medical benefit came in the form of a diagnostic reagent that detects bacterial infection of human blood by clotting in the presence of bacteri-

Figure 21.17 Horseshoe crab shell on a Massachusetts beach. (© Diana Summerhayes.)

Figure 21.18 Sunlight streaming through a kelp forest. Giant kelp grows up to 0.6 m/day and reaches lengths of 65 m. (© P. Glendell, Avon, England.)

al endotoxins. It can detect diseases, such as gonorrhoea, but is more widely used to test pharmaceutical products for the presence of contaminating endotoxins, thereby improving the quality of drugs and biological products for intravenous injection.

Biotechnology: Industrial Chemicals

Aside from medicines, marine organisms have given rise to many other kinds of products. The kelp forests of Asia and Europe, for instance, have long-been harvested as a source of fertiliser, potash, and food for livestock. Since 1929, kelp has also been harvested, off California, for a unique substance, algin, which controls the properties of mixtures containing water[29] (*Figure 21.18*). The primary industrial applications of algin are paper-coating and sizing, textile printing, and welding rod coatings, but it is also used in pharmaceutical and cosmetic products, including dental impression compounds, binding for tablets, and anti-acid formulations. Algin is used in a wide range of foods, including milk-shake mixes, and gels; it is used as an emulsifier and stabiliser in salad dressings, to enhance texture and moisture retention in bakery products, and to stabilise beer foam!

Marine organisms provide a marketable insecticide, produced in Japan since 1966 from an extract called nereistoxin from marine worms (annelids); it is active against the rice-stem borer and other insect pests[7].

In yet another kind of application, it has been found that specific compounds of bacterial origin, for example those related to melanin, can promote settlement and development of oysters and other marine invertebrates, including those of commercial value.

In contrast, there is considerable interest in using antifouling compounds made by marine organisms as a natural means of preventing the build-up of barnacles, algae, and other organisms on underwater surfaces, such as pipelines, structures, and the hulls of boats[7,12]. Success would eliminate the use of highly toxic and environmentally unfriendly antifouling compounds like tributyl tin (TBT). Much work on the possibility of bringing natural antifoulants to the market is underway at the Marine Biotechnology Institute in Japan, where one promising compound has been found in a bryozoan.

Energy from the Sea

As population grows inexorably, demand for energy rises. Oil and gas, like coal, are finite, non-renewable resources. They are also dirty, their combustion tending to pollute the air we breath and change the temperature of the planet. Knowing that they will not last forever, more effort is now being put into the development of alternative sources of energy that are clean and renewable.

The ocean has them in abundance, in its winds, waves, tides, and currents, and in the differences in heat between the surface layers and the deeps. Paradoxically, although a library could be filled with writings on tapping ocean energy, by 1993 only one sizeable project had ever been carried through to completion. It seems unlikely that ocean energy will flourish while the price of oil stays low, except in places difficult to supply with oil, natural gas, or electricity, such as capital-poor, labour-intensive sites in less-developed countries.

Wind

Wind has long-been a source of power at sea, but with the advent of the steamer its use has declined. Now, however, new ships are being built with aluminium sails to supplement the power supplied by engines. Ocean winds are a potential source of energy for coastal communities and small island nations[36]. Sweden is planning to utilise marine windmills to generate electricity, and the US has installed a wind turbine on the coast of Hawaii[40].

21.19a

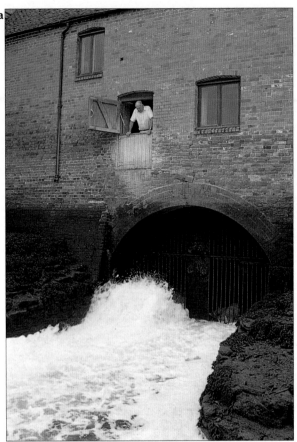

Figure 21.19 (a) The only surviving tide mill in the world still producing wholemeal flour is the Eling Tide Mill at Totton, near Southampton. The present building was erected around 1785 on the site of a tide mill known to exist since at least the mid-thirteenth century, when it was purchased by the Bishop of Winchester. The mill ran commercially until 1946, and was restored between 1975 and 1980 (photo, M. Conquer; © IOSDL). (b) The mill operates as shown. A causeway blocks the River Bartley to create a tide pool (the mill pond, MP), which is filled twice a day by Southampton's unusual double high tide (the tidal range is about 4 m). At high tide the mill pond is full (i), and the wheel stopped. A sluice gate retains water in the mill pond until the ebbing tide drops the level of water in the estuary to below the axis of the mill wheel. The sluice gate is then opened and the head of water in the mill pond drives the mill wheel until the next rising tide stops the wheel turning and replenishes the pond. (ii) 2 hours into the ebb tide – water leaving the sluice gate starts to turn the wheel, which moves slowly while part of it remains submerged. (iii) Low tide – the wheel turns at full speed, driven by water leaving the sluice gate as the level of the mill pond falls. The wheel turns for about 7.5 hours, and at its maximum efficiency for about 4 hours (based on an undergraduate project by Green[21]; redrawn at the Institute of Oceanographic Sciences Deacon Laboratory, Wormley, England.)

21.19b   (i)   (ii)   (iii)

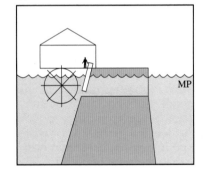
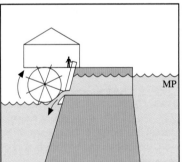
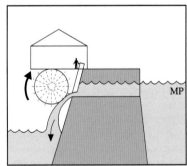

Tides

Wherever tides are strong, humans have found the means to harness their energy. For hundreds of years tidal mills supplied mechanical power up and down the European seaboard by using either the vertical rise and fall of the tides or the flood and ebb of tidal currents[4] (*Figure 21.19*). In the seventeenth century the Dutch took tidal mills to America[4]. Tidal energy is traditionally put to use by lifting a mass that carries out work while falling back into place, or by turning a paddle wheel. The more modern approach is to dam an estuary, like that of the Rance in northern Brittany, and equip it with turbines that generate electricity when driven by both the flood and ebb tides[3] (*Figure 21.20*). The 1 km long Rance barrage, built in 1966, generates half a million kilowatts of power on each tide, and doubles as a road. It cost $100 million to build; its operating costs are lower than that of any power station in France; its fuel is free; and there are no waste products[4,6]. A tidal range exceeding 10 m seems to be a prerequisite; such ranges exist in many other parts of the world, including the Bay of Fundy and the Severn Estuary (see *Box 21.1*), and barrages have been built in the former Soviet Union, in Canada, and in China[4,6,40].

21.20

Figure 21.20 The Rance barrage near St Malo in Brittany, northern France, showing water pouring through the sluice gates from upstream at low tide. The Rance barrage is equipped with 24 turbo-alternator bulb sets; chromium–nickel steel is used for the turbine blades to prevent corrosion. There are six sluice gates to empty and fill the estuary above the barrage. Operators may use the tide in one direction (as in a tide mill) or in both directions, and can use pumps to top up the estuary. The total capacity (600 GWh) has been available since 1982. Contrary to expectation, the estuary above the dam has not silted up; flooding of the estuary above the dam has greatly expanded pleasure boating; and the causeway over the dam provides a road that has contributed to the economic development of the region[34]. (Photo, Ronald Toms; © Oxford Scientific Films.)

Box 21.1 The Severn Barrage Development Project

Tidal power barrages have certain environmental attractions compared with conventional power stations, notably the lack of contribution to acid rain, greenhouse gases, or radioactive waste. The commercial feasibility of harnessing the extreme tidal conditions in the Severn Estuary to generate electricity has been under consideration for some time in the UK, where a detailed report was presented to Parliament on 23rd October 1989. Before a barrage could be built, a full environment-impact assessment needs to be carried out over several years. The design for the 16 km long barrage, which would carry a public highway from near Weston-Super-Mare in England to near Cardiff in Wales, includes 216 turbo-generators of 9 m diameter, each rated at 40 MW, giving a total capacity of 8640 MW. The barrage would be cut by 166 sluice gates, a major lock for shipping, and a lesser lock for small craft. The annual output would amount to 17 TW hours of electricity, equivalent to burning 8 million tonnes of coal and supplying 7% of the electrical demand of England and Wales. Early environmental-impact studies suggest that the water above the barrage will be less turbid, hence clearer, than it is now, allowing more light penetration and stimulating more photosynthetic activity. Mud flats exposed at low water will be less disturbed by currents, and hence more biologically productive. Thus, the estuary is likely to support more fish and birds in total, though there may be some changes in the population supported.

21.21a

21.21b

Figure 21.21 (a) A schematic cross-sectional cutaway of the prototype shore-line wave energy plant on the Scottish island of Islay, demonstrating the operating principle: (1) waves oscillate the water column; (2) upward water motion in the chamber forces air through the turbine, driving the generator to produce electricity; (3) the turbine converts reciprocal air flow into high speed uni-directional rotation. The plant has a 75 kW capacity, an energy cost of 7 p/kWh, and an energy output of 300 MWh/yr. It indicates a future potential of 1000 kW, with an energy cost of 3–4 p/kWh. This amount of production would make wave energy plants economic for the islands (courtesy of AEA Technology, Harwell, England). (b) The Islay Wave Power Station showing the wave chamber (foreground) and turbine housing (rear) (courtesy of Don Lennard, England and Australia).

21.22a

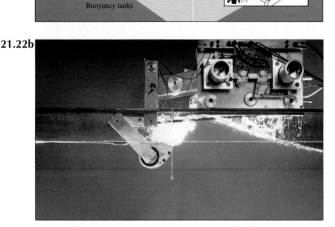

21.22b

Figure 21.22 (a) Salter's ducks are hollow floats made of reinforced concrete, with a cone-shaped cross-section. Each duck is about 33 m long and 20 m across. Named after their inventor, Professor Stephen Salter of the University of Edinburgh, several would be connected in a line by a hollow shaft attached to land or a fixed platform at one end. Incoming waves tilt each float, or duck, upward, absorbing the waves' energy and leaving calm water in its lee. Within the oscillating duck, gyroscopes (inset) activated by the rocking motion, drive a high-pressure hydraulic system; the circulating oil drives a turbine that generates electricity. Power from each duck would flow to shore through a central cable. Compared with other such plants, the ducks are extremely effective, providing the highest output per metre of sea. A pilot plant has been tested and research continues. (© ETSU, Harwell, England.) (b) A model duck at about 1/150th scale in a narrow tank, being tested for response to freak waves. The duck (centre, in water) is being subjected to a 50-year design wave coming from the wavemaker (off-stage to the right). The rig above water level, to which the duck is attached by angled struts, constrains the axis or 'spine' of the single duck to move as if it were part of a 'duck string' of 50 or more such devices restrained by a compliant mooring system, while allowing the duck–spine surge and heave reaction forces to be studied (© J. Taylor, EUWP, Edinburgh University, Scotland).

Waves

Where tides are not powerful enough to warrant the construction of barrages to generate electricity, waves may be an alternative source of energy[36]. The power potential of an average wave per kilometre of beach is around 40 MW. It has been estimated that a substantial part of Britain's energy needs could be met by putting wave energy to work[3].

Wave energy can be harnessed by fixed or floating devices. A fixed device is operational on the Isle of Islay, on the western coast of Scotland, where the wave energy offshore is equivalent to between 50–77 kW per metre of shore-line per year[3], decreasing to 20 kW/m at the device. Here, Britain's first shore-line wave power station, delivering 75 kW, was constructed by a team from the Queen's University of Belfast in 1988 and commissioned in 1991 (*Figure 21.21*). It works on the principle of an oscillating water column, with waves pushing air through a turbine[3]. The construction of this pilot plant proved the concept and opened the way to the construction of a commercial demonstrator in the 0.5–1.5 MW range. Power stations of this type would be ideal for many small island nations, and for local needs in larger countries (such as China) where the electrical grid systems are not yet fully developed. Two large systems have been installed on the coast of Norway.

Floating devices function on the principle that wave motion can be converted into reciprocal motion[6]. Some of the systems currently being refined, such as Salter's nodding duck (*Figure 21.22*), seem to hold considerable promise as energy sources for the future[3,42].

Currents

Although ships' captains have used ocean currents for decades to make speedy passage, the rivers in the sea have only once been harnessed to generate electricity, off northwest Iceland. The potential is there – the Gulf Stream[4], for example, carries 30 million m^3/s of water past Miami with a velocity of 2.5 m/s. The problem is that all this power, five times larger than the flow of all the world's rivers, is too diffuse to harness easily by conventional means. Scientists of the Woods Hole Oceanographic Institution have, nevertheless, calculated that an array of turbines stretched across the current in the Florida Straits would produce 1000 million W/day, as much as two conventional nuclear power stations[4].

Thermal energy

Closer to reality is the conversion of the ocean's thermal energy into electricity[3,6,27]. The principle of ocean thermal-energy conversion (or OTEC) is simple, and has been around for over 100 years.

OTEC uses the difference in temperature between warm surface water and cold deep water to power a turbine and generate electricity (*Figure 21.23*). The potential for applying this principle is greatest in the tropical regions, between 10°N and 10°S, where warm surface waters (25–30°C) overlie cold waters (4–7°C) situated at depths of 500–1000 m.

The first pilot OTEC plant was built by the French OTEC pioneer, Georges Claude, in Cuba in 1930. Pilot plants have since been built or proposed in the Ivory Coast, in Tahiti, in Nauru, in Japan, and elsewhere. The first closed-cycle Mini-OTEC plant to produce power was built in Hawaii in 1979. Although this successful experimental plant generated an electrical output of 50 kW, about 80% of this energy was needed to pump up the cold water for the system[23], leaving a net output of 10–15 kW. Plants like this would be doing well to produce an annual return on capital of 1%, which is unlikely to be commercially attractive to investors.

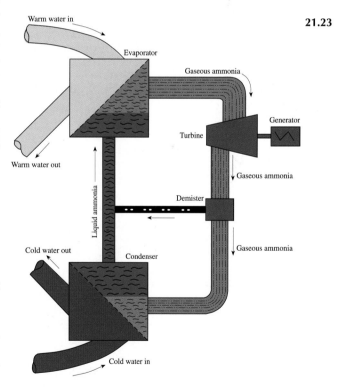

21.23

Figure 21.23 Sketch of the workings of an OTEC plant. Most modern OTEC plants are closed-cycle systems, like that shown here, in which warm surface water vaporises an intermediate fuel, such as ammonia; the vapour powers a turbine and is then condensed back into liquid by the cool waters pumped up from the depths. These are called closed-cycle systems because the working fluid is recirculated. (© Living Tapes Ltd.)

21.24

Figure 21.24 The Natural Energy Laboratory of Hawaii Authority's plant at Keahole Point, where there is an experimental 210 kW open-cycle OTEC plant. In an open-cycle plant, sea water is evaporated in near-vacuum conditions to create low-pressure steam, which drives a turbine; once the steam condenses the resulting water is discarded (compare with closed-cycle OTEC, as described in *Figure 21.23*). On the site, pipes can be seen for accessing cold ocean water (6°C from 600 m) and warm surface water (25–28°C); the large round tanks are ponds for growing fish and algae (kelp). (© Greg Vaughn, Natural Energy, Laboratory of Hawaii Authority, US.)

However, because of the considerable promise inherent in OTEC, especially for regions such as oceanic islands lacking in conventional energy sources, research is active[3]. Between 1974 and 1994 the US Department of Energy invested around $250 million in OTEC research and development. Japan and Taiwan have also spent considerable sums on it, along with some European countries[3,23,27].

Luckily, there is another way of proving the commerciality of OTEC plants, which involves considering the value of the cold deep water itself and what can be done with it[23]. Because water from depths of 100–2000 m is much richer in nutrients than the surface water (see Chapter 11), it can be used to encourage the growth of plankton to support mariculture (e.g., the production of shellfish or fish-based protein foods), or to grow algae (such as kelp) to provide biomass for feed, fertiliser, or methanol.

Since the water is cold, it can be used for air conditioning or be piped through the ground in tropical areas to cool the soil so that fruits (e.g., strawberries) that would not normally grow in the tropics can be grown year-round[3]. Through condensation or desalination the cold water can be a source of fresh water for drinking or for irrigation; through electrolysis it can be a source of hydrogen and oxygen. Combining OTEC and aquaculture, and deriving a plethora of by-products, is the vision of the future for commercial success in converting the ocean's thermal energy into electricity in tropical regions[3,23]. Work along these lines has begun at the Natural Energy Laboratory of Hawaii, where there is an open-cycle experimental OTEC plant (*Figure 21.24*).

Energy from the Sea Bed

First, there was wood for the cooking fires of early humans; then came coal, to fuel the industrial revolution; and now we live in the age of energy from oil and gas that occur in the ground. Although oil and gas are more or less ubiquitous in sedimentary rocks, in most places they are dispersed in such tiny amounts that they cannot be extracted economically. Here and there, however, conditions were such that the oil and gas have been able to migrate out of the fine-grained mud rocks in which they were formed, and into coarser-grained strata. In some places, folds, domes, faults, and other structures have formed natural traps that prevented oil and gas from escaping from the coarser strata. Traps full of oil or gas have become reservoirs, and targets for the drill (*Figure 21.25*).

Where the sandy strata reach the surface, or are cut by faults that offer avenues to the surface, much of the oil and gas escapes by natural seepage – such natural seeps on land were exploited by ancient peoples. American Indians used oil for fuel and medicine; the Chinese found natural gas while drilling for salt, and used it as a fuel as far back as 1000BC; ancient Egyptians coated mummies with pitch to help preserve them; and Nebuchadnezzer paved the streets of Babylon with asphalt around 600BC. Natural seeps at the coast are still being exploited off Coal Oil Point in California[16].

All commercial accumulations of oil and natural gas come from the decay of the buried remains of marine and terrestrial plants deposited in sedimentary basins[15]. While some of these basins were depressions on land and accumulated their sediments in lakes, most were along the drowned mar-

21.25

Figure 21.25 North Sea production platform complex in BP's Bruce gas field. (© British Petroleum.)

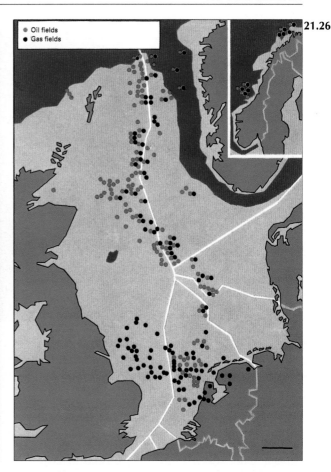

21.26

Figure 21.26 Map of distribution of oil and gas fields in the northeast Atlantic area, showing national borders. Britain and Norway have the largest offshore sectors – note the concentration of production along the boundary between the British and Norwegian sectors, above the deeply buried rift valley that forms the Viking Graben. Most fields in this area produce oil; most in the southern North Sea produce gas. The scale bar is 100 km; the light blue sea is shallower than 200 m. Based on information from Institut Français du Pétrole and redrawn from Le Monde Diplomatique, *Les Mers, Avenir de l'Europe*, 1992, p. 47.

gins of the continents and accumulated their sediments beneath the sea. Over the aeons of geological time, many marine sedimentary basins have become part of the landmass; it was in these that the easily reachable oil, not far below the surface, was discovered by the early drillers. The oil business grew and prospered from this 'easy' oil, most of which has now been found. What remains is much more difficult to find, either because it is buried deep, or because it is in places that are difficult to get to and work in, like the Arctic coasts or beneath the ocean.

Given that much oil forms from the decay of marine planktonic remains in sedimentary basins, it is hardly surprising that oil reservoirs should exist in the basins still beneath the sea. Seeps of oil at the coast (for instance, at Coal Oil Point, California, and Kimmeridge Bay, England) are indications of the likelihood of oil offshore – indeed, there are many seeps offshore[39]. Oil was first produced from beneath the sea in 1894, off California, by drilling from wooden wharves[4]. Production was costly compared with that on land, and it was not until

1936 that operations began in the Gulf of Mexico, culminating in the discovery of the Creole field in 1938. By 1948 the first offshore platform out of sight of land had been completed off the Louisiana coast, along with the first offshore pipeline[4]. Offshore oil had come of age.

Things have moved apace since then. Hydrocarbon exploration has been extended to the waters of every major coastal nation. Oil is produced in substantial amounts off the coast of California, in the Gulf of Mexico, in the North Sea (*Figure 21.26*), in the Black Sea, off Angola, off Brazil, off southeastern Australia, in the East Indies,

Figure 21.27 The IXTOC-1 blow out of 1979–1980 in the Bahia de Campeche, Gulf of Mexico; blow outs are one of the hazards of the offshore oil industry. The well, drilled in 50 m of water, blew out on June 3rd, 1979, and was capped 290 days later after a loss of between 0.5–1.6 million tonnes of oil. Drilling had reached 3625 m when gas under high pressure forced its way up the drill pipe and onto the platform; the gas caught fire, the fire and explosion destroying the platform and catching the sea alight (actually burning gas and light hydrocarbons as they reached the sea surface). The platform sank to the bottom and damaged the well casing at the sea bed, beginning the largest marine oil spill in the history of oil exploration. Wind and current carried the oil northwest, where some of it fouled the beaches of northeastern Mexico and southeast Texas; most of it evaporated or sank to the bottom of the Gulf. The blow out was stopped by drilling a relief well into the side of the original well. (© Prof. O Linden, Stockholm University, Sweden.)

in the Persian Gulf, and to a lesser extent in many other places besides[9,16]. Natural gas is also produced in substantial amounts offshore, the North Sea being a major source (*Figures 21.25* and *21.26*).

As the shallow-water oil has been found, explorers with advanced drilling technology have moved steadily into deeper and deeper water. The deepest production wells are now located down the continental slope in the Gulf of Mexico, where they have progressed from 312 m in 1978 to 914 m in early 1994. Plans are afoot for production from 1101 m off the Philippines in 1996. In deep water, production is usually not from the conventional drilling and production platform, but from 'subsea systems', in which the well-head is at the sea bed and connects directly to a pipeline[9,16]. Safety is a primary consideration to avoid blow outs, like that at the IXTOC-1 well in the Gulf of Mexico (*Figure 21.27*).

According to the UK's Institute of Petroleum, annual world-wide production of oil and natural gas liquids reached 3.2 billion tonnes in 1992, representing 65 million barrels of oil per day (1 tonne = about 7.3 barrels). Of this, about two-fifths came from offshore. Offshore production is gradually increasing, as onshore oil production peaks. Almost every year, it seems, we are given forecasts that there will be enough oil to last another 40 years; then, because of technical advances, more is found, or it becomes possible to extract more than was thought possible from existing reservoirs. Total world reserves stood at 137 billion tonnes in 1992,

enough for 43 years at current rates of extraction.

We do know that there are no large reserves of oil in the deep ocean beyond the continental slopes, because sedimentary rocks need to be buried to certain depths and cooked by the flow of heat through the Earth to produce oil, and deep-sea sediments are not buried deeply enough for the right conditions to have been met[15]. However, natural gas, or methane, is not subject to these same constraints; much of it is created early on in the decomposition of organic matter and it can readily be trapped in deep ocean sediments. The trapping process is somewhat unusual. The bottom waters of the deep ocean are so cold, and under so much pressure, that within the pore waters trapped in the underlying sediments, mixtures of water and methane form gas hydrates, or clathrates, cages of water molecules that surround gas molecules[25]. With depth in the sediment, the flow of heat from the interior of the Earth is sufficient to melt these ice cages, freeing the gas, which may then be trapped beneath the hydrate layer. It has been suggested that there are 100,000 trillion cubic feet (3×10^{15} m^3) of clean natural gas trapped in the deep ocean in this way, which is as much as has been discovered by drilling to date in conventional exploration and production wells. Clathrates are concentrated on continental slopes and rises deeper than 300 m, so cover around 10% of the total ocean area[25]. If some way can be found of extracting this deep-ocean gas, present resource estimates will have to be revised.

Figure 21.28 Gravel being pumped aboard an offshore dredger. (© United Marine Dredging Ltd, England.)

Figure 21.29 Piles of sand and gravel from offshore on the quayside awaiting distribution. (© United Marine Dredging Ltd, England.)

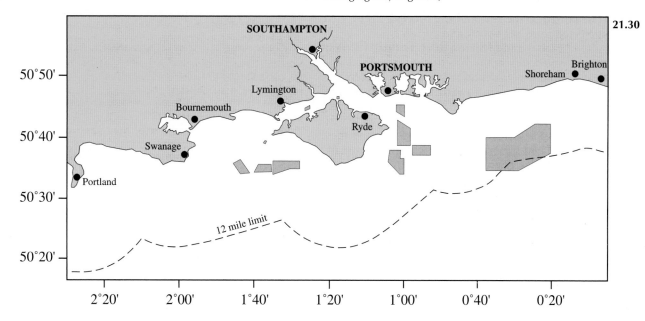

Figure 21.30 Dredging areas in the English Channel. (Courtesy of the Institute of Oceanographic Sciences Deacon Laboratory, Wormley, England.)

Minerals from the Sea Bed

Aggregates

Beaches are beautiful, but like many other things in life they may interest people for reasons other than the aesthetic. They are made of sand or gravel, the raw materials of the building trade, and so they have been used for generations in some parts of the world to feed the construction industry. This particular use of beaches is on the decline, because with the galloping growth of tourism, even in the remotest spots the amenity value of most beaches is considered to be worth more than their value as building materials.

Use of the beach for pleasure has moved the search for aggregates offshore, to tap deposits that formed during the rise and fall of sea-level that accompanied the ice ages. At least half of the surface of the world's continental shelves is covered with sands and gravels from fossil beaches, dunes, and river channels (*Figure 21.28*). This is a great boon to those countries that are highly industrialised and densely populated, and where land is scarce. Dredging for these bulk materials is a growing industry off their coasts (*Figure 21.29*). About 25% of the sand and gravel for construction in southeastern England comes from offshore (*Figure 21.30*), but about 50% of the world's offshore aggregate production is from Japan[10,16]. Where the deposits are close to shore, care must be taken to ensure that the natural offshore supplies of sand for beaches are not destroyed by dredging; this is why around the UK, for instance, dredging is regulated, being confined to specified concession areas[2], and why little offshore dredging is allowed around the US, even though there are substantial resources of sand and gravel off the US coasts[10]. Most dredging is from depths of less than 45 m, which may be extended to 50–60 m over the next few years.

Mineral ores

Most sand and gravel is made of mixtures of quartz and feldspar, the main rock-forming minerals of the continents. But the precise mineralogy of a beach depends on the mineral make-up of the local rocks, the local climate (which dictates how the rocks weather), how the eroded minerals were transported to the sea, and the processes they were subjected to by sea waves and currents.

On beaches and offshore, the continual action of waves and tides helps to concentrate hard and heavy minerals in deposits known as placers[10,16]. Placers generally form toward the base of offshore sand deposits, at the interface with the underlying rock, and in depressions and channels. A wide variety of types of dredgers can be used to mine these unconsolidated mineral ores, which can be of many different types depending on the local geology. Beaches around the world have been mined for many minerals, including diamonds (Namibia), gold (Alaska and Nova Scotia), and chromite (Oregon). Offshore, diamond placers are mined off Namibia, cassiterite (for tin) off Malaysia, Indonesia, and Thailand, and (in the past) off Cornwall, England[8,10,16]. Other minerals, such as chromite (for chromium), rutile (for titanium), ilmenite (for iron and titanium), magnetite (for iron), zircon (for zirconium), monazite (for rare earths), and scheelite (for tungsten), have been or are currently being dredged in various places around the world, like Sri Lanka and Australia[10,16]. Most dredging is from depths of less than 50 m, but the Japanese have dredged cassiterite from up to 4000 m deep. Beaches have even been mined for their most basic mineral, quartz; pure quartz is the basic constituent of glass sand, mined from the beach in western North Island, New Zealand, for instance[8].

On tropical islands fringed with coral reefs, the white sands consist not of quartz but of coral fragments, made of calcium carbonate, the basic constituent of cement. Calcium carbonate in the form of shell remains is also common in places on the continental shelves, and has been dredged offshore (e.g., off Iceland) for the production of cement[10]. Extraction of carbonate by the demolition of coral reefs on some tropical islands helps build holiday

Figure 21.31 Phosphatic limestone with a P_2O_5 content of around 15–18% dredged from the continental shelf off Morocco (maximum dimension 12 cm). This example is a conglomerate of pebbles of phosphatised limestone and phosphorite set in a phosphorite matrix, in which are embedded dark green to black, sand-sized grains of the iron–potassium aluminosilicate mineral glauconite. The rock is believed to have formed in Miocene times. Being solid, rather than pelletal, and having a moderate P_2O_5 content, it is a rather low-grade resource of fertiliser phosphate, so is uncompetitive with the extensive high-grade pelletal phosphorite deposits mined onshore in Morocco. Miocene glauconitic conglomeratic phosphatic limestones like this are also common off South Africa and California. All three sites are today washed by nutrient-rich upwelling currents; similar, but richer, currents may have prevailed at these sites during the Miocene in response to a global reorganisation of oceanic nutrients. (Photo, M. Conquer; © IOSDL.)

hotels to house tourists and boost local income. Paradoxically, these short-term gains increase the risk of damage from severe storms, by wrecking the islands' natural coastal protection (see also Chapter 17).

Some minerals have been mined from beneath the sea bed, through offshore extensions of mines that began on land, including coal off the east coast of the UK, scheelite (for tungsten) off Tasmania, iron ore off Newfoundland, and tin off Cornwall.

Other minerals are deposited on or just beneath the sea bed, and could be mined from the sea bed if economic conditions were right. Among these is calcium carbonate–fluorapatite, a phosphate mineral that forms the main component of the phosphatic rock, phosphorite (*Figure 21.31*). Sand-grain size pellets and pebble- to boulder-size nodules of phosphorite abound in certain areas, usually off the west coasts of continents in mid-latitudes where surface waters are exceptionally well supplied with phosphate[10] (e.g., Peru, California, Morocco, South Africa, and Namibia). Usually they occur where the

21.32a  21.32b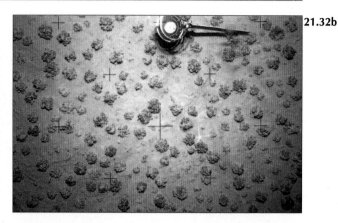

Figure 21.32 Manganese nodules carpet the sea bed at a depth of around 4100 m in the Peru Basin, as seen from a photo-sledge during RV *Sonne* cruise S079; the diameter of the outer ring of the compass is 25 cm. (a) From 8°35.0′S, 90°42.5′W, showing the sea bed densely covered with nodules; to one side of the compass a benthic organism has pushed the nodules apart in its search for food from the underlying sediment. (b) From 8°43.0′S, 90°43.0′W, showing less dense coverage by larger nodules, with the trails of benthic organisms obvious between the nodules to the side and a star fish near the centre. Most nodules look like small black potatoes or tennis balls; these ones look more like round cauliflower heads. (Courtesy of Dr von Stackelberg, Bundesanstalt für Geowissenschaften und Rohstoffe, Germany.)

water is 100–1000 m deep, so they are potentially accessible with the right mining gear. Similar deposits on land are mined for most of the world's phosphate fertiliser. Where the economics are right, offshore deposits could provide the basis for a local phosphate fertiliser industry. But land reserves are abundant and cheap to mine, so it seems unlikely that offshore phosphorites will become commercially viable before 2000AD. A likely first contender is actually an east-coast deposit, on the Chatham Rise off New Zealand. There are rich deposits also off the east coast of the southeastern US, from Florida to the Carolinas.

Elsewhere on the sea bed, iron–manganese oxide minerals are deposited in the form of nodules or crusts[10]. In certain places on the deep-ocean floor (4000–5000 m deep), where the rate of sedimentation is extremely slow, the sea bed is carpeted with small potentially mineable manganese nodules the size and shape of potatoes or tennis balls (*Figure 21.32*). Nobody is very interested in their manganese content; the attractive feature is their high content of combined copper, nickel, and cobalt (averaging 2.4% in places[18]). In shallower waters (around 1000–2000 m), the flanks of volcanic islands and seamounts may be encrusted with manganese oxides that are particularly rich in cobalt[10] (up to about 2%, *Figure 21.33*). Constraints on utilising these various deposits include the high costs of mining in deep water, the variability of ore grades, the distance to market, and the competing costs of mining on land. Deep-sea metal-mining will stay unattractive as

21.33

Figure 21.33 Manganese encrustation on pillow basalt from Nod Hill, an abyssal hill rising from the western part of the Madeira Abyssal Plain close to the eastern flank of the Mid-Atlantic Ridge. (1) Basalt core; (2) oxidised glassy surface of pillow basalt; (3) 1.6 cm thick manganese encrustation. Crust has been removed to show the smooth oxidised surfaces in the glassy section. Encrustations[41] can reach 25 cm. (Photo, M. Conquer; © IOSDL.)

21.34

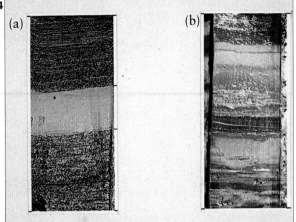

Figure 21.34 Multicoloured layers representing different minerals deposited from metal-rich Red Sea brines. (a) A grey–green section from 125–145 cm in CHAIN core 119K in DISCOVERY DEEP. It comprises detrital sediments (foraminiferal tests made of high magnesium calcite) in the middle (pale, greenish grey, and homogeneous), and a finely laminated mixture of these sediments with iron sulphides at the top and bottom. (b) A red, brown, and yellow section from 125–165 cm in CHAIN core 161K from the saddle between the CHAIN DEEP and the ATLANTIS II DEEP. It comprises a mixture of detrital sediments (foraminiferal tests of high magnesium calcite) and an orange–yellow mix of goetthite and amorphous 'limonite' (both are iron oxides). Precipitation of the goetthite–amorphous material requires oxidation of dissolved ferrous iron[14]. (© D A Ross, Woods Hole Oceanographic Institution, US.)

long as metal prices are depressed, which is likely to persist until early in the twenty-first century[10,16,18]. The constraints have not stopped international consortia of companies from Japan, the UK, France, Germany, and the US, and the Indian Government, from exploring for manganese nodules and/or developing the means to harvest them. The prime sea-bed sites are in the eastern Pacific between Hawaii and California at about 12°N[10,16].

We have known about manganese and phosphorite nodules since their discovery during the late nineteenth century. Many were recovered during the first global oceanographic expedition, by HMS *Challenger* in 1872–1876 (see Chapter 1). Much more recently two new kinds of mineral deposit have been discovered on the deep-sea floor: metal-rich muds and hydrothermal mounds.

In the early 1960s hot brines were discovered in deeps in the central graben of the Red Sea[14]. Investigations in 1965 showed that beneath these were muds rich in metal (iron, manganese, zinc, copper, cadmium, lead, and silver; *Figure 21.34*). The brines are hot hydrothermal fluids that have leached salt and metals from the underlying rocks (which include evaporites). The metals precipitate as minerals of different kinds[14] (oxides, sulphides, sulphates, or carbonates). Hydrothermal activity is concentrated along the central graben because it is here that Africa and Arabia are pulling apart from one another. German researchers, funded by Saudi Arabia and the Sudan, have established that the muds are rich enough in metals to be of economic potential, but the general depression in metal prices makes them non-viable at present. Of all the deep-sea metal deposits, these are the ones most likely to be mined first[10,16].

The second discovery, in 1975, was that hydrothermal fluids discharging in the axial valley of the East Pacific Rise were also associated with metal-rich sulphide deposits[10,16,31]. We now know that hydrothermal activity and associated massive sulphide deposits rich in copper and zinc are common along the crests of the world-encircling mid-ocean ridge system, where the several plates of the Earth's rigid outer shell are pulling apart from one another (see Chapter 10). The Red Sea is one end-member of this ridge system, where not much sea floor has yet been created, where the sediment supply is high, and where deeply buried evaporite deposits make the hydrothermal fluids exceptionally salty, forming brines. Hydrothermal systems and associated mineral deposits also occur where sea-floor spreading takes place in the marginal basins of the Pacific[10].

A typical mid-ocean ridge hydrothermal field consists of several hydrothermal vents, each exiting through a tube of sulphide minerals (see Chapter 10). Individual tubes may be anything up to 200 m high, or perhaps more. Drilling shows that massive sulphide deposits permeate the rock below the sea bed in these locations. Because the fields are small and hard to find, and because most of them consist of hard rock and are far from land, mining them before 2005 seems out of the question. However, studying them gives us valuable clues as to how to prospect more accurately for similar deposits now raised up on land.

General References

Attway, D.H. and Zaborsky, O.R. (eds) (1993), *Marine Biotechnology, Vol. 1, Pharmaceutical and Bioactive Natural Products*, Plenum, New York, 500 pp.

Barnaby, F. and Barnaby, W. (1990), *Oceans of Wealth*, Living Tapes Ltd, London, 94 pp.

Borgese, E.M. (1975), *The Drama of the Oceans*, Abrams, New York, 258 pp.

Couper, A. (ed.) (1983), *The Times Atlas of the Oceans*, Times Books, London, 272 pp.

Cronan, D.S. (1992), *Marine Minerals in Exclusive Economic Zones*, Chapman and Hall, London, 209 pp.

Earney, F.C.F. (1990), *Marine Mineral Resources*, Routledge, New York, 387 pp.

Elder, D. and Pernetta, J. (1991), *Oceans*, Mitchell Beazley Pub., London, 200 pp.

Faulkner, D.J. (1992), Biomedical uses for natural marine chemicals, *Oceanus*, **35**(1), 29–35.

Penny, M. (1990), *Exploiting the Sea*, Wayland Pub., Hove, 32 pp.

Porter, G. and Brown, J.W. (1991), *Global Environmental Politics*, Westview Press, Oxford, 208 pp.

Rogers, D. (1991), *Food from the Sea*, Wayland Pub., Hove, 32 pp.

Simpson, S. (1990), *The Times Guide to the Environment*, Times Books, London, 224 pp.

Whipple, A.B.C. (1983), *Restless Oceans*, Time-Life Books, Amsterdam, 176 pp.

References

1. Anderson, D.H. (1993), Efforts to ensure universal participation in the United Nations Convention on the Law of the Sea, *Int. Comp. Law Quarterly*, **42**, 654–664.

2. Ardus, D.A. and Harrison, D.J. (1990), The assessment of aggregate resources from the UK continental shelf, in *Ocean Resources, Vol. 1, Assessment and Utilisation*, Ardus, D.A. and Champ, M.A. (eds), Kluwer Acad. Pub., Dordrecht, pp 113–128.

3. Barnaby, F. and Barnaby, W. (1990), *Oceans of Wealth*, Living Tapes Ltd, 94 pp.

4. Borgese, E.M. (1975), *The Drama of the Oceans*, Abrams, New York, 258 pp.

5. Borgese, E.M. (1992), Ocean mining and the future of world order, in *Use and Misuse of the Sea Floor*, Hsü, K.J. and Thiede, J. (eds), Wiley and Sons, Chichester, pp 117–126.

6. Charlier, R.H. and Justus, J.R. (1993), *Ocean Energies*, Elsevier Oceanographic Series, No. 56, Elsevier, Amsterdam, 534 pp.

7. Colwell, R.R. (1984), The industrial potential of marine biotechnology, *Oceanus*, **27**(1), 3–12.

8. Cook, P.J., Fannin, N.G.T., and Hull, J.H. (1992), The physical exploitation of shallow seas, in *Use and Misuse of the Sea Floor*, Hsü, K.J. and Thiede, J. (eds), J. Wiley and Sons, Chichester, pp 157–180.

9. Couper, A. (ed.) (1983), *The Times Atlas of the Oceans*, Times Books, London, 272 pp.

10. Cronan, D.S. (1992), *Marine Minerals in Exclusive Economic Zones*, Chapman and Hall, London, 209 pp.

11. Cushing, D.H. (1982), *Climate and Fisheries*, Acad. Press, New York, 373 pp.

12. Dayton, L. (1994), Sea's green clean for boats, *New Scientist*, **May**, 21.

13. Dickson, R.R. and Brander, K.M. (1993), Effects of a changing windfield on cod stocks of the North Atlantic, *Fish Oceanogr.*, **2**(3/4), 124–153.

14. Degens, E.T. and Ross, D.A. (1969), *Hot Brines and Recent Heavy Metal Deposits in the Red Sea*, Springer-Verlag, New York, 600 pp.

15. Demaison, G. and Murris, R.J. (1984), *Petroleum Geochemistry and Basin Evaluation*, Am. Assoc. Petrol. Geol. Memoir, No 35, AAPG, Tulsa, 426 pp.

16. Earney, F.C.F. (1990), *Marine Mineral Resources*, Routledge, New York, 387 pp.

17. Elder, D. and Pernetta, J. (1991), *Oceans*, Mitchell Beazley Pub., London, 200 pp.

18. Exon, N.F., *et al.* (1992), What is the resource potential of the deep ocean?, in *Use and Misuse of the Sea Floor*, Hsü, K.J. and Thiede, J. (eds), Wiley and Sons, Chichester, pp 7–28.

19. FAO (1993a) *FAO Yearbook*, Vol. 72 (1991), FAO Fisheries Series No. 40, Catches and Landings, United Nations Food and Agriculture Organisation, Rome, 654 pp.

20. FAO (1993b) *FAO Yearbook*, Vol. 73, (1991), FAO Fisheries Series No. 41, Commodities, United Nations Food and Agriculture Organisation, Rome, 395 pp.

21. Green, M. (1990), *Physics of a Tide Mill*, Oceanography Department, University of Southampton Student Report (unpublished), 12 pp.

22. Hinds, L. (1992), World marine fisheries: management and development problems, *Marine Policy*, **16**(5), 394–403.

23. Johnson, F.A. (1990), Energy from the oceans: a small land based ocean thermal energy plant, in *Ocean Resources, Vol. 1, Assessment and Utilisation*, Ardus, D.A. and Champ, M.A. (eds), Kluwer Acad. Pub., Dordrecht, pp 201–211.

24. Kunzendorf, H. (1986), *Marine Mineral Exploration*, Elsevier Oceanography Series No 41, Elsevier, Amsterdam, 300 pp.

25. Kvenvolden, K.A., Ginsburg, G.D., and Soloviev, V.A. (1993), Worldwide distribution of subaquatic gas hydrates, *Geo-Marine Lett.*, **13**, 32–40.

26. Lee, C.M. (1984), Surimi gel and the US seafood industry, *Oceanus*, **27**(1), 35–39.

27. Lennard, D.E. (1990), OTEC developments out of Europe, in *Ocean Resources, Vol. 1, Assessment and Utilisation*, Ardus, D.A. and Champ, M.A. (eds), Kluwer Acad. Pub., Dordrecht, pp 191–200.

28. McNeil, W.J. (1984), Salmon ranching: a growing industry in the North Pacific, *Oceanus*, **27**(1), 27–31.

29. McPeak, R.H. and Glantz, D.A. (1984), Harvesting California's kelp forests, *Oceanus*, **27**(1), 19–26.

30. Novitsky, R.H. (1984), Discovery to commercialisation: the blood of the Horseshoe Crab, *Oceanus*, **27**(1), 13–18.

31. Oceanus (1984), Deep sea hot springs and cold seeps, *Oceanus*, Vol. 27, No.3, Woods Hole Oceanographic Institution, Massachusetts, 100 pp.

32. Penny, M. (1990), *Exploiting the Sea*, Wayland Pub., Hove, 32 pp.

33. Porter, G. and Brown, J.W. (1991), *Global Environmental Politics*, Westview Press, Oxford, 208 pp.

34. Rodier, M. (1992), The Rance tidal power station: a quarter of a century, in *Tidal Power: Trends and Developments*, Clare, R., Price, R., Madge, B., Binnie, C.J.A., and Wilson E.A. (eds), Proceedings of the Fourth Conference on Tidal Power, London, Thomas Telford Pub., London, pp 301–309.

35. Rogers, D. (1991), *Food from the Sea*, Wayland Pub., Hove, 32 pp.

36. Senior, A.G. (1990), Renewable energy from the sea, in *Ocean Resources, Vol. 1, Assessment and Utilisation*, Ardus, D.A. and Champ, M.A. (eds), Kluwer Acad. Pub., Dordrecht, pp 171–181.

37. Sharp, G.D. and McLain, D.R. (1993), Fisheries, El Niño-Southern Oscillation and upper-ocean temperature records: an eastern Pacific example, *Oceanogr.*, **6**(1), 13–22.

38. Smith, A.J. (1990), The potential resources of the sea areas around the remaining dependencies of the United Kingdom, in *Ocean Resources, Vol. 1, Assessment and Utilisation*, Ardus, D.A. and Champ, M.A. (eds), Kluwer Acad. Pub., Dordrecht, pp 21–48.

39. Spiess, R.B. (1983), Natural submarine petroleum seeps, *Oceanus*, **26**(3), 24–29.

40. Vadus, J.R., Bregman, R., and Takahashi, P.K. (1992), The potential of ocean energy conversion systems and their impact on the environment, in *Use and Misuse of the Sea Floor*, Hsü, K.J. and Thiede, J. (eds), Wiley and Sons, Chichester, pp 373–402.

41. von Stackelberg, U., Kunzendorf, H, Marchig, V., and Gwozdz, R. (1984), Growth history of a large ferro-manganese crust from the equatorial North Pacific nodule belt, *Geologisches Jahrbuch*, **A**75, 213–215.

42. Whipple, A.B.C. (1983), *Restless Oceans*, Time-Life Books, Amsterdam, 176 pp.

CHAPTER 22

Waste Disposal in the Deep Ocean

M.V. Angel

Introduction

Satellite pictures of the Earth clearly show that far more of its surface is covered with sea than with earth. The oceans cover just over seven-tenths of its surface, but what the satellite images fail to show is that half is covered by ocean deeper than 3000 m. At present, apart from the transit of shipping and a small amount of oceanic fishing, this half of the world's surface remains largely unused by humans, albeit modified to some extent by anthropogenic contaminants (as, for example, clinker, see Chapter 13). Contrast this with the startling data on how little land is now left for natural systems. Allowing for the land area which is covered with ice, or is desert, or is used either agriculturally or for managed forestry, less than one hectare of land per capita of the population (less than 30% of the whole land surface) remains for natural ecosystems (*Figure 22.1*). This is at the world's present population size of just over 5 billion. The most optimistic – that is to say, the lowest – forecast of world population is for it to double to 10 billion by 2100; more pessimistic forecasts are for this doubling to occur by 2050! Even using the more optimistic forecast, there will be growing pressure to utilise every usable square metre of land, including most National Parks and Nature Reserves, within a few decades.

It is possible that the demand for agricultural land may be kept within bounds by better agricultural practices. The 'Green Revolution' of the 1980s saw considerable increases in the amounts of rice grown in developing countries through the use of improved strains of rice. So, being optimistic, if one assumes that the agronomists and biological engineers can repeat their past successes, then the next major foreseeable problem is what to do with the vast increases in waste that this growing population will generate. Every year every person in the developed world generates about 2 tonnes of waste. In developing countries, the average per capita production is only about a quarter of a tonne, but these nations have ambitions to develop their industries, material wealth, and life-styles, and in so doing their waste production will increase. How are we to manage our waste streams to prevent them causing ecological havoc?

The Nature of Waste

What are the sorts of waste with which mankind is littering and contaminating the environment? The detritus of twentieth-century life-styles and the debris of industries include both natural and artificial products. The natural substances, such as sewage sludge, can be dealt with by environmental systems so long as these systems are not overloaded. There are bulky, more-or-less inert wastes, such as mining tailings, fly ash from power stations, and dredged spoils from harbours and estuaries. There are the gases we discharge: carbon dioxide, sulphur dioxide, and oxides of nitrogen. There is a great variety of chemical wastes, some of which are highly dangerous not only to ourselves, but also to other life, and which threaten the ecological processes that keep the Earth habitable.

Among the most dangerous of these are some of the man-made organic compounds, a few of which have been purposely designed to kill. Being selected for their high toxicity and persistence in the environment, they are tailor-made to create environmental problems. Being synthetic, they are resistant to degradation by micro-organisms which, never

22.1

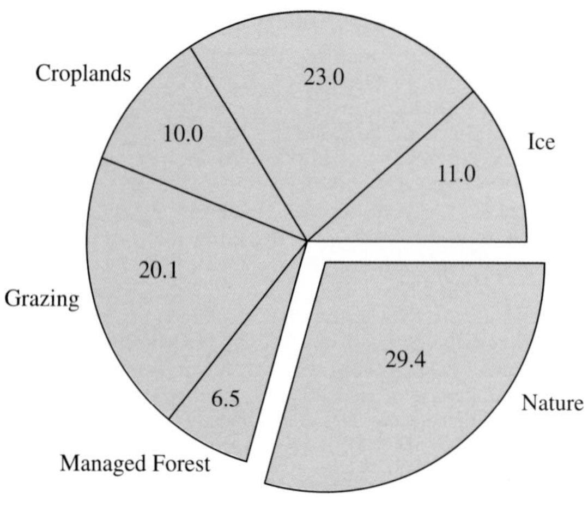

Figure 22.1 Percentage distribution of usage of land space globally (based on McAllister[3]).

having encountered them before, have not evolved the biochemical systems to cope with them. Many of these substances, despite their hazardous properties, are considered to be 'essential' not only for industry to compete more effectively, but also to control weeds in lawns and keep rats from infesting the sewers.

Another major source of waste is the social demand in developed countries that all goods be highly packaged, partly to maintain their quality, but more importantly to make them increasingly irresistible to the purchaser. As a result, massive quantities of paper, plastic, aluminium foil, glass, and cans are discarded daily. We have become too busy or too lazy to clean our own vegetables and yet over-concerned about the slightest contamination of food. Economic considerations are used as an excuse to neglect the recycling of garbage and the adequate processing of sewage. In Britain alone, about 100 million tonnes of domestic waste are generated annually, to which must be added industrial wastes – mining tailings, dredge spoils, and fly ash – plus the 200 million tonnes or so of carbon dioxide that are released into the atmosphere. Estimates of the waste stream in the US for 1988 amounted to 180 million tonnes of municipal solid waste, 300 million tonnes of sewage sludge (wet), 400 million tonnes of dredged material, and 400 million tonnes of industrial waste (wet and solid), but did not include waste gases. How are humans to ensure that the environment is kept healthy?

Options for Waste Management

The most sensible approach is to reduce the amounts of waste produced. However, while this is feasible for some materials, it is not for others. For example, the amount of sewage produced is directly proportional to the size of the population, and the more sewage that is treated, the more sludge has to be disposed of. More wastes can and should be recycled, but for some recycling can have high environmental costs and is not necessarily always the best option. It may prove better to incinerate organic materials to generate energy, rather than recycle them. For some of the persistent organic compounds, like polychlorobenzenes (PCBs), destruction by properly controlled incineration is probably the only safe way of disposal. Other materials, such as some radioactive wastes, have to be placed in properly engineered repositories where they can be stored until they become safe, which, in all practical terms, for isotopes with long half-lives is for ever.

At present, about 70% of Britain's domestic refuse is disposed of into land-fill sites. In the short term, this is a cheap and easy option, but one that in the longer term is creating problems. The numbers of sites suitable for land-fill are rapidly dwindling. Not only is there a serious risk of ground waters becoming contaminated, but also sites continue to generate gases, like methane, long after disposal has ceased. Even if such difficulties can be overcome, is it really sensible to continue to squander one of the most basic of resources – land space – in this way? The more land area that becomes restricted in its use, the greater will become the pressure to exploit the dwindling wilderness. When will such losses of wilderness begin to compromise the ability of natural systems to keep the planet habitable?

There is no single universal solution to these problems. Even if human life-styles can be changed so that less waste is generated, there needs to be more recycling and reductions in the manufacture of the more damaging synthetic chemicals like PCBs and chlorofluorocarbons (CFCs).

As mentioned, some of the more toxic materials can be destroyed by incineration. But some residues will always remain a problem – the toxic residues from sewage treatment, the fly ash from incinerators, the dredge spoils that are too contaminated to use for land reclamation, and our legacies of contaminated land from the industrial revolution and stockpiles of nuclear wastes. Despite some innovative ideas, such as the creation of artificial reefs using blocks of fly ash (see Chapter 20), bulky wastes of mixed origin present some of the most awkward problems. Where can these be put with the minimum environmental damage? Is the deep ocean a suitable place?

Marine Disposal – The Background

At present, the trend in international law is to move toward the banning of all waste disposal into the ocean. This attitude originated from the real need to clean up coastal seas which, since the industrial revolution, have been increasingly subjected to environmental abuse. In 1984, prior to the Interministerial North Sea Conference, the UK dumped just over 9 million tonnes of sewage sludge into her coastal seas (*Figure 22.2*). Added to this were numerous outfalls discharging smaller amounts, which were often highly contaminated with industrial and oily wastes. Many bathing beaches had become contaminated with unacceptably high levels of sewage-derived micro-organisms, and there was a real fear that many people using the sea for recreation were being unnecessarily exposed to pathogens. The sewage was a small, but still significant, source of various heavy metals entering UK coastal waters each year (e.g., accounting for 5% of 98 tonnes of cadmium, 3.5% of 38 tonnes of mercury, and 7% of 2370 tonnes of lead). Most of the discharges of sludge were into

22.2

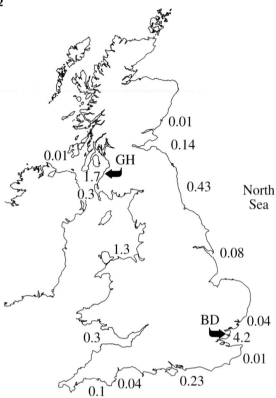

Figure 22.2 Localities and the quantities (in units of millions of tonnes) at which sewage sludge was dumped in UK coastal waters in 1984 (BD, Barrow Deep in the Thames Estuary; GH, Garroch Head in the Firth of Clyde).

Bight near Heligoland, in the North Sea. Eutrophication is more readily induced in warm tropical seas, because less oxygen dissolves in warmer water and also the higher temperatures increase the metabolic activity of bacteria. If anoxic conditions persist, sulphur bacteria begin to flourish. These bacteria gain their energy by oxidising organic material through reducing sulphate ions to sulphide ions. Dissolved sulphide ions form hydrogen sulphide, which not only makes the water stink of bad eggs, but also is lethal to most animals. When water rich in hydrogen sulphide is stirred up to the surface, there is no escape for the fish and other animals, so massive mortalities ensue.

Mortalities can also be the result of another effect of eutrophication. The highly productive conditions favour opportunistic species; some of the phytoplankton species which flourish under these enriched conditions form dense blooms and secrete powerful toxins. Such massive blooms of toxic algae (many of which are dinoflagellates) are referred to as 'Red Tides' (see Chapter 6). In 1988, *Chrysochromulina polylepis* (not a dinoflagellate) bloomed in the Skaggerak and was then carried by the currents northward along the coast of Norway. This bloom killed almost every pelagic and benthic organism in its path, and threatened to destroy the stocks of the many fish farms along the Norwegian coast (in Norway, production of farmed salmon now exceeds its land-based agricultural production, so this was a major threat to its economy).

At one site in the UK, off Garroch Head in the Clyde, sewage sludge was being dumped in an accumulative regime involving about 1.7 million tonnes each year (*Figure 22.2*). The impact of this heavy enrichment (*Figure 22.3*) resulted in the abundances of bottom-living animals increasing more than tenfold at the centre of the dump site. Biomass showed a five-fold increase. The numbers of species per unit area increased around the fringes of the dump site, but then decreased very sharply toward the centre. At the centre the populations were dominated by vast numbers of a very few, small, opportunistic species, mostly polychaete and nematode worms. The impact of the dump appears to have been restricted to about 50–75 km² of sea bed. Not all the effects were negative; the dump site attracted large numbers of fish and so became the centre of a thriving local fishery. The fish showed no evidence of having accumulated heavy metals or human pathogens. Although conditions in sediments in the centre of the dump site were highly reducing, the

tidally dynamic regimes where it was very rapidly dispersed. So, in the Barrow Deep in the Thames Estuary, the biggest dump of all, little trace of any dumped material remained after just a single tidal cycle. However, there were reports of heavy metals accumulating in local shellfish, and marked increases in the frequency of skin lesions in local fish. Biochemical and physiological characteristics of fish caught in the region showed them to be under some physiological stress, but these signs become more apparent in the centre of the North Sea, where residual tidal currents probably lead to the accumulation of contaminants from both the sewage dumps and riverine discharges of contaminants from all the coastal countries.

Coastal seas that are continually enriched with nutrients or additional organic material suffer from eutrophication. This results in a few species with very high growth and reproductive rates becoming dominant, with a consequent reduction of available oxygen in the bottom sediments because of bacterial oxidation of the excess organic matter. In regions where the tidal mixing is low and the waters are density-stratified, animals inhabiting the sea bed die as the oxygen level diminishes; in extreme cases, even those inhabiting the overlying water are killed. Recently, such anoxic conditions have occurred almost every summer in the German

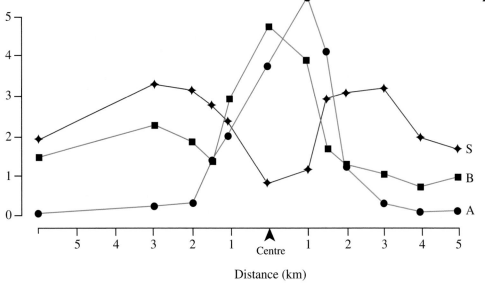

Figure 22.3 Characteristics of the benthic communities across the Garroch Head sewage sludge dump site: A, abundances of animals per cm²; B, biomass in 100 g wet weight per m²; S, numbers of species per 0.1 m². The arrow indicates the centre of the dump site (redrawn from Pearson[4]).

overlying water remained well oxygenated. The criteria used to judge the acceptability of such a dump site include the absence of deleterious effects on living resources and on other legitimate uses of the sea (shipping, recreation, and the exploitation of non-living resources), and an absence of transmission of organic and inorganic contaminants and pathogens back to the human population. Despite the rather benign effects of the Garroch Head dump, coastal dumping of sewage has generally been considered to be unacceptable by European coastal countries, and so the UK agreed to phase out marine disposal by 1997.

The Deep Ocean Option

Internationally, such unacceptable use of shallow seas has been banned under the London Dumping Convention, which is managed by the International Maritime Organisation (IMO); these embargoes have automatically been extended to deep-ocean disposal as well. Has this been wise? Other chapters demonstrate the fundamental importance of the integrity of oceanic processes to the maintenance of the global environment. Does this mean that no use can be made of the oceans in seeking solutions to some of the problems of waste management; for example, by discharging large-bulk wastes directly onto the sea bed at abyssal depths of around 4000 m, where the ensuing environmental problems will be less?

There are a number of factors which suggest

that the deep ocean may prove to be environmentally acceptable, with due care:

- First, the vast volume of the oceans, around 1370 x 10⁶ km³, will serve to dilute concentrations of any toxins to insignificance. On the abyssal plains, living biomass is too sparse (see Chapter 13) for bio-accumulation significantly to enhance transfers back to humans.

- Second, no living resources are presently exploited from the abyssal depths, nor is there any likelihood of any such resource being discovered.

- Third, at temperate latitudes the annual deposition of sedimenting material scavenges metals from the water column (Chapter 7).

- Fourth, heavy metals may not present as serious an environmental problem as they do in shallow seas. We now know that hydrothermal vents emit substantial quantities of heavy metals into the sea water overlying the mid-oceanic ridges (Chapter 10), and yet these plumes of dissolved metal remained undetected until very recently. They are very rapidly diluted and, more importantly, the geochemical processes within the water column result in the rapid redeposition of the metals back onto the sea floor. If the high concentrations of dissolved and particulate heavy metals that are normal around vents were found in inshore waters they would, quite rightly, cause great concern. But around the vents, the rich communities of animals have adapted to cope with them.

22.4

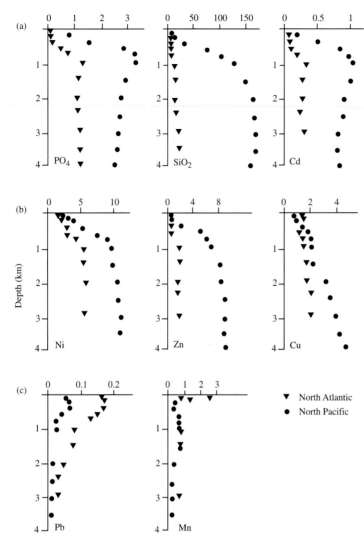

Figure 22.4 Concentration profiles (nmol/kg) of dissolved substance at two stations (t) at 34°N 66°W in the North Atlantic and (l) at 32°N 145°W in the North Pacific. (a) Marked reductions in concentration in the near-surface waters as a result of biological utilisation; these are categorised as bio-limited substances. (b) Some reduction in concentration in the near-surface waters, but the influence of biological processes on the profiles is relatively small; these are categorised as bio-intermediate substances. (c) No reduction in concentration in the surface waters, and biological processes play little or no role in determining the shape of the profiles; these are categorised as bio-unlimited substances (redrawn from Kester et al.[1]).

In ocean waters, chemical constituents can be classified by their concentration profiles into being bio-limited, bio-intermediate, and bio-unlimited (*Figure 22.4*). If wastes contaminated with metals, such as copper, cobalt, nickel, lead, and even mercury, were to be discharged directly onto the deep-ocean floor, even with the remarkable sensitivity of modern chemical analytical techniques it is doubtful if increases in dissolved concentrations of these metals would be detectable, and so there would be no biological impact. This hypothesis can easily tested by a serendipitous 'experiment' that has already been conducted. Prior to 1983, when all ocean dumping of radioactive waste was banned by international agreement, drums containing low-level radioactive waste were routinely dumped over abyssal plains at one or two licensed sites, one being in the Bay of Biscay. A few of the drums have been recovered and found to be leaking, but no contamination of the surrounding sediments was detected. Biological studies revealed little change in the communities around the drums, nor was it evident whether these minimal changes had been induced by the physical presence of the drums or by the impact of any leakage of radioisotopes. While not advocating the resumption of the disposal of radioactive wastes in the ocean, a fuller investigation of this old low-level dump site could provide much clearer evidence of the environmental acceptability of disposing other types of waste materials into the deep ocean at a few licensed sites.

Similarly, any pathogens in the waste would need to survive for two or three centuries if they were to return back into the surface waters by the slow stirring of thermohaline circulation. High hydrostatic pressures certainly prevent the metabolism of micro-organisms in sewage, but studies are needed to establish how long the pathogens remain viable. Even if some pathogens do remain viable,

22.5

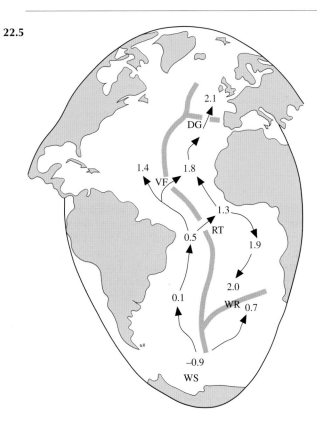

Figure 22.5 Temperatures of bottom water in the Atlantic, showing how from its source in the Weddell Sea (WS) it flows northward up the western side, its northward flow in the east being blocked by the Walvis Ridge (WR) off southwest Africa. Water flows along the Mid-Atlantic Ridge through the Romanche Trench (RT) on the equator and the Vema Fracture Zone (VF) at 10°N. Eventually, the water flows into the Western European Basin through Discovery Gap (DG), opposite the Straits of Gibraltar, at 1–2 Sverdrups (i.e., 1–2 x 10⁶ m³/s), by which time the geothermal flux from the Earth and mixing has warmed it to 2.1°C. The hatching indicates the approximate position of the mid-ocean ridges.

the other major oceans. The greater age of these bottom waters is reflected in their relatively low dissolved oxygen content. So, the deep waters of the North Pacific and Indian Oceans are more vulnerable to oxygen stress. However, microbial degradation rates are slowed by the cool *in situ* temperatures and high hydrostatic pressures, so if the waste piles up, much of the organic matter will remain buried and will not be oxidised.

If we assume the worst-case scenario, that the full chemical oxygen demand is realised, will the supply of oxygen in deep water be sufficient to cope with the demand? First, we need to know how much reduction in oxygen availability can be tolerated by the abyssal ecosystems. If, for example, the biological systems at abyssal depths in the northeastern Atlantic can tolerate a reduction of 10% (i.e., the *in situ* oxygen concentrations being reduced from 5.5 to 5.0 ml/l), then the supply of bottom water will allow the disposal of 150 million tonnes of the type of sludge dumped in Barrow

they would have to be concentrated by many orders of magnitude to be infectious and to be transferred by some as yet unknown process back to humans.

Oxygen Demand

Would oxygen concentrations be reduced significantly if the deep ocean were to be used for the disposal of organically enriched waste, such as sewage sludge? Most deep waters in the oceans are rich in dissolved oxygen; the bottom waters are formed at the surface of polar seas, when sea ice is forming. Sea water freezes at −1.9°C and, since the colder the water the more gas will dissolve in it, where the bottom water is being formed in the Weddell Sea and to the west of Greenland, the dissolved oxygen concentrations exceed 8 ml/l. The flow of bottom water northward in the Atlantic can be traced by temperature (*Figure 22.5*), where the bottom topography plays a key role in determining the patterns of flow. The concentrations of dissolved oxygen reflect the age of the deep water (*Figure 22.6*). These concentrations are enhanced in the northwestern Atlantic by the bottom water formation off Greenland, but in the northeastern Atlantic concentrations over the Porcupine Abyssal Plain to the west of Ireland remain at about 5.5 ml/l. In the North Atlantic, the bottom water mixes with the overlying waters to form North Atlantic deep water; this water mass supplies bottom waters to

22.6

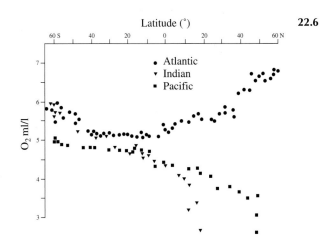

Figure 22.6 Latitudinal distribution of dissolved oxygen (ml/l) in bottom waters of the western regions of the three main oceans (modified from Mantyla and Reid[2]). Note the rise in oxygen concentration in the bottom waters of the northwestern Atlantic resulting from bottom water formation off Greenland. There is no such poleward rise in the oxygen concentrations in the northeastern Atlantic.

22.7

Figure 22.7 *Eurythenes gryllus*, a large amphipod species, which can grow to lengths of over 12 cm and is one of the most abundant and voracious species attracted to baits on the sea bed. These sorts of scavengers will probably become more abundant around an organically enriched dump site, and may prove to be an important vehicle for the dispersal of some of the contaminants. (© Heather Angel.)

Deep in the Thames Estuary. During the Quaternary (i.e., the last 2 million years), the concentrations of oxygen dissolved in ocean bottom-water fluctuated beyond this range as a result of the climatic oscillations between glacial and interglacial conditions.

Biological Impact

How much damage might be caused to the deep-living communities and would it be a serious loss of biodiversity? If dumped in one pile, the material would totally destroy the community beneath it, just as everything natural is destroyed at a land-fill site. Such an impact would be trivial compared with the effects of major geological events (Chapter 9), and would probably be preferable to dispersing the waste widely. The waste would probably be invaded by large numbers of opportunistic species, as happened in the shallow waters off Garroch Head (*Figure 22.3*), and in turn these dense concentrations of 'detritivores' would attract large numbers of predators and scavengers, in much the same way that scavengers are attracted to baited cameras and traps (*Figures 22.7* and *22.8*; see also Chapter 13). These mobile animals may play a greater role in dispersing the contaminants than would the water currents, which at abyssal depths are generally weak, except in regions where meso-scale eddies generate benthic storms (Chapter 4).

Oceanographers have still not sampled even a hundred millionth of the total area of the ocean bed, so we still know very little about the broad-scale distribution patterns of deep-living species. However, preliminary data for the larger animals imply that most benthic species have widespread distributions, both within and between the ocean basins. There seems to be no reason to doubt the assumption that the smaller species are equally

widespread, but this needs to be verified. However, there are some obvious exceptions, such as the specialised inhabitants of hydrothermal vents and seeps (Chapters 10 and 13), and in hadal environments (>6000 m) at the bottom of ocean trenches. The proximity of these and any other special environments would have to be accounted for if any selection is made of sites for such waste disposal.

The Future

The question that urgently needs to be addressed is whether continuing with the extensive use of land-fill sites is likely to be environmentally more damaging and ultimately more expensive than introducing other waste-management options, including the use of a limited number of sites on the ocean floor. What should not be disposed of in the ocean?

- Anything that can be considered to be a resource.
- Any persistent synthetic organic substance which is totally unknown in the natural marine or terrestrial environment (for example, organohalo-

22.8

Figure 22.8 The remains of a mackerel used as bait in a trap set on the sea bed for 2 days at a depth of 4500 m off the Cape Verde Islands. Three species of amphipods, including *Eurythenes gryllus*, were in the trap and had taken part in the rapid consumption of the bait. (Courtesy of the Southampton Oceanography Centre, UK.)

gens and PCBs) and is truly essential to manufacture should be chemically destroyed.

- Radioactive isotopes? Maybe there is a case for reconsidering how the isotopes produced for weapons and peaceful purposes are to be contained and disposed of. Some of the present repositories are vulnerable to natural catastrophes and terrorist acts.

- How about carbon dioxide? Can humans slow down the rate of increase in the atmosphere (which promises to change our climate) by discharging carbon dioxide into the deep ocean, where it will dissolve? The Japanese are actively developing technologies to do so, but there are potential dangers which require research. Adding more carbon dioxide to deep waters will lower their pH. If these more acidic waters erode and destabilise calcareous sediments at the base of the continental slopes, the result might be massive failures of the continental margins (Chapter 9). There may be substantial biological impact – an effective way of anaesthetising aquatic animals is to squirt carbonated water into their container. Would the disposal of so much carbon dioxide into the deep ocean put benthic communities to sleep permanently, and what would be the scale of this impact – over 10, 100, or even a 1000 km? And what scale of impact might we consider to be acceptable in the context of even greater problems associated with global warming? The scale of such impacts is of serious concern, because terrestrial ecologists repeatedly find it impossible to extrapolate the results of experiments conducted in plots of a few metres to forecast what will happen at scales of tens and hundreds of kilometres.

These are complex issues which need further research and large-scale experiments to assess. But will these be conducted before the urgency to find solutions to waste problems becomes so overwhelming that decisions have to be made (because of socio-economic pressures on the land) in the absence of proper scientific evaluations based on sound ecological and biogeochemical principles?

General References

Kullenberg, G. (ed.) (1986), *The Role of the Oceans as a Waste Disposal Option*, NATO ASI Series, D. Reidel Publishing Co, Dordrecht, 725 pp.

Spencer, D.W. (1991), *Report of a Workshop to Determine the Scientific Research Required to Assess the Potential of the Abyssal Ocean as an Option for Future Waste Management*, Woods Hole Oceanographic Institution, 111 pp.

References

1. Kester, D.R., Burt, W.V., Capuzzo, J.M., Park, P.K., Ketchum, B.H., and Duedall, I.W. (1985), *Wastes in the Ocean*, Vol. 5, *Deep-Sea Waste Disposal*, Wiley Interscience, 346 pp.
2. Mantyla, A.W. and Reid, J.W. (1983), Abyssal characteristics of the world ocean waters, *Deep Sea Res.*, **30A**, 805–833.
3. McAllister, D.E. (1993), How much land is there on Earth? For people? For nature?, *Global Biodivers.*, 3, 6–7.
4. Pearson, T.H. (1986), Disposal of sewage in dispersive and non-dispersive areas: contrasting case histories in British coastal waters, in *The Role of the Oceans as a Waste Disposal Option*, Kullenberg, G. (ed.), NATO ASI Series, D. Reidel Publishing Co, Dordrecht, pp 577–595.

Some Commonly Used Words and Terms

Depth Zones

Abyssal. A subdivision of the benthic zone encompassing the ocean floor between a depth of 2–6 km.

Abyssopelagic. Open-ocean (oceanic) environment below 4 km depth.

Aphotic. The dark region of the ocean that lies below sunlit surface waters.

Benthic. That part of the ocean adjoining the sea bed.

Epipelagic. The upper region of the ocean extending to a depth of about 200 m.

Euphotic. The surface layer of the ocean that receives enough light to support photosynthesis.

Eutrophic. That with an abundance of nutrients.

Littoral. The benthic zone between the highest and lowest normal water marks; the intertidal zone.

Neritic. The water that overlies the continental shelf, generally of water depth less than 200 m.

Oceanic. The waters beyond the shelf break, generally of water depth greater than 200 m.

Pelagic. All water in the oceans, including the neritic zone and oceanic zones.

Sublittoral. That portion of the benthic environment extending from low tide to a depth of 200 m, often taken as the surface of the continental shelf.

Subneritic. The benthic environment extending from the shoreline across the continental shelf to the shelf break.

Plankton, Bacteria, and Marine Animals

Aerobic bacteria. Bacteria that undergo respiration in the presence of free oxygen (O_2).

Anaerobic bacteria. Bacteria that undergo respiration in the absence of free oxygen (O_2).

Autotroph. Plants and bacteria that synthesise food from inorganic nutrients.

Benthos. Organisms that live on or within the sea bed.

Coccolithophores. Microscopic, single-celled plant plankton having exo-skeletons composed of tiny, calcareous plates or discs called coccoliths.

Demersal organisms. Organisms which rest on the sea bed, but swim and feed in the water column.

Diatoms. Microscopic, unicellular phytoplankton possessing silica valves.

Dinoflagellates. Microscopic unicellular phytoplankton that propel themselves using tiny, whip-like flagella.

Epifauna. Animals that live on the sea bed, either attached or moving freely over it.

Epiflora. Plants that live in contact with the sea bed.

Eutrophication. The process whereby water becomes anoxic through decomposing organic matter.

Foraminifera. Planktonic and benthic protozoans that have a skeleton or shell composed of calcium carbonate ($CaCO_3$).

Holoplankton. Plants and animals that are plankton for their entire life.

Infauna. Animals that live within or burrow through the substrate (sand or mud).

Macroplankton. Large plankton (such as jellyfish and Sargassum weed).

Meiofauna. Small species of animals that live in the spaces among particles in a marine sediment.

Meroplankton. Planktonic larval forms of organisms that are members of the benthos or nekton as adults.

Microplankton. Plankton of length 0.06–1 mm.

Nanoplankton. Plankton of length <50 µm.

Nektobenthos. Those members of the benthos that are active swimmers and spend much time off the bottom.

Nekton. Pelagic animals, such as adult squids, fish, and mammals, that are active swimmers to the extent they can determine their position in the ocean by swimming.

Phytoplankton. Plant plankton, the primary producers of the oceans.

Picoplankton. The smallest plankton, with a body length of <2 µm.

Plankton. Organisms that float or have weak swimming abilities.

Sulfur-oxidizing bacterium. Any bacteria which use energy released by oxidation to synthesise organic matter chemosynthetically.

Ultraplankton. Plankton with body length <5 µm.

Zooplankton. Animal plankton.

Further definitions may be found in the Glossaries of:

- Pinet, P.R. (1992), *Oceanography. An Introduction to the Planet Oceanus*, West Pub. Co., New York, 570 pp.
- Thurman, H.V. (1994), *Introductory Oceanography*, 7th edn, Macmillan, New York, 550 pp.

Acronyms

ADCP — Acoustic Doppler Current Profiler
AGCM — Atmosphere General Circulation Model
ALACE — Autonomous Lagrangian Circulation Explorer
ARCS — Autonomous Remote Controlled Submersible
ASP — Amnesiac Shell Poisoning
ATOC — Acoustic Thermometry of Ocean Climate
ATSR — Along-Track Scanning Radiometer
AVHRR — Advanced Very High Resolution Radiometer
BP — Time in years Before Present day
CASI — Compact Airborne Spectrographic Imaging
CCD — Calcite Compensation Depth
CFCs — Chlorofluorocarbons
CLIMAP — Climate: Long-Range Interpretation, Mapping, and Prediction
CTD — Conductivity, Temperature, and Depth measuring Instrument
CZCS — Coastal Zone Colour Scanner
DGPS — Differential Global Positioning System
DIC — Dissolved Inorganic Carbon
DMS — Dimethyl Sulphide
DOBS — Digital Ocean Bottom Seismometer
DOC — Dissolved Organic Carbon
DSP — Diarrhoetic Shellfish Poisoning
DSV — Deep Submergence Vehicle
EEZ — Exclusive Economic Zone
EKE — Eddy Kinetic Energy
EM — Electromagnetic
ENSO — El Niño Southern Oscillation
EPSONDE — This is *not* an acronym, but the name of an instrument derived from running together the words 'epsilon' (the Greek letter normally used to denote the rate of dissipation of turbulent kinetic energy – which is what the probe is used to measure) with 'sonde', the German word for probe.
ERBE — Earth Radiation Budget Experiment
ERS — Earth Resources Satellite
FAD — Fish Attracting Device
FAO — United Nations Food and Agriculture Organisation
FCA — Floc Camera Assembly
FLIP — Floating Instrument Platform
FRAM — Fine Resolution Antarctic Model
GCOS — Global Climate Observing System
GEOSECS — Geochemical Ocean Sections Programme
GLORIA — Geological Long Range Inclined Asdic
GOOS — Global Ocean Observing System
GPS — Global Positioning System
ICES — International Council for the Exploration of the Sea
IFOV — Instantaneous Field of View
IMO — International Marine Organisation
IOSDL — Institute of Oceanographic Sciences, Deacon Laboratory, UK
IPCC — Intergovernmental Panel on Climate Change
ITCZ — Intertropical Convergence Zone

JAMSTEC — Japan Marine Science and Technology Centre
MAFF — Ministry of Agriculture, Food and Fisheries, UK
MODE — Mid-Ocean Dynamics Experiment
MORT — Mean Ocean Residence Time
MYRTLE — Multi-Year Tide and Sea Level Equipment
NAEC — Normal Atmosphere Equilibrium Concentration
NEADS — North East Atlantic Dynamics Study
NSP — Neurotoxic Shellfish Poisoning
ODP — Ocean Drilling Programme
OSCR — Ocean Surface Current Radar
OTEC — Ocean Thermal Energy Conversion
PCBs — Polychlorinated Biphenyls
POC — Particulate Organic Carbon
POM — Particulate Organic Matter
PSP — Paralytic Shellfish Poisoning
RAFOS — SOFAR backwards – to indicate that the RAFOS system works in the opposite mode to conventional acoustically tracked floats
ROV — Remotely Operated Vehicles
SA — Selective Ability
SAR — Synthetic Aperture Radar
SCUBA — Self-Contained Underwater Breathing Apparatus
SCV — Sub-mesoscale Coherent Vortex
Sigma t (s_t) — A convenient measure, commonly used in place of density, equal to $1000[(\rho/\rho_m) - 1)]$ where ρ is the density and $\rho_m = 999.975$ kg/m^3, the maximum density of pure water [see Gill, A.E. (1982), *Atmosphere–Ocean Dynamics*, Academic Press, London, 662 pp]
SMBA — Scottish Marine Biological Association (now the Scottish Association of Marine Science, SAMS)
SOC — Southampton Oceanography Centre, UK
SOFAR — Sound Fixing and Ranging
SST — Sea Surface Temperature
STD — Salinity, Temperature, and Depth measuring instrument
TAG — Trans-Atlantic Geotravers
TBT — Tributyl Tin
TEP — Transparent Exopolymer Particles
TM — Thematic Mapper
TOBI — Towed Ocean Bottom Instrument
TOC — Total Organic Carbon
TOGA — Tropical Ocean–Global Atmosphere Experiment
UML — Upper Mixed Layer
UOR — Undulating Oceanographic Recorder
WHOI — Woods Hole Oceanographic Institution, USA
WOCE — World Ocean Circulation Experiment

Index

A **bold** number indicates major information on an entry (including relevant illustrations); an *italic* number indicates an illustration or caption (and any relevant text).